CRC Handbook
of
Microbial Iron
Chelates

Editor

Günther Winkelmann

Professor
Department of Microbiology and Biotechnolgoy
University of Tubingen
Tubingen, Germany

CRC Press
Taylor & Francis Group
Boca Raton London New York

CRC Press is an imprint of the
Taylor & Francis Group, an **informa** business

PREFACE

The present handbook seeks to bring into one book theoretical and practical information on the various aspects of microbial iron chelates which are now collectively called siderophores. While an earlier comprehensive treatise on *Iron Transport in Microbes, Plants and Animals*, edited by Günther Winkelmann, Dick van der Helm, and Joe B. Neilands in 1987 emphasized the different biological systems, the present handbook focuses on the various iron chelating compounds. Thus, the reader will find important physicochemical and biological data of the presently known natural siderophores and of many synthetic analogues which have been synthesized in recent years. Few branches of natural science have been so rapidly altered by recent advances as has microbiology. Since the field of microbial iron chelates in particular has increased tremendously during the past five years, an updated compilation of the available data in the form of a handbook seemed to be necessary.

It was the intention of the editor to create a reference book for both the chemists involved in analysis and synthesis of natural compounds, as well as for the biologists interested in aspects of biosynthesis, function, and genetic regulation of microbial iron transport compounds. The first part of the handbook deals with the isolation procedures, the various siderophores, their producing organisms as well as with physiological and genetic aspects of siderophore transport in microorganisms. The second part focuses on the physicochemical data, such as formation constants, electrochemistry, and other spectroscopic properties of natural and synthetic iron chelates. Moreover, some chapters describe in detail the methods of characterization and structure elucidation. The last part of the book is concerned with synthetic aspects of siderophores which may open the possibility to tailor particular siderophores for scientific and commercial purposes. Last but not least, the potential of several natural and synthetic compounds for clinical use is discussed, which will encourage establishment of closer links between those working on basic biological aspects and those involved in aspects of therapeutical treatments of iron disorders in human health. This brief introduction touches on only a few of the topics covered in this handbook. I believe that it is good practice to compile the data for those entering the field and also for those who want to see what is happening in other fields. I hope that those entering the study of microbial iron chelates will find it as fascinating and rewarding as I do.

I would like to thank all my colleagues who have contributed to this handbook. The friendly cooperation that exists between the "iron people" made it easy and pleasant to edit this comprehensive handbook on microbial iron chelates.

Günther Winkelmann
September 1990

THE EDITOR

Günther Winkelmann, Dr. rer. nat. is Professor of Microbiology at the Department of Microbiology & Biotechnology, University of Tübingen, Germany.

Dr. Winkelmann obtained his diploma and doctoral degree at the University of Hamburg in Biology and Biochemistry in 1967 and 1969, respectively. After his habilitation in 1976 he was appointed as a Professor of Microbiology in 1980 at the University of Tübingen.

Dr. Winkelmann is a member of the German Society for Microbiology (VAAM), the American Society for Microbiology (ASM), and the International Society for Human and Animal Mycology (ISHAM).

He has been the recipient of research grants from the Deutsche Forschungsgemeinschaft (DFG). Most of his papers and books focus on microbial iron transport compounds and antimicrobial agents. He is the Editor-in-Chief of the journal *Biology of Metals*. His current major research interests are in isolation and characterization of microbial products, aspects of molecular recognition, and microbial degradation of natural products.

CONTRIBUTORS

Mohamed A. Abdallah, D.Sc., D.I.C.
Department of Chemistry
University of Strasbourg 1
Strasbourg, France

Raymond J. Bergeron, Ph.D.
Professor of Medicinal Chemistry
Departments of Medicinal Chemistry,
  Medicine and Chemistry
University of Florida
Gainesville, Florida

Volkmar Braun, Ph.D.
Professor of Microbiology
Department of Microbiology
University of Tubingen
Tubingen, Germany

Alvin L. Crumbliss, Ph.D.
Professor
Department of Chemistry
Duke University
Durham, North Carolina

John B. Dionis, Ph.D.
Assistant Professor
Department of Pediatrics
New England Medical Center
Tufts University School of Medicine
Boston, Massachusetts

Klaus Hantke, Ph.D.
Professor
Department of Microbiology
University of Tubingen
Tubingen, Germany

Dick van der Helm, Ph.D.
Research Professor
Department of Chemistry and
  Biochemistry
University of Oklahoma
Norman, Oklahoma

Mahbubul A. F. Jalal, Ph.D.
Scientist
Metabolism and Development Studies
The Plant Cell Research Institute
Dublin, California

Hans-Beat Jenny, Ph.D.
Head of Analytical Laboratories
Pharmaceuticals Research
Biotechnology Subdivision
CIBA-GEIGY Limited
Basel, Switzerland

Jacqueline Libman, Ph.D.
Senior Staff Scientist
Department of Organic Chemistry
Weizmann Institute of Science
Rehovot, Israel

Joey D. Marugg, Ph.D.
Research Manager
Gene Technology and Fermentation
Unilever Research Laboratory
Vlaardingen, The Netherlands

Berthold F. Matzanke, Ph.D.
Associate Professor
Department of Biology
University of Tubingen
Tubingen, Germany

James S. McManis, Ph.D.
Research Associate
Department of Medicinal Chemistry
University of Florida
Gainesville, Florida

Kayoko Nakamura, Ph.D.
Associate Professor
Department of Radiology
Keio University
Tokyo, Japan

J. B. Neilands, Ph.D.
Professor
Department of Biochemistry
University of California
Berkeley, California

Heinrich H. Peter, Ph.D.
Head of Microbial Chemistry Section
Pharmaceuticals Research
Biotechnology Subdivision
CIBA-GEIGY Limited
Basel, Switzerland

Abraham Shanzer, Ph.D.
Associate Professor
Department of Organic Chemistry
Weizman Institute of Science
Rehovot, Israel

Peter. J. Weisbeek, Ph.D.
Professor
Department of Molecular Cell Biology
University of Utrecht
Utrecht, The Netherlands

Günther Winkelmann, Ph.D.
Professor
Department of Microbiology and
  Biotechnology
University of Tubingen
Tubingen, Germany

TABLE OF CONTENTS

DETECTION, DETERMINATION, ISOLATION, CHARACTERIZATION AND REGULATION OF MICROBIAL IRON CHELATES

J. B. Neilands and Kayoko Nakamura

HISTORICAL

It is not a simple matter to identify the first microbial iron chelates to appear in the research literature. This distinction is usually accorded to the ferrichromes since these were the compounds first prepared by low iron growth of the microorganisms and attention was focused initially on their capacity to bind iron as a biological activity. When applied to selected species of both bacteria and fungi, growth under low iron stress disclosed the synthesis of ferric chloride positive material to be a response fairly general for microorganisms growing under such cultural conditions. Propagation of the smut fungus, *Ustilago sphaerogena,* at low iron gave high yields of the ligands of ferrichrome and, especially, ferrichrome A. Addition of excess iron salt then afforded the ferric complexes in amounts of the better part of a gram per liter. In the course of chemical characterization of the ferrichromes, it was discovered that the ferric ion is held in a trihydroxamate center. A search of the literature revealed that a microbial product, aspergillic acid, had been characterized from *Aspergillus flavus* in 1947. Other hydroxamic acid or ferric hydroxamate compounds which have to be regarded as contemporaries of the ferrichromes include coprogen, mycobactin, griseinalbomycin, nocardamin, mycelianamide, and the ferrioxamines.

Meanwhile the purple ferric chloride positive product of iron-stressed *Bacillus subtilis* was identified as the 2,3-dihydroxybenzoate conjugate of glycine. These two chelating groups, the hydroxamic acid and the catechol, are still the most commonly found among the microbial iron chelates although, as expected, the structural diversity of these compounds, which have assumed importance in fields as diverse as clinical medicine and agriculture, continues to expand.

The apparent general ability, with some notable exceptions, of microorganisms to synthesize ferric chelating agents required the invention of a new generic name. The earlier designations, siderochromes, sideramines, and sideromycins, were replaced by the single term siderophores. This was suggested by Lankford[1] who, prophetically, reasoned that not all of these ferric chelates might be colored and that the inclusion of "chrome" in the name hence could both be inappropriate and misleading.

Siderophores are technically defined as virtually ferric ion specific binding compounds produced by bacteria and fungi growing under low iron stress.

To date probably a hundred different siderophores have now been described at some level of detail ranging from an initial detection of a biological activity through a full, three-dimensional X-ray structure to a molecular mechanism of regulation of synthesis by iron. However, many more siderophores remain to be characterized, new bio-chelation centers for Fe(III) will be discovered and the present, primitive level of understanding of the regulatory mechanism will be refined, This, then, is the rationale and justification for the present survey of these aspects of the field of microbial iron chelates.

DETECTION AND DETERMINATION

LOW IRON MEDIA

Because of a tight control by iron on their biosynthesis, production of excess siderophore cannot be expected until the culture is actually stressed for the metal ion. Thus regardless of the type of medium adopted, the microorganisms employed, or the method chosen for siderophore production, the addition of only a small amount of a pure iron salt, generally ferrous sulfate, should result in a *faster* growth rate and a *higher* cell density. An excessively stringent removal of iron may cut the growth rate to zero, in which case the back addition of an appropriate level of iron should give the desired result of an intermediate growth velocity and yield of cells. In the case of enteric bacteria, the optimal "added" iron will usually be in the range 0.1 to 5 μM, with the latter figure representing a considerable excess over that required to stop siderophore production. Tests for microbial mass by measurement of turbidity at 500 to 600 nm and probes for siderophore production should be carried out at periodic intervals. Flasks equipped with sidearms are convenient for this purpose. Siderophore synthesis in excess cannot be expected until the culture actually runs out of iron and hence the curve of growth will always precede the curve of chelator production.

The two general types of media that may be employed can be described as either "minimal" or "complex". The former is by far the simpler to use if the micorroganism, as is the case for the enteric bacteria, lacks complex growth requirements. However, even minimal media may contain enough iron to repress siderophore synthesis and hence some measure may be necessary to further restrict the iron level. The review by Lankford,[1] the most critical and comprehensive yet prepared on bacterial iron nutrition, gives a good account of the various methods available for deferration of media. A common practice is to shake out the medium with a solution of 1% 8-hydroxyquinoline in chloroform until the organic layer comes away clear. The ferric complex of 8-hydroxyquinoline, which is also soluble in chloroform, has a greenish-grey-black color. As a final step, the medium can be extracted with pure chloroform to get rid of any residue of chelating agent that might linger in the aqueous phase.

It is important to realize that 8-hydroxyquinoline is a general chelating agent for heavy metals and hence it may be necessary to re-supplement the medium with factors other than iron if a reasonable level of growth is to be achieved.

An addition point to bear in mind is that only that certain ingredient making the medium "complex", for example, yeast extract or casamino acids, may have to be deferrated. These may be the sole ingredients contaminated with a significant amount of iron. This avoids the difficulty of extraction of a very large volume of liquid medium.

FUNCTIONAL GROUP REAGENTS
Ferric Chloride Reaction

The classical probe for ferric ion reactive substances is the so-called ferric chloride reaction in which a red to purple color is generated upon coordination of the metal ion. The reagent is 5 mM Fe(ClO$_4$)$_3$ dissolved in 0.1 M HClO$_4$; the perchlorate is used rather than the chloride in order to discourage binding of the anion and hence the solution should be essentially colorless.[2] The excess acid prevents hydrolysis and precipitation of the ferric ion as its hydroxide. A 0.2 ml volume of the cell-free culture supernatant is mixed with 1.0 ml of reagent and the optical density read at the wavelength of maximum absorbancy. The resulting charge-transfer bands will have relatively weak absorbancies, in the range of 1000 to 3000 for molar extinctions, and hence a corrected reading of 0.1 corresponds, at most, to 0.03 μmol of material per milliliter assay solution. The ferric ion reacts with enols, phenols, catechols and hydroxamic acids to give products with maximum absorbancies in the 400 to 600 nm range. The reaction is mainly of historical interest since the lack of specificity and the relative insensitivity have resulted in the adoption of more effective methods.

The Csaky Test for Bound Hydroxylamine

The Csaky test for hydroxylamine, widely used as a standard method for detection and assay of hydroxamic acid type siderophores, is a modification of the Blom iodine oxidation method.[3] The end product actually assayed is nitrite. The reaction was evaluated by Gillam et al.,[4] who controlled the time and temperature of initial hydrolysis, the pH of oxidation and who applied N-(1-naphthyl)ethylenediamine as coupling agent for formation of the azo dye. These workers found the method to be the most sensitive of four tried and to be capable of detecting less than 0.02 μmol of hydroxylamine N. The method is sensitive, is specific for hydroxamate type siderophores, but it suffers from the disadvantage that it requires a number of steps and involves the use of rather toxic reagents.

Periodate Oxidation of Hydroxamic Acids

In the course of chemical characterization of the ferrichromes, it was observed that the metal-free ligands reacted with HIO_4 to give a product with a sharp, intense absorbancy at 264 nm.[5] The product was determined to be the cis-nitroso alkane dimer,

$$\overset{O}{\underset{R}{\nwarrow}} \underset{N=N}{} \overset{O}{\underset{R}{\nearrow}}$$

Free hydroxylamine does not give the reaction since it is oxidized to nitrous acid; the test is applicable to the ligands of hydroxamic acid type siderophores since these are probably all derived from a secondary series bearing an alkyl substituent (R) on the hydroxylamino N. Catechols do not react with periodate. Although the extinction of the cis-nitroso dimer is close to 10,000 M^{-1} cm^{-1}, the fact that two moles are needed for its formation and the yield from anything other than simple N-methylhydroxylamine is unknown means that the method has been used only occasionally.[6] By working so deep in the ultraviolet it is difficult to have confidence that the material being measured is actually the desired dimer. In theory the reaction could detect as little as 0.02 μmol of a secondary hydroxamic acid.

Arnow Reaction for Catechols

The method of choice for quantitation of catechol type siderophores is the reaction introduced in 1937 by Arnow in which the molecule is treated in succession with nitrous acid, molybdate and alkali to yield a pink chromogen with maximum absorbancy at 515 nm.[7] In the slightly miniturized adaption used in this laboratory, 1.0 ml of sample is mixed with 0.1 ml of 5 N HCl, 0.5 ml of reagent containing 10 g each of $NaNO_2$ and $Na_2MoO_4 \cdot 2H_2O$ in 50 ml water and 0.1 ml of 10 N NaOH. This gives an absorbancy for 2,3-dihydroxybenzoic acid of about 0.59, which means that considerably less than 0.02 μmol can be determined with adequate precision. The sample should not be too strongly buffered since the function of the HCl is to generate nitrous acid from the sodium nitrite, the nitrous acid being required for nitrosation of the catechol ring. The purpose of the excess alkali is to convert the initially formed yellow complex to one absorbing at longer wavelengths. Again, if the sample contains too much of an ester, such as ethylacetate, the latter will consume the alkali and the reaction will remain yellow in color. Catechol type siderophores frequently contain up to three moles of 2,3-dihydroxybenzoic acid and it is assumed that each ring is more-or-less capable of quantitative reaction with the Arnow reagents, although this point has apparently not been investigated specifically. A positive reaction requires adjacent hydroxyls and lack of substitution at specific sites elsewhere on the aromatic ring.

Specialized Reactions and Tests

Mention should be made of the capacity of hydroxylamines to reduce alkaline tetrazolium in the cold. This reaction was used by Snow for determination of the structure of the unique amino acid of mycobactin, N^6-hydroxy-L-lysine, and was said to be as sensitive as ninhydrin.[8] Prior hydrolysis of the hydroxamic acid bond, which should have a stability in acid intermediate between that of amides and esters, is required in order to release the hydroxylamine. The hydroxyamino N has a pKa of about 5 and exposure of this functional group for long periods of time in air in the free base form is not advised. Similarly, hydrolysis of the ferri-form of the siderophore in strong acid or alkali leads to extensive degradation and the generation of many artifacts of hydrolysis.

Catechol-type siderophores display a three-banded electronic absorption spectrum with maxima at about 318, 250 and 210 nm with increasing intensity into the ultraviolet. The extinction for the first of these, about 3200 M^{-1} cm^{-1}, is sufficiently strong to be useful for analysis and detection, especially if the spectrophotometric measurements are made in an alcoholic or other suitable solvent to avoid the strong carbonyl absorption associated with, for example, ethyl acetate. Further, the catechol type siderophores will have the potent blue ultraviolet fluorescence associated with di-hydroxybenzoyl compounds. This fluorescence is quenched upon the formation of the ferric complex.

The family of siderophores known as the pseudobactins or pyoverdines is the only presently known case in which the deferri form of the molecule is colored. This can be attributed to the presence of the substituted quinoline moiety, which is both fluorescent and chromogenic.

UNIVERSAL CHEMICAL TESTS

From the foregoing it is apparent that reasonably good tests are available for the two common types of functional groups found in siderophores, namely, hydroxamic acids and catechols. However, the isolation of a line of phytosiderophores from plants, of which mugineic acid is the prime example,[9] containing neither of these groupings suggests the need for a general and universal chemical method for detection and determination of siderophores. This need was further underlined with the characterization of rhizobactin from *Rhizobium meliloti* DM4, a siderophore which like the phytosiderophores, is devoid of catechol and hydroxamate functionalities.[7,10] Microbiological tests, which are sensitive and which may be specific, require that a viable organism be kept in stock and this is sometimes difficult in a chemical laboratory. For these reasons we have sought to develop a purely chemical method that is independent of the particular structure of the siderophore.

Some attention has been paid to methods whereby the amount of ferric ion solubilized by the low-iron supernatant fluid of a culture could be assessed directly. In theory, the maximum amount of Fe^{3+} held in solution in water at pH 7.0 should be in the range of 10^{-17} to 10^{-18} M, and so it should be possible to spin out insoluble iron and then measure the soluble iron by atomic absorption or, after reduction, as the ferrozine complex. Unfortunately, attempts to follow this protocol have been frustrated by the refusal of ferric hydroxide to precipitate quantitatively from neutralized, spent media.

The best we have been able to do is the Chrome Azurol S (CAS) test, which is based on the removal of ferric ion from an intensely pigmented complex by a competing ligand, namely, a siderophore.[11] This test has been rather widely used in spite of the fact that it has been in the literature only a few years. The main advantages of the CAS assay, apart from its generality, are its sensitivity and its limited applicability to the search for mutants on plates. The reaction is sensitive since the molar extinction of the midnight blue ferric Chrome Azurol S at 630 nm is in excess of 100,000 M^{-1} cm^{-1}, which means that as little as 0.002 μmol of siderophore can be determined.

Unfortunately, the CAS assay suffers from a number of drawbacks. Phosphate is rather good at stripping iron out of the blue complex to yield the intensely colored gold form of the deferrated dye. For this reason we restrict the total phosphate to 0.03%, which still suffices for maximum growth rate of *Escherichia coli*. Chelating agents other than siderophores have to be avoided and rich media may contain interferring substances. In the original formulation we used 1,4-piperazinediethanesulfonic acid (PIPES) as the main buffer, but we subsequently found that 3-(N-morpholino)propanesulfonic acid (MOPS), which is substantially cheaper, could as well be used.

The use of CAS for direct plating of microorganisms requires several comments. The detergent that must be used to disperse the dye is rather toxic to Gram-positive bacteria and fungi. A zwitterionic detergent, 3-(dimethyldodecylammonium)propanesulfonate, is less toxic to all species, Gram-positive bacteria and fungi, lacking an outer membrane, but in this case the iron complex is brilliant green and the iron-free dye a bright yellow in color. The toxicity of the standard detergent, hexadecyltrimethylammonium bromide (HDTMA), could be mitigated by topping the plates with a solution of perchlorate or a suspension of a polystyrene cation exchanger. These treatments appear to bind the excess HDTMA and make it less toxic to the fungi and Gram-positive bacteria. The HDTMA can be mixed with the CAS reagent directly if trouble is experienced with precipitation of the dye during preparation of the assay solution or blue plates.

Regarding direct plating of mutants, it should in principle be possible to select biosynthetic, regulatory and transport defective strains from the same plate. Biosynthetic mutants should be able to survive on their low affinity uptake system and should give no halo, whereas both regulatory and transport mutants should give an extra large halo. These latter two classes should be distinguishable on the basis of their response to the available iron supply. It should be possible to shrink the halo of the transport mutants by introduction of a higher level of iron in the medium. In addition, application of a competing ligand not utilized by the bacteria, such as ethylenediamine-N,N'-bis(2-hydroxyphenylacetic acid (EDDA), should severely and selectively retard the growth rate of the transport mutants. This technique has been applied to *Rhizobium meliloti* 1021 and has resulted in the isolation of all three types of mutants.[12] It is a simple matter to pick up putative regulatory mutants on agar surfaces but many of these will be found to be "stress" mutants in which siderophore synthesis is affected by factors other than iron.

MICROBIOLOGICAL METHODS
General

Any microbiological method for detection and determination of siderophores should be at least one order of magnitude more sensitive than even the CAS test. As siderophores represent a high affinity, chelate-mediated uptake pathway, provision must usually be made to block out the low affinity channels so that overgrowth of the test organism does not result. This can be achieved by addition of either Fe(III) or Fe(II) binding agents. Among the former, EDDA, deferriferrichrome A or conalbumin are commonly used. The last named has essentially the same affinity for Fe(III) as transferrin but is much cheaper. Naturally, these proteins cannot stand the heat exposure of the aforementioned organic chelating agents, which may be added to the hot medium. The exact level to be used will depend on both the amount of iron to be complexed, which will be a function of the complexity of the medium, and the toxicity of the scavenging agent. The desirable feature of deferriferrichrome A is that it is virtually ferric specific; however, it is not available in most laboratories. Bipyridyl is a general divalent metal binding agent and is capable of drawing all except heme iron into coordination in the presence of any source of electrons as reducing agent. The safest procedure is to use the least concentration of additive that will suppress low affinity iron nutrition, especially in view of the well known ability of phenanthrolines and bipyridyls to complex with DNA.

Arthrobacter flavescens JG-9 (ATCC 29091)

As its name suggests, this is a yellow-pigmented soil bacterium. It appears to be relatively deficient in capacity to use anything save hydroxamate siderophore chelated iron but will respond to synthetic chelating agents at certain levels.[13] Catechol-type siderophores, however, are not utilized.

Amazingly, *A. flavescens* can be cultured in a complex medium only following the addition of a trace level of siderophore. The medium suggested by ATCC contains 1 g each of Bacto Peptone and Difco Yeast Extract, and 0.2 g K_2HPO_4 (Medium No. 424) per 100 ml. When supplemented with 2 mg of a hydroxamate siderophore, such as ferrichrome or rhodotorulic acid — the latter will be saturated with iron immediately in this rich medium — adequate growth will be realized following incubation at 30°C. The assay can be performed in liquid media or on agar surfaces. Most hydroxamate type siderophores are relatively resistant to thermal decomposition, but in the interests of total recovery the test samples should be sterilized by filtration.

Salmonella typhimurium LT-2 *enb-1* and *enb-7*

A series of "iron mutants" of *S. typhimurium* isolated by Ames by chemical mutagenesis were found defective in growth on high citrate media unless supplemented with iron salts. These mutants were eventually found to be blocked at some stage in the biosynthesis of a siderophore indigenous to virtually all enteric bacteria and known as enterobactin or enterochelin. The siderophore arises from a spur on the aromatic pathway originating with chorismate and proceeds though 2,3-dihydroxybenzoate as iintermediate. Class I mutants, as exemplified by *enb-1*, are blocked between 2,3-dihydroxybenzoic acid and enterobactin, while Class II, such as *enb-7*, are blocked before the catechol and after chorismate. Thus, while both will respond to enterobactin, only the Class II type are activated by 2,3-dihydroxybenzoate on the citrate media. Citrate plays a double role in serving as a carbon source for the growth of the bacteria and it also binds low affinity iron in a form not transported by *S. typhimurium*. By contrast, *E. coli* cannot use free citrate for growth and energy production but does have a inducible outer membrane system for recognition of ferric citrate.

Class I and II mutants of *S. typhimurium* use a number of so-called "exogenous" siderophores, such as ferrichrome. The complete list of hydroxyamate siderophores utilized by *S. typhimurium* has not been determined, but is probably as extensive as the corresponding series in *E. coli* (see below).

Escherichia coli RW193 (ATCC 33,475) and B18 (ATCC 33,476)

Both of these strains are derived from the K12 line of *E. coli* and both are *ent*A, namely, blocked between chorismate and 2,3-dihydroxybenzoate. Mutant B18 is, additionally, unable to synthesize the 81,000 MW outer membrane receptor for ferric enterobactin. Use of both of these mutants confirms that the activity measured is, in fact, ferric enterobactin. Very high levels of the siderophore may give some activity on B18 via a non-receptor mediated supply of iron. Both strains are auxotrophic for several amino acids, which can be supplied at a level of 40 μg/ml. This is unnecessary when the bacteria are grown on complex nutrient media in the presence of deferriferrichrome A, EDDA, or conalbumin as complexing agents for excess iron. Of these, the first-named is routinely used in this laboratory since it appears to be completely specific for Fe(III).

The most convenient assay is performed on agar surfaces with the test samples pipetted onto 6 mm paper discs which are then placed on the surface of the agar. Difco nutrient broth powder (0.4 g) and 0.75 g agar are suspended in 50 ml water, swirled and heated to dissolve the agar, at which point 0.5 ml of 0.1 mM deferriferrichome A is added to complex adventitious iron.[14] The solution is divided into two equal parts, brought to 40°C and each inoculated with two drops of either RW193 or B18 culture which had been grown up overnight in the same medium without deferriferrichrome A. Each disc, about 5 or 6 of which can be placed on one plate, can be impregnated with up to 20 ul of solution containing, at most, a few hundred picomoles of enterobactin. The diameter of exhibition of growth of RW193 should correlate approximately with the amount of enterobactin supplied.

Escherichia coli **LG1522**

This strain of *E. coli* carries, among its several markers, the *iuc* (iron uptake chelate) mutation on its pColV-K30 and, on its chromosome, *fepA*. Thus the biosynthetic pathway to aerobactin on the pColV-K30 which it harbors is blocked, but not the outer membrane receptor for ferric aerobactin. Although the strain can produce enterobactin, the absence of the outer membrane receptor for its ferric complex means that it cannot respond to this siderophore and is hence useful for assay of aerobactin in the presence of enterobactin.

For assay on agar surfaces, an overnight culture of LG1522 in Luria broth is sedimented, washed with sterile saline and 3×10^7 cells seeded into Tris medium containing 1% sodium succinate and the required amino acids at 40 μg and thiamine at 2.5 μg per ml.[15] The samples to be tested for aerobactin can be applied via paper disc or wells. Following incubation at 37°C overnight, the amount of aerobactin applied is approximately proportional to the diameter of halo of exhibition of growth.

E. coli has the capacity to use, via outer membrane receptors, ferric enterobactin, ferrichrome, ferric aerobactin, ferric coprogen-rhodotorulic acid, and ferric citrate. Thus, the use of a genetically altered strain is required if any conclusion is to be drawn regarding the nature of the activity measured by a simple growth test.

ISOLATION

GENERAL CONSIDERATIONS

A decision early on in the isolation procedure has to be made regarding the form, whether free ligand or metal complex, to be isolated. It is worthwhile to bear in mind that the iron complex cannot be analyzed directly in the NMR spectrometer owing to the paramagnetism of the ferric ion. Basically, a wild type or a mutant cell may be employed, depending on the level of siderophore produced. A mutant blocked in the uptake of the siderophore has an inherent advantage since such a cell is easily iron-starved and, furthermore, the siderophore cannot be taken inside to be metabolized. Naturally, the cleaner the medium to start with the easier it will be to come out with the pure siderophore.

A few initial probes with the paper electrophoresis apparatus, using the CAS liquid assay reagent as spray, should reveal the charge on the unknown siderophore as a function of pH. This information is necessary in order to learn the pH at which there is the most likely prospect of driving the siderophore into an organic solvent, which is the best way to effect its purification. Obviously, if the siderophore has both + and − charges, there is no pH at which it will show much solubility in an organic phase.

We record here condensed versions of the isolation of four siderophores, of which two are hydroxamate and two catechol. While this does not cover all of the possibilities that may be encountered, it does, nonetheless, span many of those commonly met with in bacteria and fungi. Specialized structures, such as the rhizobactins, will require the elaboration of a particular method developed for that purpose.

FERRICHROME A

The following procedure has been used by several generations of undergraduate students in the Biochemistry Department at Berkeley, who have never failed to obtain crystals of ferrichrome A, and on occasion, crystals of ferrichrome (Figure 1).

The sterile medium contains, per liter, 1 g K_2SO_4, 3 g each K_2HPO_4 and ammonium acetate, 1 g of citric acid and 20 g of sucrose. This requires about 0.7 ml of concentrated ammonium hydroxide for neutralization to pH 6.8. The medium is then supplemented with 2 mg thiamine, 0.005 mg Cu^{++}, 0.035 mg Mn^{++}, 2 mg Zn^{++}, and 80 mg Mg^{++} per liter and dispensed in 0.21 volumes in 1 liter erlenmeyer flasks. After a 1% inoculum from a young culture of *Ustilago sphaerogena*, the culture is shaken with vigorous aeration at 30°C for 1 week.

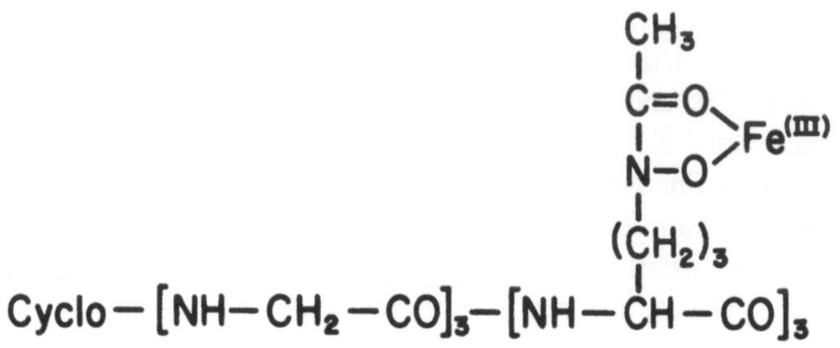

$$\text{Cyclo}-[\text{NH}-\text{CH}_2-\text{CO}]_3-[\text{NH}-\overset{\displaystyle |}{\underset{\displaystyle |}{\text{CH}}}-\text{CO}]_3$$

FIGURE 1. Structure of ferrichrome, a siderophore common to many fungal species and utilized, although apparently not synthesized, by several bacterial species. In ferrichrome A, which is generally much less active as a siderophore, the triglycyl peptide is replaced by the sequence seryl-seryl-glycyl and the acyl part of the hydroxamic acid bond is trans-β-methyl glutaconic acid rather than acetic acid.

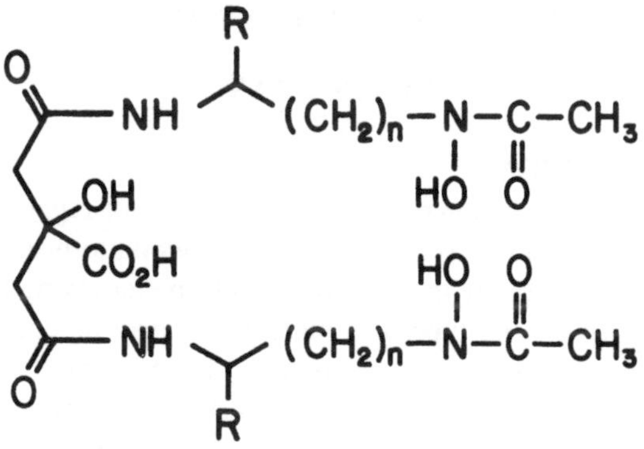

FIGURE 2. Basic structure of the citrate-hydroxamate family of siderophores. The best studied member, aerobactin (R = COOH, n = 4), is the product of *Aerobacter aerogenes* 62-I and of clinical isolates of enteric bacteria, including *Escherichia coli*.

The cells are separated by centrifugation, 1 g $FeSO_4 \cdot 7H_2O$ is added and dissolved by shaking, the pH is brought to 3 with dilute sulfuric acid, the solution is saturated with ammonium sulfate, and the color taken into benzyl alcohol in three aliquots. Any emulsion can be broken by centrifugation. After the addition of three volumes of diethyl ether to the pooled benzyl alcohol extracts, the color is back-extracted into a small volume of water in three successive steps. The combined water extracts are swirled with ether to remove any residual benzyl alcohol and then set aside in a shallow evaporation dish covered with a filter paper. The yield is about 100 mg of ferrichrome A tetrahydrate.

A very small amount of ferrichrome, generally not enough for crystallization, may be obtained by neutralizing the mother liquor from the ferrichrome A crystallization and extracting it with benzyl alcohol. The organic phase is diluted with ether and the ferrichrome extracted into water in the usual way.

This method is applicable to the ferrichromes since they hold the iron tightly at pH 3, have no charge at this pH, and when neutral they can be driven into benzyl alcohol.

AEROBACTIN

The original source of aerobactin (Figure 2) *Aerobacter aerogenes* 62-I, is an effective source of the siderophore. However aerobactin is also the common product of *Escherichia*,

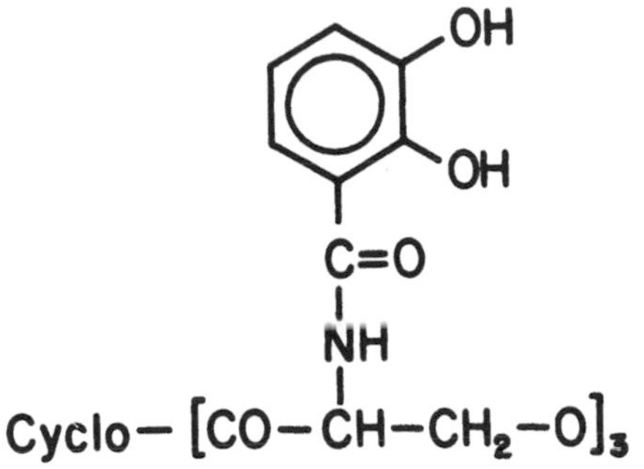

FIGURE 3.   Enterobactin, also called enterochelin, a siderophore common to many enteric
bacterial species.

Salmonella, Shigella, and other enteric genera. *A. aerogenes* 62-I is inoculated into one liter of Tris-buffer medium containing sodium succinate at 1% in place of glucose as carbon source and after 2 to 3 d of growth the cells are separated by centrifugation. After adjusting the pH to 6 with HCl, the supernatant is passed through a 150 × 25 mm column containing AG-1X2 equilibrated with 0.4 M NH$_4$Cl. The column is washed with 0.4 M NH$_4$Cl and the aerobactin eluted with a straight gradient up to 1.0 M NH$_4$Cl. The CAS-positive fractions are pooled, the pH adjusted to 2, ammonium sulfate added to saturation, and the solution extracted with benzyl alcohol. The siderophore is returned to a small volume of water, as in the procedure for the ferrichromes, and lyophilized. The product is dissolved in a small volume of water and passed through a 0.5 m × 15 mm column of Biogel P-2. Lyophilization yields 25 to 50 mg of hygroscopic material which can be converted to the more stable salt by passage through a small column of sodium form of CM-25. The ferric complex is formed by reaction with one equivalent of ferrous sulfate followed by chromatography on the sodium form of Sephadex CM-25. The reddish-brown band, representing the sodium salt of ferric aerobactin, is lyophilized to afford about 35 mg of siderophore per liter.

ENTEROBACTIN

The following method employs a *fepA* mutant of *Escherichia coli* and avoids the use of the ferric complex. Although the latter has a 3-negative charge and adheres nicely to anion exchangers, the iron cannot be removed without extensive degradation of the siderophore. The medium contains 5.8 g NaCl, 1.1 g NH$_4$CL, 0.1 g MgCl$_2$6H$_2$O, 0.142 g Na$_2$SO$_4$, 0.272 g KH$_2$PO$_4$, 12.1 g Tris base, 40 mg each proline, leucine and tryptophane, 25 mg thiamine chloride hydrochloride, 100 mg casamino acids and 17 mg of MnSO$_4$H$_2$O per liter. The pH is adjusted to 7.4 using an electrode designed for Tris, sterilized, supplemented with 5 g of glucose and set on the shaker at 37°C. This culture is used to inoculate a 10 l carboy of the same medium fitted with a bubbler for vigorous aeration. When the Arnow reaction has reached a maximum the cells are spun off and the supernatant fluid extracted with ethyl acetate.[7] The organic phases are pooled, washed with a small amount of pH 5.5 citrate solution, dried over MgSO$_4$, concentrated to about 5 ml and the enterobactin (Figure 3) precipitated with excess hexane. The crude enterobactin is dissolved in a few ml of ethyl acetate and chromatographed over a short column of Mallinkdrodt 100 mesh silicic acid. Brownish oxidation and polymerization products are retained by the column. The effluent is concentrated and the enterobactin crystallized by addition of hexane. The yields are variable but should be in the range of several hundred milligrams from the 10 l.

CHRYSOBACTIN

Erwinia chrysanthemi 4098, a transport-defective strain of the wild type *E. chrysanthemi* 3937 causing systemic disease in *Saintpaulia* plants is grown up in 0.5 l batches in Fernbach

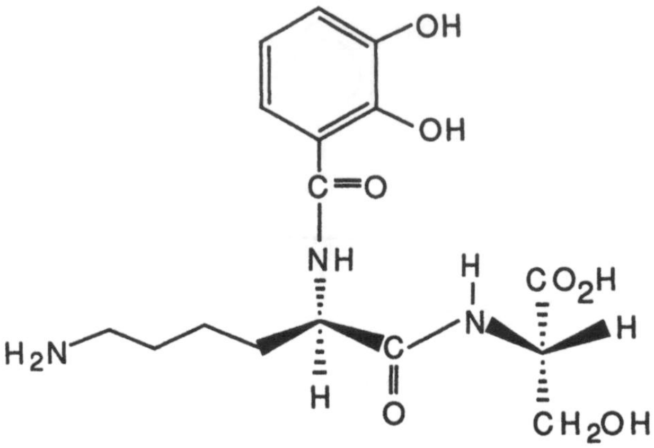

FIGURE 4.   Structure of chrysobactin, a siderophore from the bacterial phytopathogen *Erwinia chrysanthemi*. The lysyl and seryl residues have the D and L configuration, respectively.

flasks containing modified M9 medium (MM9). This is the standard M9 medium containing the single phosphate, KH_2PO_4, reduced to 0.03%.[11] The medium is supplemented with 0.1 *M* Tris, pH 7.4, and 0.2% glucose is added as carbon source. After incubation at 30°C for about 2 d, the cell-free supernatant is percolated through a 5 × 60 cm bed of non-ionic polystyrene XAD-4 and the column washed with water. Chrysobactin (Figure 4) was eluted with 1:1 water-methanol and detected by the Arnow reaction. This siderophore does not react readily with the CAS reagent. The Arnow positive material was concentrated, filtered through Sephadex G-25 at pH 5.5, lyophilized, dissolved in the least volume of water and injected into an HPLC equipped with a reverse phase C-18 column. Separation was performed at pH 5.5 with the catechol peaks determined via their absorbancy at 315 nm. The maximum yield was 0.1 mmol chrysobactin per liter of culture. The structure of chrysobactin was determined to be N-α-(2,3-dihydroxybenzoyl)-D-lysyl-L-serine.[16]

CHARACTERIZATION

PRELIMINARY STEPS

A flat-bed electrophoresis apparatus has been in use in this laboratory for several years as the primary piece of equipment for investigation of new siderophores. The working space is such that a section of Whatman No. 1 or No. 3 paper of standard 55 cm length will reach across the glass surface and dip into the buffer compartments on each end. For any siderophore or its hydrolysis product that can acquire a positive charge, 4% formic acid is a suitable buffer since it is both volatile and capable of separation of most of the usual 20 "protein" amino acids. This analysis affords a good estimate of the size and charge on the molecule since the rate of migration in an electric field is dependent primarily on net charge and only secondarily on mass. Thus a solution of ferrichrome A left in absolute methanol will display a pattern of four redbrown bands moving toward the anode and corresponding to the tri-, di-, and mono-esters and to the native molecule. These are equally spaced since the mass is in each case virtually identical and the separation is achieved on the basis of charge.

After a preliminary analysis it should be possible to decide if the siderophore is new or known. A fairly comprehensive list of already characterized siderophores, and their structures, has been published by Hider.[17]

If the siderophore can be crystallized, and this is usually best achieved with the ferric complex, an X-ray diffraction structure gives the most information. However, only about a third of all known siderophores can be crystallized and hence resort must be made to alternative modes of analysis.

If crystallization cannot be achieved, then resort must be made to NMR and MS analysis, which are mutually supportive and best carried out in tandem. The ^1H spectrum is most informative since the chemical shifts are very sensitive to the environment. The ^{13}C spectra are also useful since in this case the shifts are very large and it is possible, in effect, to add up the number and types of carbon atoms in the molecule.

When the 2,3-dihydroxybenzoyl or salicyl moiety is conjugated to a β-hydroxy amino acid, such as serine or threonine, the formation of an oxazoline ring can be expected. The N atom in this ring, which is a good ligand for Fe(III), can be charged up below pH 3 and this will result in a characteristic spectral change in the ultraviolet.

THE METAL BINDING CENTER

A plot of abundance of atoms relative to silicon vs. atomic mass shows a sharp peak centered at 56, which represents iron.[18] The nucleus ^{56}Fe is the most stable in the Periodic Table. In terms of its abundance on the surface of the earth, iron ranks fourth among all elements and, among the metals, it is surpassed only by aluminum. The aqueous chemistry of iron is dominated by the exceedingly low solubility of the oxyhydroxy polymers of Fe(III). Thus the estimated solubility product constant of 10^{-38} for ferric hydroxide will limit the concentration of free ferric ions to about $10^{-18} M$ at pH 7.4. This limitation does not apply to the ferrous ion, which is soluble to the extent of 100 mM at neutral pH. The biological utility of iron is dependent mainly on the degree to which its redox potential can be altered by changes in the environment in which the metal ion is coordinated. Standard redox potentials range from -0.4 V for certain ferredoxins to $+0.4$ V for cytochrome c oxidase. Regardless of oxidation state, iron prefers to be six-coordinate and to complex O, N, or S in a roughly octahedral geometry. The equilibrium (stability) constant and binding patterns follow the Irving-Williams order and Pearson's rules. Thus Fe(III) has a higher charge and smaller radius than Fe(II) and hence ranks as a harder acid which will prefer to bind to an atom classified as a harder base, such as O^-. Stability constants for a wide variety of ligands have been measured by different methods, of which potentiometic titration is the most general, and published in tabular form.

What this means is that siderophores, which are intended to dissolve ferric iron in the environment, will offer mainly oxygen atoms as electron donors to the metal ion. Accordingly, the functional groups bound to ferric ion are generally, although not exclusively, hydroxamic acid and catechols. Unloading of the iron from the ferri-siderophores is most efficiently accomplished by reduction since the relatively soft Fe(II) has little affinity for oxygen.

It was once thought that siderophore activity was dependent on a very large formation constant for ferric ion, namely, of the order of those measured in enterobactin and the ferrichromes. It is now recognized that a much lower affinity will suffice since many siderophores have been found to carry only a single catechol moiety for binding the iron.

The only metal ion with affinity for the siderophore ligand comparable to that of Fe(III) is gallium, which has the same charge and a very similar ionic radius. Since Ga(III) is diamagnetic, it may be used as a substitute for Fe(III) when NMR is planned. Also, as Ga(III) has a single oxidation state, the use of the gallium analog in metabolic studies constitutes a probe for the possible presence of a reduced form of the metal. Similarly, the Cr(III) complex may be used to screen for possible separation of the ligand and metal since in this case the ion is kinetically inert and it can only be inserted into the ligand via an input of energy. Although the redox potential of ferri-siderophores will be low owing to the differential affinity for Fe(III) and Fe(II), cells do possess the capability to remove the iron via a reductive step.

A study of the relationship between pH and the electronic absorption spectrum in the range 400 to 550 nm will reveal at once if a ferric hydroxamate siderophore contains more than one hydroxamic acid ring around the Fe(III). At neutral pH ferrichrome and ferric acetohydroxamate will be red-brown in color and will have absorption maxima at about 425 nm. However, at pH 3 the extinction of the ferric acetohydroxamate will fade to about a third of its value at neutral pH and the solution will assume a purple color with an absorption maximum over 500 nm. At pH 3 the 3:1 ferric complex in ferrichrome will remain intact, as will its color and spectrum. Rhodotorulic acid, a dihydroxamate, will display intermediate behavior.

Siderophores are coordination compounds and hence both geometrical and optical isomers are possible. In the ferrichrome and enterobactin type cycles the sidechains do not have sufficient flexibility to allow trans isomers and hence only the cis form will result. Optical activity in the ligand will confer a favored disposition in space of the metal-binding atoms thus affording either the right-handed delta (Δ) or the left-handed lambda (Λ) optical isomer. In *E. coli* the natural enterobactin made from L-serine is Δ-cis; a synthetic Λ-cis prepared from D-serine is not just inactive, but it complexes iron in a form not utilized by the cell.[19]

REGULATION

A plausible mechanism has been published for the regulation by iron of siderophore synthesis in *E. coli*. According to this mechanism, when the internal iron (II) concentration reaches a critical level the metal ion binds to a specific protein (Fur = *f*erric *u*ptake *r*egulation) which then associates with a region upstream from iron regulated genes and operons to block their transcription.[20]

Two methods were used independently for acquiring *fur* mutants of *E. coli* K-12. The first were obtained by constitutive derepression of a fhuA :: Mu d(Ap-lac) fusion.[21] In the second, *lacZ* was inserted as reporter gene in the aerobactin operon and constitutive mutants then acquired by mutagenesis of the chromosome with Tn5.[20] Both mutants mapped at 15 to 16 min on the K-12 chromosome.

The *fur* gene was sequenced and shown to code for a 17 kDa protein containing a high percent histidine.[22] The protein is stimulated by a number of divalent heavy metal ions to act as a repressor, binding an "iron box" operator with the sequence 5′-GATAATGA-TAATCATTATC. To date three promoters of Fur regulated genes and operons have been footprinted by the repressor and shown to fall into one of two classes. In the first, as illustrated by the aerobactin operon,[23] Fur has multiple, overlapping sites of attachment radiating out from both directions from the initial point of contact of the holorepressor with the operator. In the second type, as exemplified by the gene for the Fur protein and the cir gene, there is but a single operator site.[24,25]

The Fur protein appears to exist in solution mainly as a dimer with molecular weight 34 kDa. This was determined by use of a Hitachi Gel Pack column W-530 preceded by guard column GL-F500. Volumes of 10 µl of Fur or standards were injected in 20 m*M* Tris-HCl (pH 7.0) in 150 m*M* NaCl as the mobile phase at a flow rate of 0.8 ml/min. The detector was set at 275 nm. As may be seen from the data recorded in Table 1, the retention time of Fur was identical to that of the myoglobin dimer, namely, 11.5 min. This corresponds to a molecular weight of 34 kDa. In other experiments in this laboratory, performed with gravity gel filtration by Mark Coy, the presence of some tetramer could be seen. This was not found by HPLC analysis. In conclusion, it appears that Fur exists in an oligomeric form that is at least a dimer. The state of oligomerization does not seem to be radically affected by dilution or by the presence of EDTA or DTT.

The *fur* mutants of *E. coli* are not only constitutive in synthesis of siderophores and the cognate outer membrane receptors for their ferric complexes, but are altered in several other ways, including a failure to produce the iron-containing B form of superoxide dismutase.[26] The reason for the pleiotropic character of the *fur* mutation is not understood at present.

TABLE 1
Determination of Molecular Weight of Fur by High Pressure Liquid Chromatography

Proteins	Molecular weight	Retention time (min)
Bovine serum albumin	66.0 K	10.2
Egg albumin	45.0 K	10.9
Myoglobin dimer	35.0 K	11.5
Fur		11.5
Carbonic anhydrase	29.0 K	12.0
α-Lactalbumin	14.2 K	12.7
Cytochrome c	12.4 K	14.9
Aprotinin	6.5 K	15.1

REFERENCES

1. **Lankford, C. E.**, Bacterial assimilation of iron, *Crit. Rev. Microbiol.*, 2, 273, 1973.
2. **Atkin, C. L., Neilands, J. B., and Phaff, H. J.**, Rhodotorulic acid from species of *Rhodospirillum, Rhodotorula, Sporidiobolus and Sporobolomyces, J. Bacteriol.*, 103, 722, 1970.
3. **Csaky, T. Z.**, On the estimation of bound hydroxylamine in biological materials, *Acta Chem. Scand.*, 2, 450, 1948.
4. **Gillam, A. H., Lewis, A. G., and Andersen, R. J.**, Quantitative determination of hydroxamic acids, *Anal. Chem.*, 53, 841, 1981.
5. **Emery, T. and Neilands, J. B.**, Further observations concerning the periodic acid oxidation of hydroxylamine derivatives, *J. Org. Chem.*, 27, 1075, 1962.
6. **Holzberg, M. and Artis, W. M.**, Hydroxamate siderophore production by opportunistic and systemic fungal pathogens, *Infect. Immun.*, 40, 1134, 1983.
7. **Arrow, L. E.**, Colorimetric determination of the components of 3,4-dihydroxyphenylalanine-tyrosine mixtures, *J. Biol. Chem.*, 118, 531, 1937.
8. **Snow, G. A.**, Mycobactins: iron chelating growth factors from *Mycobacteria, J. Bacteriol. Rev.*, 34, 99, 1970.
9. **Sugiura, Y. and Nomoto, K.**, Phytosiderophores, *Struct. Bonding*, 58, 107, 1984.
10. **Smith, M. J., Shoolery, J. N., Schwyn, B., Holden, I., and Neilands, J. B.**, Rhizobactin, a structurally novel siderophore from *Rhizobium meliloti* DM4, *J. Am. Chem. Soc.*, 107, 1739, 1985.
11. **Schwyn, B. and Neilands, J. B.**, Universal chemical assay for detection and determination of siderophores, *Anal. Biochem.*, 160, 47, 1987.
12. **Gill, P. and Neilands, J. B.**, Cloning a genomic region required for a high affinity iron uptake system in *Rhizobium meliloti* 1021, *Mol. Microbiol.*, 1991.
13. **Neilands, J. B.**, Methodology of siderophores, *Struct. Bonding*, 58, 1, 1984.
14. **Emery, T. and Neilands, J. B.**, Contribution to the structure of the ferrichrome compounds, *J. Am. Chem. Soc.*, 82, 3658, 1960.
15. **McDougall, S. and Neilands, J. B.**, Plasmid and chromosome-coded aerobactin synthetis in enteric bacteria, *J. Bacteriol.*, 159, 300, 1984.
16. **Persmark, M., Expert, D., and Neilands, J. B.**, Isolation, characterization and synthesis of chrysobactin, a compound with siderophore activity from *Erwinia chrysanthemi, J. Biol. Chem.*, 264, 3187, 1989.
17. **Hider, R. C.**, Siderophore mediated absorption of iron, *Struct. Bonding*, 58, 25, 1984.
18. **Bowen, H. J. M.**, *Trace Elements in Biochemistry*, Academic Press, New York, 1966.
19. **Neilands, J. B., Erickson, T. J., and Rastetter, W. H.**, Stereospecificity of the ferric enterobactin receptor of *Escherichia coli* K-12, *J. Biol. Chem.*, 256, 3831, 1981.
20. **Bagg, A. and Neilands, J. B.**, Molecular mechanism of siderophore-mediated iron assimilation, *Microbiol. Rev.*, 51, 509, 1987.
21. **Hantke, K.**, Regulation of ferric iron transport in *E. coli*: isolation of a constitutive mutant, *Mol. Gen. Genet.*, 182, 288, 1981.
22. **Schaffer, S., Hantke, K., and Braun, V.**, Nucleotide sequence of the iron regulatory gene, *fur, Mol. Gen. Genet.*, 201, 204, 1985.

23. **de Lorenzo, V., Giovannini, F., Herrero, M., and Neilands, J. B.,** Metal ion regulation of gene expression, *J. Mol. Biol.,* 204, 875, 1988.
24. **de Lorenzo, V., Herrero, M., Giovannini, F., and Neilands, J. B.,** Fur (*ferric uptake regulation*) protein and CAP (*catabolite activator protein*) modulate transcription of *fur* gene in *Escherichia coli, Eur. J. Biochem.,* 173, 537, 1988.
25. **Griggs, D. W. and Konisky, J.,** Mechanism for iron-regulated transcription of the *Escherichia coli cir* gene, *J. Bacteriol.,* 171, 1048, 1989.
26. **Niederhoffer, E. C., Naranjo, C. M., and Fee, J. A.,** Relationship of the superoxide dismutase genes, *sod*A and *sod*B, to the iron uptake (*fur*) regulon in *Escherichia coli* K-12, in *Metal Ion Homeostasis,* Hamer, D. H. and Winge, D. R., Eds., Alan R. Liss, New York, 1988, 149.

STRUCTURES, COORDINATION CHEMISTRY AND FUNCTIONS OF MICROBIAL IRON CHELATES

Berthold F. Matzanke

INTRODUCTION

A great variety of low molecular weight (0.5 to 1.5 kDa) iron-chelating agents are synthesized by microbes under various conditions.[1-4] The majority of these ligands have been shown to serve as iron solubilizers. The chelates are iron transport agents for microbes under iron-limited growth conditions. The ligands are called siderophores (from the Greek: σιδεροσ = iron: φορευζ = carrier); they exhibit a wide range of different structural building blocks and they will be briefly characterized in the next section. In addition, many ferric and ferrous iron chelating agents of microbial origin have been detected for which the biological role is yet unclear. One group of these compounds, the anthracyclines, have gained increasing importance as anticancer agents.

Siderophore uptake in microorganisms is in general a receptor-dependent process. However, siderophores may exhibit both optical and geometrical isomers. Synthesis of Cr^{3+} or Rh^{3+}-siderophores enables isolation and characterization of these isomers. Stereochemically well-characterized isomers are an indispensible prerequisite to study the specificity of siderophor mediated uptake.

The metal centers of siderophores have also been characterized by Mössbauer spectroscopy and EPR. These results combined with literature data on cytochromes, Fe-S-proteins and ferritins have enabled *in vivo* Mössbauer spectroscopic investigations on the time course of iron assimilation and have revealed that siderophores are sometimes iron storage compounds in microbial systems. Moreover, these analyses allow a detailed description of the redox state, abundance and nature of the main components of microbial iron metabolism.

STRUCTURES OF MICROBIAL IRON CHELATING AGENTS

SIDEROPHORES

Iron, the fourth most abundant element in the earth's crust, is present under aerobic conditions at nearly neutral pH in the form of extremely unsoluble minerals like hematite, goethite, and pyrite or as polymeric oxidehydrates, -carbonates, and -silicates which severely restricts the bioavailability of this metal. Thus, microorganisms have evolved siderophores, exhibiting extraordinary high complex formation constants for ferric iron.[5] The major role of siderophores is extracellular solubilization of iron from minerals or organic substrates under conditions of cellular iron deprivation and specific transport of Fe^{3+} into microbial cells. The biological rationale for the production of siderophores by microorganisms is the irreplaceable role of iron in nearly all oxidation and reduction processes of cells, combined with the extreme insolubility of ferric hydroxide at physiological pH.

Despite the considerable structural variation found in the siderophores, their common feature is to form six-coordinate octahedral complexes. The classical chelating groups are catecholates, phenolates, and hydroxamates. Catecholates and phenolates were assumed to be typical of bacteria, whereas hydroxamates would be characteristic of *Actinomycetes* species and fungi. However, these criteria for siderophore systematics became obsolete with the detection of novel siderophores containing diverse functional binding groups like oxazoline nitrogen, α-hydroxy carboxylates, complexone-like structures and even hydrazide. Moreover, biosynthesis and excretion of hydroxamates was also found to occur in bacteria,[6-15] in a fungal system (*Phycomyces*) large amounts of gallic and protocatechuic acid were produced.[16] As with secondary metabolites multivarious and complex chemical structures are typical within the siderophores preventing their unequivocal and universal classification. Nevertheless, many siderophores can be classified into families, determined by a similar or identical backbone.

Catecholate-type siderophores

Compared to hydroxamate siderophores, catecholates display only little structural variations. In 1958 the first representative of this group of iron chelating agents, 2,3-dihydroxy-*N*-benzoyl-glycine (DHB-glycine) (1) was detected in supernatants of *Bacillus subtilis* cultures.*[17] Other mono-catecholates include DHB-threonine from *Klebsiella oxytoca*,[18] DHB-serine from *E. coli*[19] and *Salmonella typhimurium* (1), and DHB itself.[20-23] The latter is excreted by many bacteria. Very recently, chrysobactin (2) was isolated from the phytopathogenic bacterium *Erwinia chrysanthemi*.[24] It was characterized as *N*-[N^2-(2,3-dihydroxybenzoyl)-D-lysyl]-L-serine.

Myxochelin A, a bis-catecholate (3), was isolated from the culture broth of the myxobacterium *Angiococcus disciformis*.[25] Its structure, *N,N*-bis-(2,3-dihydroxybenzoyl)-lysinol is related to azotochelin, *N,N*-bis-(2,3-dihydroxybenzoyl)-L-lysine, found in culture fluids of iron-deficient cultures of *Azotobacter vinelandii* growing on nitrate.[26] An unnamed bis-catecholate (4), N^2,N^8-bis-(2,3-dihydroxybenzoyl)spermidine is synthesized by *Micrococcus denitrificans* (now called *Paracoccus denitrificans*).[27]

In 1970 the first tricatechol siderophore, named enterochelin (5), was isolated from culture fluids of *E. coli* and *Aerobacter aerogenes*.[28] It is the cyclic triester of DHB-serine. Independently this compound was obtained from *Salmonella typhimurium* by a second group and termed enterobactin, which is now the most commonly used trivial name for this siderophore.[29] All other known tris catecholate siderophores exhibit a linear backbone based on spermidine or nor-spermidine. Parabactin, 2-hydroxybenzene-3-methyl-oxazoline-2-carboxamidyl-N^4,N^1,N^8-bis-(2,3-dihydroxybenzoyl)spermidine, is excreted by *Paracoccus denitrificans*. Exposure to acid destroys the oxazoline ring producing a threonyl-moiety. This compound is termed parabactin A (4).[27,30] Agrobactin (6) and agrobactin A, isolated from *Agrobacterium tumifaciens* are homologous to parabactins. The phenol group on N^4 is a catechol group in agrobactin.[30] From low-iron cultures of *Vibrio cholerae* vibriobactin (7) was obtained. Vibriobactin, like agrobactin, exhibits three 2,3-hydroxybenzoyl residues, but also contains two residues of threonine per molecule, both of which are present as oxazoline rings.[31] The polyamine backbone proved to be *N*(3-aminopropyl)-1,3-diaminopropane, trivially named norspermidine.

In addition, various catecholate containing siderophores have been reported for which the chemical structures are not yet resolved. These include spirilobactin,[32] amonobactin P and T,[33] and others.[34] The phenolate salicylic acid and its derivatives were found in culture liquids from *Mycobacteria* species and *Thermoactinomyces vulgaris*.[21,35] Like DHB, salicylic acid can be either a precursor or a degradation product of siderophores, or might itself serve as a siderophore.

Hydroxamates

One large hydroxamate siderophore family are the ferrioxamines (8). They are excreted by *Actinomycetes* (Table 1). Ferrioxamines occur as both linear and cyclic compounds containing 1-amino-5-hydroxylaminopentane, in some cases 1-amino-5-hydroxylaminobutane, and succinic acid as building blocks.[36-40] The first representative was isolated and characterized in 1960. Desferrioxamine B, trade-named Desferal,® has become the drug of choice for the treatment of transfusional iron overload.[4,36] Ferrioxamine E (nocardamine)

---

* (Structures (1)-(44) are compiled in the appendix).

TABLE 1
Residues of Ferrioxamines and of Ferrioxamine Derived Antibiotics
Corresponding to Structure (8)

X	m	n	y	Name	Ref.
NH$_2$	5	4	–CH$_3$	Desferrioxamine A$_1$	37
NH$_2$	4	4	–CH$_3$	Desferrioxamine A$_2$	37
NH$_2$	5	5	–CH$_3$	Desferrioxamine B	36
CH$_3$–C(O)NH	5	5	–CH$_3$	Desferrioxamine D$_1$	
NHcyclo	5	4	–(CH$_2$)$_2$-CO-cyclo	Desferrioxamine D$_2$	37
NH$_2$	5	5	–(CH$_2$)$_2$-COOH	Desferrioxamine G	39
NHcyclo	5	5	–(CH$_2$)$_2$-CO-cyclo	Desferrioxamine E	38
NHR^a	5	5	–CH$_3$	Desferrimycin A$_1$	39
OH	5	5	–(CH$_2$)$_2$-COOR^b	Desferridanomycin A	39
OH	5	5	–(CH$_2$)$_2$-COOR^c	Desferridanomycin B	39
OH	5	5	–(CH$_2$)$_2$-COOH	Desferridanoxamine	39

Note: R^a = structure (9), R^b = C$_{13}$H$_{23}$NO$_7$OCONH$_2$ (sugar), and R^c = C$_{13}$H$_{24}$NO$_8$ (sugar).

and D$_2$ are cyclic compounds whereas all other ferrioxamines are linear. Certain derivatives of the ferrioxamines such as ferrimycin A$_1$, danomycin A, and danomycin B (Table 1) show antibiotic activity.[39,41,42] Bisucaberin (10), a siderophore, sensitizing tumor cells to macrophage-mediate cytolysis, was isolated from a culture broth of the marine bacterium *Alteromonas haloplanktis*.[7] This molecule is a cyclic dimer of succinyl-(*N*-hydroxycadaverine). It is closely related to ferrioxamine E, the trimer of the same moiety.

Another large family of hydroxamate siderophores are the ferrichromes (11) (Table 2) found in low-iron cultures of many fungi. The ancestor of this group, ferrichrome, was isolated in 1952 by Neilands from culture fluids of *Ustilago sphaerognena*.[43] With the exception of tetraglycyl-ferrichrome, a heptapeptide,[44] all ferrichromes possess cyclic hexapeptide backbones containing a tripeptide sequence of *N*$^\delta$-acyl-*N*$^\delta$-hydroxy-L-ornithine.[45-50] The variety of the ferrichromes is caused by variations in the remaining tripeptide sequence and the *N*-acyl substituents. DDF[des(diserylglycyl)] ferrirhodin (12) and some antibiotics, termed albomycins (12 + 13), are related to the ferrichromes. Their structures contain, as a central building block, linear trimers of *N*$^\delta$-OH-ornithine.[51,52]

A biologically active compound produced by *Penicillium* species and *Neurospora crassa* was first described by Hesseltine in 1952, who termed it coprogen (14).[53-55] Recently a variety of siderophores belonging to the coprogen family was isolated from *Fusarium dimerum*,[56] *Epicoccum purpurascens*,[57,58] *Curvularia subulata*, and *Alternaria lonipes*.[58-61] Coprogens are linear trihydroxamates composed of *N*$^\delta$-acyl-*N*$^\delta$-hydroxy-L-ornithine, anhydromevalonic acid, acetic acid and additional variable building blocks (14) (Table 3). Foroxomithine (15) is the solitary bacterial representative of the coprogen family.[62] It is produced by *Streptomyces nitrosporeus* and acts as an inhibitor of an angiotensin-converting enzyme. Foroxomithine contains a seryl building block instead of an anhydromevalonoyl group and the terminal *N*-hydroxy groups are formylated.

Building blocks composed of *N*$^\delta$-hydroxy-ornithine acylated with cis-5-hydroxy-3-methylpentenoic-2-acid are found in fusarinines (fusigens) (16) isolated from *Aspergillus, Penicillium,* and *Fusarium* species.[63-67] These building blocks are esterified head-to-tail to build the various linear and cyclic fusarinines. Ferric neurosporin isolated from cultures of *Neurospora crassa* has been identified as the cyclic triester of *N*$^\delta$-acetyl-*N*$^\delta$-hydroxy-*N*$^\delta$(3-hydroxybutyryl)-ornithine.[68] A siderophore function of this fusarinine-like molecule is not known.

Unlike the other common siderophores which form 1:1 complexes with Fe^{3+}, rhodotorulic acid (RA) (17) and dimerum acid (DA) (18) are only tetradentate and form 2:3

TABLE 2
Structural Variations of Ferrichrome — Substituents X_1 X_2
and R are Shown in Structure (11)

X_1	X_2	R	Siderophore	Ref.
–H	–H	–CH$_3$	Ferrichrome	47
–CH$_2$OH	–CH$_2$OH	–CH$_3$	Ferrichrysin	46
–CH$_2$OH	–CH$_2$OH	A	Ferrichrome A	154
–CH$_2$OH	–CH$_2$OH	B	Ferrirubin	47
–CH$_2$OH	–CH$_2$OH	C	Ferrirhodin	47
–CH$_2$OH	–CH$_2$OH	2B's and –CH$_3$	Asperochrome B$_1$	50
–CH$_2$OH	–CH$_2$OH	2B's and –CH$_3$	Asperochrome B$_2$	50
–CH$_2$OH	–CH$_2$OH	1B and 2 –CH$_3$	Asperochrome D$_1$	50
–CH$_2$OH	–CH$_2$OH	1B and 2 –CH$_3$	Asperochrome D$_2$	50
–CH$_2$OH	–CH$_2$OH	1B and 2 –CH$_3$	Asperochrome D$_3$	50
–CH$_2$OH	–CH$_2$OH	2B's and 1 D	Asperochrome C	49
–CH$_3$	–CH$_3$	B	Asperochrome A	49
–CH$_3$	–CH$_2$OH	–CH$_3$	Sake colorant A	1
–CH$_3$	–H	–CH$_3$	Ferrichrome C	1
–CH$_2$OH	–H	–CH$_3$	Ferricrocin	45
–CH$_3$	–H	–CH$_2$OH	Malonichrome	48

Note: Where there are different substituents, the sequence is not specified.

complexes with iron at pH 7. Two molecules of N^δ-(res)-N^δ-hydroxyornithine are cyclized to form a diketopiperazine ring. The terminal residues are acetyl groups in RA and anhydromevalonyl groups in DA. RA is synthesized in large quantity by *Rhodotorula* species[69,70] and DA by *Fusarium dimerum*.[56]

Mixed Ligands

Pseudomonas species and *Azotobacteria* produce a variety of fluorescent chromopeptide siderophores, termed pseudobactins,[71-74] pyoverdins,[75-77] and azotobactins.[76,78] The chromophores are derived from 2,3-diamino-6,7-dihydroxyquinoline depicted in (19-21). The peptide sequences are summarized in Table 4. Metal binding is accomplished by dihydroxyquinoline, hydroxamate, and, depending on the peptide chain, by a second hydroxamate or an α-hydroxycarboxylate. Pyoverdin Pa, pyoverdin Pa A, and pyoverdin Pa C[76] differ only in the chromophoric part of the structure. Pseudobactin A is not fluorescent since the chromophore is partially hydrogenated.[72]

Mycobactins (22) from *Mycobacteria* and the closely related nocobactins (22) from *Nocardia* represent a series of lipid-soluble siderophores located in the lipid-rich boundary layers of these bacteria.[79-81] Iron-binding is achieved by two hydroxamates, a phenolate group, and oxazoline nitrogen. Mycobactins and nocobactins exhibit the same main nucleus with particular substituents at various points of the structure.

The plasmid-related bacterial siderophore anguibactin (23), isolated from the fish pathogen *Vibrio anguillarum,* shows a unique molecular composition. Its structure has been identified as o-N-hydroxy-o-N-[[2'-(2'',3''-dihydroxyphenyl)thiazolin-4'-yl]histamine.[82] Also very unusual is pyochelin (24) from low-iron cultures of *Pseudomonas aeruginosa* and its structure has been assigned as 2-(2-o-hydroxyphenyl-2-thiazoline-4-yl)-3-methylthiazolidine-4-carboxylic acid.[83]

TABLE 3
Structural Variations Within the Coprogen Siderophores — Positions of Residues R_{1-4} are Shown in Structure (14)

R_1	R_2	R_3	R_4	Siderophore	Ref.
–H	$COCH_3$	A	A	Coprogen	56
–H	$COCH_3$	–CH_3	A	Neocoprogen I (isotriornicin)	59
–H	$COCH_3$	A	–CH_3	Isoneocoprogen I (triornicin)	58
–H	$COCH_3$	–CH_3	–CH_3	Neocoprogen II	59
–CH_3	–CH_3	A	A	N^α-Dimethylcoprogen	60
–CH_3	–CH_3	–CH_3	A	N^α-Dimethylneocoprogen	60
–CH_3	–CH_3	A	–CH_3	N^α-Dimethylisoneocoprogen	60
–H	$COCH_3$	A	B	Hydroxycoprogen	61
–H	$COCH_3$	–CH_3	B	Hydroxyneocoprogen I	61
–H	$COCH_3$	A	–CH_3	Hydroxyisoneocoprogen I	61

TABLE 4
Peptide Sequence of Various Pyoverdins, Pseudobactins and Azotobactin D. Corresponding to Structures (19)–(21)

Peptide sequence	Name	Ref.
CHR₁–Lys–*threo*–β–OH–Asp–Ala–*allo*–Thr–Ala–cOHOrn	Pseudobactin	71
CH(hydro)–Lys–*threo*–β–OH–Asp–Ala–*allo*–Thr–Ala–cOHOrn	Pseudobactin A	72
CHR₁–Ser–Ala–Gly–Ser–Ala–*threo*–β–OH–Asp–*allo*–Thr–OHOrn	PseudobactinA214	74
CHR₅–O–cyclo[Ser–Ser–*threo*–β–OH–Asp–Thr–Ser–OHOrn–Ala–Gly]	Pseudobactin7SR1	73
CH–Asp–Ser–Hse–Gly–*threo*–β–OH–Ser–Cit–Hse–OHOrn–HSe	Azobactin D	76

where $R_1 = $ succinamide for CHR₁, etc., the bracketed structure:

CHR₁ ⎫
CHR₂ ⎬–Ser–Arg–Ser–OHOrn –Lys—Thr
CHR₃ ⎭ | |
           OHOrn–Thr

CHR₄

Pyoverdin Pa		270—273
Pyoverdin Pa A		270—273
Pyoverdin Pa C		270—273
Pyoverdin Pa B		270—273

Note: CH [structure (19)], CHR₁ [structure (20)] represent chromophores and CH–hydro a hydrochromophore [structure (21)]. $R_1 = $ succinamide, $R_2 = $ succinic acid, $R_3 = $ succinate methyl ester, $R_4 = $ α-keto-glutarate, $R_5 = $ hydrogen. OHOrn and cOHOrn are abbreviations for N^6-hydroxyornithine and its cyclized form; Hse = homoserine, cit = citrullin.

From culture supernatants of *Actinomadura madurae* a further novel siderophore was isolated and termed maduraferrin (25).[84] It is an oligopeptide composed of salicylic acid, β-alanine, glycine, serine, N^6-hydroxy-N^α-methylornithine, and hexahydropyridazine-3-carboxylic acid. Ferric ion is complexed by salicylamide, hydroxamate, and the unique acid hydrazide group.

Citrate-hydroxamates exhibit citric acid as a common building block and its α-hydroxy-carboxylate group participates in iron binding. Representatives of this family are aerobactin (26), first isolated from cultures of *Aerobacter aerogenes*,[14] arthrobactin (26), also known as terregens factor, from *Arthrobacter* strains,[13] and schizokinen (27), excreted from *Bacillus megaterium*.[12] In these siderophores two molecules of N^6-acetyl-N^6-hydroxylysine, 1-amino-5-(N-acetyl-N-hydroxy)amino pentane, or 1-amino-3-(N-hydroxyamino)propane, respectively, are attached via an amide bond to citric acid. Recently another uncommon mixed ligand siderophore was obtained from cultures of a marine species of *Alcaligenes denitrificans* and named alcaliligin (28).[8] Alcaliligin represents a cyclic dimer of succinyl-*N*-hydroxy-C-hydroxy-putrescin.

A completely novel class of siderophores possesses neither hydroxamate nor phenolate iron chelating groups. Rather, ferric ion binding is gained exclusively by α-hydroxy carboxylates and carboxylates obviously exhibiting excellent iron binding properties. These compounds are colorless requiring therefore increased efforts for screening and isolation (CAS-test and iron nutrition bioassays).[85] A very hydrophilic iron complex was isolated from *Staphylococcus hyicus* exhibiting siderophore activity to the producer and 37 other staphylococci.[86] This compound, termed staphyloferrin A (29) is composed of two moles of citrate linked by ornithine. *Rhizobium meliloti,* capable of fixing atmospheric nitrogen when symbiotically associated with certain legumes, excretes and utilizes the siderophore rhizobactin (30).[87] The compound, N^2-[2-[(1-carboxymethyl)amino]ethyl]-N^6(3-carboxy-3-hydroxy-1-oxopropyl)lysine, contains ethylenediaminedicarboxyl and α-hydroxycarboxyl moieties.

FERRIC AND FERROUS ION COMPLEXING AGENTS WITHOUT A SIDEROPHORE FUNCTION

There exists a number of hydroxamate-type iron complexing agents isolated from microorganisms whose role as siderophores is not clear or unlikely: *N*-phenylacetohydroxamic acid (31) is produced in low-iron cultures of *Pseudomonas mildenbergii,* actininonin (32) by streptomyces species, mycelianamin (33) by *Penicillium griseofulvum* and astechrome (34) by *Aspergillus terreus.*[1,9,88-90] Hadacidin (*N*-formyl-*N*-hydroxyglycine), a compound with antitumor and anticancer activity was originally isolated from *Penicillium frequentans,*[91] and trichostatin A from *Streptomyces hygroscopicus.*[92] Tris (*N*-methylthioformohydroxamato-Fe(III)(fluopsin F) has been found in culture supernatant fluids of *Pseudomonas* and *Streptomyces* species (S. Miyamura, Japan Patent 87087, 1973) and displays broad-spectrum antibiotic activity against both Gram-positive and Gram-negative bacteria and fungi.[93,94] *Pseudomonas cepacia* excretes, under iron deficient conditions, pyochelin and another low molecular weight compound strongly chelating Fe^{3+}.[95] This latter compound has been named cepabactin (35) and identified as 1-hydroxy-5-methoxy-6-methyl-2(1H)-pyridinone. This pyridinone has already been described as an antibiotic produced by *Pseudomonas alcaligenes.*[96] Cepabactin shows structural similarities to aspergillic acids (36) from *Aspergillus* species and pulcherriminic acid (37) from *Candida pulcherrima.*[88] An unusual orange-brown ferric ion complex has been isolated from culture supernatants of *Streptomyces antibioticus.*[97] This compound contains a 3-hydroxy-pyridine and a thiazoline moiety and has been termed ferrithiocin (38). It should be mentioned that tropolone-derived compounds (39) synthesized by various *Penicillium* species also form stable complexes with ferric ion exhibiting colors ranging from red to green.[98-100]

The ability of anthracyclines and, to some extent, of tetracyclines, to bind strongly ferric ions has long been neglected.[101-108] The basic structure of all anthracyclinones is the 7,8,9,10-tetrahydro-5,12 naphatacenquinone. Over 200 derivates of the basic anthracyclinone structure are known. They are obtained from various microorganisms, by microbial transformation by partial synthesis and by total synthesis.[109,110] Typical representatives are the antitumor antibiotics adriamycin (ADR) and daunomycin (DM) (40), secondary metabolites produced by *Streptomyces peucetius* strains. These compounds are of particular interest because of their widespread application in cancer chemotherapy. Over 2 million patients have been treated with these drugs since the early 1970s. Recently it was shown that both ADR and DM bind iron with a very high complex formation constant ($K_m = 10^{28.4}$).[103-105] In fact, a growing body of investigations suggests that the ferric anthracycline complex could account for all the known cytotoxic effects of the drug. The complex may also be less cardiotoxic than the metal-free drug.[111-115]

Whereas siderophore mediated ferric iron uptake in microorganisms is well characterized, little is known about the possibility of ferrous ion uptake.[116-118] Due to the involvement of ferric and ferrous iron in intracellular radical reactions it is very unlikely that $[Fe(H_2O)_6]^{2+}$ can pass freely through the cellular membrane. To date, however, no ferrous ion siderophore has been identified. Nevertheless a variety of microbial compounds were characterized forming stable complexes with ferrous iron. From cultures of plant pathogenic *Erwinia rhapontici* a pink pigment was isolated and shown to be the ferrous complex of ferrosamine A (41), 2(2-pyridyl)-1-pyrroline-5-carboxylic acid.[119] The same compound has previously been isolated from *Pseudomonas* species and called ferropyrimine.[120,121] Under acidic conditions the pyrroline opens to the zwitterion of the α-amino acid and this no longer complexes ferrous ions (42). Closely related to ferrosamine A are siderochelins A—C isolated from *Nocardia* and *Actinomycetes* which have been described as 3,4 dihydro-4-hydroxy-5-(3-hydroxy-pyridinyl)-4-alkyl-2H-pyrrole-2 carboxamides.[122-124] A series of para-substituted ortho-nitrosophenols are found in iron-rich culture supernatants of *Streptomyces* cultures. These microbial iron chelators, termed ferroverdin, viridomycins and actinoviridins (44), form extraordinary stable, intensively green colored, low-spin complexes with Fe^{2+}.[125-128] Finally, bleomycins, a family of basic glycopeptides, were isolated from *Streptomyces* species. These compounds form complexes with bivalent metal ions such as Zn^{2+}, Cu^{2+}, Co^{2+}, and in particular, Fe^{2+}. Moreover, bleomycins show antitumor activity. The ultimate agent of cell damage are oxygen radicals, produced as a consequence of oxidation of Fe^{2+}-bleomycin to Fe^{3+} in a quaternary DNA-bleomycin-iron-oxygen complex.[129]

METAL CENTER SYMMETRIES OF SIDEROPHORES AND CHROMIC DESFERRISIDEROPHORES

Many siderophores are hexadentate ligands with three asymmetrical bidentate functional units attached to an asymmetrical backbone. Upon metal complexation a more or less distorted coordination octahedron is formed at the metal site by three 5-membered chelate rings. These complexes exhibit optical and geometrical isomers.

According to IUPAC rules the optical isomers are defined as follows:[130] Looking down the pseudo C_3 axis, Δ-isomers exhibit a right-handed propeller configuration, whereas the Λ-isomers have a left-handed propeller configuration (Figure 1). The absolute configuration of the metal center in siderophores is determined unambiguously using the Bijvoet method for anomalous dispersion of Cu-K_α radiation by the ferric ion in the crystalline solids. Assignments in solution can be achieved by CD spectroscopy (vide infra).

Eight enantiomeric pairs of geometrical isomers are theoretically possible for the metal complexes (Figure 1). A nomenclature for the geometrical isomers of hydroxamate tris bidentates was outlined by Leong and Raymond.[4,131,132] (1) No general rule can be applied for an absolute assignment of the chelate ring sequence. However, if the structure exhibits a unique functional group, this group can be utilized to define the chelate ring sequence. In the case of ferrioxamine B or D_1 this group is the N terminus. In the case of the coprogens, the unique group is a diketopiperazine ring placed between rings one and two. (2) Looking down the C_3 axis the sequence of the chelate rings 1, 2, and 3 corresponds to the rotation direction, i.e., clockwise for Λ-isomers and counterclockwise for Δ-isomers. (3) If the ring 1 has the carbon atom of the hydroxamate group below the nitrogen, it is denoted "C". If the reverse is true, it is called "N". (4) For rings 2 and 3 each is called *cis* or *trans* depending upon whether it has the same or opposite relative orientation with respect to the coordination axis as does ring one. As mentioned previously, siderophores achieve iron chelation mainly through two functional groups: hydroxamates and catecholates. Unlike hydroxamate, catecholate is a symmetric, bidentate ligand. Thus, there are no geometrical isomers of simple tris(catecholate) metal complexes. However, all siderophore catecholates are substituted asymmetrically on the catecholate ring, so that geometric isomers may, in principle, exist.

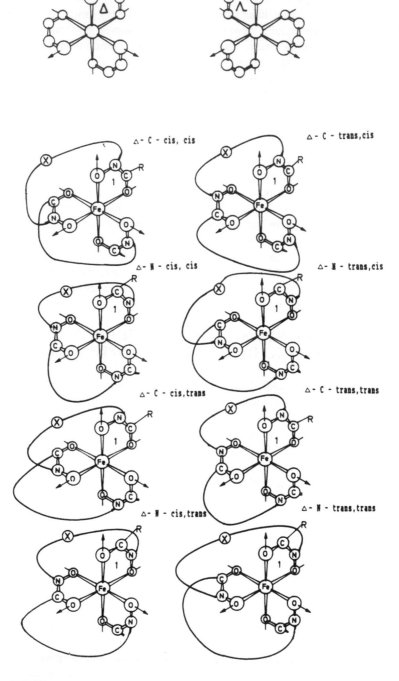

FIGURE 1.   Λ and Δ optical configuration at the metal center (top) and eight geometrical isomers of a d ferric tri-chelate complex, involving three asymmetrical bidentate ligands attached to an asymmetric backbone. Not shown is the set of eight Λ diastereomers. R, X correspond to unique functional groups, e. g., amino-terminus of ferrioxamine B (R) or diketopiperazine ring of coprogen (X).

METAL SUBSTITUTED MICROBIAL IRON CHELATES

As shown by EPR-, Mössbauer and magnetic susceptibility measurements, all ferric ion complexes of microbial iron chelates, investigated so far, are high-spin d^5.[135-144] The d^5 electronic configuration rules out any crystal field stabilization energy (CFSE) and makes the complexes relatively labile with respect to isomerization and ligand exchange in aqueous solution. Furthermore, high spin Fe^{3+} has no spin-allowed d-d transitions. Therefore the UV/VIS and CD spectra of ferric microbial iron chelates are due to charge-transfer transitions, which are not as readily interpreted as are ligand-field (d-d) transitions. This is a disadvantage for any stereochemical investigation. It has been circumvented by substituting the feric ion with kinetically more inert d^3 Cr^{3+} or d^6 Rh^{3+} ions. Since the substituted ions have almost the same ionic radii and the same charge as Fe^{3+}, their complexes show a high degree of structural similarity with the corresponding Fe^{3+} complexes, as demonstrated by the crystal structures of model compounds. The d-electron configurations of Cr^{3+} and Rh^{3+} grant significant CFSE for kinetic inertness and provide well characterized d-d transitions with distinct UV/VIS and CD spectra.

Preparation of Desferrisiderophores and of Cr^{3+} Substituted Chelates

Most siderophores are produced in a chemically defined medium without added iron. In principle, metal-free siderophores can be directly isolated from these culture suspensions. Siderophores which are commonly isolated in their desferri-form are rhodotorulic acid,[69] pyochelin,[83] agro-,[30] vibrio-,[31] and enterobactin.[28] However, for the direct isolation of des-ferri-compounds, iron depleted media and glassware are required. Moreover, desferrisider-ophores are in general colorless. Therefore, a common procedure for isolation of siderophores from a culture is to add ferrous sulfate to the culture filtrate and to purify the siderophores as the colored ferric ion complexes. Ferric hexadentate hydroxamates display colors ranging from brown to orange which are maintained over a wide pH range. Bidentate and tetradentate hydroxamates show pH-dependent color changes. Ferric catecholates undergo pH dependent protonation leading to a color shift from red to purplish-blue or green with decreasing pH. In order to extract the metal from these complexes two strategies have been used. One procedure is based on acidification of the aqueous siderophore solution and transfer of the uncharged desferrisiderophore into an organic phase (e.g., ethyl acetate). This route is successful only, if (1) the ligand protonation constants are significantly higher than the pH of the solution and (2) the ligand is sufficiently stable at these acidic conditions. A second procedure is based on ligand exchange with a chelator whose ferric iron complex is insoluble in water. Generally 8-hydroxyquinoline is employed whose grey black ferric ion oxinate precipitates in aqueous solution. Both the oxinate and excess quinoline are soluble in a chloroform phase. To avoid co-extraction of the desferrisiderophore a pH value should be employed which forces the desferrisiderophores to be charged.

With regard to the formation of kinetically inert complexes, there arises an immediate paradox. The very property of inertness that is to be exploited would seem to make the preparation of the exchange-inert compound difficult, since the ligand exchange reaction needed to introduce the metal is, by definition, slow. A strategy to overcome this problem is to introduce the metal as a complex which is much more labile than the complex to be formed. Cr^{3+} has been used most frequently for the synthesis of kinetically inert siderophore complexes. However, in aqueous solutions Cr^{3+} quickly forms aquo complexes of very low reactivity. Therefore, in many cases the synthesis of Cr^{3+} substituted siderophores is per-formed under anhydrous conditions. The violet tetrahydrofuranate $CrCl_3 \cdot 3THF$ is a partic-ularly useful material for the preparation of chromium compounds of desferrisiderophores, as it is soluble in organic solvents. Chromium complexes are unstable in light. Therefore, darkness is required for the reaction and storage of chromium complexes. Once formed, light sensitivity and dissociation in water are sufficiently slow allowing separation of isomers and uptake kinetics in microorganisms.

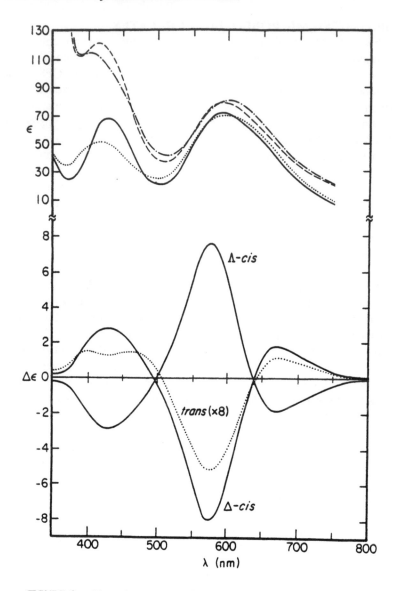

FIGURE 2.   Absorption spectra of Cr(benz)$_3$ in 17% CH$_3$OH-CHCl$_3$ solution, and both absorption and CD spectra of Cr(men)$_3$ in 3% CH$_3$OH-CHCl$_3$ solution: *cis*-Cr(benz)$_3$ (— — — — —), *trans*-Cr(benz)$_3$ (— · —), *cis*-Cr(men)$_3$ (———) and *trans*-Cr(men)$_3$ (····). The CD spectrum of the mixture of *trans* isomers has been multiplied by eight. (With permission of American Chemical Society.)

STEREOCHEMISTRY OF CHROMIC AND FERRIC SIDEROPHORE COMPLEXES

For bidentate ligands with nonequivalent ligating groups, four isomers are possible, Λ-*cis*, Δ-*cis*, Λ-*trans*, Δ-*trans*. Raymond and co-workers have synthesized and separated optical and geometrical isomers of simple hydroxamate and phenolate Cr^{3+} complexes and assigned the absolute configuration of these isomers.[145] Tris(benzhydroxamate)chromium (III),[Cr(benz)$_3$] was separated in *cis* and *trans* geometrical isomers, as confirmed by X-ray structural analyses.[146] Three fractions of optically active tris(N-methyl-1-methoxyacetohydroxamato) chromium(III), [Cr(men)$_3$], were separated by silica gel column chromatography. These were assigned to be a Λ-*cis* fraction, a Δ-*cis* fraction, and a mixture of Λ,Δ-*trans* isomers (Figure 2, Table 5). The *cis* isomers exhibit a higher dipole moment than the *trans*

TABLE 5
UV/Vis (λ_{max}) and CD Spectral Parameters (in nm, Θ in Parenthesis ($M^{-1}cm^{-1}$) of Various Chromic Siderophores and Model Compounds in Solution

Cr-Complex	Δ-cis	Δ-trans	λ-trans	λ-cis	Ref.
Rhodotorulic acid	415 (+1.05)	425(+1.24) −445	405—440 (−0.4)		152
		576(−4.35)	572(+1.92)		
	668(0.95)	665(+1.05)	637—677 (−0.3)		
λ_{max}	418(67)	416(49.01)	414(50.03)		152
	582(68.1)	588(66.6)	584(67.5)		
Coprogen		414(+0.05)		418(−0.73)	153
		567(−1.77)		561(+1.61)	
		666(+0.47)		654(−1.01)	
λ_{max}		406(82.7)		406(88.2)	153
		588(85.3)		584(82.1)	
Cr(men)$_3$	425(+2.95)	395(+0.21)		429(−2.82)	145
	574(−8.26)	461(+0.21)		573(+7.83)	
	671(+1.91)	574(−0.62)		670(−1.85)	
		678(+0.17)			
λ_{max}	424(66.7)	416(49.6)		425(65.5)	145
	593(70.2)	596(69.6)		591(70.3)	
Ferrichrysin				431(−2.48)	273
				570(+7.31)	
				659(−1.86)	
Ferrichrome				432(−2.48)	273
				571(+7.47)	
				661(−1.74)	
K$_3$[Cr(cat)$_3$]	435(+0.93)			435(−0.91)	259
	582(−2.0)			582(+2.3)	
	663(0.43)			663(−0.49)	
λ_{max}	425(104)				259
	592(78)				
[NH$_4$]$_3$[Cr(ent)]	420(+1.6)				259
	574(−7.0)				
λ_{max}	425(60)				
	586(80)				

Ferrioxamine B (absorption maxima of Vis spectra; all isomers racemic)			150
N-cis,cis	C-cis,cis	trans	
420(67.6)	420(74.7)	412(50.4)	
586(70.2)	582(71.1)	592(71.8)	

isomers. Therefore, they bind more strongly to sorbents (e.g., silica gel, ion exchange resin) yielding bands with lower R_f values as compared to *trans* isomers. This parallels the results obtained for the geometrical isomers of a number of similar compounds.[147-149]

The visible spectra of the Cr^{3+} geometrical isomers exhibit two spin-allowed d-d transitions for octahedral symmetry, namely, $^4A_{2g} \rightarrow {}^4T_{1g}$ at higher energy and $^4A_{2g} \rightarrow {}^4T_{2g}$ at lower energy. The absorption maximum of the *trans* isomers is considerably lower at the high energy transition manifold, and in addition the peak maxima are shifted to higher and lower energy as compared to the *cis* isomers (Figure 2, Table 5). The CD spectrum for *cis* tris hydroxamate complexes consists of three bands, due to E_a, A_2 and E_b transitions. The lambda configuration in the model compounds was found to have a positive CD band for the dominant energy transition at 500 to 600 nm (E_a), the delta isomer had a negative CD band in this region. For the *trans* isomer, which has no symmetry (C_1) the E_b transition at high energy is split, providing an additional criterion for differentiating between *cis* and *trans* isomers (Figure 2, Table 5). The electronic features are discussed in more detail in a recent review article.[4]

Based on these assignments the absolute configuration of chromium(III) complexes of various siderophores could be determined. The corresponding UV/VIS-spectral data and CD-spectroscopic parameters are shown in Table 5. Whereas chromic desferriferrichromes exclusively show lambda chirality at the metal center (Table 5), delta chirality of the co-ordination octahedron is observed for chromic desferrienterobactin. Chromic complexes of linear chiral ligands such as chromic desferricoprogen and chromic rhodotorulate represent mixtures of Λ- and Δ-isomers with an overall preference for delta chirality at the metal site, as demonstrated by separation of various geometrical and optical isomers of chromic linear siderophores. Geometrical *cis* and *trans* isomers could be separated for linear chromic hydroxamate siderophores, including ferrioxamine B, coprogen and rhodotorulic acid (see Table 5).[131,150-153] For the cyclic siderophores, the existence of *trans* isomers has not been detected. Apparently, much larger macrocycle ring dimensions or longer side chains bearing hydroxamate groups are required to allow sufficient flexibility for *trans* isomer formation.

The X-ray crystallographic analysis of all ferrichrome siderophores yielded Λ,-C-*cis*,*cis* configurations of the coordination octahedron.[51,154,155] Neocoprogen I adopts a Δ-C,*trans*,*trans* configuration in the crystal structure.[59] N,N',N''-triacetylfusarinine crystallizes as either Δ- or Λ-isomer, depending on the solvent system used.[156] Achiral siderophores such as ferrioxamine E and ferrioxamine D_1 crystallize as racemic mixtures of Δ- and Λ isomers. The determination of the metal center chirality of Fe^{3+}-complexes in solution is possible through comparison of the solution and solid-state CD spectra. However, the correlation of the rotary power with left- or right-handed helical stereochemistry requires an absolute assignment based on crystal structure data. This correlation could be established for triacetylfusarinine and neurosporin.[68,156] This correspondence extends into the other Fe^{3+} complexes and shows that the CD spectra of the ferric complexes in solution can be used for determinations of the metal center chirality. Fe^{3+} complexes will have the Λ configuration (at least predominantly) if the CD band in the region of the absorption maximum (400 to 500 nm for hydroxamates) has a positive sign (see Table 6).

Both $\Lambda \leftrightarrows \Delta$ and *cis* $\leftrightarrow$ *trans* equilibria of siderophore complexes can exist in solution. The chirality of the ligand can impose a preferred metal center chirality. In addition, the degree of this preference depends on the stereochemical rigidity of the ligand. In principle, the magnitude of the molar circular dichroism can be used as a measure for diastereoisomeric equilibria based on a comparison of the solid state and solution ellipticity. Nevertheless, predictions of metal center chiralities require theoretical calculations. For example, empirical-force-field calculations of ferric enterobactin show that the delta orientation at the metal center is more stable than the lambda by 0.5 kcal/mol which is consistent with the CD-spectra.[157] The flexibility of the lactone ring is probably the reason for the energy difference between the Λ and Δ orientations of the complex to be small. Consequently, the possibility that certain changes of solvent and temperature may modify the complex's Δ-*cis* preference cannot be excluded. On the other hand, there exists the possibility that a preorientation of the Δ conformation in desferrienterobactin dictates the path of complexation toward the Δ-*cis* conformation of the complex.

In linear siderophores increasing the chain lengths will obviously cause greater flexibility resulting in smaller energy difference between both optical and geometrical isomers. These small energy differences allow intramolecular rearrangement reactions of tris chelate complexes of labile metal ions, such as Mn^{2+}, $Fe^{3+}Al^{3+}$, and Ga^{3+}.[158-160] For ferrioxamine B the kinetics of isomerization have been analyzed in detail by NMR spectroscopy.[161] In these experiments Fe(III) was substituted by Ga(III) and Al(III) to circumvent the problems associated with paramagnetic line broadening. The isomerization reactions were slow enough to allow detection of two geometrical isomers and to measure the rearrangement kinetics by NMR spectroscopy employing the hydroxamate carbonyls as sensitive probes for metal center symmetries. In D_2O the isomerization is essentially proton independent over the range where

TABLE 6
Visible and CD Spectral Characteristics of Aqueous Siderophores

	λ_{max}	$\lambda_{max},\Delta\epsilon$ (M^{-1}cm^{-1}) (Δ)	$\lambda_{max},\Delta\epsilon$ (M^{-1}cm^{-1}) (Λ)	Ref.
Ferrichrome	425 (2895)		360 (−3.7)	154
			465 (2.4)	
Ferrichrome A	440 (3360)		330 (−3.9)	274, 275
			365 (−2.7)	
			465 (3.2)	
Ferrichrysin	430 (3020)			276
Ferricrocin	434 (2460)		290 (−3.78)	172
			360 (−1.62)	
			450 (+2.47)	
Ferrioxamine B	428 (2800)			
Ferrioxamine E	430 (2750)			275
Coprogen	434 (2820)	375 (+2.1)		172
		474 (−1.26)		
Neocoprogen I	428 (2720)	365 (+1.77)		132
		461 (−1.21)		
Neocoprogen II	422 (2612)	372 (+1.82)		132
		456 (−1.16)		
N,N',N''-triacetylfusarinine		370 (+3.25)		156
		467 (−2.04)		
Mycobactin P	445 (3780)			275, 276
Fe$_2$RA$_3$	425 (2700)	372 (+2.73)		15, 277
pH 7		464 (−1.41)		
Fe(benz)$_3$[a]	435 (4910)	350 (+2.3)	350 (−2.8)	274
		452 (−1.5)	455 (+1.1)	
Aerobactin	398 (2170)		415 (−0.12)	278
pH 7			574 (+0.25)	
			668 (−0.10)	
Enterobactin	495 (5600)	Δ cis		259
Agrobactin	505 (4100)		Λ-cis	279
Parabactin	512 (3300)		Λ-cis	279
Pseudobactin	400 (15,000)		400 (+2.0)	71
			436 (−0.8)	
			502 (+0.3)	
Pseudobactin A	400 (2000)			72
Neurosporin		360 (−4.8)		68
		465 (+4.5)		

[a] In acetone solution.

the complex is stable ($\approx$pH 2 to 10). This rules out mechanisms that require protonation of one of the hydroxamate arms. The enthalpies of activation ($\Delta H = 13$ [kcal/mol] in D$_2$O and 17 [kcal/mol] in methanol-d$_4$) are indicative for small activation barriers. These observations favor an intramolecular rearrangement process that neither involves explicit solvent association nor metal-hydroxamate dissociation in the transition state.[161]

SIDEROPHORE UPTAKE IN MICROORGANISMS PROBED WITH MODIFIED SIDEROPHORES

A wealth of experimental data demonstrates that siderophore uptake in microorganisms is both a receptor- and an energy-dependent process.[162,163] Moreover, many receptor-siderophore interactions are very specific.[4,149-153,164-171] Recognition of siderophores may be dependent on different parts and structural features of the molecule: (a) the molecule as an entity, (b) the geometry and chirality of the backcone, (c) chirality or geometry at the metal center, and (d) peripheral groups. Since X-ray structure determinations of siderophore receptor proteins are still lacking, the structural elements responsible for specific recognition must be traced out by using compounds which are related to the natural siderophore. That is, by employing homologous siderophores, synthetic enantiomers or synthetic analogs.[4,150-153,164-171] Moreover, diastereoisomers or geometrical isomers might affect such discriminations. Therefore, the metal center symmetry is a very important aspect in the analysis of siderophore recognition by membrane receptors. Substitution of ferric iron with kinetically inert ions such as Cr^{3+} or Rh^{3+} enables separation of isomers and studies of their receptor specificity *in vivo*.

The complete siderophore complex can be taken up by cells, or the metal may be absorbed by other means of transportation, preceded by ligand exchange or reduction at the cell membrane. The redox potentials of gallic and chromic desferrisiderophores are outside the range observed for biological reductants. If these complexes show lack of intracellular accumulation, this could imply a reductive step during uptake. If Ga is transported and Cr is not, the uptake process should involve a ligand exchange step because gallic complexes are kinetically labile.

Based on metal substituted siderophores and separated geometrical isomers the stereospecificity and mechanisms of uptake were studied in some detail in *E. coli*,[164-166] *Rhodotorula minuta*,[152,171] *Streptomyces pilosus*,[150,168] and *Neurospora crassa*.[153] *E. coli* is treated separately in the next section. Stereospecific recognition in *Rhodotorula minuta* was probed with separated Λ-*trans*, Δ-*trans*, Δ-*cis* chromic RA as well as with *enantio*-RA (Figure 3).[152] The RA uptake system discriminates between RA and its enantiomeric form. It also discriminates between Λ-*trans* and Δ-*trans* isomers of $Cr_2(RA)_3$ as shown by inhibition of ferric RA uptake[152] and, finally, changes in the backbone have only little effect on uptake rates.[171] Comparison of metal uptake mediated by gallic and chromc RA disclosed an uptake mechanism which involves ligand exchange.[171] In *Streptomyces pilosus* the racemic C-*cis*,*cis*, N-*cis*,*cis*, and the *trans* isomers of Cr^{3+}-desferriferrioxamine B (see Table 5) inhibited equally well the [55]Fe-ferrioxamine B uptake, suggesting that no differentiation between *cis* and *trans* geometrical isomers occurs in this system.[150] Coprogen is the species specific siderophore synthesized by *Neurospora crassa*.[172] For Cr^{3+}-desferricoprogen two fractions were separated by HPLC, a *cis* complex (20% excess Λ) and a mixture of *trans*-complexes (20% excess Δ) (see Table 5).[153] Although this separation of optical and geometrical isomers was not complete, inhibition studies suggest a predominant recognition of the Δ-*trans* isomers. In summary, these investigations revealed that the metal center chirality and the shape of the metal center portions of siderophores are of central importance for molecular recognition by receptors in the cell envelope.

MONITORING THE INTRACELLULAR FATE AND FUNCTION OF SIDEROPHORES BY *IN VIVO* MÖSSBAUER SPECTROSCOPY OF [57]Fe-LABELED SIDEROPHORES

The siderophore uptake process in microorganisms has been studied intensively by means of radiolabel techniques and by employing kinetically inert and reduction inert metal complexes. However, the pathways of iron metabolization and the intracellular roles of siderophores are much less understood. In order for iron to be available for microbial metabolism, the siderophore-bound metal ion must be decomplexed either by enzymatic reduction or ligand destruction or by agents which exhibit thermodynamically as well as kinetically significant higher affinities to ferric ion than the transporting siderophore.

Moreover, knowledge of the intracellular iron metabolism of both higher and microbial organisms is fragmentary. Therefore, the following problem areas are to be addressed: (1) at a qualitative level: are siderophores major constituents of iron metabolism and what are

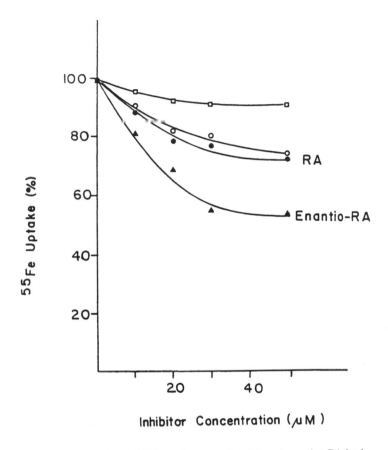

FIGURE 3. Inhibition of ^{55}Fe uptake assayed as RA and enantion-RA by increasing concentrations of geometrical or optical isomers of Cr_2RA_3. Fe_2RA_3 + Δ-*trans*-Cr_2RA_3 (●); *enantio*-Fe_2RA_3 + Δ-*trans*-Cr_2RA_3 (▲); Fe_2RA_3 + Δ-*cis*-Cr_2RA_3 (○); Fe_2RA_3 + Λ-*trans*-Cr_2RA_3 (□). (With permission of American Society of Biological Chemistry.)

the main components of this process? (2) Quantitatively: to what extent do these components contribute to the total cellular iron pool? (3) Dynamics: what kind of compositional changes occur with time or when metabolic changes occur? These questions require a systematic approach enabling simultaneous discrimination of siderophores, cytochromes, Fe-S-proteins, ferritins, and other chelates and the determination of their contribution to the total iron pool. Of course, radioactive label techniques or spectrophotometric analyses are completely unsuitable for solving these problems. However, EPR and Mössbauer spectroscopy should allow, in principle, nondestructive investigations of iron metabolism *in vivo*, if the spectra are not too complex, i.e., if only a few main metabolites are visible and discernible in the spectra. In fact, a wealth of information about the redox states of the metal centers, their spin configurations (high- or low-spin), symmetries of the ligand fields, types of ligands (e.g., O, N, S) can be derived in general from Mössbauer spectra.[173-175] Mössbauer spectroscopy is radiation specific for ^{57}Fe (2% naturally abundant). A drawback of *in vivo* Mössbauer spectroscopy is the need for a relatively high resonance nuclide concentration (minimum 20 μg ^{57}Fe per sample), and as a consequence large sample volumes (1 to 2 cm^3) and long counting times (from 2 days to 4 weeks) are used. In contrast, the advantages of *in vivo* EPR spectroscopy are relatively high sensitivity (concentrations as low as 50 μ*M*), small sample volumes (200 μl), and little time required for measurements (minutes). However, metal ions with an even number of electrons (e.g., Fe^{2+}) are EPR silent. Moreover,

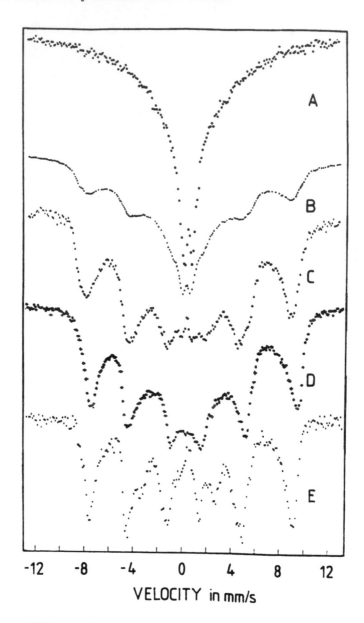

FIGURE 4.   Temperature (295 K: A; 77 K: B, C; 4.2 K: D,E) and concentration dependent Mössbauer spectra of coprogen. Measurements with amorphous coprogen (A, B, D) or freeze-dried powders diluted with BSA (C, E). BSA minimizes relaxation and simulates the protein rich intracellular environment.

paramagnetic ions other than iron and organic radicals are EPR active and may mask iron signals. Therefore, Mössbauer spectroscopy is the method of choice for *in vivo* investigations of siderophore mediated iron assimilation.

MÖSSBAUER SPECTROSCOPIC PROPERTIES OF MICROBIAL IRON CHELATES

As shown in Figure 4, Mössbauer spectra of siderophores display a strong temperature and concentration dependence. This can be understood in terms of relaxation phemomena. Two types of relaxation processes have to be considered: spin-spin and spin-lattice relaxation.[138,139,176] Spin lattice relaxation depends on temperature, whereas spin-spin relaxation

TABLE 7
Mössbauer and EPR Parameters of Siderophores

Siderophore	δ (mm s^{-1})	ΔE_Q (mm s^{-1})	H_{int} (T)	D	λ	Ref.
Enterobactin(methanol)	0.28	−0.38	55.5	0.48	0.46	135
	0.51		53.6			134
Enterobactin	0.51		53.6			134
Mycobactin				0.34	0.27	142
Ferrichrome A		0.64	−53.75	−0.29	0.25	139
		0.94	−55.0	1.0	0.25	138
Ferrirubin				0.44	0.3	141
Ferrichrysin				0.42	0.267	141
Ferricrocin	0.65	−0.60	−56.0	0.46	0.31	137
Coprogen	0.65	−0.60	−55.0	0.47	0.32	137
RA-analog (n = 10)	0.5	−0.63	−55.0	−0.8	0.33	140

Note: Mössbauer analyses have been performed of spectra measured at 4.2 K.

depends on the concentration of the magnetic ions, but is in most cases independent of the temperature. Fortunately, only little residual relaxation is observed in siderophore spectra at 4.2 K if the material is magnetically diluted, for instance, with bovine serum albumine (BSA). In addition, BSA simulates reasonably well the protein-rich environment to be expected in the cytoplasm of microorganisms. In some cases even water, glycerol or methanol are suitable dilutants.[133-135]

Although the diamagnetically diluted siderophore spectra are well resolved at 4.2 K, the absorption line pattern is complex. The complexity of the spectra arises from the magnetic interaction of the nucleus with unpaired shell electrons, which results in a splitting of the nuclear states and hence, in a multitude of transitions. For high-spin ferric iron as many as 288 transitions might be observed in the most general case $[(2S + 1)^2(2I_e + 1)(2I_g + 1)]$. However, transitions between different electronic spin states have not been observed by Mössbauer spectroscopy in such systems. Moreover, due to Kramers' rule all ions with odd number of electrons exhibit energy levels which are at least twofold degenerate in the absence of magnetic fields. Thus, the three Kramers' doublets of the $^6S_{5/2}$ state of ferric high-spin and the coupling of these states with the nuclear ground- ($I_g = 1/2$) and nuclear excited state ($I_e = 3/2$) yield Mössbauer spectra which are superpositions of three six-line subspectra arising from the three Kramers' doublets. For a complete description see Reference 4.

Various siderophores have been analyzed by EPR and Mössbauer spectroscopy. The parameters obtained by a spin Hamiltonian analysis of the experimental spectra are listed in Table 7. All siderophores have some features in common. The spectra are superposition of three magnetically split sub-spectra (vide supra). Isomer shifts and quadrupole splittings are typical for high spin Fe^{3+}. The internal field is of the order of −55T, indicating a high degree of ionicity typical for an isolated FeO_6 octahedral configuration. The λ-values near 0.333 of all siderophores listed in Table 7 reflect a nearly complete rhombic distortion of the FeO_6-octahedron arising from crystal fields at the nucleus with symmetries lower than C_3. The low symmetry observed in Mössbauer and EPR spectra of siderophores seems to be a general feature of these ferric ion complexes.

Besides siderophores, Mössbauer spectroscopic data of ferritins,[177-185] bacterioferritins,[186-189] cytochromes[175,190-192] and Fe-S-proteins[193-196] are indispensible prerequisites for the *in vivo* analysis of siderophore metabolism in microorganisms. Cooperative phenomena are typical for ferritins and bacterioferritins and enable a relatively facile Mössbauer spectroscopic estimation of ferritin contributions in cell spectra.[176,177] At high temperatures the Mössbauer spectrum will display a quadrupole-split two-line pattern. Below the blocking temperature a six-line spectrum with broadened lines is expected whereas siderophores show relaxation effect from room temperature to 4.2 K. The internal field and temperature-dependent spectral pattern of ferritins, bacterioferritins, and siderophores differ significantly enabling their discrimination in cell spectra. In general, a comparison of experimental and literature data enables a discrimination of most iron proteins.

SIDEROPHORE UPTAKE IN MICROORGANISMS AND IRON METABOLISM
Methodical Aspects

Mössbauer spectroscopic studies have been performed with enzyme complexes like photosystem I[197] and II,[198] membranes,[199] mitochondria,[200] various tissues,[79,200-202] erythrocytes,[203] and with whole microbial cells.[178,186-188,205-218] Even Mössbauer spectra of whole cells measured at room temperature have been reported.[219] However, low f-factors, line broadening due to thermal motion and sample instability, caused by metabolic activity, preclude a detailed analysis at room temperature. Thus, cell samples are in general prepared by quick-freezing.[220] Although there are some problems associated with freezing, frozen cells assume a proper environment compared to isolated enzymes or cell fractions, and justify the term *in vivo* for such preparations.

Resonance Nuclide Concentration

Because of the relatively small iron content in cells, the problem of achieving good signal-to-noise ratio in a Mössbauer experiment of reasonable duration (3 days) is often severe. A decrease of the resonance nuclide concentration by a factor of F requires an increase of counting time by factor of F^2 in order to obtain an equivalent signal-to-noise ratio. The Gram-negative bacterium *E. coli* will be chosen here as a concrete example. Growth of this organism in a ferrichrome supplied mineral salt medium to an $OD_{578} = 0.8$ yields an overall accumulation of 0.5 μmol iron per gram cells. Approximately 2 g of cells can be filled into a conventional cylindrical sample holder (1.5 cm diameter, 0.9 cm height). This corresponds to 2×10^{17} iron atoms/cm² or 4×10^{15} ^{57}Fe/cm². The peak resonant cross section for γ-absorption is approximately 2×10^{-18} cm². The effective cross section is reduced by a factor of two by the convolution of finite source and absorber line width. The recoil-free fraction (the fraction of nuclear transitions which occur without recoil) for iron in proteins is of the order of 0.7 at 4.2 K. Moreover, the effective intensity observed depends on the number of iron species in the sample. Quadrupole splitting further decreases the intensity of any individual absorption by a factor of two. Magnetic hyperfine interaction can broaden a spectrum by an effective factor of ten or more. As an example only two quadrupole doublets will be assumed for simplicity, i.e., an additional factor of four has to be taken into account. The predicted absorption dips would thus have a strength of 0.07% of the 14.4 keV γ-ray intensity. With a source strength of 50 mCi, a counting rate of 25,000/sec, and the use of 256 channels, about 100 counts per channel and per second are obtained. Under these conditions, at least 24 days of accumulation are required to obtain a fractional uncertainty of 1/100 for the signal. Finally, the absorption pattern of an *in vivo* spectrum is expected, in most cases, to be much more complex than is assumed in the example above and this increases the counting time considerably. Thus, the resonance nuclide concentration of specimens must be significantly increased. This cannot be achieved by increasing the sample volume. A larger sample thickness results in considerable loss of resonance absorption. Due to the geometry of Mössbauer experiments, larger diameters of sample holders do not increase resonance absorption restricting the effective sample volume to 2 to 2.5 cm³. The most efficient, albeit expensive, method for increasing the signal-to-noise ratio is the enrichment of the Mössbauer samples with ^{57}Fe. ^{57}Fe (95% purity, about $20/mg) is commercially available and complexes with desferrisiderophores are easily formed. The counting time can also be reduced by the application of a stronger Mössbauer source. This requires detectors with a high counting efficiency. In addition, there are problems posed by self-absorption of the source. This can be solved by software by spectral subtraction if the spectral pattern associated with self-absorption is known. The upper limit is a source strength of 100 mCi. γ-cross sections increase rapidly with atomic number. High concentration of salts may, therefore, drastically increase the opacity of the sample. Solvents containing sulfur or chlorine are strongly absorbing and severely restrict sample thickness.

Experimental Design

For *in vivo* Mössbauer studies of siderophore uptake and iron metabolism four different experimental routes are important.

1. In longtime studies ^{57}Fe-labeled siderophores are added with the inoculum to a low-iron culture. Depending on the organism the final ^{57}Fe-siderophore concentration is 2 to 3 μM for bacteria, and 10 to 30 μM for fungi and yeasts. These conditions allow iron-sufficient growth since high amounts of siderophores remain in the culture supernatant after many cycles of cell division. For example, culture supernatants of *E. coli* supplemented with ferricrocin (2 μM at $OD_{578} = 0.01$) still exhibit a ferricrocin concentration of 0.9 μM at $OD_{578} = 0.8$. Cells are then harvested at various cell densities. Any changes observed under these conditions are governed by growth phase effects.

2. Kinetics of siderophore uptake and metabolization are most easily analyzed in cultures grown for many generations under iron-deficient conditions (e.g., late logarithmic growth phase). After adding ^{57}Fe siderophores, at least 10 minutes of uptake is required to obtain a cellular resonance nuclide concentration sufficiently high for measurements. After this incubation time, the kinetics of iron uptake and metabolization can be observed with a time resolution which is limited only by the time required for sample preparation, i.e., the time required to cool down a cell suspension of approximately 21 volume to 0°C. This is a temperature at which no further metabolic activity is expected. Depending on the experimental set-up the time resolution can be as low as 2 minutes. Machines to be designed for quick cooling of cell suspensions with volumes on a liter scale could certainly permit even a much better time resolution.

3. Analysis of microbial iron storage or detoxification compounds requires cell growth in mineral salt or complex media strongly enriched with ^{57}Fe (50 to 100 μM). This preparation procedure leads to a build-up of iron storage compounds inside the cell, whereas iron containing enzyme systems will be masked under such conditions.

4. Based on experimental routes 1 or 2, an *in vivo* analysis of iron-containing enzymes can be performed, if the intracellular concentration of the enzyme system of interest is sufficiently high. An enhanced expression of enzymes can be in part achieved by metabolic changes, i.e., by changing the carbon and energy source, the terminal electron acceptor etc., or by employing mutants which are lacking or overexpressing selected enzyme complexes.

Spectral Analysis

Since fitting procedures of cell spectra containing siderophore contributions are difficult to perform, it is convenient to start a spectral analysis with subtraction of the siderophore subspectra. This procedure is called stripping. In order to keep statistical scattering of stripped spectra as low as possible care must be taken not to take the experimental siderophore spectra for the subtraction procedure unless the signal-to-noise ratios of siderophore and cell spectra are excellent. Moreover, theoretical envelopes based on spin Hamiltonian analyses often do not match the experimental spectra accurately enough to be employed for stripping.[4] In our experience, theoretical envelopes based on nonphysical fits, yield the best results (see Figures 5 and 6). Due to relaxation effects, Mössbauer spectra of undiluted siderophores are generally not suited for the analysis of cell spectra.

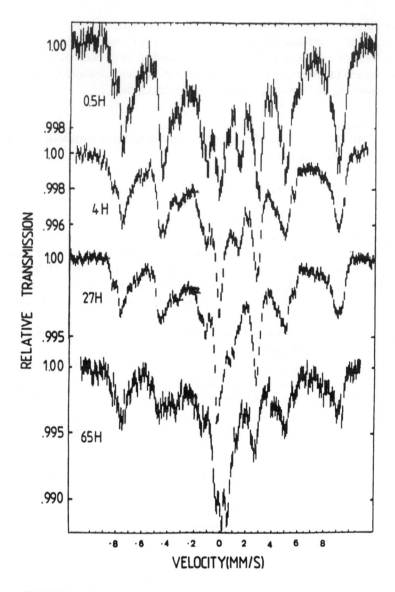

FIGURE 5.   Mössbauer spectra of frozen *N. crassa arg-5 ota aga* cells at 4.2
K in a magnetic field H_{app} = 20 mT perpendicular to the γ-rays. Sample preparation
is described in the section on methodical aspects of siderophore uptake in microorganisms and iron metabolism. Cells were grown for 72 h under low-iron stress
and subsequently incubated for 1 h with ^{57}Fe-labeled coprogen. Metabolization
was stopped by freezing samples after 0.5, 4, 27, and 65 h of additional growth.

Depending on the internal fields, different siderophores can be discriminated. The internal fields of the siderophores coprogen and ferricrocin, for example differ by 1T (see
Table 7). In terms of the energy scale, the spectrum of ferricrocin spans a range from -7.9
mm s^{-1} to $+9.5$ mm s^{-1}, whereas the outermost resonance absorption of coprogen is near
-7.7 mm s^{-1} and $+9.0$ mm s^{-1}. This difference is sufficiently large to discriminate these
siderophores in *in vivo* cell spectra. Similarly, ferrichrome exhibits the same internal field
as ferrocrocin whereas H_{int} of ferrichrome A is 2T smaller (see Table 7). The effective fields
of ferritins are significantly lower than those of siderophores enabling an unequivocal distinction between siderophores and ferritins. By least-squares fits of Lorentzian lines to the
residual spectra, Mössbauer parameters and percentage of the total absorption area of all
other cellular metabolites can be obtained.

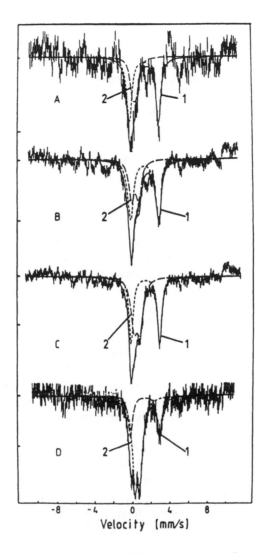

FIGURE 6. Mössbauer spectra of frozen *N. crassa arg-5 ota aga* cells after stripping. Experimental conditions as in Figure 5: 0.5 h (A), 4 h (B), 27 h (C), and 65 h (D). Contributions of coprogen and, when necessary, of ferricrocin (spectra C and D) were subtracted from the cell spectra. The resulting metabolite spectra were fitted with two quadrupole doublets of Lorentzian line shape (see text). The corresponding percentages of the absorption areas are listed in Table 8.

Intracellular Fate of Siderophores
Requirement for Intracellular Iron Binding Agents

All of the iron used in life-sustaining processes is potentially harmful when released from its natural environment. Iron may participate in the oxidative damage of cells by activating oxygen.[221-223] The resultant free oxygen radicals have been hypothesized to be causative factors in aging,[224] carcinogenesis,[224,225] and radiation injury and to be a contributory factor in tumor promotion.[226] In fact, it was demonstrated that oxygen free-radicals generated by Fe^{2+} in aqueous solution are mutagenic.[226]

One of the simplest mechanisms for the generation of oxygen free-radicals involves the reduction of H_2O_2 by divalent metal ions, particularly Fe^{2+}, with the formation of hydroxyl radicals (Fenton reaction):

$$H_2O_2 + Fe^{2+} \rightarrow Fe^{3+} + OH^- + OH* \tag{1}$$

H_2O_2 is a normal metabolite in the cell and its steady state concentration is in the range 10^{-8} to 10^{-9} M.[227] Instead, the hydroxyl radical reacts with almost every type of molecule found in living cells: sugars, amino acids, phospholipids, DNA bases, and organic acids.[223,228] The hydroxyl radical attacks molecules with a rate constant usually of the order of 10^9 M^{-1} s^{-1}.[229]

When the Fe^{2+} concentration is limiting, Fe^{3+} reduction is required for the continued formation of hydroxyl radicals. This can be accomplished by the superoxide anion which functions, in addition, as a chain carrier of the radical reaction:

$$O_2^-* + Fe^{3+} \rightarrow Fe^{2+} + O_2 \tag{2}$$

The net reaction of (1) and (2) has been termed the iron-catalyzed Haber-Weiss reaction.

The solubility of ferric ion at physiological pH is so low ($\approx 10^{-18}$ M) and the tendency of both ferrous and ferric iron to form coordination compounds is so high, that it is fair to assume that iron is tied in complexes under these conditions. Given the opportunity to reach equilibrium, it will be complexed to the ligands which it binds most tightly. Nevertheless, iron complexes do not preclude reactions with activated oxygen species. Some chelates (e.g., the Fe-EDTA complex) are reactive in the iron catalyzed Haber-Weiss reaction, whereas others (e.g., ferrioxamine) are not.[223,230-232] Ferrous NADH and NADPH complexes also appear to act as Fenton reagents,[233] as do AMP, ADP, and ATP complexes.[233-235] Therefore, virtually all organisms, growing either aerobically or anaerobically, must have evolved well designed iron containing enzymes, as well as low or high molecular weight iron transport and storage compounds which are sufficiently inert with respect to iron catalyzed Haber-Weiss-Fenton reactions. The potential toxicity, but also the importance of iron is taken into account in most organisms by the synthesis of iron storage compounds like ferritin and bacterioferritin. Moreover, stable intracellular iron carriers must be postulated enabling sheltered iron transfer. Such carriers still remain to be discovered. Although experimental facts on iron catalyzed oxygen free radical formation *in vitro* and *in vivo* are well documented, it should be mentioned that the outdated paradigm on the occurrence of bulk iron to occur, intracellularly in a free accessible form, e.g., as $Fe(H_2O)_6^{2+}$, still exists.

Siderophore Transport and Metabolization in the Ascomycet Neurospora Crassa

Under conditions of iron deficiency various siderophores are synthesized by the fungus *Neurospora crassa*. The major component being excreted is desferricoprogen (14), Table 3.[172] Relatively high amounts of a second siderophore, ferricrocin (11), Table 2, can be extracted from iron-supplied mycelia.[236] Mere traces of ferricrocin are detectable in the extracellular space. However, *N. crassa* displays comparable uptake rates for both ferricrocin and coprogen.[169]

In order to disclose the intracellular role of coprogen and ferricrocin, the iron uptake and metabolization in wild-type *Neurospora crassa* 74A and in the siderophore-free mutant *N. crassa arg-5 ota aga* were compared by Mössbauer spectroscopy.[208,211,212] A typical series of spectra is shown in Figure 5. For comparison Figure 6 displays the same spectra after stripping the siderophore contributions. The corresponding contributions to the total absorption area of the cell spectra are listed in Table 8. In the series of experiments no ferritins with large particle size can be detected in the spectra suggesting that this type of storage is not present in *N. crassa*. All Mössbauer spectra, after coprogen uptake, exhibit coprogen contributions indicating that this siderophore is accumulated inside the cell as an entity. The siderophore is continuously metabolized; however this process is much faster in the wild type than in the mutant. Moreover, after 3 days of iron starvation the rate of metabolization

TABLE 8
^{57}Fe-Coprogen and ^{57}Fe-Ferricrocin Uptake in *N. crassa* 74A and in Mutant Strain *N. crassa arg-5 ota aga*, Which is Unable to Synthesize Siderophores

Time	Percentage of intracellular iron species			
	Coprogen	Ferricrocin	Fe^{2+}	Fe^{3+} (?)
A. ^{57}Fe-Coprogen Uptake in *N. crassa* 74A				
0.5 h	65%	28%	—	7%
4 h	36%	47%	?	17%
27 h	0	73%	9%	18%
B. ^{57}Fe-Coprogen Uptake in *N. crassa* "Ota"				
0.5 h	85%		15%	
4 h	68%		16%	16%
27 h	55%	12%	18%	15%
65 h	23%	26%	16%	35%
C. ^{54}Fe-Coprogen Uptake in *N. crassa* "Ota" (1 Day Def.)				
4 h	80%		12%	8%
30 h	75%		12%	13%
D. ^{57}Fe-Ferricrocin Uptake in *N. crassa* "Ota"				
0.5 h	—	99%	1%	?
4 h	—	89%	?	11%
65 h		90%	?	10%
E. ^{57}Fe-Ferricrocin Uptake in *N. crassa* 74A				
0.5 h		95%	3%	2% (?)
4 h		93%	—	7%
27 h		90%	—	10%

Note: Cells were grown for 24 h (C) and 72 h (A, B, D, E) in iron depleted minimal medium and incubated for one hour with ^{57}Fe-siderophore. The percentages of subspectra contributing to the total absorption area of cell spectra in Figure 5 were obtained by stripping (coprogen, ferricrocin) and least squares fit procedures of the residual spectra (Figure 6).

is much higher than after one day of iron starvation. From these observations it can be concluded that metal release from coprogen is regulated by the iron requirement of the cell. In wild-type mycelia, the main component of iron metabolism is a different siderophore, ferricrocin. Even in the mutant strain ferricrocin is detected. This is unexpected, since *N. crassa arg-5 ota aga* is unable to synthesize ornithine. It is supposed that ferricrocin is synthesized by the use of coprogen degradation products. Coprogen is obviously only a transient intracellular iron pool of *N. crassa*. In an additional set of Mössbauer experiments, ferricrocin uptake was monitored in both 74A and *ota*. Surprisingly metabolization of ^{57}Fe-ferricrocin did not exceed 11% of the total absorption area (see Table 8).

The large amount of metal transfer from coprogen to ferricrocin and the scarcity of metabolization of Fe^{3+} bound to ferricrocin indicates a quasi inert iron storage function of this siderophore in *N. crassa*. Moreover, we have shown that ferricrocin acts as an iron storage compound in spores of *N. crassa*. These observations point to a very important intracellular function of ferricrocin in sporulation. In fact, many nutritional factors contribute to the induction of fungal sporulation and this process is associated with tremendous metabolic activity. Macromolecules must be synthesized to provide sporulation-specific structures. It is known that a number of fungal spores have prepackaged RNA and/or proteins that are required for the preliminary stages of spore germination.[237-239] In addition, it has been suggested that mycelial vesicles of *Neurospora crassa* which contain basic amino acids serve as a reservoir for precursors required during sporulation.[240] In this context, the intracellular accumulation of ferricrocin seems to warrant an iron pool sufficiently large for sporulation. Thus, the rationale for high amounts of ferrocrocin and the relative inertness of ferricrocin in mycelia of *N. crassa* has to be seen from the viewpoint of propagation.

These experiments nicely demonstrate how the time course of intracellular ligand exchange can be traced out and how cellular main components of iron metabolism can be characterized by Mössbauer spectroscopy. Mössbauer spectroscopy revealed that the two siderophores, coprogen and ferricrocin, have different functions in *N. crassa*. Whereas coprogen acts as an iron transport agent and transient intracellular iron pool, ferricrocin is a quasi inert intracellular iron storage compound.

Uptake of Ferric Rhodotorulate in the Yeast Rhodotorula Minuta

The biosynthesis of the dihydroxamate rhodotorulic acid (RA), (17), is characteristic of heterobasidiomycetous yeasts.[241] Transport of ferric RA in *Rhodotorula minuta* var. ex *texensis* (formerly termed *Rhodotorula pilimanae*) has been studied by means of radioactive labeled ^{55}Fe- and ^{67}Ga-complexes.[152,171] In addition, inhibition studies with chromium substituted isomers enabled a mapping of the features of siderophore receptor recognition. It is important to note that in *Rhodotorula* the $Fe_2(RA)_3$ complex is not transported across the cell membrane.[152,171] Rather, ligand exchange occurs. To illuminate the nature of this unknown ligand and of the intracellular targets, *in vivo* Mössbauer spectroscopy was employed.[215]

Mössbauer measurements of frozen aqueous Fe_2RA_3 yield a typical siderophore spectrum (Figure 7A). In Figure 7B a Mössbauer spectrum is depicted of frozen *Rhodotorula minuta* cells, grown in a salt medium supplied with 15 μM $^{57}Fe_2(RA)_3$ and harvested at $OD_{660} = 0.85$. The spectrum contains a magnetically split subspectrum which closely resembles $Fe_2(RA)_3$. From this spectrum the experimental ferric rhodotorulate spectrum (Figure 7A) was stripped yielding a $Fe_2(RA)_3$ contribution to the envelope spectrum of 55 (± 5)%. The identity of RA was confirmed by isolation of $Fe_2(RA)_3$ from cells grown in the same way as in the Mössbauer experiment. The remaining metabolite spectrum after stripping is shown in Figure 7C. Two metabolites can be discerned: a Fe^{2+} high spin species ($\delta = 1.36(8)$ mm/s, $\Delta E_Q = 2.84(9)$ mm/s) and an additional component, representing Fe^{3+} in a high-spin state ($\delta = 0.43(5)$ mm/s, $\Delta E_Q = 1.07(6)$ mm/s). The contribution to the envelope spectrum is 18(± 3)% for Fe^{2+} and 27(± 3)% for Fe^{3+}. Prolonged growth to $OD_{660} = 1.5$ yields a similar pattern: $Fe_2(RA)_3$ 55(± 6)%, Fe^{2+} 34(± 2)%) and Fe^{3+} 11(± 4)%. If $^{57}Fe_2(RA)_3$ (15 μM) is added to iron depleted cultures of *R. minuta* at $OD_{660} = 0.8$, 60% of the metal is accumulated within 30 min. Figure 7D represents the corresponding Mössbauer spectrum. The $Fe_2(RA)_3$-contribution is 68%. The residual contribution is solely due to an Fe^{2+} high-spin species.

No ferritin-like iron pools could be detected. Surprisingly, the main intracellular iron component in *Rhodotorula minuta* is the siderophore RA. As mentioned already, since energy dependent ligand exchange occurs at the membrane level, ligand exchange is not a simple exchange between $Fe_2(RA)_3$ and excess desferri-RA.[242] Rather, the process of metal transfer from extracellular RA to intracellular iron storage RA requires an additional mediator. Because the major intracellular source of iron in this yeast is a siderophore and ferritin-like structures are not at all present we assign to RA an additional function of intracellular main iron storage compound.

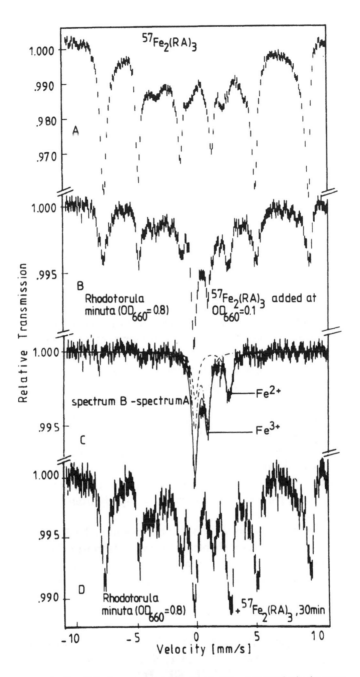

FIGURE 7. Mössbauer spectra of ferric rhodotorulate uptake in the yeast *Rhodotorula minuta*. A correponds to frozen aqueous $Fe_2(RA)_3$. Spectrum B corresponds to frozen *Rhodotorula minuta* cells, grown in a salt medium supplied with 15 μM $^{57}Fe_2(RA)_3$ and harvested at $OD_{660} = 0.85$. The magnetically split subspectrum is due to $Fe_2(RA)_3$. 1C shows 1B after stripping 1A. This metabolic pattern is similar to those found in Gram-negative bacteria (see Figure 8). If $^{57}Fe_2(RA)_3$ (15 μM) is added to iron-depleted cultures of *R. minuta* at $OD_{660} = 0.8$ 60% of the metal is accumulated within 30 min. Figure 1D represents the corresponding Mössbauer spectrum.

The siderophore accumulation process is slightly different in the yeast *Ustilago sphaerogena*.[215] Similar to RA, the transport siderophore, ferrichrome A, (11), Table 2, is not accumulated. The intracellular storage siderophore is not ferrichrome A, but ferrichrome. Based on these findings, it is concluded that siderophores replace ferritins not only in certain ascomycetes but also in heterobasidiomycetes. This further extends the biological importance of siderophores.

Ferrioxamine E Uptake in the Gram-Negative Bacterium Erwinia herbicola

Whereas Enterobacteriaceae typically synthesize enterobactin under iron deficient conditions,[28] the species specific siderophore of *Erwinia herbicola* K4 is ferrioxamine E, (8), Table 1.[6] Ferrioxamine E mediated iron uptake and metabolization of the metal was analyzed by Mössbauer spectroscopy.[216] In Figure 8 dynamics of iron metabolization after ^{57}Fe-ferrioxamine E uptake is depicted. Mössbauer parameters are summarized in Table 9. A comparison of cell spectra with the spectrum of ferrioxamine clearly demonstrates that ferrioxamine E is not accumulated in the cell indicating a fast metal transfer. Thirty minutes after uptake, 85% of the internalized metal corresponds to a ferrous ion compound and 15% to a ferric iron species. Thus, the metal transfer involves a reductive process. Only two main components of iron metabolism could be detected, differing from bacterioferritin, cytochromes and iron-sulfur proteins. All other iron requiring cellular activities are at least one order of magnitude less in concentration. Figure 9 shows cell spectra of *Erwinia herbicola* at various cell densities. As cell growth proceeds increasing amounts of the ferrous species are converted into the ferric component which therefore seems to represent the acceptor for processed ferrous iron (see Table 9). Mössbauer parameters and autoradiograms of cell extracts indicate that these main components correspond to the same low molecular weight proteins which have been detected in and isolated from *E. coli*.[209,213,214] In an additional study, evidence was provided for a linkage between these intracellular iron proteins and their redox states and the ferric iron uptake regulation protein Fur of *e. coli*.[217] Fur participates in both in the regulation of siderophore mediated iron uptake and in the intracellular regulation of the iron redox-states.[217] In *Pseudomonas aeruginosa*, also two intracellular main species are found[218] the Mössbauer parameters of which differ slightly from those found in *Erwinia* and *E. coli*.

From the studies presented in this section the conclusion can be drawn, that siderophores in Gram-negative bacteria are quickly shunted into iron metabolism. Two main components are detectable which differ from bacterioferritin, cytochromes and Fe-S proteins.[214] However, in siderophore producing fungi and yeasts the transport siderophore or a second siderophore represents the main intracellular iron storage compound the contribution of which hardly declines below 50% of the total cellular iron pool.

CHARACTERIZATION OF ENTEROBACTIN AND ENTEROBACTIN MEDIATED IRON TRANSPORT IN *E. COLI* A CASE STUDY

Enterobactin (5), synthesized by *E. coli* and other enteric bacteria,[28,29] is one of the most thoroughly studied siderophore systems. The complex and redox chemistry of enterobactin is complicated and has been a challenge for many scientists. The same holds for the mechanism of enterobactin mediated Fe-transport which still awaits full clarification. Much of the recent work in this field was performed by Raymond and co-workers.

CHARACTERISTICS OF ENTEROBACTIN AND OF ITS MODEL COMPOUND MECAM

Whereas trihydroxamate siderophores form stable complexes with the ferric ion over a large pH regimen, tricatecholate complex structures are strongly pH dependent. The pH

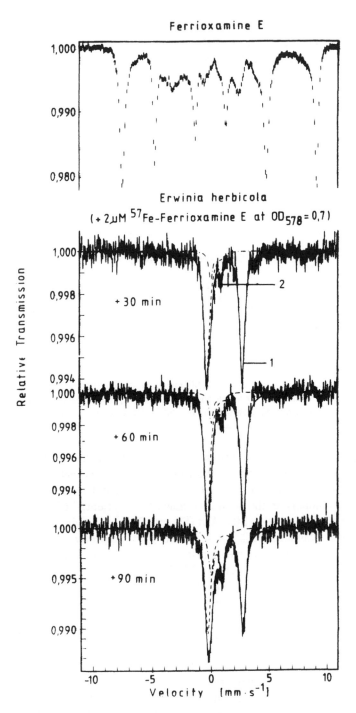

FIGURE 8. Mössbauer spectra of Ferrioxamine E uptake in the Gram-negative bacterium *Erwinia herbicola*. Dynamics of iron metabolization after ^{57}Fe-ferrioxamine E uptake is depicted. Mössbauer parameters are summarized in Table 9. A comparison of cell spectra (Figure 8B to D) with the spectrum of ferrioxamine E clearly demonstrates that ferrioxamine E is not accumulated in the cell indicating a fast metal transfer. Thirty minutes after uptake, 85% of the internalized metal corresponds to a ferrous ion compound and 15% to a ferric iron species. Thus, the metal transfer involves a reductive process. Only two main components of iron metabolism could be detected, differing from bacterioferritin, cytochromes and iron-sulfur proteins.

TABLE 9
Mössbauer Parameters of the Iron Species in Cell Spectra of *Erwinia herbicola* Obtained by Least-Squares Fits

Time of metabolization	δ (mm/s)	ΔE_Q (mm/s)	Percentage of total absorption area
Component 1 [Fe³⁺ (O)₆]			
30 min	0.47 (4)	0.85 (5)	13.6 (33)
60 min	0.53 (4)	0.88 (4)	13.3 (25)
90 min	0.49 (4)	0.91 (4)	29.6 (16)
$OD_{578} = 0.6$	0.46 (2)	0.96 (2)	42.2 (11)
$OD_{578} = 1.0$	0.43 (2)	0.97 (2)	50.1 (7)
$OD_{578} = 1.3$	0.44 (3)	1.02 (3)	52.8 (7)
Component 2 [Fe²⁺ (O)₆]			
30 min	1.28 (2)	3.10 (2)	86.4 (14)
60 min	1.24 (2)	3.06 (2)	86.7 (18)
90 min	1.25 (2)	3.04 (2)	70.4 (12)
$OD_{578} = 0.6$	1.26 (2)	2.95 (2)	57.8 (13)
$OD_{578} = 1.0$	1.23 (2)	2.86 (3)	49.9 (9)
$OD_{578} = 1.3$	1.26 (4)	2.88 (5)	47.2 (10)

Note: Ferrioxamine E was added to a M9 minimum medium ([Fe]≤0.5 μM) either with the inoculum for monitoring growth phase dependent changes of cellular iron or at midlog growth to follow uptake kinetics and initial intracellular reaction on a short time scale. For comparison see Figures 8 and 9.

dependence reflects changes of the iron environment. In enterobactin, the metal is coordinated by three catecholate dianions at pH 7. With decreasing pH the catecholate oxygens are protonated stepwise concomitant to a bond shift from catecholate to a salicylate mode of binding resulting eventually in the neutral complex $[Fe^{3+}(H_3ent)]^0$ (Figure 11).[243] For enterobactin and the synthetic analog MECAM (Figure 10) the following complex protonation constants have been evaluated: $K_{MHL} = 4.89$ and 7.08, respectively, $K_{MH2L} = 3.15$ and 5.6 respectively.[243,244] It should be mentioned that the sequential protonation and the concomitant bond shift were called in question.[245,246] However, spectrophotometric titrations,[243] pH dependent Mössbauer spectra,[134] an infrared study,[247] and very recently a ²H-NMR investigation[248] corroborated the initial proposal. In the Mössbauer study it was shown that the six-line pattern is replaced by a doublet when the pH is lowered indicating changes at the metal center (Figure 11). The sextet-doublet ratio of the Mössbauer spectra can be exploited as a rough probe for the pH of a sample. Measurable quantities of the doublet species require at least a pH of 6.5 or lower for enterobactin, and a pH of 8 for MECAM. Direct involvement of the α-carbonyl group in coordination to the iron has been probed using FTIR. The carbonyl stretching frequency (in the region 1593 to 1640 cm⁻¹) is affected by both ligand deprotonation and metal iron coordination. The study showed that all catechoyl arms remain associated with the metal ion when protonated (Figure 12, Scheme 1). Additional structural information on the protonated forms of tricatecholamides was obtained by monitoring the NMR paramagentic shift of the CD₃ resonance in Fe(1,3,5-tris-[[[4-(deuteromethyl)-2,3 dihydroxybenzoyl]amino]methyl]benzene), a synthetic enterobactin analog. Large shifts have been seen in these types of complexes, resulting from the π delocalization of unpaired spin density in a dominant contact shift interaction. Comparison with ²H-NMR spectra of deuteromethylated ferric salicylate established that protonation of ferric enterobactin and related model compounds results in a shift of the catechoylamides from a catecholate to a salicylate mode of bonding.

Erwinia herbicola

$(+2\mu M\ ^{57}Fe\text{-Ferrioxamine E at }OD_{578}=0.01)$

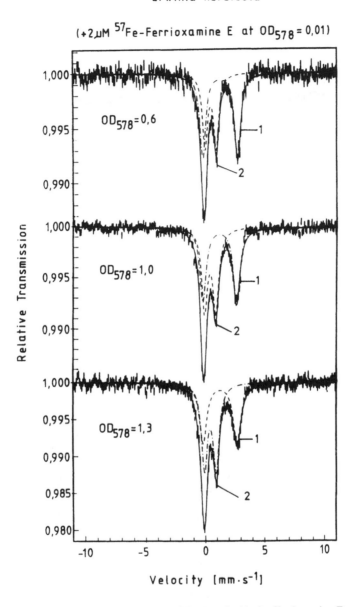

FIGURE 9. Mössbauer spectra of *Erwinia herbicola*. Ferrioxamine E was added with the inoculum. Cells were harvested at various cell densities. As cell growth proceeds increasing amounts of the ferrous species (1) are converted into the ferric component (2) (see Table 9).

Enterobactin has the highest complex formation constant ever observed for natural Fe(III) complexing chelators ($K_f \approx 10^{52}$). The enormous stability of the ferric complex, its high specificity for Fe^{3+} and low affinity for Fe^{2+}, results in the Fe^{3+}/Fe^{2+} formal potential of enterobactin to be highly negative. The full coordination of Fe^{3+} by all three catechol groups requires the loss of six protons. Thus there is a strong dependence of the stability of the ferric enterobactin complex, and its reduction potential on pH. The formal electrochemical potentials for Fe(III/II) enterobactin are (at pH noted) -0.99 (>10.4), -0.79 (7.4), -0.57 (6.0), $+0.17$ (4.0) V (vs. NHE).[134,250] The redox equilibria have also been discussed controversially for some years but there now seems to be agreement on the formulation of

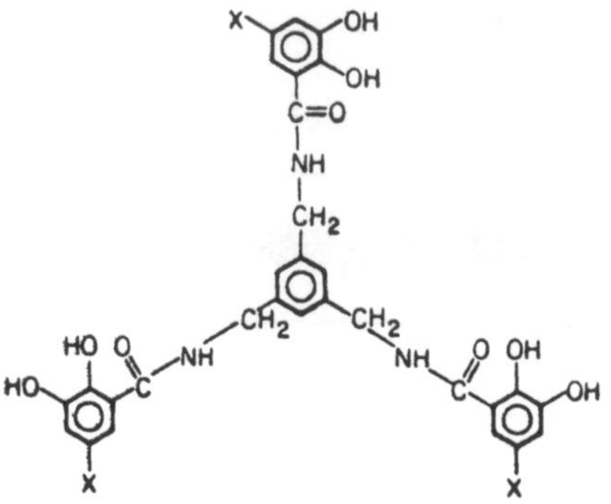

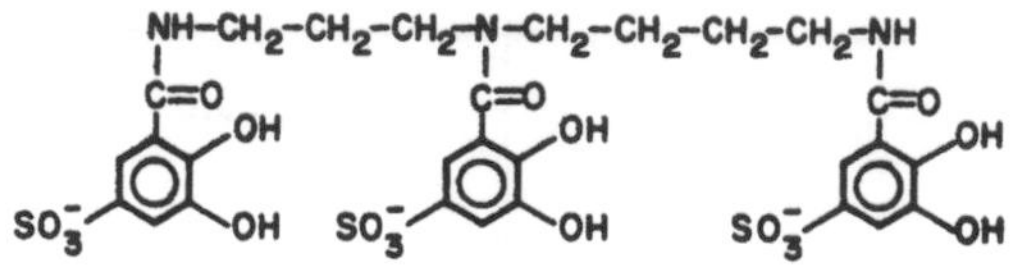

FIGURE 10.   Structures of synthetic catecholate-type siderophore analogs. (A): N,N',N'',-tris(2,3-dihydroxy-5-benzoyl)-1,3,5-tris(aminomethyl)benzene (ME-CAM), (X = H); (B):N,N',N'',-tris(2,3-dihydroxy-5-sulfobenzoyl)-1,3,5-tris(aminomethyl)benzene (MECAMS), (X = SO_3^-); (C): N,N',N'',-tris(2,3-dihydroxy-5-sulfobenzoyl)-1,5,10-triazadecane (LICAMS); (D): N,N',N'',-tris(2,3-dihydroxy-5-benzoyl)-1,3,5-tricarbamoylbenzene (TRIMCAM: X = H; TRIM-CAMS: X = SO_3^-); (E): N,N-dimethyl-2,3-dihydroxybenzamide (DMB)

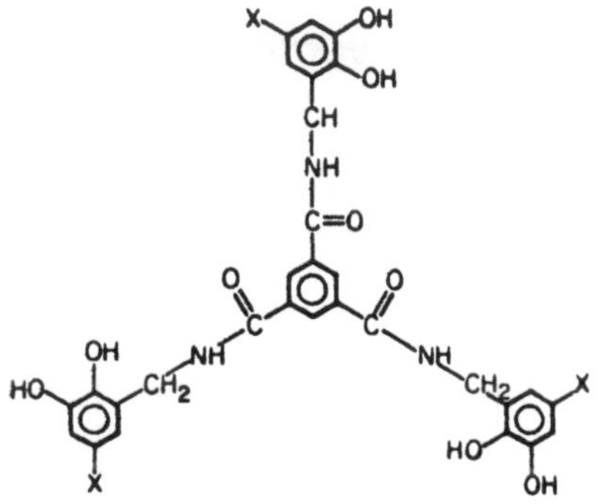

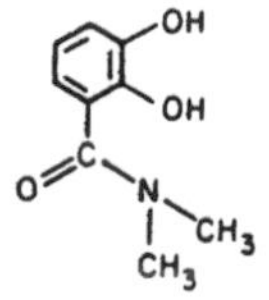

the redox chemistry.   Below pH 6 protonation of the ferric complex and dissociation of the ferrous become significant and the electrochemical reaction is not well defined.   Therefore, at pH 4 the potential has been estimated assuming no complexation of the ferrous ion by enterobactin at this pH.[134] Mössbauer spectra revealed that in a methanolic solution ferric enterobactin undergoes protonation and an intrinsic redox reaction with the formation of a semiquinone/Fe(II) couple under acidic conditions.[133] The reaction becomes irreversible if the solution is allowed to stand under aerobic conditions.   The mechanism is described as follows:

$$[Fe^{3+}(ent)]^{3-} \xrightleftharpoons{\ nH^+,\ methanol\ } [Fe^{3+}(H_nent)]^{n-3} \qquad (3)$$

$$[Fe^{2+}(H_nent)]^{n-3} \xrightarrow{\ O_2\ } polymeric\ quinones\ +\ Fe^{2+} \qquad (4)$$

ENTEROBACTIN UPTAKE IN *ESCHERICHIA COLI*

Three problem areas are to be addressed with respect to enterobactin uptake in *E. coli*: (1) recognition of the siderophore by the cell, (2) the transport mechanism, and (3) the intracellular metal release from the complex.

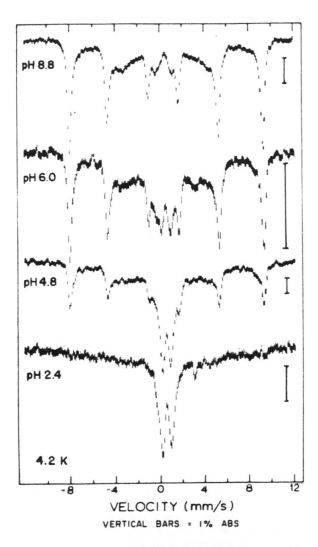

VELOCITY (mm/s)

VERTICAL BARS = 1% ABS

FIGURE 11. Mössbauer spectra at 4.2 K of ferric enterobactin in water as a function of pH (H_{app} = 60 mT parallel). The vertical bars indicate 1% absorption. With decreasing pH a fast relaxing Fe(III) high-spin doublet arises at about δ = 0.5 mm s^{-1}. (With permission of American Chemical Society.)

It is now well established that enterobactin uptake in *E. coli* is energy-dependent and necessitates the enantioselective and stereospecific recognition of the complex by the FepA outer membrane receptor.[164,166,167,251-256] As with many other siderophore receptors the molecular nature of stereospecific recognition is not well characterized. Certainly, the crystal structure of the receptor proteins will unravel this puzzle but no structure is available yet.[257] Thus variation of the siderophore structural elements and application of homologous compounds in uptake kinetics and inhibition studies should map, on a molecular level, some of the functionalities involved in recognition. In fact, the ferric complexes of some synthetic analogs of enterobactin can act as iron sources for *E. coli* and detailed kinetic and inhibition studies enabled a more precise description of the molecular groups essential for recognition by FebA.[164,166,167]

Synthetic ferric *enantio*-enteriobactin is not transported in *E. coli*.[258] Enterobactin uptake is effectively inhibited by the structural analog Fe-MECAM and Rh-MECAM. However,

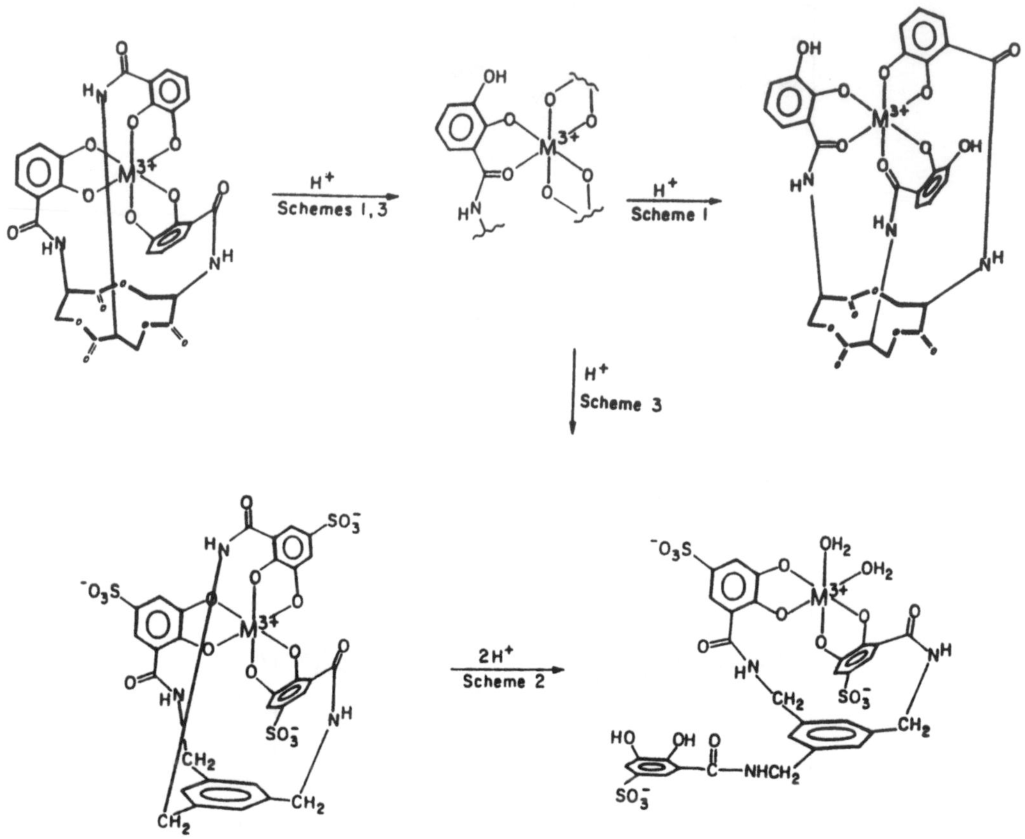

FIGURE 12. Possible protonation schemes of tris(catecholate) metal complexes. In Scheme 1 the metal complex undergoes a series of two overlapping one-proton steps to generate a mixed salicylate-catecholate coordination. Further protonation results in the precipitation of a tris(salicylate) complex (e.g., enterobactin, MECAM). This differs from Scheme 2, in which a single two-proton dissociated one arm of the ligand to form a bis(catecholate) chelate. Scheme 3 incorporates features of Schemes 1 and 2. In this model the metal again undergoes a series of two overlapping one-proton reactions. However, unlike the case of Scheme 1, the second proton displaces a catecholate arm, which results in a bis(catecholate) metal complex.

substrates which are modified at the aromatic rings (MECAMS, LICAMS, structure, see Figure 10) or at the amid bond connecting the 2,3 hydroxybenzoyl-groups with the backbone (TRIMCAM) are ineffective inhibitors (Figure 13).[164] From these experiments it was concluded that the part of the molecule recognized by the receptor is the metal-binding end, but the ligand to which these functionalities are attached is not important in receptor recognition. The exact meaning of "metal-binding end" and "ligand platform" was probed by Rh^{3+}-substituted catecholate model compounds. Rh-complexes, which are among the most ligand-exchange-inert complexes, were used because the corresponding chromic complexes are very oxidation sensitive.[166,259] In contrast to Rh-MECAM, the simple tris(catecholate) complex of Rh^{3+}, $[Rh(cat)_3]^{3-}$, is not an inhibitor of ferric-enterobactin uptake (see Figure 14A). When the Rh^{3+} tris ligand complex of 2,3-dihydroxy-*N,N* dimethylbenzamide, DMB, is used, the result is strikingly different. This complex, which has amide functionalities on the ligands is very effective in inhibiting ^{55}Fe-ent uptake (Figure 14B). In addition, replacement of the amide protons of MECAM with methyl groups does neither change the iron binding properties of the ligand nor receptor specificity. Thus the carbonyl group is essential for recognition but the proton on the amide nitrogen is not. These studies clearly show that modified siderophores and synthetic models are excellent tools in order to determine stereospecificity of enterobactin uptake on a molecular level.

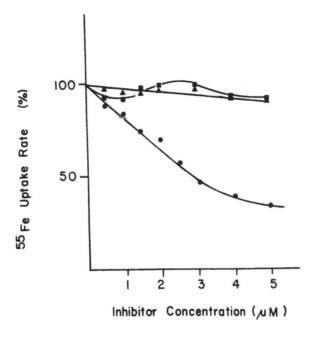

FIGURE 13.   Inhibition of [^{55}Fe]enterobactin uptake by structural analogs. The [^{55}Fe]enterobactin uptake rate (36 pmol/mg per min at 0.5 μM complex concentration) without any inhibitor is defined as 100%. Compounds added as potential inhibitors were Fe-ME-CAM, Fe-TRIMCAM, and Fe-3,4-LICAMS. (With permission of American Society of Microbiology.)

Very recently, uptake and inhibition studies have been published with the labile Al, Ga, In, and Sc complexes of enterobactin.[164,260,261] These complexes ranged from relatively effective inhibitors of ferric-enterobactin uptake to ineffective inhibitors in the series Fe > Sc > In ≫ Al > Ga. Ga complex uptake rates as low as 15% of the corresponding Fe(ent) uptake were measured. However, this process is energy dependent and the absorbed complex is exchangeable with excess of both unlabeled Ga(ent) and Fe(ent).[161] These results are neither consistent with a total lack of specific uptake nor with merely unspecific binding. The underlying rationale for these observations is unknown and still awaits clarification. It might well be that differences of the various enterobactin complexes in their coordination and pH dependent protonation and redox chemistry affect their affinity to the outer and inner membrane transport systems.

Intracellular iron utilization requires removal of the metal from enterobactin. Three possible routes for metal removal are (1) ligand exchange, (2) reductive ligand exchange via reductases, and (3) ligand degradation followed by route 1 or 2

As a result of its complex stability and redox potential which is outside the range available from biological reductants, iron can only be extracted from the enterobactin complex, at physiological pH, if the ligand is degraded. In fact, ligand degradation steps have been postulated. An esterase was found in the cell extracts of *E. coli* cleaving the ester backbone of ferric enterobactin.[262-265] In addition, it was demonstrated that the hydrolysis product (di-hydroxybenzoylserine, DHBS) is not an intermediate in the biosynthesis of enterobactin and therefore cannot be reused. The redox potential of the Fe(DHBS)$_3$ complex is −350 mV and within the range available from biological reductants.[263] Therefore a mechanism of enterobactin-mediated iron uptake was proposed which involved an initial ester cleavage in order to metabolize iron.[262] However, the esterase activity was reexamined and it was found that the esterase acts on both the free ligand and the iron complex, but with a 2.5-fold higher

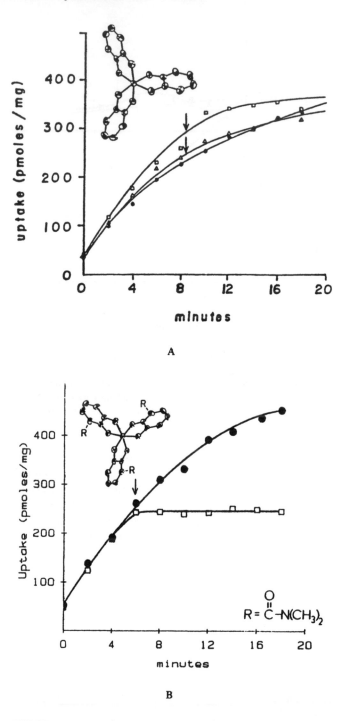

A

B

FIGURE 14. (A) Inhibition of 2 μM [^{59}Fe]enterobactin by [Rh(cat)$_3$]$^{3-}$. Control (closed circles) represents uptake of 1 μM label with no inhibitor added. Inhibition experiments: K$_3$[Rh(cat)$_3$]$^{3-}$ added at t = 6 min in 10-fold excess (open squares) or 100-fold excess (open triangles). (B) Inhibition of 2 μM [^{59}Fe]enterobactin by [Rh(DMB)$_3$]$^{3-}$. Control (closed circles) represents uptake of 1 μM label with no inhibitor added. K$_3$[Rh(DMB)$_3$]$^{3-}$ added at t = 6 min in 50-fold excess (open squares). (With permission of American Chemical Socitey.)

TABLE 10
^{57}Fe(ent) and ^{57}Fe(MECAM) Uptake in *E. coli* K-12 RW193

Total incubation time (min)	^{57}Fe (siderophore) (%) (sextet)	Fe^{2+} (%) (S = 2)	Fe^{3+} (%) (Fast relaxing)
Enterobactin	^{57}Fe-ent		
20	66	20	14
50	62	23	15
80	65	25	10
140	50	40	10
MECAM	^{57}Fe-MECAM		
5	90	Some	10
20	70	10	20
50	90	?	10
110	90	Some	10

Note: Cells grown in iron depleted minimum medium were harvested in the mid-log phase, resuspended in fresh minimum medium and incubated for 20 min with 2 μM ^{57}Fe-siderophore. The cells were then washed (0°C) and grown for an additional period of time. The data are derived from stripping procedures and least-squares fits of the corresponding *in vivo* Mössbauer spectra (Reference 206).

rate for the free ligand.[265] Moreover, uptake studies with synthetic analogs which lack ester units and which are not susceptible to hydrolysis showed growth response by *B. subtilis* and by *E. coli*.[251,266] Thus the esterase mechanism for *in vivo* iron release from ferric enterobactin became uncertain. The redox properties of enterobactin in a nonaqueous environment or at low pH (below pH 6) provide alternative pathways for metal transfer. The cell membranes represent nonaqueous environments, and an acidic compartment of the cell is the space intervening outer and cytoplasmic membrane, the periplasmic gel,[267-269] where such a reductive removal can occur.

An *in vivo* Mössbauer study of ^{57}Fe(ent) and ^{57}Fe(MECAM) uptake in the *aro*-mutant *E. coli* K12 RW193 has helped to address this difficult problem. Whereas both Fe(ent) and Fe(MECAM) are taken up as the intact complex, their rates of metabolism are drastically different (see Table 10). Depending on the experimental batch, 50 to 70% of enterobactin-bound iron is transferred into various Fe^{2+} and Fe^{3+} species within 140 min. Only traces of Fe^{2+} were detectable after 2 h of Fe(MECAM) uptake. This is clear evidence that ferric MECAM is not specifically metabolized. This is consistent with the original esterase mechanism of iron removal from enterobactin. Since no appreciable amounts of Fe(DHBS)$_3$ could be detected in the enterobactin experiments, any esterase activity and reductive ligand exchange must be closely correlated. The positive growth promotion with Fe(MECAM) uptake must be interpreted as being due to a small amount of nonspecifically metabolized complex which is still sufficient for normal growth rates of *E. coli*. The fast relaxing central quadrupole doublets found in cell spectra after ^{57}Fe-enterobactin and ^{57}Fe-MECAM uptake fit well with the doublets correlated with protonated catecholates and might indicate a distribution of Fe(ent) and Fe(MECAM) in either cell compartment, the acidic periplasm and the cytoplasm.

REFERENCES

1. **Neilands, J. B.,** Microbial iron transport compounds, in *Handbook of Microbiology,* Vol. IV, 2nd ed., Laskin, A. I. and Lechevalier, H. A., Eds., CRC Press, Boca Raton, FL, 1982, 565.
2. **Winkelmann, G.,** Iron complex products in *Biotechnology,* Vol. 4, Rehm, H.-J. and Reed, G., Eds., VCH Verlagsgesellschaft, Einheim, FRG, 1986, 215.
3. **Hider, R. C.,** Siderophore mediated absorption of iron, *Struct. Bonding,* 58, 25, 1984.
4. **Matzanke, B. F., Müller-Matzanke, G., and Raymond, K. N.,** Siderophore mediated iron transport, in *Iron Carriers and Iron Proteins,* Loehr, T. M., Ed., VCH, New York, 1989. 1.
5. **Raymond, K. N., Müller, G. I., and Matzanke, B. F.,** Complexation of iron by siderophores. A review of their solution and structural chemistry and biological function, *Topics Curr. Chem.,* 123, 49, 1984.
6. **Berner, I., Konetschny-Rapp, S., Jung, G., and Winkelmann, G.,** Characterization of ferrioxamine E as the principal siderophore of *Erwinia herbicola (Enterobacter agglomerans), Biol. Met.,* 1, 51, 1988.
7. **Takahashi, A., Nakamura, H., Kameyama, T., Kurasawa, S., Naganawa, H., Okami, Y., Takeuchi, S., and Umezawa, H.,** Bisucaberin, a new siderophore, sensitizing tumor cells to macrophage-mediated cytolysis. II. Physico-chemical properties and structure determination, *J. Antibiot.,* 40, 1671, 1987.
8. **Nishio, T., Tanaka, N., Hiratake, J., Katsube, Y., Ishida, Y., and Oda, J.,** Isolation and structure of the novel dihydroxamate siderophore alcaliligin, *J. Am. Chem. Soc.,* 110, 8733, 1988.
9. **Hulcher, F. H.,** Isolation and characterization of a new hydroxamic acid from *Pseudomonas mildenbergi, Biochemistry,* 21, 4491, 1982.
10. **Paoletti, L. C. and Blakemore, R. P.,** Hydroxamate production by *Aquaspirillum magnetotacticum, J. Bacteriol.,* 167, 73, 1986.
11. **Rabsch, W., Paul, P., and Reissbrodt, R.,** A new hydroxamate siderophore for iron supply of *Salmonella, Acta Microbiol. Hung.,* 34, 85, 1987.
12. **Mullis, K. B., Pollack, J. R., and Neilands, J. B.,** Structure of schizokinen, an iron transport compound from *Bacillus megaterium, Biochemistry,* 10, 4894, 1972.
13. **Linke, W. D., Crueger, A., and Diekmann, H.,** Stoffwechselprodukte von Mikroorganismen. 106. Mitteilung. Zur Konstitution des Terregensfaktors, *Arch. Mikrobiol.,* 85, 44, 1972.
14. **Gibson, F. and Magrath, D. I.,** The isolation and characterization of a hydroxamic acid (aerobactin) formed by *Aerobacter aerogenes* 62-I, *Biochim. Biophys. Acta,* 192, 175, 1969.
15. **Meyer, J.-M., and Abdallah, M. A.,** The siderochromes of non-fluorescent pseudomonas: production of nocardamin by *Pseudomonas stutzeri, J. Gen. Microbiol.,* 118, 125, 1980.
16. **Bernhard, K. and Albrecht, H.,** Stoffwechselprodukte des Mikroorganismus *Phycomyces blaakesleeanus* in glucosehaltiger Nährlösung und Untersuchungen über das Wachstum dieses Schimmelpilzes bei verschiedenen Stickstoffquellen, *Helv. Chim. Acta,* 30, 627, 1947.
17. **Ito, T. and Neilands, J. B.,** Products of "low fermentation" with *Bacillus subtilis.* Isolation, characterization and synthesis of 2,3 dihydroxybenzoyl glycine, *J. Am. Chem. Soc.,* 80, 4645, 1958.
18. **Korth, H.,** Über das Vorkommen von 2,3-Dihydroxybenzoesäure und ihren Aminoderivaten in Kulturmedien von *Klebsiella oxytoca, Arch. Mikrobiol.,* 70, 297, 1970.
19. **O'Brien, I. G., Cox, G. B., and Gibson, F.,** Biologically active compounds containing 2,3-dihydroxybenzoic acid and serine formed by *Escherichia coli, Biochim. Biophys. Acta,* 201, 453, 1970.
20. **Ratledge, C.,** Relationship between the products of aromatic biosynthesis in *Mycobacterium smegmatis* and *Arobacter aerogenes, Nature,* 203, 428, 1964.
21. **Ratledge, C. and Chaudrey, M. A.,** Accumulation of iron-binding phenolic acids by actinomycetales and other organisms related to mycobacteria, *J. Gen. Microbiol.,* 66, 7, 1971.
22. **Dyer, J. R., Heding, H., and Schaffer, C. P.,** Phenolic metabolite of "low-iron fermentation" of *Streptomyces griseus.* Characterization of 2,3 dihydroxybenzoic acid, *J. Org. Chem.,* 29, 2802, 1964.
23. **Leong, S. A. and Neilands, J. B.,** Siderophore production by phytopathogenic microbial species, *Arch. Biochem. Biophys.,* 218, 351, 1982.
24. **Persmark, M., Expert, D., and Neilands, J. B.,** Isolation, characterization, and synthesis of chrysobactin, a compound with siderophore activity from *Erwinia chrysanthemi, J. Biol. Chem.,* 264, 3187, 1989.
25. **Kunze, B., Bedorf, N., Kohl, W., Höfle, G., and Reichenbach, H.,** Myxochelin A, a new iron-chelating compound from *Angiococcus disciformis* (myxobacteriales), *J. Antibiot.,* 42, 14, 1989.
26. **Corbin, J. L. and Bulen, W. A.,** The isolation and identification of 2-*N*,6-*N*-di-(2,3-dihydroxybenzoyl)-L-lysine formed by iron-deficient *Azotobacter vinelandii, Biochemistry,* 8, 757, 1969.
27. **Tait, G. H.,** The identification and biosynthesis of siderochromes formed by *Micrococcus dinitrificans, Biochem. J.,* 146, 191, 1975.
28. **O'Brien, I. G. and Gibson, F.,** The structure of enterochelin and related 2,3-dihydroxy-*N*-benzoylserine conjugates from *Escherichia coli, Biochim. Biophys. Acta,* 215, 393, 1970.
29. **Pollack, J. R. and Neilands, J. B.,** Enterobactin, an iron transport compound from *Salmonella typhimurium, Biochem. Biophys. Res. Commun.,* 38, 989, 1970.

30. **Ong, S. A., Peterson, T., and Neilands, J. B.**, Agrobactin, a siderophore from *Agrobacterium tumifaciens*, *J. Biol. Chem.*, 254, 1860, 1979.

31. **Griffith, G. L., Sigel, S. P., Payne, S. M., and Neilands, J. B.**, Vibriobactin, a siderophore from *Vibrio cholerae, J. Biol. Chem.*, 259, 383, 1984.

32. **Bachhawat, A. K. and Ghosh, S.**, Iron transport in *Azospirillum brasilense:* role of the siderophore spirilobactin, *J. Gen. Microbiol.*, 133, 1759, 1987.

33. **Barghouthi, S., Young, R., Olson, M. O., Arceneaux, J. E. L., Clem, L. W., and Byers, B. R.**, Amonobactin, a novel tryptophan- or phenylalaninine-containing phenolate siderophore in *Aeromonas hydrophyla, J. Bacteriol.*, 171, 1811, 1989.

34. **Skorupska, A., Choma, A., Derylo, H., and Lorkiewicz, Z.**, Siderophore containing 2,3-dihydroxy-benzoic acid and threonine formed by *Rhizobium trifoli, Acta Biochim. Pol.*, 35, 121, 1988.

35. **Ratledge, C. and Winder, F. G.**, The accumulation of salicylic acid by mycobacteria during growth on an iron-deficient medium, *Biochem. J.*, 84, 501, 1962.

36. **Bickel, H., Hall, G. E., Keller-Schierlien, W., Prelog, V., Vischer, E., and Wettstein, A.**, 261. Stoffwechselprodukte von Actinomycten. Über die Konstitution von Ferrioxamin B, *Helv. Chim. Acta*, 43, 2129, 1960.

37. **Keller-Schierlein, W., Mertens, P., Prelog, V., and Walser, A.**, 77. Stoffwechselprodukte von Mikroorganismen. Die Ferrioxamine A_1, A_2 und D_2, *Helv. Chim. Acta*, 48, 710, 1965.

38. **Keller-Schierlein, W. and Prelog, V.**, 245. Stoffwechselprodukte von Actinomycten. Über das Ferrioxamin E; ein Beitrag zur Konstitution des Nocardamins, *Helv. Chim. Acta*, 44, 1981, 1961.

39. **Keller-Schierlein, W., Huber, P., and Kawaguchi, H.**, Chemistry of danomycin, an iron-containing antibiotic, in *Natural Products and Drug Development*, Alfred Benzon Symposium, Krogsgaard-Larsen, P., Christensen, S. B., and Kofod, H., Eds., Copenhagen, 1984, 213.

40. **Adapa, S., Huber, P., and Keller-Schierlein, W.**, 179. Stoffwechselprodukte von Mikroorganismen. 216. Mitteilung. Isolierung, Strukturaufklärung und Synthese von Ferrioxamin H, *Helv. Chim. Acta*, 65, 1818, 1982.

41. **Knüsel, F. and Zimmermann, W.**, Sideromycins, in *Antibiotics. Mechanisms of Action of Antimicrobial and Antitumor Agents*, Vol. III, Corcoran, J. W. and Hahn, F. E., Eds., Springer-Verlag, Berlin, 1975, 653.

42. **Bickel, H., Gäumann, E., Nussberger, G., Reusser, P., Vischer, E., Voser, W., Wettstein, A., and Zähner, H.**, 259. Stoffwechselprodukte von Actinomyceten. 25. Mitteilung. Über die Isolierung der Ferrimycine A_1 und A_2, neuer Antibiotika aus der Sideromycin-Gruppe, *Helv. Chim. Acta*, 43, 2105, 1960.

43. **Emery, T. F. and Neilands, J. B.**, Structure of ferrichrome compounds, *J. Am. Chem. Soc.*, 83, 1626, 1961.

44. **Deml, G., Voges, K., Jung, G., and Winkelmann, G.**, Tetraglycylferrichrome — the first heptapeptide ferrichrome, *FEBS Lett.*, 173, 53, 1984.

45. **Llinas, M. and Neilands, J. B.**, Structure of ferricrocin, *Bioinorg. Chem.*, 2, 159, 1972.

46. **Norrestam, R., Stensland, B., and Branden, C. I.**, On the conformation of cyclic iron-containing hexapeptides: the crystal and molecular structure of ferrichrysin, *J. Mol. Biol.*, 99, 501, 1975.

47. **Keller-Schierlein, W.**, Stoffwechselprodukte von Mikroorganismen. 45. Mitteilung über die Konstitution von Ferrirubin, Ferrirhodin und Ferrichrom A, *Helv. Chim. Acta*, 46, 1920, 1963.

48. **Emery, T.**, Malonichrome, a new iron chelate from *Fusarium roseum, Biochim. Biophys. Acta*, 629, 382, 1980.

49. **Jalal, M. A. F., Mocharla, R., Barnes, C. L., Hossain, M. B., Powell, D. R., Eng-Wilmot, D. L., Grayson, S. I., Benson, B. A., and van der Helm, D.**, Extracellular siderophores of *Aspergillus ochraceus, J. Bacteriol.*, 158, 683, 1984.

50. **Jalal, M. A. F., Hossain, M. B., and van der Helm, D.**, Structure of ferrichrome-type siderophores with dissimilar N^δ-acyl groups: asperochrome B_1, B_2, B_3, D_1, D_2 and D_3, *Biol. Met.*, 1, 77, 1988.

51. **Jalal, M. A. F., Galles, J. L., and van der Helm, D.**, Structure of des(diserylglycyl)ferrirhodin, DDF, a novel siderophore from *Aspergillus ochraceus, J. Org. Chem.*, 50, 5642, 1985.

52. **Benz, G., Schröder, T., Kurz, J., Wünsche, Ch., Karl, W., Steffens, G., Pfitzner, J., and Schmidt, D.**, Konstitution der Desferriform der Albomycin δ_1, δ_2 and ϵ, *Angew. Chem.*, 94, 552, 1982.

53. **Hesseltine, C. W., Pidacks, C., Whitehill, A. R., Bohonos, N., Hutchings, B. L., and Williams, J. H.**, Coprogen, a new growth factor for coprophilic fungi, *J. Chem. Soc.*, 74, 1362, 1952.

54. **Zähner, H., Keller-Schierlein, W., Hutter, R., Hess-Leisinger, K., and Deer, A.**, Stoffwechselprodukte von Mikroorganismen. 40. Mitteilung. Sideramine aus Aspergillaceen, *Arch. Mikrobiol.*, 45, 119, 1963.

55. **Keller-Schierlein, W. and Diekmann, H.**, Stoffwechselprodukte von Microorganismen. 85. Mitteilung. Zur Konstitution des Coprogens, *Helv. Chim. Acta*, 53, 2035, 1970.

56. **Diekmann, H.**, Stoffwechselprodukte von Mikroorganismen. 81. Mitteilung. Vorkommen und Strukturen von Coprogen B und Dimerumsäure, *Arch. Mikrobiol.*, 73, 65, 1970.

57. **Frederick, C. B., Bentley, M. D., and Shive, W.**, The structure of the fungal siderophore, isotriornicin, *Biochem. Biophys. Res. Commun.*, 105, 133, 1982.

58. **Frederick, C. B., Bentley, M. D., and Shive, W.,** Structure of triornicin, a new siderophore, *Biochemistry,* 20, 2436, 1981.

59. **Hossain, M. B., Jalal, M. A. F., Benson, B. A., Barnes, C. L., and van der Helm, D.,** Structure and conformation of two coprogen-type siderophores: neocoprogen I and neocoprogen II, *J. Am. Chem. Soc.,* 109, 4948, 1987.

60. **Jalal, M. A. F., Love, S. K., and van der Helm, D.,** N^α-Dimethylcoprogens. Three novel trihydroxamate siderophores from pathogenic fungi, *Biol. Met.,* 1, 4, 1988.

61. **Jalal, M. A. F. and van der Helm, D.,** Siderophores of highly phytopathogenic *Alternaria longipes.* Structures of hydroxycoprogens, *Biol. Met.,* 2, 11, 1989.

62. **Umezawa, H., Aoyagi, T., Ogawa, K., Obata, T., Iinuma, H., Naganawa, H., Hamada, M., and Takeuchi, T.,** Foroxomythine, a new inhibitor of angiotensin-converting enzyme, produced by actinomycetes, *J. Antibiot.,* 38, 1813, 1985.

63. **Diekmann, H.,** Stoffwechselprodukte von Mikroorganismen. 56. Mitteilung. Fusigen-ein neues Sideramin aus Pilzen, *Arch. Mikrobiol.,* 58, 1, 1967.

64. **Sayer, J. M. and Emery, T.,** *Biochemistry,* 7, 184, 1968.

65. **Diekmann, H.,** Stoffwechselprodukte von Mikroorganismen. 68. Mitteilung. Die Isolierung und Darstellung trans-5-Hydroxy-3-methylpenten-(2)-säure, *Arch. Mikrobiol,* 62, 322, 1968.

66. **Diekmann, H. and Krezdorn, E.,** Stoffwechselprodukte von Mikroorganismen. 150. Mitteilung. Ferricrocin, Triacetylfusigen und anderes Sideramine aus Pilzen der Gattung *Aspergillus,* Gruppe *Fumigatus, Arch. Microbiol.,* 106, 191, 1975.

67. **Moore, R. E. and Emery, T.,** N^δ-Acetylfusarinines: isolation, characterization, and properties, *Biochemistry,* 15, 2719, 1976.

68. **Eng-Wilmot, D. L., Rhaman, A., Mendenhall, J. V., Grayson, S. L., and van der Helm, D.,** Molecular structure of ferric neurosporin, a minor sideophore-like compound containing, N^δ-hydroxy-D-ornithine, *J. Am. Chem. Soc.,* 106, 1285, 1984.

69. **Atkin, C. L. and Neilands, J. B.,** Rhodotorulic acid, a diketopiperazine dihydroxamic acid with growth factor activity. I. Isolation and characterization, *Biochemistry,* 7, 3734, 1968.

70. **Atkin, C. L., Neilands, J. B., and Phaff, H. J.,** Rhodotorulic acid from species of *Leucosporidium, Rhodosporidium, Rhodotorula, Sporidiobulus* and *Sporobolomyces,* and a new alanine-containing ferrichrome from *Cryptotoccus melibiosum, J. Bacteriol.,* 103, 722, 1970.

71. **Teintze, M., Hossain, M. B., Barnes, C. I., Leong, J., and van der Helm, D.,** Structure of ferric psuedobactin, a siderophore from plant growth promoting *Pseudomonas, Biochemistry,* 20, 6446, 1981.

72. **Teintze, M. and Leong, J.,** Structure of pseudobactin A, a second siderophore from plant growth promoting *Pseudomonas* B10, *Biochemistry,* 20, 6457, 1981.

73. **Yang, C.-C. and Leong, J.,** Structure of pseudobactin 7SR1, a siderophore from a plant-deleterious *Pseudomonas, Biochemistry,* 23, 3534, 1984.

74. **Buyer, J. S., Wright, J. M., and Leong, J.,** Structure of pseudobactin A214, a siderophore from a bean-deleterious *Pseudomonas, Biochemistry,* 25, 5492, 1986.

75. **Wendenbaum, S., Demange, P., Dell, A., Meyer, J. M., and Abdallah, M. A.,** The structure of pyoverdine Pa, the siderophore of *Pseudomonas aeruginosa, Tetrahedron Lett.,* 24, 4877, 1983.

76. **Demange, P., Wendenbaum, S., Bateman, A., Dell, A., and Abdallah, M. A.,** Bacterial siderophores: structure and physicochemical properties of pyoverdins and related compounds, in *Iron Transport in Microbes Plants and Animals,* Winkelmann, G., van der Helm, D., and Neilands, J. B., Eds., VCH, Weinheim, 1987, 167.

77. **Cody, Y. S. and Gross, D. C.,** Outer membrane protein mediating iron uptake via pyoverdin$_{pss}$, the fluorescent siderophore produced by *Pseudomonas syringae* pv. *syringae, J. Bacteriol.,* 169, 2207, 1987.

78. **Philson, S. B. and Llinas, M.,** Siderochromes from *Pseudomonas fluorescens, J. Biol. Chem.,* 257, 8086, 1982.

79. **Ratledge, C.,** Mycobactins and nocobactins, in *Handbook of Microbiology,* Vol. IV, 2nd ed., Laskin, A. I. and Lechevalier, H. A., Eds., CRC Press, Boca Raton, FL, 1982, 575.

80. **Snow, G. A.,** Mycobactins: iron-chelating growth factors from mycobacteria, *Bacteriol. Rev.,* 34, 99, 1970.

81. **Ratledge, C., Patel, P. V., and Mundy, J.,** Iron transport in *Mycobacterium smegmatis:* the location of mycobactin by electron microscopy, *J. Gen. Microbiol.,* 128, 1559, 1982.

82. **Jalal, M. A. F., Hossain, M. B., van der Helm, D., Sanders-Loehr, J., Actis, L. A., and Crosa, J. H.,** Structure of anguibactin, a unique plasmid-related bacterial siderophore from the fish pathogen *Vibrio anguillarum, J. Am. Chem. Soc.,* 111, 292, 1989.

83. **Cox, C. D., Rinehart, K. L., Moore, M. L., and Cook, J. C., Jr.,** Pyochelin, novel structure of an iron-chelating growth promoter for *Pseudomonas aeruginosa, Proc. Natl. Acad. Sci. U.S.A.,* 78, 4256, 1981.

84. **Keller-Schierlein, W., Hagmann, L., Zähner, H., and Huhn, W.,** 167. Stoffwechselprodukte von Mikroorganismen. Maduraferrin ein neuartiger Siderophor aus *Actinomadura madurae, Helv. Chim. Acta,* 71, 1528, 1988.

85. **Schwyn, B. and Neilands, J. B.**, Universal chemical assay for the detection and determination of siderophores, *Anal. Biochem.*, 160, 47, 1987.

86. **Meiwes, J., Fiedler, H.-P., Haag, H., Koetschny-Rapp, S., and Jung, G.**, Isolation and characterization of staphyloferrin A, a compound with siderophore activity from *Staphylococcus hyicus* DSM 20459, *FEMS Microbiol. Lett.*, 67, 201, 1990.

87. **Smith, M. J., Schoolery, J. N., Schwyn, B., Holden, I., and Neilands, J. B.**, Rhizobactin, a structurally novel siderophore from *Rhizobium meliloti*, *J. Am. Chem. Soc.*, 107, 1739, 1985.

88. **Neilands, J. B.**, Microbial iron transport compounds, in *Inorganic Biochemistry*, Eichhorn, G. I., Ed., Elsevier, Amsterdam, 1973, 167.

89. **Hulcher, F. H.**, Isolation and characterization of a new hydroxamic acid from *Pseudomonas mildenbergi*, *Biochemistry*, 21, 4491, 1982.

90. **Arai, K., Sato, S., Shimizu, S., Nitta, K., and Yamamoto, Y.**, Metabolic products of *Aspergillus terreus*. VII. Astechrome: an iron-containing metabolite of the strain IFO 6123, *Chem. Pharm. Bull.*, 29, 1510, 1981.

91. **Eng-Wilmot, D. L., Hossain, M. B., and van der Helm, D.**, The structures of the mono- and disodium salts of hadacidin, an antibiotic hydroxamic acid, *Acta Crystallogr.*, B37, 1207, 1981.

92. **Eng-Wilmot, D. L. and van der Helm, D.**, Structure of trichostatin, a primary hydroxamic antibiotic, *Acta Crystallogr.*, B37, 718, 1981.

93. **Itoh, S., Inuzuka, K., and Suzuki, T.**, New antibiotics produced by bacteria grown on n-paraffin (mixture of C_{12}, C_{13} and C_{14} fractions), *J. Antibiot. (Tokyo)*, 23, 542, 1970.

94. **Martinez-Molina, E. Del Rio, L. A., and Olivarez, J.**, Copper and iron as determinant factors of antibiotic production by *Pseudomonas reptilivora*, *J. Appl. Bacteriol.*, 41, 69, 1974.

95. **Meyer, J.-M., Hohnadel, D., and Hallé, F.**, Cepabactin from *Pseudomonas cepacia*, a new type of siderophore, *J. Gen. Microbiol.*, 35, 1479-1487, 1989.

96. **Barker, W. R., Callaghan, C., Hill, L., Nobel, D., Acred, P., Harper, P. B., Sowa, M. A., and Fletton, R. A.**, A new cyclic hydroxamic acid antibiotic, isolated from culture broth of *Pseudomonas alcaligenes*, *J. Antibiot.*, 32, 1096, 1979.

97. **Naegeli, H.-U. and Zähner, H.**, 147. Stoffwechselprodukte von Mikroorganismen. Ferrithiocin, *Helv. Chim. Acta*, 63, 1400, 1980.

98. **Oxford, A. E., Raistrick, H., and Smith, G.**, Anti-bacterial substances from moulds. VI. Puberilic acid $C_8H_6O_6$ and puberilonic acid $C_8H_4O_6$, metabolic products of a number of species of Penicillium, *Chem. Ind.*, 485, 487, 1942.

99. **Segal, W.**, 571. Stipitatonic acid. A new mold tropolone from *Penicillium stipitatum*, *J. Chem. Soc.*, 2847, 1959.

100. **Bryant, R. W. and Light, R. J.**, Stipitatonic acid biosynthesis. Incorporation of (formyl ^{14}C)-3-methylorcylaldehyde and ^{14}C-stipitaldehydic acid, a new tropolone metabolite, *Biochemistry*, 13, 1516, 1974.

101. **Garnier-Suillerot, A.**, Metal anthracycline and anthracenedione complexes as a new class of anticancer agents, in *Anthracycline and Anthracenedione-Based Anti-cancer Agents*, Lown, J. W., Ed., Book series on bioactive molecules, Elsevier, Amsterdam, 1989.

102. **Gosalvez, A., Blanco, M. F., Vivero, C., and Valles, F.**, Quelamycin, a new derivative of adriamycin with several therapeutic advantages, *Eur. J. Cancer*, 14, 1185, 1978.

103. **Beraldo, H., Garnier-Suillerot, A., Toso, L., and Lavelle, F.**, Iron(III)-adriamycin and iron(III)-daunorubicin complexes: physicochemical characteristics, interaction with DNA, and antitumor activity, *Biochemistry*, 24, 284, 1985.

104. **Fantine, E. O. and Garnier-Suillerot, A.**, Interaction of 5-iminodaunorubicin with Fe(III) and with cardiolipin-containing vesicles, *Biochim. Biophys. Acta*, 856, 130, 1986.

105. **Fiallo, M. M. L. and Garnier-Suillerot, A.**, Physicochemical studies of the iron(III)-carminomycin complex and evidence of the lack of stimulated superoxide production by NADH dehydrogenase, *Biochim. Biophys. Acta*, 840, 91, 1985.

106. **Lambs, L., Venturini, M., Decock-Le Révérend, Kozlowski, H., and Berthon, G.**, Metal ion-tetracycline interactions in biological fluids. Part 8. Potentiometric and spectroscopic studies on the formation of Ca(II) and Mg(II) complexes with 4-dedimethylamino-tetracycline and 6-desoxy-6-demethyl-tetracycline, *J. Inorg. Biochem.*, 33, 193, 1988.

107. **Zweier, J. L., Gianni, L., Muindi, J., and Myers, C. E.**, Differences in O_2-reduction by the iron complexes of adriamycin and daunomycine: the importance of the sidechain hydroxyl group, *Biochim. Biophys. Acta*, 884, 326, 1986.

108. **Gelvan, D. and Samuni, A.**, Reappraisal of the association between adriamycin and iron, *Cancer Res.*, 48, 5645, 1988.

109. **Neidle, S. and Waring, M. J.**, *Molecular Aspects of Anti-Cancer Drug Action*, VCH, Weinheim, 1983.

110. **Fujiwara, A. and Hoshino, T.**, Anthracycline antibiotics, *CRC Crit. Rev. Biotechnol.*, 3, 133, 1986.

111. **Sinha, B. K., Katki, A. G., Batist, G., Cowan, K. H., and Myers, C. E.**, Differential formation of hydroxyl radicals by adriamycin in sensitive and resistant MCF-7 human breast tumor cells: implications for the mechanism of action, *Biochemistry*, 26, 3776, 1987.

112. **Bellamy, W. T., Dorr, R. T., Dalton, W. S., and Alberts, D. S.,** Direct relation of DNA lesions in multidrug-resistant human myeloma cells to intracellular doxorubicin concentration, *Cancer Res.*, 48, 6360, 1988.
113. **Hasinoff, B. B. and Davey, J. P.,** The iron(III)-adriamycin complex inhibits cytochrome c oxidase before its inactivation, *Biochem. J.*, 250, 827, 1988.
114. **Mimnaugh, E. G., Dusre, L., Atwell, J., and Myers, C. E.,** Differential oxygen radical susceptibility of adriamycin-sensitive and -resistant MCF-7 human breast tumor cells, *Cancer Res.*, 49, 8, 1989.
115. **Gianni, L., Vigano, L., Lanzi, C., Niggeler, M., and Malatesta, V.,** Role of Daunosamine and hydroxyacetyl side chain in reaction with iron and lipid peroxidation by anthracyclines, *J. Natl. Cancer Inst.*, 80, 1104, 1988.
116. **Hantke, K.,** Ferrous iron transport mutants in *Escherichia coli* K12, *FEMS Microbiol. Lett.*, 44, 53, 1987.
117. **Lodge, J. S. and Emery, T.,** Anaerobic iron uptake by *Escherichia coli*, *J. Bacteriol.*, 160, 801, 1984.
118. **Bezkorovainy, A., Solberg, L., Poch, M., and Miller-Catchpole, R.,** Ferrous iron uptake by *Bifidobacterium bifidum* var *pennsylvanicus:* the effect of metals and metabolic inhibitors, *Int. J. Biochem.*, 19, 517, 1987.
119. **Feistner, G., Korth, H., Ko, H., Pulverer, G., and Budzikiewitz, H.,** Ferrosamine A from *Erwinia rhapontici, Curr. Microbiol.*, 8, 239, 1983.
120. **Pouteau-Thouvenot, M., Choussy, M., Barbier, M., and Viscontini, M.,** Sur le role métabolique de la proferrosamine A produite par *Pseudomonas roseus fluorescens* J. C. Marchal, *Helv. Chim. Acta*, 52, 2392, 1937.
121. **Shiman, R. and Neilands, J. B.,** Isolation, characterization and synthesis of pyrimine, an iron (II) binding agent from *Pseudomonas* H. G., *Biochemistry*, 4, 2233, 1965.
122. **Liu, W.-C., Fisher, S. M., Wells, J. S., Ricca, C. S., Principe, P. A., Trejo, W. H., Bonner, D. P., Gougoutos, J. Z., Toeplitz, B. K., and Sykes, R. B.,** Siderochelin, a new ferrous-ion chelating agent produced by *Nocardia, J. Antibiotics*, 34, 791, 1981.
123. **Okuyama, D., Nakamura, H., Nganawa, H., Takita, T., Umezawa, H., and Iitaka, Y.,** Isolation, racemization and absolute configuration of siderochelin A, *J. Antibiot.*, 35, 1240, 1982.
124. **Mitscher, L. A., Högberg, T., Drake, S. D., Burgstahler, A. W., Jackson, M., Lee, B., Sheldon, R. I., Gracey, H. E., Kohl, W., and Theriault, R. J.,** Isolation and structural determination of siderochelin C, a fermentation product of an unusual *Actinomyocytes* sp., *J. Antibiot.*, 37, 1260, 1984.
125. **Blinova, I. N., Egorova, S. A., Marchenko, I. V., Saulina, L. I., Blionov, N. O., and Khokhlov, A. S.,** New iron containing antibiotics—Isolation and properties of viridomycin A, B and C, *Chem. Nat. Compd. (USSR)*, 11, 506, 1976.
126. **Chain, E. B., Tonolo, A., and Carilli, A.,** Ferroverdin, a green pigment containing iron produced by a streptomycete, *Nature*, 176, 645, 1955.
127. **Khokhlov, A. S. and Blinova, I. N.,** Structure of viridomycin A a new iron containing antibiotic, *Dokl. Akad. Nauk. SSSR*, 215, 1493, 1974.
128. **Kurobane, I., Dale, P. L., and Vining, L. C.,** Characterization of new viridomycins and requirements for production in cultures of *Streptomyces griseus, J. Antibiot.*, 40, 1131, 1987.
129. **Povirk, L. F.,** Bleomycin, in *Molecular Aspects of Anti-Cancer Drug Action*, Neidle, S. and Waring, M. J., Eds., VCH, Weinheim, 1983, 157.
130. Tentative proposals for nomenclature of absolute configurations concerned with six-coordinated complexes based on the octahedron, *Inorg. Chem.*, 9, 1, 1970.
131. **Leong, J. and Raymond, K. N.,** Coordination isomers of biological iron transport compounds. IV. Geometrical isomers of chromic desferrioxamine B, *J. Am. Chem. Soc.*, 97, 293, 1975.
132. **Hossain, M. B., Jalal, M. A. F., Benson, B. A., Barnes, C. L., and van der Helm, D.,** Structure and conformation of two coprogen-type siderophores: neocoprogen I and neocoprogen II, *J. Am. Chem. Soc.*, 109, 4948, 1987.
133. **Bock, J. L. and Lang, G.,** Mössbauer spectroscopy of iron chelated by deferoxamine, *Biochim. Biophys. Acta*, 264, 245, 1971.
134. **Pecoraro, V. L., Wong, G. B., Kent, T. A., and Raymond, K. N.,** Coordination chemistry of microbial iron transport compounds. 22. pH-dependent Mössbauer spectroscopy of ferric enterobactin and synthetic analogues, *J. Am. Chem. Soc.*, 105, 4617, 1983.
135. **Spartalian, K., Oosterhuis, W. T., and Neilands, J. B.,** Electronic state of iron in enterobactin using Mössbauer spectroscopy, *J. Chem. Phys.*, 62, 3538, 1975.
136. **Oosterhuis, W. T.,** The electronic state of iron in some natural iron compounds: determination by Mössbauer and ESR spectroscopy, *Struct. Bonding*, 20, 59, 1974.
137. **Matzanke, B. F.,** Dissertation, Universität Tübingen, Tübingen, FRG, 1982.
138. **Wickman, H. H., Klein, M. P., and Shirley, D. A.,** Paramagnetic hyperfine structure and relaxation effects in Mössbauer spectra: Fe[57] in ferrichrome A, *Phys. Rev.*, 152, 345, 1966.
139. **Hoy, G. R., Corson, M. R., and Balko, B.,** Nonadiabatic, stochastic model for the classic relaxing paramagnet ferrichrome A: theory and experiment, *Phys. Rev. B*, 27, 2652, 1983.

140. **Barclay, S. J., Huynh, B. H., and Raymond, K. N.**, Coordination chemistry of microbial iron transport compounds. 27. Dimeric iron (III) complexes of dihydroxamate analogues of rhodotorulic acid, *Inorg. Chem.*, 23, 2011, 1984.

141. **Dowsing, R. D. and Gibson, J. F.**, Electron spin resonance of high-spin d^5 systems, *J. Chem. Phys.*, 50, 294, 1969.

142. **van der Ven, N. S., Spartalian, K., Oosterhuis, W. T., and Ashkin, J.**, ESR and Mössbauer studies of Fe^{3+} in Mycobactin P, *Bull. Am. Phys. Soc.*, 19, 373, 1974.

143. **Spartalian, K., Oosterhuis, W., and Window, B.**, Mössbauer effect studies of iron storage and transport proteins, in *Mössbauer Effect Methodology*, Vol. 8, Gruverman, I. J., Ed., Plenum Press, New York, 1973, 137.

144. **Mielczarek, E.**, personal communication.

145. **Leong, J. and Raymond, K. N.**, Coordination isomers of biological iron transport compounds. I. Models for the siderochromes. The geometrical and optical isomers of tris(N-methyl-1-methoxyacetohydroxamato) chromium (III), *J. Am. Chem. Soc.*, 96, 1757, 1974.

146. **Abu-Dari, K. and Raymond, K. N.**, Coordination chemistry of microbial iron transport compounds. 20. Crystal and molecular structures of two salts of cis- and trans-tris(benzohydroximato)chromate(III), *Inorg. Chem.*, 19, 2034, 1980.

147. **Raymond, K. N., Abu-Dari, K., and Sofen, S. R.**, Stereochemistry of microbial iron transport compounds, in *ACS Symposium, Ser. 119, Stereochemistry of Optically Active Transition Metal Compounds*, Douglas, B. E. and Saito, Y., Eds., American Chemical Society, 1980, 7.

148. **Raymond, K. N.**, Kinetically inert complexes of the siderophores in studies of microbial iron transport, in *Adv. in Chem. Ser., No. 162. Bioinorganic Chemistry II*, Raymond, K. N., Ed., American Chemical Society, 1977, 33.

149. **McCaffery, A. J., Mason, S. F., and Ballard, R. E.**, Optical rotary power of coordination compounds. III. The absolute configurations of trigonal metal complexes, *J. Chem. Soc.*, 2883, 1965.

150. **Müller, G. and Raymond, K. N.**, Specificity and mechanisms of ferrioxamine-mediated iron transport in *Streptomyces pilosus*, *J. Bacteriol.*, 160, 304, 1984.

151. **Carrano, C. J. and Raymond, K. N.**, Coordination chemistry of microbial iron transport compounds. 10. Characterization of the complexes of rhodotorulic acid, a dihydroxamate siderophore, *J. Am. Chem. Soc.*, 100, 5371, 1978.

152. **Müller, G., Isowa, Y., and Raymond, K. N.**, Stereospecificity of siderophoremediated iron uptake in *Rhodotorula pilimanae* as probed enantiorhodotorulic acid and isomers of chromic rhodotorulate, *J. Biol. Chem.*, 260, 13921, 1985.

153. **Chung, T. D. Y., Matzanke, B. F., Winkelmann, G., and Raymond, K. N.**, Inhibitory effect of the partially resolved coordination isomers of chromic desferricoprogen on coprogen uptake in *Neurospora crassa*, *J. Bacteriol.*, 165, 283, 1986.

154. **van der Helm, D., Baker, J. R., Loghry, R. A., and Ekstrand, J. D.**, Structure of alumichrome A and ferrichrome A at low temperatures, *Acta Crystallogr.*, B37, 323, 1981.

155. **Barnes, C. L., Eng-Wilmot, D. L., and van der Helm, D.**, Ferricrocin ($C_{29}H_{44}FeO_{13}N_9x7H_2O$), an iron(III) binding peptide from *Aspergillus versicolor*, *Acta Crystallogr.*, C40, 922, 1984.

156. **Hossain, M. B., Eng-Wilmot, D. L., Loghry, R. A., and van der Helm, D.**, Circular dichroism, crystal structure, and absolute configuration of the siderophore ferric N, N',N''-triacetylfusarinine, $FeC_{39}H_{57}N_6O_{15}$, *J. Am. Chem. Soc.*, 102, 5766, 1980.

157. **Shanzer, A., Lobman, J., Lifson, S., and Felder, C. E.**, Origin of the Fe^{3+}-binding and conformational properties of enterobactin, *J. Am. Chem. Soc.*, 108, 7609, 1986.

158. **Eaton, S. S., Hutchinson, J. R., Holm, R. H., and Muetterties, E. L.**, Intramolecular rearrangement reactions of tris-chelate complexes. III. Analysis of the rearrangements of tris(α-isopropyltropolonato)aluminum(III) and -cobalt(III). Examples of stereochemically nonrigid aluminum(III) and cobalt (III) complexes, *J. Am. Chem. Soc.*, 94, 6411, 1972.

159. **Muetterties, E. L.**, Stereochemically nonrigid structures, *Acc. Chem. Res.*, 3, 266, 1970.

160. **Rorabacher, D. B. and Margerum, D. W.**, Multidentate ligand kinetics. VI. The exchange of ethylenediaminetetraacetate ion with triethylenetatramine nickel(II) and tetraethylenepentaminenickel(II), *Inorg. Chem.*, 3, 382, 1964.

161. **Borgias, B., Hugi, A. D., and Raymond, K. N.**, Isomerization and solution structures of desferrioxamine B complexes of Al^{3+} and Ga^{3+}, *Inorg. Chem.*, 28, 3538, 1989.

162. **Winkelmann, G., van der Helm, D., and Neilands, J. B.**, *Iron Transport in Microbes Plants and Animals*, VCH, Weinheim, 1987.

163. **Braun V. and Winkelmann, G.**, Microbial iron transport structure and function of siderophores, *Progr. Clin. Biochem. Med.*, 5, 67, 1987.

164. **Ecker, D. J., Matzanke, B. F., and Raymond, K. N.**, Specificity of ferric enterobactin transport in *E. coli*, *J. Bacteriol.*, 167, 666, 1986.

165. **Matzanke, B. F., Müller, G. I., and Raymond, K. N.**, Stereospecific uptake of hydroxamate siderophores and synthetic enantio-rhodotorulic acid in *E. coli*, *Biochem. Biophys. Res. Commun.*, 121, 922, 1984.

166. **Ecker, D. J., Loomis, L. D., Cass, M. E., and Raymond, K. N.**, Substituted complexes of enterobactin and synthetic analogues as probes of the ferric-enterobactin receptor in *Escherichia coli*, *J. Am. Chem. Soc.*, 110, 2457, 1988.

167. **Raymond, K. N., Ecker, D. J., Loomis, L. D., and Matzanke, B. F.**, Structure recognition and transport of ferric enterobactin in *E. coli* in *Frontiers of Bioinorganic Chemistry*, Xavier, A. V., Ed; VCH, Weinheim, 1986.

168. **Müller, G., Matzanke, B. F., and Raymond, K. N.**, Iron transport in *Streptomyces pilosus* mediated by ferrichrome type siderophores, rhodotorulic acid and enantio-rhodotorulic acid, *J. Bacteriol.*, 160, 313, 1984.

169. **Huschka, H., Naegeli, H.-U., Leuenberger-Ryf, H., Keller-Schierlein, W., and Winkelmann, G.**, Evidence for a common siderophore transport system but different siderophore receptors in *Neurospora crassa*, *J. Bacteriol.*, 162, 715, 1985.

170. **Huschka, H., Jalal, M. A. F., van der Helm D., and Winkelmann, G.**, Molecular recognition of siderophores in fungi: role of iron-surrounding N-acyl residues and the peptide backbone during membrane transport in *Neurospora crassa*, *J. Bacteriol.*, 167, 1020, 1986.

171. **Müller, G., Barclay, S. J., and Raymond, K. N.**, The mechanism and specificity of iron transport in *Rhodotorula pilimanae* probed by synthetic analogs of rhodotorulic acid, *J. Biol. Chem.*, 280, 13916, 1985.

172. **Wong, G. B., Kappel, M. J., Raymond, K. N., Matzanke, B., and Winkelmann, G.**, Coordination chemistry of microbial iron transport compounds. 24. Characterization of coprogen and ferricrocin, two ferric hydroxamate siderophores, *J. Am. Chem. Soc.*, 105, 810, 1983.

173. **Gütlich, P., Link, R., and Trautwein, A.**, *Mössbauer Spectroscopy and Transition Metal Chemistry*, Springer-Verlag, Berlin, 1978.

174. **Dickson, D. P. E., and Johnson, C. E.**, Physiological and medical applications, in *Applications of Mössbauer Spectroscopy*, Vol. II, Academic Press, New York, 1980, 209.

175. **Spartalian, K. and Lang, G.**, Oxygen transport and storage materials, in *Applications of Mössbauer Spectroscopy*, Vol. II, Academic Press, New York, 1980, 249.

176. **Mørup, S.**, *Paramagnetic and Superparamagentic Relaxation Phenomena Studied by Mössbauer Spectroscopy*, Polyteknik Forlag, Kopenhagen, 1980.

177. **Harrison, P. M. and Lilley, T. H.**, Ferritin, in *Iron Carriers and Proteins*, Loehr, T. M., Ed., VCH, New York, 1989, 123.

178. **Spartalian, K., Oosterhuis, W. T., and Smarra, N.**, Mössbauer effect studies in the fungus *Phycomyces*, *Biochim. Biophys. Acta*, 399, 203, 1975.

179. **Rimbert, J. N., Dumas, F., Kellershohn, C., Girot, R., and Brissot, P.**, Mössbauer spectroscopy of iron overloaded livers, *Biochimie*, 67, 663, 1985.

180. **Bell, S. H., Weir, M. P., Dickson, D. P. E., Gibson, J. F., Sharp, G. A., and Peters, T. J.**, Mössbauer spectroscopic studies of human haemosiderin and ferritin, *Biochim. Biophys. Acta*, 787, 227, 1984.

181. **Bauminger, E. R., Harrison, P. M., Nowik, I., and Treffry, A.**, Mössbauer spectroscopy study of the initial stages of iron-core formation in horse spleen apoferritin: evidence for both isolated Fe(III) atoms and oxobridged Fe(III) dimers as early intermediates, *Biochemistry*, 28, 5486, 1989.

182. **St. Pierre, T. G., Bell, S. H., Dickson, D. P. E., Mann, S., Webb, J., Moore, G. R., and Williams, R. J. P.**, Mössbauer spectroscopic studies of the cores of human, limpet and bacterial ferritins, *Biochim. Biophys. Acta*, 870, 127, 1986.

183. **Yang, C.-Y., Meagher, A., Huynh, B. H., Sayers, D. E., and Theil, E. C.**, Iron (III) bound to horse spleen apoferritin: an X-ray absorption and Mössbauer spectroscopy study that shows that iron nuclei can form on the protein, *Biochemistry*, 26, 497, 1987.

184. **Watt, G. D., Frankel, R. B., and Papaefthymiou, G. C.**, Reduction of mammalian ferritin, *Proc. Natl. Acad. Sci. U.S.A.*, 82, 3640, 1985.

185. **Frankel. R. B., Papaefthymiou, G. C., and Watt, G. D.**, Binding of Fe^{2+} by mammalian ferritin, *Hyperf. Interact.*, 33, 233, 1987.

186. **Bauminger, E. R., Cohen, S. G., Labenski de Kanter, F., Levy, A., Ofer, S., Kessel, M., and Rottem, S.**, Iron storage in *Mycoplasma capricolum*, *J. Bacteriol.*, 141, 378, 1980.

187. **Dickson, D. P. E. and Rottem, S.**, Mössbauer spectroscopic studies of iron in *Proteus mirabilis*, *Eur. J. Biochem.*, 101, 291, 1979.

188. **Bauminger, E. R., Cohen, S. G., Dickson, D. P. E., Levy, A., Ofer, S., and Yariv, J.**, Mössbauer spectroscopy of *Escherichia coli* and its iron-storage protein, *Biochim Biophys. Acta*, 623, 237, 1980.

189. **Watt, G. D., Frankel, R. B., Papaefthymiou, G. C., Spartalian, K., and Stiefel, E. I.**, Redox properties and the Mössbauer spectroscopy of *Azotobacter vinelandii* bacterioferririn, *Biochemistry*, 25, 4330, 1986.

190. **Huynh, B. H., Lui, M. C., Moura, J. J. G., Moura, I., Ljungdahl, P. O., Münck, E., Payne, W. J., Peck, H. D., DerVartanian, D. V., and LeGall, J.**, Mössbauer and EPR studies on nitrite reductase from *Thiobacillus dentrificans*, *J. Biol. Chem.*, 257, 9576, 1982.

191. **Lipscomb, J. D., Andersson, K. K., Münck, E., Kent, T. A., and Hooper, A. B.**, Resolution of multiple heme centers of hydroxylamine oxidoreductase from *Nitrosomonas*. 2. Mössbauer spectroscopy, *Biochemistry*, 21, 3973, 1982.

192. **Geary, P. J. and Dickson, D. P. E.**, Mössbauer spectroscopic studies of the terminal dioxygenase protein of benzene dioxygenase from *Pseudomonas putida, Biochem. J.*, 195, 199, 1981.

193. **Orme-Johnson, H. and Orme-Johnson, N. R.**, Iron-sulfur proteins: the problems of determining cluster type, in *Metal Ions in Biology*, Vol. 4, Spiro, T. G., Ed., Wiley Interscience, New York, 1982, 67.

194. **Schulz, C. and Debrunner, G.**, Rubredoxin, a simple iron-sulfur protein: its spin hamiltonian and hyperfine parameters, *J. Phys. (Paris)*, 37, Colloque C6, 153, 1976.

195. **Bill, E., Bernhardt, F.-H., and Trautwein, A. X.**, Mössbauer studies on the active Fe . . . [2Fe-2S] site of putidamonoxin, its electron transport and dioxygen activation mechanism, *Eur. J. Biochem.*, 121, 39, 1981.

196. **Kent, T. A., Emptage, M. H., Merkel, H., Kennedy, M. C., Beinert, H., and Münck, E.**, Mössbauer studies of aconitase, *J. Biol. Chem.*, 260, 6871, 1985.

197. **Evans, E. H., Rush, J. D., Johnson, C. E., and Evans, M. C. W.**, Mössbauer spectra of photosystem-I reaction centers from the blue-green alga *Chloroglea fritschii, Biochem. J.*, 182, 861, 1979.

198. **Petrouleas, V. and Diner, B. A.**, Investigation of the iron components in photosystem II by Mössbauer spectroscopy, *FEBS Lett.*, 147, 111, 1982.

199. **Evans, E. H., Carr, N. G., Rush, J. D., and Johnson, C. E.**, Identification of non-magnetic iron centre and an iron-storage-or-transport material in blue-green algal membranes by Mössbauer spectroscopy, *Biochem. J.*, 166, 547, 1977.

200. **Semin, B. K., Novakova, A. A., Aleksandrov, A. Yu., Ivanov, I. I., Rubin, A. B., and Kuzmin, R. N.**, Mössbauer spectroscopy of iron metabolism and iron intracellular distribution in liver of rats, *Biochim. Biophys. Acta*, 715, 52, 1982.

201. **Kaufman, K. S., Papaefthymiou, G. C., Frankel, R. B., and Rosenthal, A.**, Nature of iron deposits on the cardiac walls in β-thalassemia by Mössbauer spectroscopy, *Biochim. Biophys. Acta*, 629, 522, 1980.

202. **Bauminger, E. R., Cohen, S. G., Ofer, S., and Bachrach, U.**, Study of storage iron in cultured chick embryo fibroblasts and rat glioma cells, using Mössbauer spectroscopy, *Biochim. Biophys. Acta*, 623, 237, 1980.

203. **Bauminger, E. R., Cohen, S. G., Ofer, S., and Rachmilewitz, E. A.**, Quantitative studies of ferritinlike iron in erythrocytes of thalassemia, sickle-cell anemia, and hemoglobin Hammersmith with Mössbauer spectroscopy, *Proc. Natl. Acad. Sci. U.S.A.*, 76, 939, 1979.

204. **Ofer, S., Cohen, S. G., Bauminger, E. R., and Rachmilewitz, E. A.**, Recoil-free absorption in thalassemic red blood cells, *J. Phys. (Paris)*, 37, Colloque C6, C6-199, 1976.

205. **Matzanke, B. and Winkelmann, G.**, Siderophore iron transport followed by Mössbauer spectroscopy, *FEBS Lett.*, 130, 50, 1981.

206. **Matzanke, B. F., Ecker, D. J., Yang, T.-S., Huynh, B. H., Müller, G., and Raymond, K. N.**, Iron enterobactin uptake in *Escherichia coli* followed by Mössbauer spectroscopy, *J. Bacteriol.*, 167, 674, 1986.

207. **Matzanke, B. F., Bill, E., Winkelmann, G., and Trautwein, A. X.**, A ^{57}Fe Mössbauer study of iron assimilation in *N. crassa* mediated by siderophores, *Hyperf. Interact.*, 29, 1415, 1986.

208. **Matzanke, B. F., Bill, E., Müller, G. I., Winkelmann, G., and Trautwein, A. X.**, Metabolization of ^{57}Fe-coprogen in *N. crassa*. An in vivo Mössbauer study, *Eur. J. Biochem.*, 162, 643, 1987.

209. **Matzanke, B. F., Bill, E., Winkelmann, G., and Trautwein, A. X.**, A novel main component of microbial iron metabolism detected by in vivo Mössbauer spectroscopy, *Rec. Trav. Chim. Pays-Bas*, 106, 258, 1987.

210. **Matzanke, B. F.**, Mössbauer spectroscopy of microbial iron uptake and metabolism, in *Iron Transport in Microbes, Plants and Animals*, Winkelmann, G., van der Helm, D., and Neilands, J. B., Eds., VCH, Weinheim, 1987, 251.

211. **Matzanke, B. F., Bill, E., Winkelmann, G., and Trautwein, A. X.**, Role of siderophores in iron storage in spores of *N. crassa* and *A. ochraceus*. *J. Bacteriol.*, 169, 5873, 1987.

212. **Matzanke, B. F., Bill, E. Trautwein, A. X., and Winkelmann, G.**, Ferricrocin functions as the main intracellular iron-storage compound in mycelia of *Neurospora crassa*, *Biol. Met.*, 1, 18, 1988.

213. **Matzanke, B. F., Bill, E., Müller, G. I., Trautwein, A. X., and Winkelmann, G.**, In vivo Mössbauer spectroscopy of iron uptake and ferrometabolism in *Escherichia coli*. Proceedings of the 3rd Seeheim Workshop on Mössbauer Spectroscopy, *Hyperf. Interact.*, 47, 311, 1989.

214. **Matzanke, B. F., Müller, G., Bill, E., and Trautwein, A. X.**, Iron metabolism of *E. coli* studied by Mössbauer spectroscopy and biochemical methods, *Eur. J. Biochem.*, 183, 371, 1989.

215. **Matzanke, B. F., Bill, E., Trautwein, A. X., and Winkelmann, G.**, Siderophores as iron storage compounds in the yeasts *Rhodotorula minuta* and *Ustilago sphaerogena* detected by in vivo Mössbauer spectroscopy, *Hyperf. Interact.*, 58, 2359, 1990.

216. **Matzanke, B. F., Berner, I., Bill, E., Trautwein, A. X., and Winkelmann, G.**, Iron uptake in *Erwinia herbicola* K4 and detection of *E. coli*-like novel main components of iron metabolism, *Forum Mikrobiol.*, 13, 53 (P17), 1990.

217. **Matzanke, B. F., Bill, E., and Trautwein, A. X.**, Fur participates in the regulation of iron redox states in *Escherichia coli, Forum Mikrobiol.*, 13, 20 (V73), 1990.

218. **Mielczarek, E. V., Royt, P. W., and Toth-Allen, J.**, Microbial acquisition of iron, *Comments Mol. Cell. Biophys.*, 6, 1, 1989.

219. **Bauminger, E. R., Cohen, S. G., Giberman, E., Nowik, I., Ofer, S. G., Yariv, J., Werber, M. M., and Mevarech, M.**, Iron in the whole bacterial cell. Exploratory investigations, *J. Phys. (Paris)*, 37, Colloque C6, C6-227, 1976.

220. **Douzou, P.**, The use of subzero temperatures in biochemistry: slow reactions, *Methods Biochem. Anal.*, 22, 401, 1975.

221. **Fridovich, I.**, The biology of oxygen radicals, *Science*, 201, 875, 1978.

222. **Dunford, H. B.**, Free radicals in iron-containing systems, *Free Radic. Biol. Med.*, 3, 405, 1987.

223. **Halliwell, B. and Gutteridge, J. M. C.**, Oxygen toxicity, oxygen radicals, transition metals and disease, *Biochem. J.*, 219, 1, 1984.

224. **Harman, D.**, The aging process, *Proc. Natl. Acad. Sci. U.S.A.*, 78, 7124, 1981.

225. **Ames, B. N.**, Dietary carcinogens and anticarcinogens, *Science*, 221, 1256, 1983.

226. **Loeb, L. A., James, E. A., Waltersdorph, A. M., and Klebanoff, S. J.**, Mutagenesis by the autoxidation of iron with isolated DNA, *Proc. Natl. Acad. Sci. U.S.A.*, 85, 3918, 1988.

227. **Filho, A. C. M. and Meneghini, R.**, In vivo formation of single-strand breaks in DNA by hydrogen peroxide is mediated by the Haber-Weiss reaction, *Biochim. Biophys. Acta*, 781, 56, 1984.

228. **Minotti, G. and Aust, S. D.**, The role of iron in the initiation of lipid peroxidation, *Chem. Phys. Lipid*, 44, 191, 1987.

229. **Dorfman, L. M. and Adams, G. E.**, Reactivity of the hydroxyl radical in aqueous solutions, National Standards Reference Data System, *Nat. Bur. Stand. Bull.*, No. 46, 1973.

230. **Dreyer, G. and Dervan, P. B.**, Sequence-specific cleavage of single-stranded DNA: oligonucleotide-EDTA·FE(II), *Proc. Natl. Acad. Sci. U.S.A.*, 82, 968, 1985.

231. **Inoue, S. and Kawanishi, S.**, Hydroxyl radical production and human DNA damage induced by ferric nitrilotriacetate and hydrogen peroxide, *Cancer Res.*, 47, 6522, 1987.

232. **Basile, L. A., Raphael, A. L., and Barton, J. K.**, Metal activated hydrolytic cleavage of DNA, *J. Am. Chem. Soc.*, 109, 7550, 1987.

233. **Rowley, D. A. and Halliwell, B.**, Superoxide dependent formation of hydroxyl radicals from NADH and NADPH in the presence of iron salts, *FEBS Lett.*, 142, 39, 1982.

234. **Rush, J. D. and Koppenol, W. H.**, Oxidizing intermediates in the reaction of ferrous EDTA with hydrogen peroxide. Reactions with organic molecules and ferrocytochrome c, *J. Biol. Chem.*, 261, 6730, 1986.

235. **Floyd, R. A. and Lewis, C. A.**, Hydroxyl free radical formation from hydrogen peroxide by ferrous iron-nucleotide complexes, *Biochemistry*, 22, 2645, 1983.

236. **Horowitz, N. H., Charlang, G., Horn, G., and Williams, N. P.**, Isolation and identification of the conidial germination factor of *Neurospora crassa*, *J. Bacteriol.*, 127, 135, 1976.

237. **Brambl, S., Dunkle, L. D., and Van Etten, J. L.**, Nucleic acid and protein synthesis during fungal spore germination, in *The Filamentous Fungi*, Smith, J. E. and Berry, D. R., Eds., Edward Arnold, London, 1978, 94.

238. **Dahlberg, K. R. and Van Etten, J. L.**, Physiology and biochemistry of fungal sporulation, *Annu. Rev. Phytopathol.*, 20, 281, 1982.

239. **Van Etten, J. L., Dahlberg, K. R., and Russo, G. M.**, Nucleic acids, in *The Fungal Spore: Morphogenetic Controls*, Turian G. and Hohl, H. R., Eds., Academic Press, New York, 1981, 277.

240. **Brody, S.**, Generic and biochemical studies on *Neurospora* conidia germination and formation, in *The Fungal Spore: Morphogenetic Controls*, Turian G. and Hohl, H. R., Eds., Academic Press, New York, 1981, 605.

241. **Deml, G. and Oberwinkler, F.**, Studies in heterobasidiomycetes. Part 22. A survey on siderophore formation in low-iron cultured anther smuts of *caryophyllaceae*, *Zbl Bakt. Hyg., I. Abt. Orig. C3*, 475, 1982.

242. **Tufano, T. P. and Raymond, K. N.**, Coordination chemistry of microbial iron compounds. 21. Kinetics and mechanism of iron exchange in hydroxamate siderophore complexes, *J. Am. Chem. Soc.*, 103, 6617, 1981.

243. **Harris, W. R., Carrano, C. J., Cooper, S. R., Sofen, S. R., Avdeef, A. E., McArdle, J. V., and Raymond, K. N.**, Coordination chemistry of microbial iron transport compounds. 19. Stability constants and electrochemical behavior of ferric enterobactin and model complexes, *J. Am. Chem. Soc.*, 101, 6097, 1979.

244. **Harris, W. R. and Raymond, K. N.**, Ferric ion sequestering agents. 3. The spectrophotomeric and potentiometric evaluation of two new enterobactin analogues: 1,5,9-N,N′,N″-tris(2,3-dihydroxybenzoyl)-cyclotriazatridecane and 1,3,5-N,N′,N″-tris(2,3-dihydroxybenzoyl)triaminomethylbenzene, *J. Am. Chem. Soc.*, 101, 6534, 1979.

245. **Hider, R., Silver, J., Neilands, J. B., Morrison, I. E. G., and Rees, L. V. C.**, Identification of iron(II) enterobactin and its possible role in (*Escherichia coli*) iron transport, *FEBS Lett.*, 102, 325, 1979.

246. **Hider, R. C., Mohd-Nor, A. R., Silver, J., Morrison, I. E. G., and Rees, L. V. C.**, Model compounds for microbial iron-transport compounds. Part 1. Solution chemistry and Mössbauer study of iron(II) and iron(II) complexes from phenolic and catecholic systems, *J. Chem. Soc. Dalton Trans.*, 609, 1981.

247. **Pecoraro, V. L., Harris, W. R., Wong, G. B., Carrano, C. J., and Raymond, K. N.**, Coordination chemistry of microbial iron transport compounds. 23. Fourier transform infrared spectroscopy of ferric catecholamide analogues of enterobactin, *J. Am. Chem. Soc.*, 105, 4623, 1983.

248. **Cass, M. E., Garrett, T. M., and Raymond, K. N.**, The salicylate mode of bonding in protonated ferric enterobactin analogues, *J. Am. Chem. Soc.*, 111, 1677, 1989.

249. **Cooper, S. R., McArdle, J., and Raymond, K. N.**, Siderophore electrochemistry: relation to intracellular iron release mechanism, *Proc. Natl. Acad. Sci. U.S.A.*, 75, 3551, 1978.

250. **Lee, C.-W., Ecker, D. J., and Raymond, K. N.**, The pH dependent reduction of ferric enterobactin probed by electrochemical methods and its implications for microbial iron transport, *J. Am. Chem. Soc.*, 107, 6920, 1985.

251. **Heidinger, S., Braun, V., Pecoraro, V. L., and Raymond, K. N.**, Iron supply to *Escherichia coli* by synthetic analogs of enterochelin, *J. Bacteriol.*, 153, 109, 1983.

252. **Cox, G. B., Gibson, R. K. J., Luke, N. A., Newton, I. G., O'Brien, I. G., and Rosenberg, H.**, Mutations affecting iron transport in *Escherichia coli*, *J. Bacteriol.*, 104, 219, 1970.

253. **Pierce, J. R., Pickett, C. L., and Earhart, C. F.**, Two *fep* genes are required for enterochelin uptake in *Escherichia coli* K-12, *J. Bacteriol.*, 155, 330, 1983.

254. **Elkins, M. D. and Earhart, C. F.**, Nucleotide sequence and regulation of the *Escherichia coli* gene for ferrienterobactin transport protein *fepB*, *J. Bacteriol.*, 172, 5443, 1989.

255. **Pugsley, A. P. and Reeves, P.**, Uptake of ferrienterochelin by *Escherichia coli*: energy-dependent stage of uptake, *J. Bacteriol.*, 130, 26, 1977.

256. **Hollifield, W. C. and Neilands, J. B.**, Ferric enterobactin transport system in *Escherichia coli* K-12. Extraction, assay, and specificity of the outer membrane receptor, *Biochemistry*, 17, 1922, 1978.

257. **Jalal, M. A. F. and van der Helm, D.**, Purification and crystallization of ferric enterobactin receptor protein, fep A, from the outer membranes of *Escherichia coli* UT5600/pBB2, *FEBS Lett.*, 243, 366, 1989.

258. **Neilands, J. B., Erickson, T. J., and Rastetter, W. H.**, Stereospecificity of the ferric enterobactin receptor of *Escherichia coli* K-12, *J. Biol. Chem.*, 256, 3831, 1981.

259. **Isied, S. S., Kuo, G., and Raymond, K. N.**, Coordination isomers of biological iron transport compounds. V. The preparation and chirality of the chromium(III) enterobactin complex and model tris(catechol)chromium(III) analogues, *J. Am. Chem. Soc.*, 98, 1763, 1976.

260. **Plaha, D. S. and Rogers, H. J.**, Antibacterial effect of the scandium complex of enterochelin. Studies of the mechanism of action, *Biochim. Biophys. Acta*, 760, 246, 1983.

261. **Rogers, H. J., Woods, V. E., and Synge, C.**, Antibacterial effect of the scandium and indium complexes of enterochelin on *Escherichia coli*, *J. Gen. Microbiol.*, 128, 2389, 1982.

262. **Langman, L. L., Young, I. G., Frost, G. E., Rosenberg, H., and Gibson, F.**, Enterochelin system of iron transport in *Escherichia coli*: mutations affecting ferric-enterochelin esterase, *J. Bacteriol.*, 112, 1142, 1972.

263. **O'Brien, I. G., Cox, G. B., and Gibson, F.**, Enterochelin hydrolysis and iron metabolism in *Escherichia coli*, *Biochim. Biophys. Acta*, 237, 537, 1971.

264. **Bryce, G. F. and Brot, N.**, Studies on the enzymatic synthesis of the cyclic trimer of 2,3-dihydroxy-N-benzoyl-L-serine in *Escherichia coli*, *Biochemistry*, 11, 1708, 1972.

265. **Greenwood, K. T. and Luke, R. K. J.**, Enzymatic hydrolysis of enterochelin and its iron complex in *Escherichia coli* K-12. Properties of enterochelin esterase, *Biochim. Biophys. Acta*, 525, 209, 1978.

266. **Lodge, J. S., Gaines, C. G., Arcenaux, J. E. L., and Byers, B. R.**, Non-hydrolytic release of iron from ferrienterobactin analogs by extracts of *Bacillus subtilis*, *Biochem. Biophys. Res. Commun.*, 97, 1291, 1980.

267. **Hobot, J. A., Carlemalm, E., Villiger, W., and Kellenberger, E.**, Periplasmic gel: new concept resulting from the reinvestigation of bacterial cell envelope ultrastructure by new methods, *J. Bacteriol.*, 160, 143, 1984.

268. **Stock, J. B., Rauch, B., and Roseman, S.**, Periplasmic space in *Salmonella typhimurium* and *Escherichia coli*, *J. Biol. Chem.*, 252, 7850, 1977.

269. **Cramer, W. A., Dankert, J. R., and Uretanis, Y.**, The membrane channel-forming antibiotic protein, colicine E, *Biochim. Biophys. Acta*, 737, 173, 1983, and references therein.

270. **Briskot, G., Taraz, K., and Budzikiwicz, H.**, Siderophore vom Pyoverdin-Typ aus *Pseudomonas aeruginosa*, *Z. Naturforsch.*, 41c, 497, 1986.

271. **Briskot, G., Taraz, K., and Budzikiwicz, H.**, Pyoverdine-type siderophores from *Pseudomonas aeruginosa*, *Liebigs Ann. Chem.*, 375, 1989.

272. **Demange, P., Wendenmaum, S., Linget, C., Mertz, C., Cung, M. T., Dell, A., and Abdallah, M. A.**, Bacterial siderophores: structural and NMR assignment of pyoverdins Pa, siderophores of *Pseudomonas aeruginosa* ATCC 15692, *Biol. Met.*, 3, 155, 1990.

273. **Leong, J. and Raymond, K. N.,** Coordination isomers of biological iron transport compounds. II. The optical isomers of chromic desferrichrome and desferrichrysin, *J. Am. Chem. Soc.,* 96, 6628, 1974.

274. **Abu-Dari, K. and Raymond, K. N.,** Coordination isomers of biological iron transport compounds. 8. The resolution of tris(hydroxamato) and tris(thiohydroxamato)complexes of high-spin iron(III), *J. Am. Chem. Soc.,* 99, 2003, 1977.

275. **Neilands, J. B.,** Naturally occurring non-porphyrin iron compounds, *Struct. Bonding,* 1, 59, 1966.

276. **Hough, E. and Rogers, D.,** The crystal structure of ferrimycobactin P, a growth factor for the *Mycobacteria, Biochem. Biophys. Res. Commun.,* 57, 73, 1974.

277. **Carrano, C. J., Cooper, S. R., and Raymond, K. N.,** Coordination chemistry of microbial iron transport compounds. 11. Solution equilibria and electrochemistry of ferric rhodotorulate complexes, *J. Am. Chem. Soc.,* 101, 599, 1979.

278. **Harris, W. R., Carrano, C. J., and Raymond, K. N.,** Coordination chemistry of microbial iron transport compounds. 16. Isolation, characterization and formation constants of ferric aerobactin, *J. Am. Chem. Soc.,* 101, 2722, 1979.

279. **Neilands, J. B., Peterson, T., and Leong, S. A.,** High affinity iron transport in microorganisms, in *Inorganic Chemistry in Biology and Medicine,* ACS Symp. Ser. 140, American Chemical Society, Washington, D.C., 1980, 263.

APPENDIX
STRUCTURES OF REPRESENTATIVE MICROBIAL IRON CHELATING AGENTS

Catecholates: (1) monocatecholates (2,3-dihydroxybenzoyl-glycine (X = H), -serine (X = CH_2OH),-threonine (X = $-CH(OH)CH_3$); (2) chrysobactin, (3) myxochelin, (4) unnamed catecholate (R = H) and parabactin A (R = 2 hydroxybenzoyl N-threonyl-), (5) agrobactin, (6) vibriobactin, and (7) enterobactin.

Hydroxamates: (8) ferrioxamines (see Table 1 for chain length and terminal groups), (9) building block of desferrimycin A_1 (Table 1), (10) bisucaberin, (11) ferrichromes (details, see Table 2), (12) DDF [X = H, R = cis-5-hydroxy-3-methylpent-2-enoyl), albomycines (X = (13), R = CH_3, R_1 = $-NCO-NH_2$ (albomycine δ_2), R_1 = NH (albomycine ϵ), R = O (albomycine δ_1)], (14) coprogens (see Table 3), (15) foroxomythine, (16) fusarinines [fusarinin, (n = 1); fusarinine A (n = 2, R = H), fusarinine B (n = 3, R = H), fusarinine C (r = 3, R = H, cyclic), triacetylfusarinine (n = 3, R = acetyl, cyclic)], (17) rhodotorulic acid, and (18) dimerum acid.

Mixed chelating groups: (19) — (21) chromophores of pyoverdins, azotobactins, and pseudobactins (peptide sequence X and residues R_1 are described in Table 4), (22) myco-bactins and nocobatins (R_1 = various alkyl chains; R_2 = R_3R_5 = CH_3 and H, R_4 = alkyl- or H, details in Reference 79), (23) anguibactin, (24) pyochelin, (25) maduraferrin, (26) aerobactin (R = CO_2H) and arthrobactin (R = H), (27) schizokinen, (28) alcaligin, (29) staphyloferrin, and (30) rhizobactin.

Other microbial iron chelators: (31) N-phenyl hydroxamic acid, (32) actinonin, (33) mycelianamide, (34) astechrome, (35) cepabactin, (36) aspergillic acids (AA) [AA (R_1 = H, R_2 = C_2H_5, R_3 = CH_3), neo-AA(R_1 = H, R_2 = $CH(CH_3)_2$, R_3 = H), meta AA (R_1 = OH, R_2 = CH_3, R_3 = CH_3), hydroxy-AA (R_1 = OH, R_2 = C_2H_5, R_3 = CH_3), neo-hydroxy-AA (R_1 = OH, R_2 = $CH(CH_3)_2$, R_3 = H)], (37) pulcherriminic acid, (38) fer-rithiocin, (39) tropolones [thujaplicin α, β, and ν (R_1, R_2, and R_3 correspond to H, or $C(CH_3)_2$)], (40) anthracyclines [daunomycin (R = H), adriamycin (R = OH)], (41) fer-rosamine A, (42) acidic form of ferrosamine A, (43) siderchelins A-C

$$\left(A: X = -C\begin{smallmatrix} OH \\ \\ CH_3 \end{smallmatrix} \quad B: -C\begin{smallmatrix} CH_3 \\ \\ OH \end{smallmatrix} \quad C: -C\begin{smallmatrix} OH \\ \\ C_2H_5 \end{smallmatrix} \right)$$, (44) viridomycins [viridomycin A (X = H), actinoviridin A (X = OH), ferroverdin (X = O-ϕ-C=CH_2)]

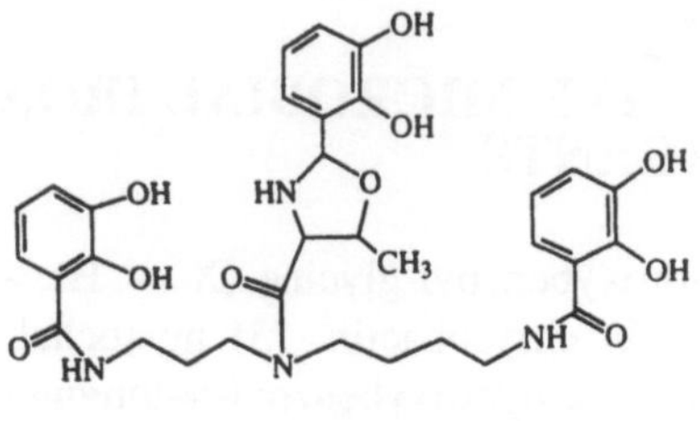

6

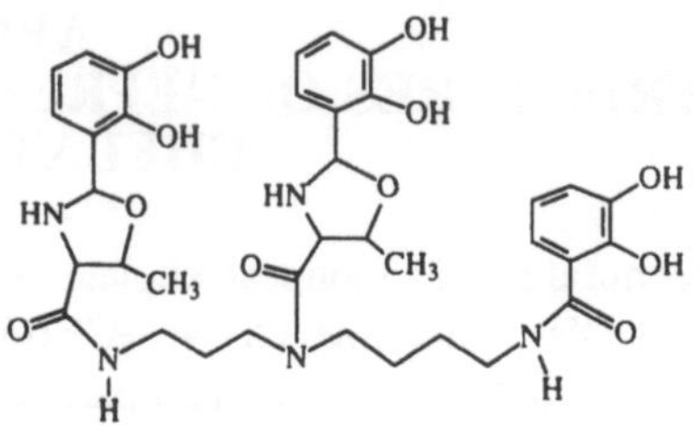

7

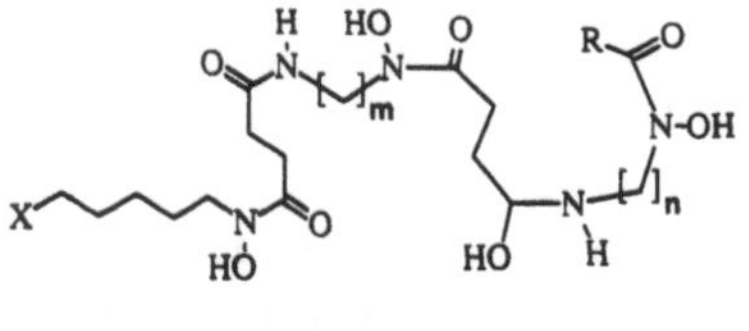

8

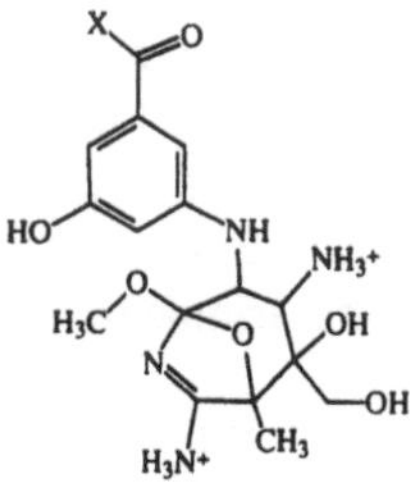

9

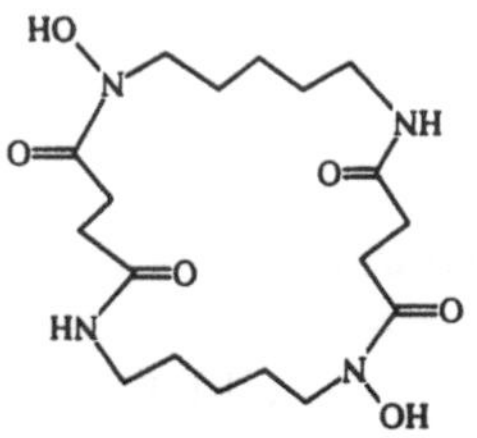

10

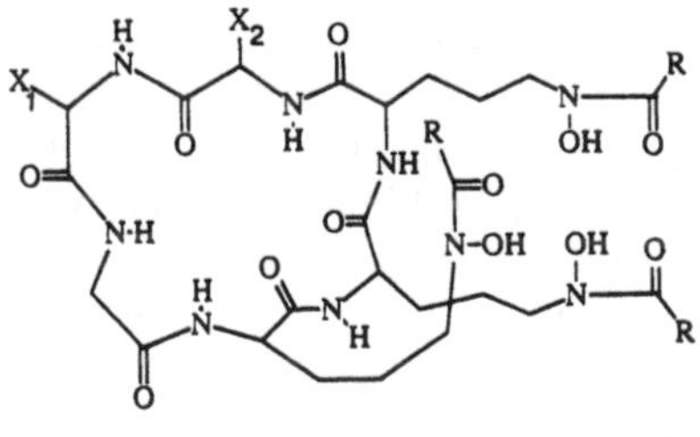

11

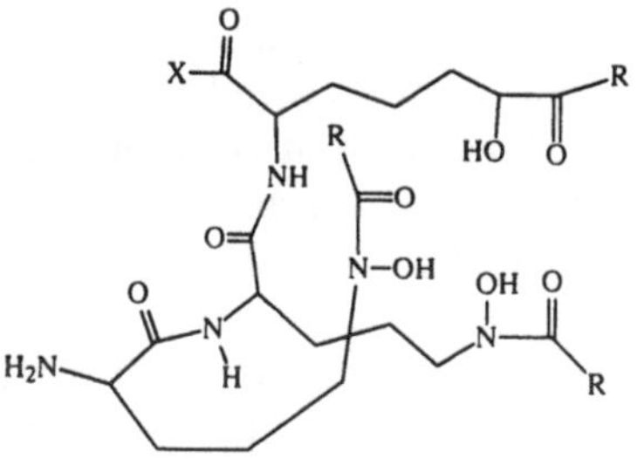

12

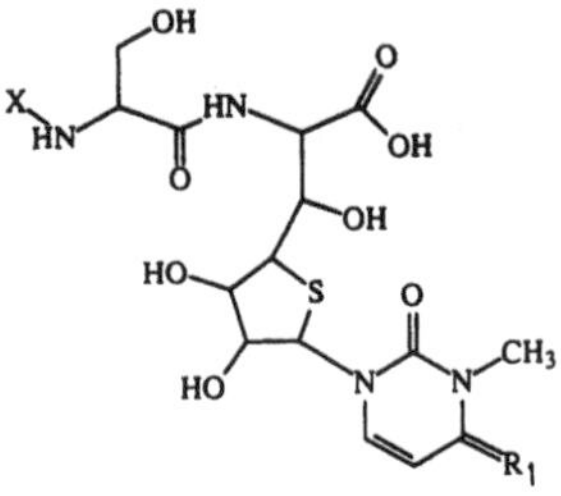

13

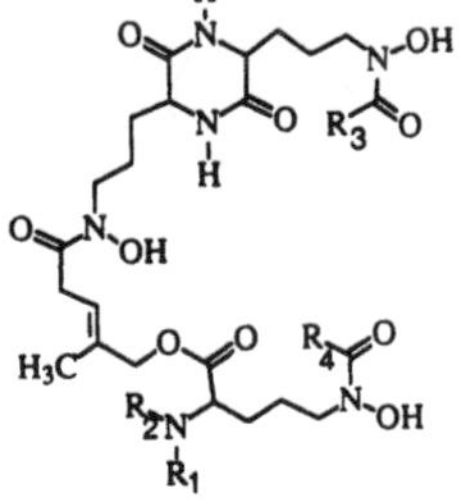

14

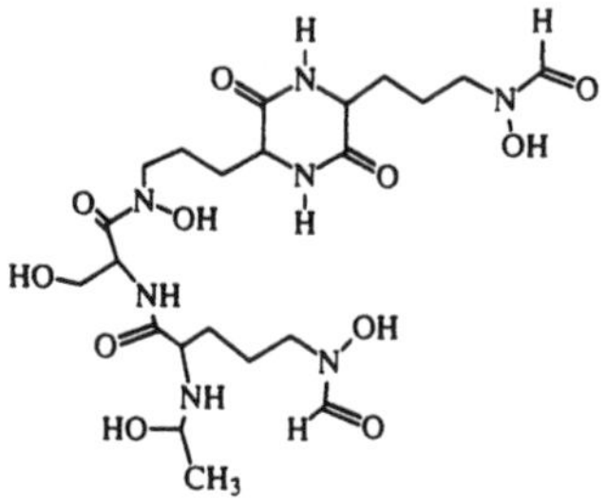

15

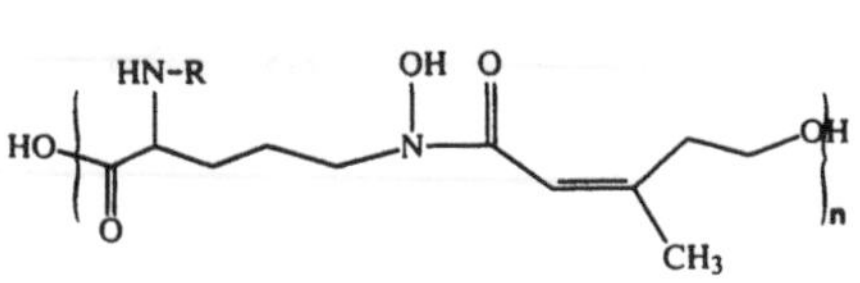

16

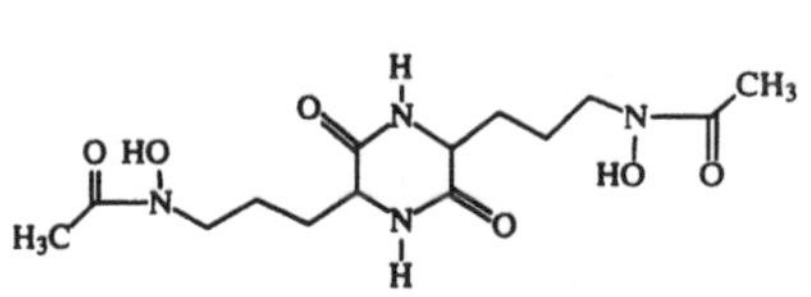

17

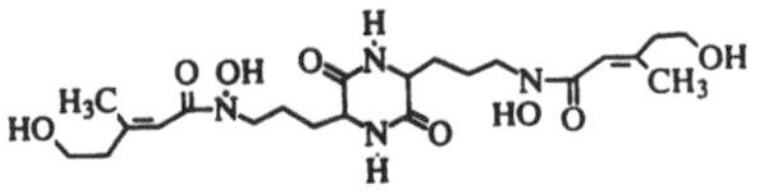

18

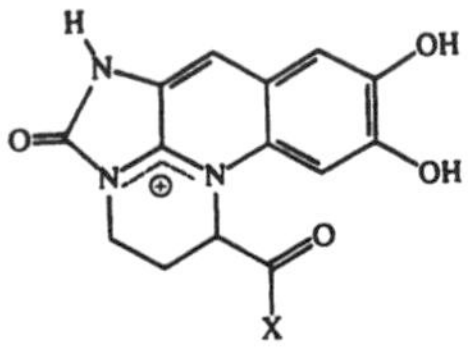

19

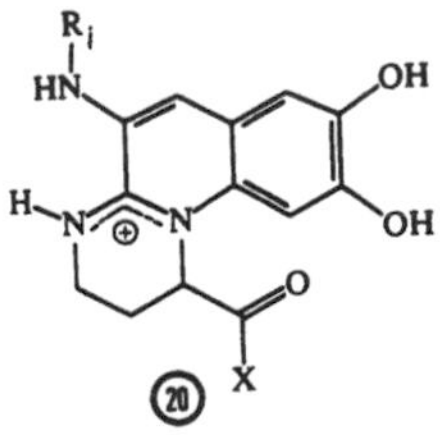

20

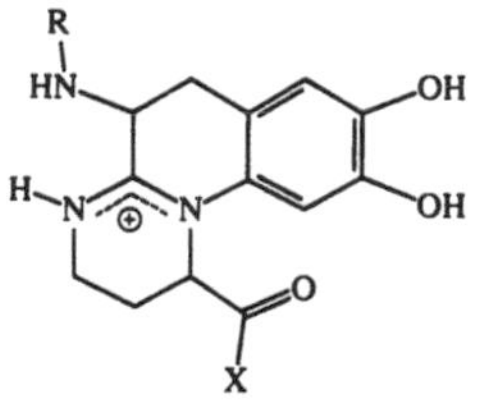

21

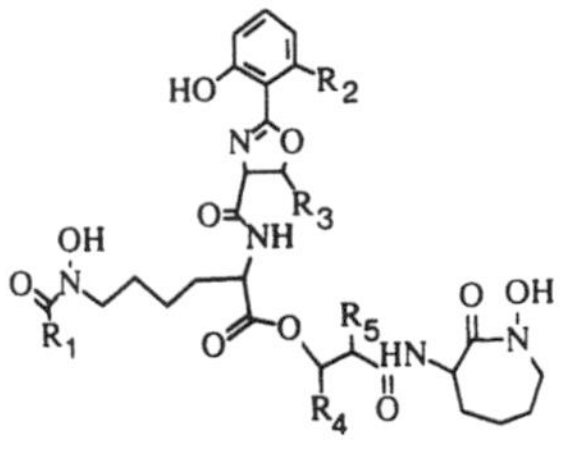

22

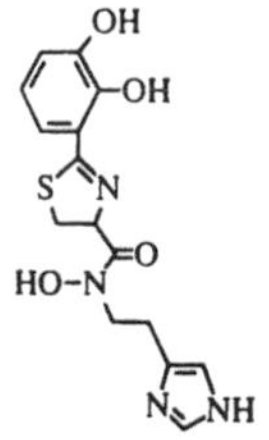

23

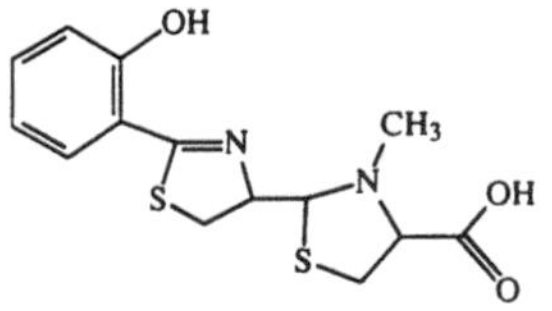

24

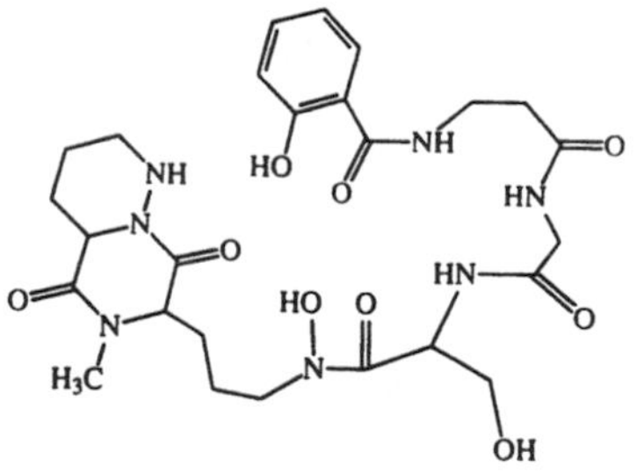

25

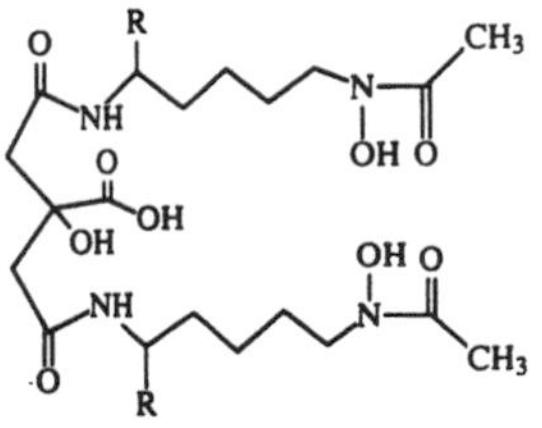

26

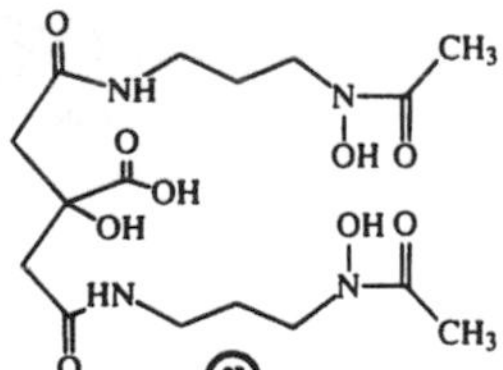

27

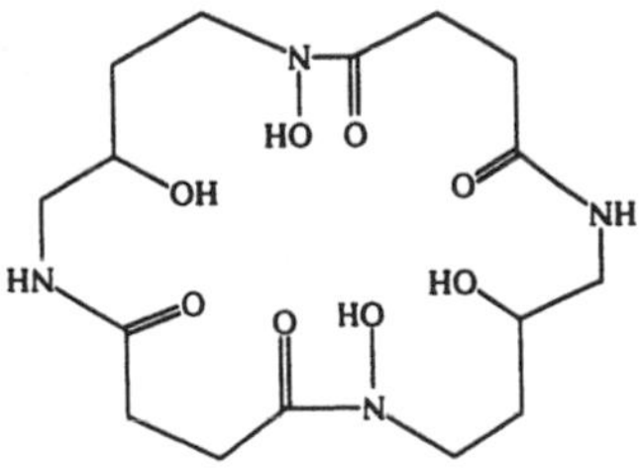

28

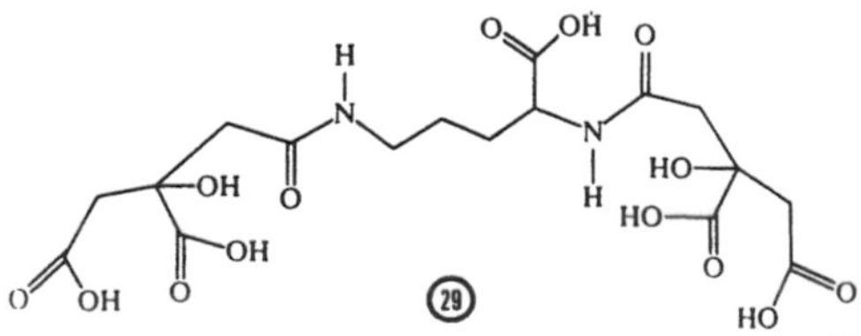

29

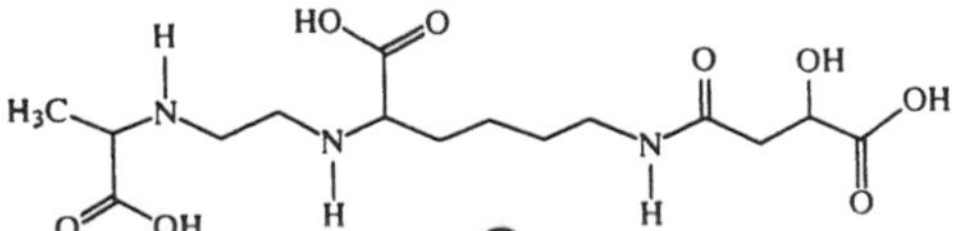

30

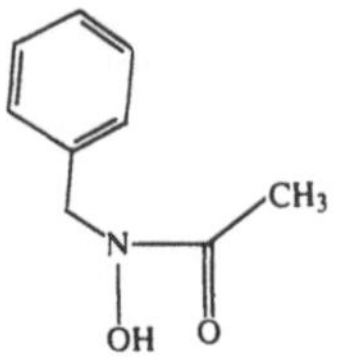

31

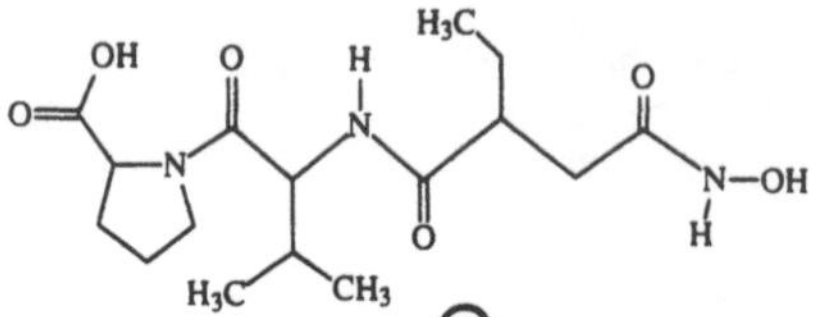

32

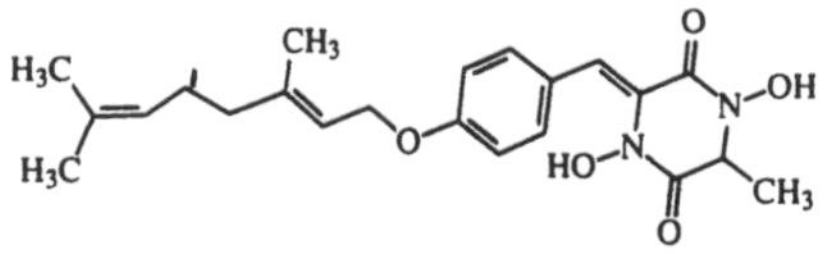

33

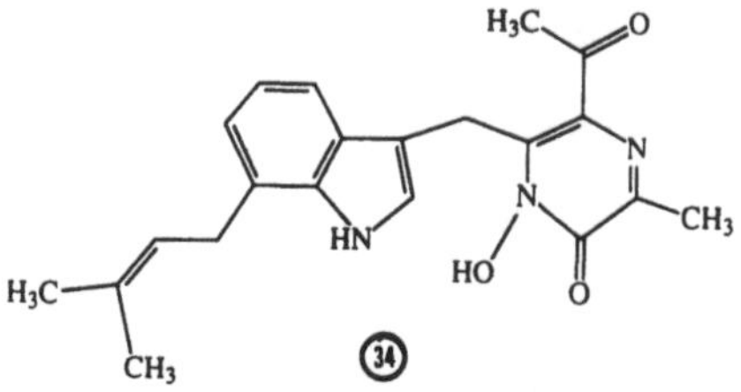

34

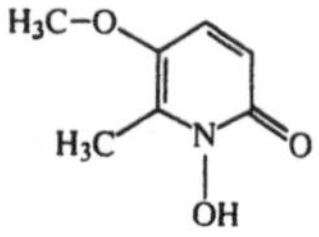

35

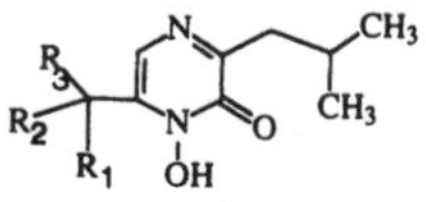

36

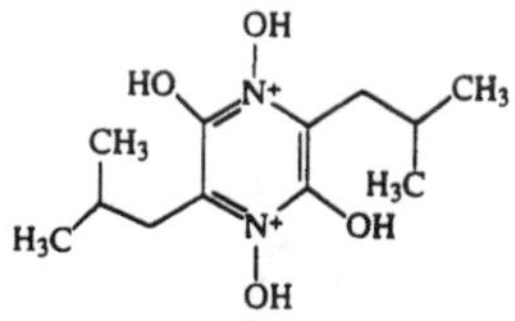

37

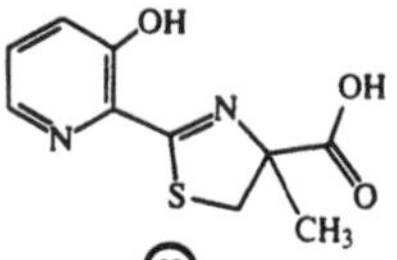

38

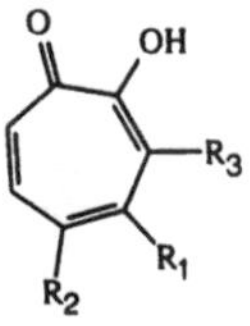

39

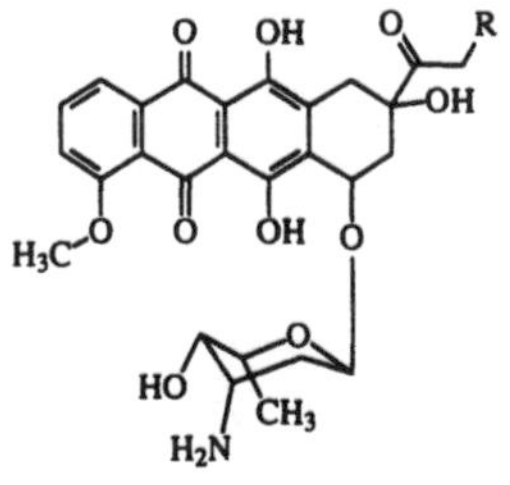

40

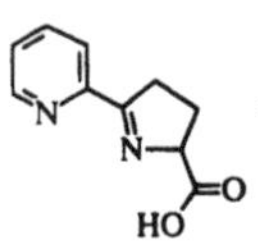

41

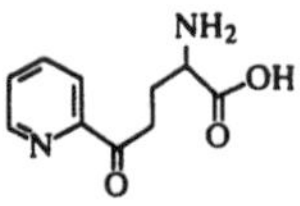

42

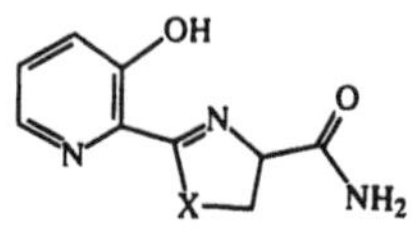

43

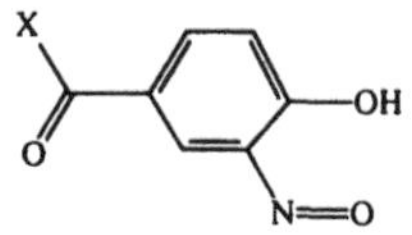

44

SPECIFICITY OF IRON TRANSPORT IN BACTERIA AND FUNGI

Günther Winkelmann

GENERAL CONSIDERATIONS

Iron transport in microbes is measured most often with isotope-labeled siderophores although other methods, like atomic absorption, ESR and Mössbauer spectroscopic measurements, have been used as alternative methods. Generally, ^{55}Fe- or ^{59}Fe-labeled iron complexes have to be prepared before iron uptake can be followed by kinetic measurements. The fate of the ligand may be studied by using ^{3}H- or ^{14}C-labeled ligands. Occasionally, simultaneous measurement of double-labeled ^{55}Fe/^{14}C-labeled siderophores is required. The method of choice for transport measurements in microorganisms is the membrane filter method, in which labeled cells are filtered off after uptake of siderophores. For protoplasts or fragile microbial cells a sedimentation procedure in high-density media is recommended. Kinetic studies may be performed either in a time-dependent or concentration-dependent manner. Time-dependent uptake is generally carried out to confirm a continuous accumulation of iron or ligand, whereas concentration-dependent uptake is the preferred method to analyze the mechanism of uptake. While time-dependent assays generally do not allow to differentiate between diffusion, binding or actual transport, concentration-dependent assays may at least in part differentiate between different mechanisms. Saturation kinetics are clear evidence for a mediated transport which by definition requires the involvement of a mediating transport protein or transport system within the membrane. However, it does not define active transport, as is sometimes inferred. In order to obtain proof for an active transport, one has to use specific metabolic inhibitors, such as cyanide, azide, or other respiratory poisons and uncouplers. Calculation of concentration gradients is also a valuable experiment, although it is not always possible to determine exactly the intracellular concentration of siderophores, since iron release may make it difficult to obtain a clear interpretation of the siderophore uptake. Generally, saturation kinetics allow a distinction between mediated transport and diffusion-like behavior of transported siderophores.

Transport data are generally recorded as the concentration of half-maximal velocity of transport (K_m) and maximal velocity of transport (V_{max}). However, when kinetics of membrane transport are discussed in terms of Michaelis-Menten constants, it is important to point out that there is only a formal conformity between enzyme and transport kinetics. The transport kinetic constant K_m is the result of several characteristic membrane interactions and the maximal transport velocity V_{max} is generally dependent on the slowest reaction step. Thus, these values represent only an approximation to simple enzyme kinetics. We know, for example, that during iron transport mediated by siderophores in Gram-negative bacteria, two membranes (outer and inner membrane) have to be penetrated. Moreover, it appears that further periplasmic mediators of siderophores are involved as well. These facts have to be considered when siderophore-iron transport into bacterial cells is determined. Similar problems of interpretation arise when siderophore transport across fungal membranes is considered. A more specific aspect of siderophore-mediated iron transport, however, is that the kinetics of both the iron and the ligand have to be followed, although in most cases the fate of iron is the more interesting one and the enrichment of iron is generally more pronounced than the uptake of the ligand. Therefore, analysis of double-labeled siderophore kinetics is the preferred method, if the mechanism of siderophore transport is to be investigated.

In certain cases it has been possible to determine association/dissociation constants K_D of siderophores with isolated membrane receptors. The K_D values often differ from the corresponding K_m values of the whole transmembrane transport, confirming the view that further permeation steps within the membrane are involved. Small K_D values indicate high binding affinity with membrane receptors or with the binding sites of transport systems. In whole cells, rapid interaction with receptors or binding sites may be inferred from high initial uptake rates followed by significantly slower uptake rates after some minutes. In this case saturation of receptors or binding sites preceeds the actual transport event. The latter may be determined from the slope of the following uptake curve.

Another interesting feature of siderophore transport is the deviation from ideal saturation kinetics which is often observed. Several workers in the field regard deviations from the ideal hyperbolic curve as proof for the existence of additional transport systems, i.e., high- and low-affinity systems. It has indeed been observed that the kinetic data in the range of high siderophore concentrations differ from those in the low concentration range. This difference may indicate that siderophores gain entrance into the cells by both unspecific and specific routes, respectively. In fact, high concentrations of siderophores often contribute to a diffusion-like behavior during saturation kinetics. The transport systems probably start to operate inefficiently or are bypassed. The diffusion-like part of saturation kinetics is easily identified in the presence of metabolic inhibitors when active uptake is no longer possible. Moreover, in order to obtain correct saturation curves, only those time intervals should be used which assure initial linear uptake rates. As long as these laws of kinetics are obeyed, the formal Michaelis-Menten kinetics can be adapted for characterizing transport phenomena.

Besides the formal kinetic aspects, the biological state and growth condition of the organism is of crucial importance for the exact determination of kinetic data. In most instances cells from the logarithmic growth phase or actively growing culture are used, although the amplification of cells during the uptake experiment should be avoided. Therefore, time-dependent kinetics are best performed with actively metabolizing cells during the smallest possible time periods. In Gram-negative bacteria an iron-deficient state is required for both the induction of desferri-siderophores and overproduction of the cognate outer membrane receptor proteins. Thus, maximal transport rates are influenced significantly by the amount of expressed membrane receptors. In fungi, iron-deficiency also induces siderophore pro-duction but overproduction of plasma membrane proteins has never been observed. However, uptake rates are enhanced at least fivefold in cells grown in low-iron media. It appears that transport of iron, mediated by siderophores, is regulated at different levels, genetically as well as metabolically. Some common principles of iron acquisition among microorganisms have been determined so far. However, there exists a great variation in siderophore structures and mechanisms of uptake. The following chapter will deal with the specificity of sider-ophore-mediated iron transport in certain bacterial and fungal genera. Although this chapter includes only a small fraction of the naturally existent microorganisms and corresponding siderophores, it represents a systematic survey of studies in which iron transport has been analyzed in greater detail.

GRAM-POSITIVE BACTERIA

ARTHROBACTER

The functions originally ascribed to the siderophores were their growth stimulating properties. In 1952 three microbial iron complexes, coprogen, ferrichrome, and arthrobactin (terregens factor) were discovered which enhanced the growth of bacteria and fungi at nanogram quantities.[1-4] The growth factor activity of these structurally very different com-pounds could not be fully explained at that time, but it seemed clear that the activity was due to the activation of iron metabolism. Neilands proposed a unified hypothesis to account for their biological activity and suggested that the growth factors may act as coenzymes for the intracellular transfer of iron in microorganisms.[5] We now know that the test organism

Arthrobacter flavescens, used to determine the biological activity, represents a siderophore auxotrophic organism, which is unable to synthesize its own siderophore and is therefore dependent on the presence of exogenous siderophores. Other *Arthrobacter* species have been shown to produce arthrobactin as a siderophore.[4] The list of organisms requiring iron transport compounds such as ferrichrome, coprogen, or arthrobactin for growth was later extended to strains of *Microbacterium lacticum*.[6-8] Both *Microbacterium* and *Arthrobacter* belong to the Gram positive, irregular bacteria and are thus related to some extent. The lag-reducing activity or division initiation activity of siderophores observed with other bacteria may likewise be regarded as a growth promotion activity which is based on the cells ability to acquire iron from siderophores for intracellular metabolism. In *Arthrobacter* the microbial iron complexes cannot be replaced by catecholate siderophores, citrate, or ethylendiamine tetraacetate, confirming the specific interaction with hydroxamate-type siderophores. Moreover, *Arthrobacter flavescens*, as well as a variety of selected transport mutants, are currently in use for the identification of new siderophores from fungi and bacteria.[9-11] With few exceptions, virtually all microorganisms studied so far have shown to produce and utilize siderophores for growth. While some of these are siderophore-auxotrophic, the majority of microorganisms are able to biosynthesize their own siderophores. *Arthrobacter* species can easily be isolated from soil.[12,13] The siderophore-auxotrophic, *Arthrobacter flavescens* JG9 (ATCC 25091), has been deposited at the American type culture collection and is now widely used to detect hydroxamate siderophores in fungi. Recently the production of hydroxamate siderophores from ectomycorrhizal fungi,[14] like *Amanita muscaria*, *Boletus edulis*, *Suillus brevipes*, *Pisolithus tinctorius*, and *Cenococcum geophilum* have been detected by the use of *Arthrobacter flavescens* JG9. Moreover, the biological activity of synthetic analogs of ferrichrome, like retrohydroxamate ferrichrome and biomimetic retro-ferrichromes possessing a C_2-symmetric molecule (triamine or tricarboxylate) as anchor has been studied by the use of *A. flavescens*.[15-17]

STREPTOMYCES

Iron complexing compounds isolated from microorganisms were originally named sideramines by Bickel et al.[18-25] This designation sounds similar to that of the vitamins and indeed the fact that both types of compounds are essential for growth is striking. The ferrioxamines (Figure 1) of *Streptomyces* were found to contain exclusively diamines and carboxylic acids. Until now the structures of the following ferrioxamines have been published: ferrioxamine A_1, A_2, B, D_1, D_2, E, G, H, and I. Because of the low available amounts of pure compounds the structures of ferrioxamines C and F have not yet been elucidated. Strains of *Streptomyces* also produce the antibiotically active sideromycins, like the ferrimycins and albomycins (Figure 2). While albomycins are derivatives of ferrichrysin, ferrimycins are derived from ferrioxamines.[18,26] Several ferrimycins have so far been described: ferrimycin A_1, A_2, A22765, and Daunomycin.[27] The growth-promoting activity of ferrioxamines is competitively antagonized in the presence of ferrimycins which led to the conclusion that both compounds enter the cells by the same transport system. Agar-diffusion bioassays for ferrimycins have been described by Zähner et al. using *Staphylococcus aureus* and *Bacillus subtilis* as indicator strains.[28] Ferrioxamines have so far been detected in the genus *Streptomyces*, and some other bacterial genera, e.g., *Arthrobacter*, *Chromobacterium*, and *Pseudomonas*.[29-31] Recently, enterobacterial strains, such as *Erwinia herbicola*, also named *Enterobacter agglomerans* and *Hafnia alvei* have been shown to produce ferrioxamines.[32,70] The siderophores of *Erwinia herbicola* consisted of ferrioxamine E as the principal siderophore with small amounts of ferrioxamine B and D_2, while *Hafnia alvei* produced mainly ferrioxamine G (the linear form of ferrioxamine E) and small amounts of ferrioxamine E.[33,70] Transport of ^{55}Fe-labeled ferrioxamine E in strain *Erwinia herbicola* A111 showed kinetic

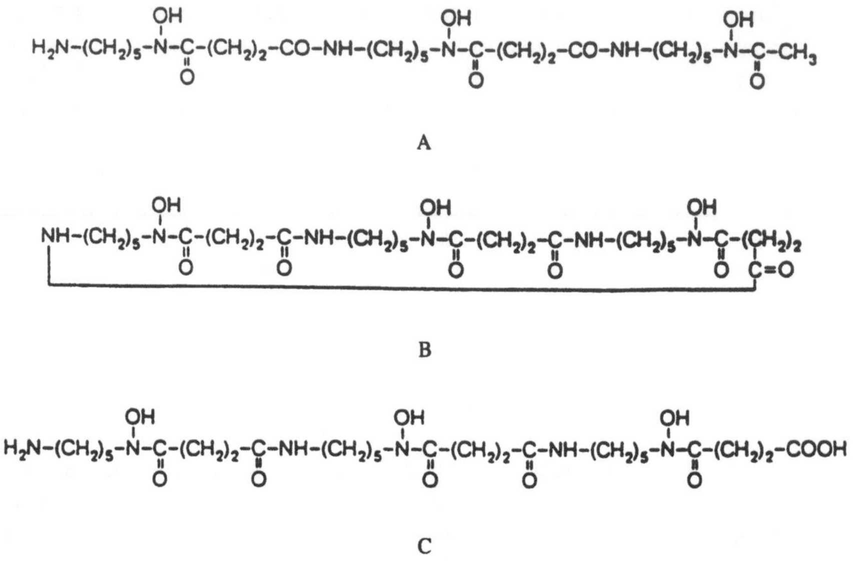

FIGURE 1. Structures of ferrioxamines (iron-free): (A) desferrioxamine B, (B) desferrioxamine E, and (C) desferrioxamine G.

data of $K_m = 0.1\ \mu M$, $V_{max} = 8$ pmol/min mg. Higher V_{max} values of 40 pmol/min mg have been found in *Erwinia herbicola* K4 which are close to the values found for *Streptomyces pilosus* (Table 1).

Iron transport mediated by ferrioxamines in *Streptomyces* has been studied using radiolabeled ^{55}Fe-ferrioxamines as well as Cr(III)-and Ga(III)-substituted ferrioxamine complexes.[34] These studies revealed that freshly germinated spores of *S. pilosus* can take up iron from ferrioxamines B, D_1, D_2, E when cultivated under iron-deficient conditions. Uptake was energy-dependent and displayed saturation kinetics. The values of $K_m = 0.1\ \mu M$ (ferrioxamine B) and $K_m = 0.2\ \mu M$ (ferrioxamines D_1, D_2, E) and the corresponding uptake rates are within the range reported for other bacterial siderophore transport systems. In *S. pilosus*, ^{67}Ga(III), Cr(III) and ^{55}Fe(III) complexed by desferrioxamine were taken up at comparable rates. Since ^{67}Ga(III) cannot be reduced, and Cr(III) is inert to ligand exchange, these results suggest that neither reduction nor decomplexation is the rate limiting step during transport, indicating that the intact complexes are taken up by the cells. Chromic isomers of desferrioxamine B have been prepared according to earlier described procedures.[36] The isomers N-*cis-cis*, C-*cis,cis*, and a mixture of *trans*-isomers showed substantially the same inhibition of ^{55}Fe-ferrioxamine B uptake. This was taken as proof that the ferrioxamine transport system in *Streptomyces* cannot discriminate between different geometrical isomers. Moreover, the chromic desferri-ferrioxamine B isomers inhibited the uptake of all ferrioxamines (D_1, D_2, E) which is consistent with a common receptor or transport system for ferrioxamines. Uptake of iron mediated by exogenous siderophores,[35] e.g., ferrichrome, ferrichrysin, and rhodotorulic acid was comparable to that mediated by the ferrioxamines, possibly indicating a common step during membrane transport. However, the existence of different receptors or a multifunctional receptor in *S. pilosus* could not be ruled out. Since the ligands of the exogenous siderophores have no structural similarity to the ferrioxamines, it was speculated that only the hydroxamate center and its direct surroundings are important for recognition and uptake. Ferrichromes, substituted at the hydroxamate carbonyl groups, like ferrirubin and ferrichrome A, were only poorly effective in supplying iron to *S. pilosus*.

MYCOBACTERIA

Mycobacterium smegmatis has been used as a model organism for studying iron me-

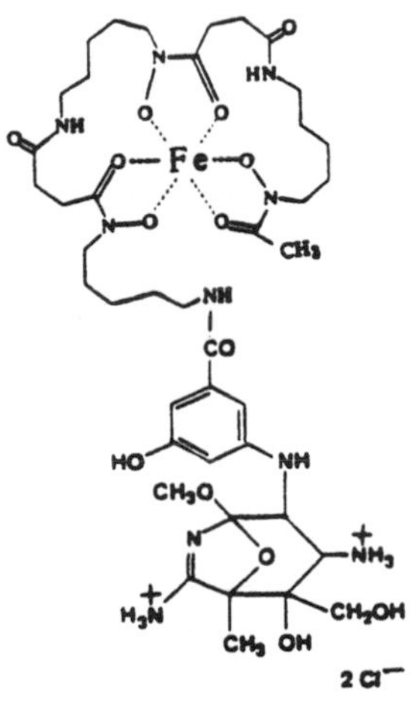

A

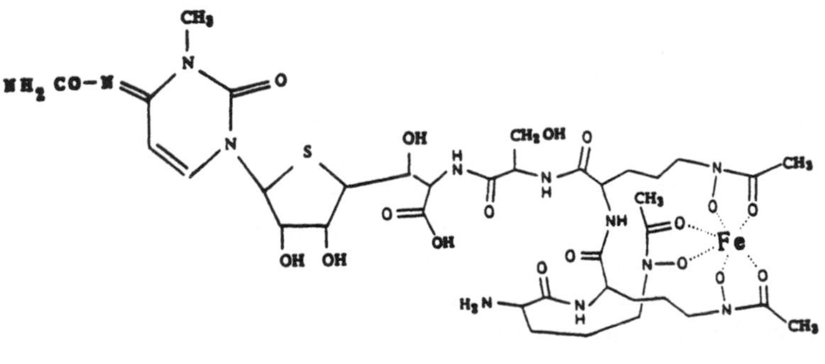

B

FIGURE 2. Structures of sideromycins: (A) ferrimycin A_1 and (B) albomycin δ_2.

TABLE 1
Kinetic Data of Siderophore Iron Transport in Bacteria

Organism	Siderophore	K_m (μM)	V_{max} (pmol/min mg)	Ref.
Streptomyces pilosus	Ferrioxamine B	0.1	30	34
Streptomyces pilosus	Ferrioxamine D_1, D_2, E	0.2	20	34
Erwinia herbicola A111	Ferrioxamine E	0.1	8	33
Erwinia herbicola K4	Ferrioxamine E	0.063	40	229
Mycobacterium smegmatis	Exochelin MS	6	—	44
Anabaena sp.	Schizokinen	0.04	10	62
Staphylococcus aureus	Staphyloferrin	0.25	82	64
	Ferrioxamine E	0.058	45	64
Escherichia coli	Enterobactin	0.23	73	85
	MECAM	0.69	68	85
	Ferricrocin	0.15	44	100
	Fe-rhodotorulate	1.4	23	100
	Enantio-Fe-rhodotorulate	0.2	3.7	100
Paracoccus denitrificans	L-parabactin (low affinity)	3.9	494	116
	L-parabactin (high affinity)	0.24	108	116
Pseudomonas WCS358	Pseudobactin 358	0.23	140	134
Azospirillum brasiliense	Spirillobactin	0.23	270	141

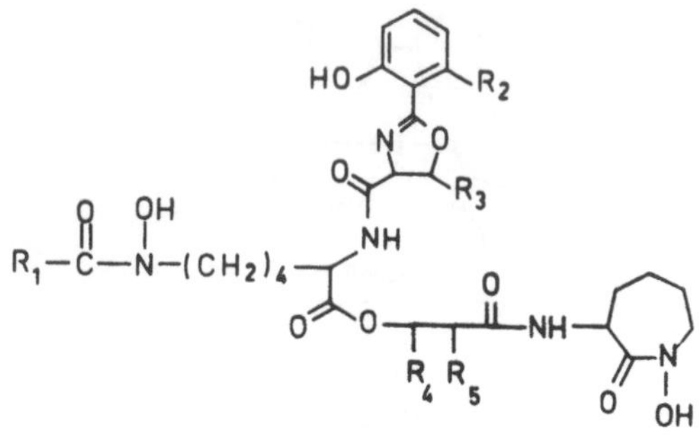

FIGURE 3.   Structure of mycobactin (R_1 is usually an unsaturated alkyl chain, R_2, R_3 and R_5 are either H or CH_3, R_4 may be a short or a long alkyl chain.

tabolism in other members of this genus which includes a variety of saprophytic mycobacteria as well as pathogens, like *M. tuberculosis*, *M. paratuberculosis* (syn. *johnei*) and *M. leprae*.[37] Mycobacteria have been shown to produce the intracellular lipid-soluble mycobactins (Figure 3) and the water-soluble exochelins.[38-45] Like other siderophores, both mycobactins and exochelins are produced in increased amounts in response to iron-limited growth. In addition, citrate can also act as an iron donor in mycobacteria.[45]

Among the fungal hydroxamate siderophores, rhodotorulic acid can serve as a siderophore in mycobacteria, but ferrichrome and ferrioxamines are ineffective.[46] A simple mechanism of iron transport based on the loading and unloading function of mycobactin has been described.[47] Thus, addition of iron into the medium of an iron-deficient culture of *M. smegmatis* resulted in a rapid conversion of desferrimycobactin into the ferri-form. According to this model ferrimycobactin functions as an iron-shuttle within the cell boundary layers of the mycobacteria. Mycobactin has been located using electron microscopy of whole cells stained with vanadate.[48] Although mycobactin was clearly seen to be surrounding the mycobacterial cell in an apparently discrete but discontinuous inner layer of the outer envelope, it was not possible to equate this layer unequivocally with the cytoplasmic membrane. The distance between this region and the outer surface of the cell was approximately 30 nm, which is about the thickness of the cell wall. In iron-sufficient cells exposed to vanadate there was a faint but continuous inner electron-dense layer at the expected position of the cytoplasmic membrane, corresponding to the position of the mycobactin in iron-deficient bacteria. From these studies it was concluded that mycobactin may be acting either as a store for iron, or as an ionophore, or possibly fulfilling both roles.

The actual iron solubilizing and transporting agents in the aqueous environment were initially unknown. Although salicylate is secreted by mycobacteria, its role as an iron carrier has been questioned and no evidence could be presented that salicylate could fulfill the function of an extracellular iron-solubilizing agent, equivalent to the hydroxamate or catecholate siderophores in other bacteria.[49] The detection of the exochelins as species-specific extracellular siderophores of the mycobacteria has allowed us to propose the mechanism of iron uptake in more detail. Exochelins from *M. smegmatis* have been purified and shown to be a mixture of at least seven compounds (including some breakdown products) which are separable by ion-exchange chromatography and gel filtration.[42,43] However, the chemical structure has not been elucidated so far. Exochelin originating from *M. smegmatis* has been designated exochelin MS whereas the one from *M. bovis* has been called exochelin MB. The uptake of iron from ferriexochelin MS (main fraction) in *M. smegmatis* at low external concentration occurs via a high-affinity system, with a K_m of approximately 6 μM as determined by ^{3}H- and ^{55}Fe-labeled exochelins.[44] Uptake of ferriexochelin was sensitive to KCN, NaN$_3$, CCCP, DNP, and HgCl$_2$, suggesting an energy-requiring and receptor-depen-

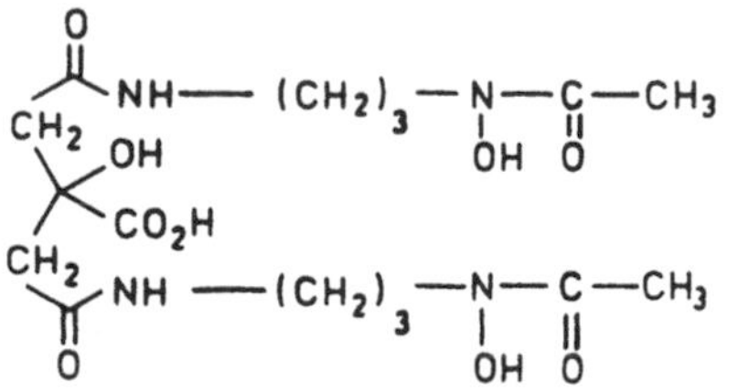

FIGURE 4. Structure of schizokinen.

dent transport mechanism. Exochelins from *M. smegmatis* can solubilize iron from ferric phosphate and ferritin, indicating their effectiveness in body fluids. Recently several iron-regulated envelope proteins have been detected in *M. smegmatis*.[46] These iron-regulated proteins were ascribed apparent molecular weights of 180, 84, 29, and 25 kDa. Antibodies were raised to each of the four proteins; the one raised to protein III inhibited exochelin-mediated iron uptake into iron-deficient grown cells. A mycobactin-deficient mutant has also been obtained which allowed the study of the uptake process in more detail.[50]

BACILLUS

The genus *Bacillus* comprises a group of aerobic, Gram-positive, spore-forming bacteria which are ubiquitous in soil and water. *B. megaterium* has been shown to produce schizokinen (Figure 4). a dihydroxamate-type siderophore,[51,52] while *B. subtilis* was shown to produce 2,3-dihydroxybenzoylglycine (DHBG) as a siderophore.[53] Iron transport studies with the hydroxamate(schizokinen)-producing *Bacillus megaterium* ATCCC 192113 and mutants derived from this strain, which cannot produce schizokinen, have led to the conclusion that iron uptake from the schizokinen-iron chelate is an active process, and represents the primary mode of iron transport in *B. megaterium*.[54] Kinetics of radioactive iron transport was examined in three strains of *Bacillus megaterium*. In strain ATCC 19213, which secretes the ferric-chelating secondary hydroxamic acid schizokinen, ^{59}Fe uptake from ^{59}FeCl$_3$ or the ferric hydroxamate ^{59}Fe-ferrioxamine B (^{59}Fe-Desferal) was rapid and reached saturation within 3 min. In strain SK11, which does not secrete schizokinen, transport from ^{59}FeCl$_3$ was markedly reduced. A strain (Ard-1) resistant to the ferric hydroxamate antibiotic A22765 (ferrimycin) was isolated from strain SK11. Strain Ard-1 failed to grow with Desferal-Fe and was unable to incorporate ^{59}Fe from this source, indicating that this mutant may be unable to recognize both the antibiotic A22765 and the structurally similar ferrioxamine B (Desferal), while retaining the capacity to utilize iron from schizokinen. The behavior of the transport mutant suggested that there exist at least two uptake routes in *B. megaterium:* one for ferric schizokinen and another one for ferrioxamines. Aerobactin was not a siderophore in SK11, indicating that even the two related citrate-dihydroxamates, Fe-schizokinen and aerobactin, are discriminated by the Fe-schizokinen transport system in *B. megaterium*.[55]

The fate of double-labeled ^{3}H/^{59}Fe-schizokinen was studied in both a wild type and a schizokinen-requiring mutant.[56] When ^{3}H-schizokinen was used for transport of ^{59}Fe in the wild type *B. megaterium*, uptake of both the ligand and metal was observed followed by a rapid discharge of ^{3}H-schizokinen within 2 min. The discharge was temperature-dependent and did not occur at 0°C. In the schizokinen-auxotroph strain *B. megaterium* SK11 similar release of ^{3}H-schizokinen occurred but only at elevated concentrations of the double-labeled chelate. These results indicated an initial temperature-independent binding of the ferric hydroxamate, followed by a temperature-dependent transport of the chelate into the cell and an enzyme catalyzed separation of iron from the chelate.

Membrane vesicles of *B. megaterium* SK11 and *B. megaterium* Ard-1, prepared from cells grown at low iron concentrations, showed temperature and energy-independent binding of both ^{59}Fe-schizokinen and ^{59}Fe-ferrioxamine B.[57] Using doubly labeled ^{3}H,^{59}Fe-schizokinen it was confirmed that uptake of the intact chelate occurred. Addition of excess unlabeled ferric schizokinen caused a rapid release of radioactive schizokinen, while the ferric forms of other chelates, such as aerobactin, rhodotorulic acid and ferrioxamine B caused no measurable release, suggesting a specificity of binding to the putative membrane receptors. Ferri-schizokinen receptors were present on membranes in greater numbers (about eightfold) compared to the ferrioxamine B receptors. Ferric aerobactin was not bound to the membranes in a specific manner and was not able to dissociate the ferric forms of schizokinen or ferrioxamine B. Moreover, these studies implied independent transport systems for ferric schizokinen and ferrioxamine B and suggested the possibility of additional ferrihydroxamate receptors on the membranes of *B. megaterium*.

Reductive transfer of iron from ferrisiderophores to the ferrous-chelating agent ferrozine was measured spectrophotometrically.[58] The ferrisiderophore reductase was NADPH dependent and displayed maximal iron reduction under anaerobic conditions. The enzyme was associated primarily with the cell soluble fraction and its activity was inhibited by oxygen, heat, protease treatment and iodoacetamide. Kinetic data yielded $K_m = 2.5 \times 10^{-4} M$ and $V_{max} = 35.7$ nmol/min mg of protein. Fractionation by gel filtration of the soluble material revealed that three peaks of different molecular weights showed ferrisiderophore reductase activity. Substrate specificity was broad but the activity was greatest for ferric schizokinen. Ferric aerobactin was also reduced indicating that the inability to use ferric aerobactin for growth was due to lack of receptors for this siderophore on the cell surface of *B. megaterium*. The mechanism of iron uptake in *Bacillus subtilis* has been analyzed by using (a) phenolic compounds which are produced under low levels of iron by this bacterium, like 2,3-dihydroxybenzoic acid and 2.3-dihydroxybenzoylglycine, and (b) by using the fungal siderophore ferrichrome.[59] The results of these experiments showed that iron uptake was mediated by phenolic compounds as well as by the hydroxamate ferrichrome.

CYANOBACTERIA

There are several reports on the occurrence of siderophores in cyanobacteria.[60-62] *Anabaena* sp. has been shown to produce schizokinen,[61] suggesting some evolutionary relationship to the genus *Bacillus*. Transport of ferric schizokinen in *Anabaena* sp. (ATCC 27898) showed saturation kinetics ($K_m = 0.04$ μM, $V_{max} = 10$ pmol/mg min). Light-driven transport was inhibited by uncouplers and ATPase inhibitors, whereas transport in dark-adapted cells was blocked by inhibitors of respiration.[62]

STAPHYLOCOCCUS

Two iron chelating compounds have been recently detected and purified from a strain of *Staphylococcus hyicus* DSM20459 and named Staphyloferrin A and B. Staphyloferrin A (Figure 5) is composed of D-ornithine linked at N^α and N^δ to citric acid residues.[63] Thus staphyloferrin is N^2,N^5-di-(1-oxo-3-hydroxo-3,4-dicarboxybutyl)-D-ornithine. Together with arthrobactin and aerobactin, staphyloferrin represents another siderophore containing citric acid as carboxylate type ligands. Ferric staphyloferrin has a slightly yellow color and the UV lacks the charge transfer bands between 420 and 500 nm which is typical for hydroxamate and catecholate type siderophores. Growth promotion tests with staphyloferrin on agar plates containing EDDA revealed siderophore activity in all *Staphylococcus* strains (37) examined.[64] Several other iron chelators served as iron donors in *Staphylococcus*, i.e., NTA, citrate, 2,3 DHBA, transferrin, lactoferrin and hemoglobin. It is interesting to note that iron from ferrioxamine E and B was also taken up by *Staphylococcus* strains. The kinetic data of transport—ferrioxamine E ($K_m = 0.058$ μM, $V_{max} = 45.5$ pmol/min mg) staphyloferrin ($K_m = 0.25$ μM, $V_{max} = 82$ pmol/min mg)—suggested the existence of at least two active and separate transport systems in *Staphylococcus*.

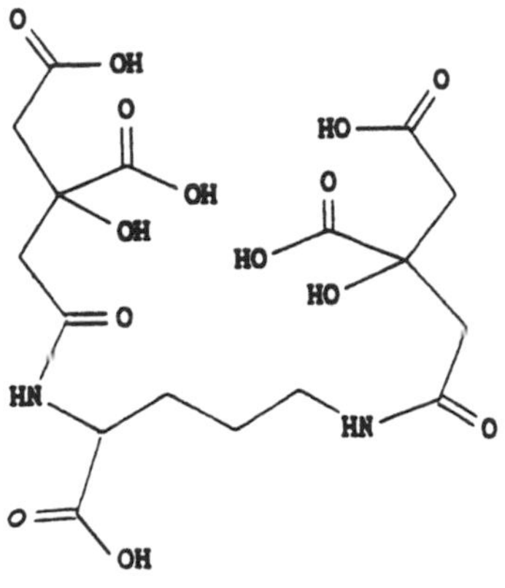

FIGURE 5. Structure of staphyloferrin A.

LACTOBACILLUS

Strains to the genus *Lactobacillus* have been shown to grow in the absence of iron, as their enzymes obviously do not require iron.[65] Although *Streptococcus mutans* could be grown without addition of iron, the addition of 0.45 μM iron stimulated growth.[66]

GRAM-NEGATIVE BACTERIA

ENTEROBACTERIA

Transport of iron in enteric bacteria deserves special attention because of its significance in medical microbiology and because of the fact that members of the Enterobacteriaceae are easily amenable to genetic analysis (see chapter on genetics of iron transport). Enterochelin, also named enterobactin (Figure 6), is the principal siderophore of *E. coli*,[67] *Salmonella*,[68] *Shigella*,[69] *Klebsiella*,[69] and other enterobacterial genera. Strains of *Enterobacter agglomerans* have recently been shown to produce ferrioxamine E, and strains of *Hafnia alvei* have been shown to produce ferrioxamine G as their principal siderophores.[32,70] Since *Proteus*, *Providencia*, *Morganella*, and *Yersinia* failed to show enterobactin production in bioassays other siderophore systems must be active.[71,72] It has been reported that several of these strains, originating from hospital infections, carry aerobactin synthesis genes on plasmids.[73] More recent preliminary results suggest that catecholate iron uptake systems are operating in *Proteus* and *Morganella*.[74]

Salmonella is known as an enterobactin producer.[68] Mutants of *Salmonella typhimurium* LT2 defective in the biosynthesis of enterobactin were isolated and named *enb*.[75] Class I mutants like *enb-1* and *enb-13* excreted the enterobactin precursor 2,3-dihydroxybenzoic acid (DHBA), while class II mutants like *enb-7* were blocked in steps before DHBA. The growth of both mutant classes was inhibited by citrate. This inhibition was reversed by enterobactin. Moreover, addition of DHBA to class II mutants restored enterobactin biosynthesis. *Enb* mutants have been reported to enable further differentiation of enterobacteriaceae by means of siderophore pattern analysis.[71] Recently a *tonB* mutant derived from an *enb-7 Salmonella* mutant has been obtained which proved useful as an indicator strain for 2,3-dihydroxybenzoic acid producing enterobacterial strains.[76] It has also been shown that *enb* mutants, blocked in the biosynthesis of enterobactin, were able to utilize various fungal siderophores, such as ferrichromes, rhodotorulic acid and coprogen or siderophores from other bacterial genera like ferrioxamines and schizokinen.[77] Host-adapted strains of *Salmonella* show often a weak enterobactin production in citrate containing media, such as Vogel-Bonner medium, suggesting the existence of additional iron transport systems.[78]

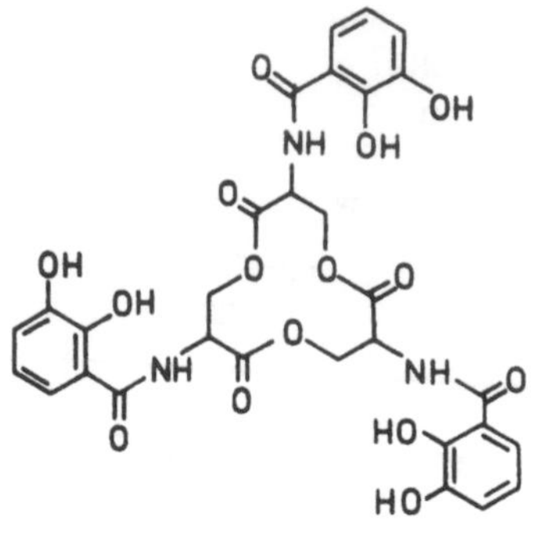

FIGURE 6.   Structure of enterobactin (=enterochelin).

Strains of *S. arizonae* (subsp. IIIa and IIIb) have been shown to produce aerobactin.[79] This fact could be confirmed when *S. arizonae* of different stock cultures were screened for aerobactin synthesis.[74]

Escherichia coli has been the subject of intensive siderophore research. Early transport studies with enterobactin (=enterochelin) in *E. coli* have shown that several genes are involved: the biosynthetic genes (*ent*), genes for the enterobactin transport (*fep*) and a gene for degradation of enterobactin, the enterobactin esterase gene (*fes*).[80] Uptake of iron by mutants, defective in one of these genes, was compared with the parent strain. Transport assays were performed in the presence of nitrilotriacetate (NTA) which inhibited nonenterobactin mediated iron uptake. In the presence of NTA the parent strain took up approximately 5 pmol $^{55}Fe/10^9$ cells within 20 min, when no enterobactin was added. After enterobactin had been added, the iron uptake reached a value of approximately 90 pmol $^{55}Fe/10^9$ cells, indicating rapid uptake of iron from the ferric enterobactin complex. At the point of saturation the intracellular concentration of ferric enterobactin was about 50 times that in the medium, suggesting the operation of an active siderophore mediated iron transport system. In contrast, the receptor-deficient mutant (*fep*$^-$) failed to transport ferric enterobactin. Although it was originally assumed that the products of cleavage of enterobactin by the esterase (linear trimer, dimer and monomer) are no longer involved in iron transport, there are reports that 2,3-dihydroxybenzoylserine and analogous monomeric catecholates containing β-lactam antibiotics enter the cells via Cir and Fiu outer membrane receptors.[81-83]

The mechanism of enterobactin uptake in *E. coli* has been the subject of some controversy. Based on the finding of an enterobactin esterase it was concluded that the ligand has to be cleaved before iron can be released. This view was supported by the highly negative reduction potential of -790 mV vs. NHE and the fact that ferric enterobactin has the highest formation constant known ($K_f = 10^{52}$). Later studies with the uncleavable synthetic enterobactin analog MECAM showed that an esterolytic cleavage is not a prerequisite for iron uptake from enterobactin, suggesting that a reductive removal of iron may well occur before the ligand is cleaved by an esterase.[84,85] Obviously the destruction of the ligand prevents reuse of the molecule. However, only 10% of iron from MECAM was really metabolized after uptake, indicating that the synthetic compound was not as effective as the natural ferric enterobactin. Two possible mechanisms of iron release from ferric enterobactin in *E. coli* are still under discussion: (1) protonation in a medium of low dielectric constant, as found in the membrane interior, followed by an internal electron-transfer reaction to yield ferrous iron, and (2) protonation concurrent with reduction of the ferric enterobactin complex in a low pH environment.[86] Recent studies have shown that protonation of ferric enterobactin and related compounds results in a shift of the catecholate to salicylate mode of bonding.[87] It has been shown that the ferric complex of the synthetic structural analog of enterobactin, MECAM, was transported with similar saturation kinetics ($K_m = 0.69$, $V_{max} = 68$ pmol/

min mg) as was ferric enterobactin (K_m = 0.23, V_{max} = 73 pmol/min mg).[85] A double-label transport assay with ^{59}Fe, ^{3}H MECAM showed that the ligand and the metal were transported across the outer membrane at nearly identical rates. The results suggested a mechanism of active transport of an unmodified coordination complex across the outer membrane with a possible accumulation in the periplasm, independent of subsequent metabolism. Thus, an intermediate storage of ferric enterobactin before its release of iron to the cell may be assumed. Based on considerations of sequential protonation of ferric enterobactin with decreasing pH and a reduction potential of -0.56 V at pH 6, a reductive removal of iron from accumulated ferric enterobactin seems the most probable mechanism.

Iron transport in *E. coli* is mediated by outer membrane proteins (receptors) designed to bind and process siderophores for iron transport across the inner cytoplasma membrane. In 1973 Guterman first suggested that the colicin B receptor may bind ferric enterobactin which subsequently made it possible to select for enterobactin receptor-deficient mutants.[88-90] Colicin B resistant mutants of *E. coli* K-12 lacked a single band in the outer membrane when analyzed by SDS polyacrylamide gel electrophoresis.[91] The same authors presented evidence for *in vitro* competition between ferric enterobactin and colicin B for the extracted receptor. Ferrichrome gave little inhibition of ferric enterobactin binding which would be expected since the ferrichrome uptake route requires a different outer membrane receptor. The FepA receptor protein has recently been isolated and crystallized using a plasmid carrying strain of *E. coli*.[92]

While natural and synthetic enterobactin (L-seryl enterobactin) functions as a siderophore, enantioenterobactin (D-seryl enterobactin) was a poor iron transporting agent in *E. coli* and did not show competition with the natural ferric enterobactin.[93] The natural enterobactin forms a Δ coordination isomer with ferric iron, while the synthetic enantioenterobactin adopts a Λ coordination about the metal center. The failure of the unnatural D-serine derived material to support growth of *E. coli* mutants indicated exclusion of the diastereomeric Λ-cis ferric enterobactin by the outer membrane receptor (FepA). Outer membrane specificity for the natural complex may be determined by the chirality of the seryl triester backbone, by the chirality of the metal center, or by both of these factors. The insensitivity of the membrane transport process to changes in the skeleton, as found in synthetic enterobactin analogs, suggests that the configuration of the metal center is the more important factor in determining enterobactin recognition by the receptor.[94,95]

Uptake of ferric enterobactin in *E. coli* was found to be strongly dependent on an energized membrane state, as demonstrated by its sensitivity to 2,4-dinitrophenol, sodium azide and potassium cyanide.[96] Since ferric enterobactin uptake was also sensitive to inhibitors of glycolysis, such as iodoacetic acid and sodium fluoride in ATPase-deficient mutants and arsenate-treated cells, the authors concluded that ferric enterobactin uptake is dependent on the presence of cellular ATP or an analogous source of phosphate bond energy. While earlier authors could not observe uptake of ferric enterobactin in cells grown anaerobically, Lodge and Emery[97] reported uptake of iron from various siderophores in *E. coli* under anaerobic conditions, although the rate and extent of uptake were less than those in aerobically grown cells. Thus, although the biosynthesis of siderophores is strongly dependent on aerobic conditions, siderophore mediated iron transport seems not to be restricted to aerobic conditions and may proceed anaerobically, i.e., in the gastrointestinal tract or in anaerobic surface layers, provided metabolic energy is available. A ferrous iron transport system in *E. coli* has also been described.[98] Mutants lacking the ferrous iron transport (feo$^-$) were obtained after nitrosoguanidine treatment, selection with streptonigrin and growth in an iron citrate medium.

In 1976 Leong and Neilands reported uptake of ^{55}Fe- and ^{3}H-labeled siderophores and their chromic analogs in the mutants *E. coli* K-12 RW193 and *Salmonella typhimurium* LT-2 *enb-7* which are both defective in the production of their native siderophore, enterobactin.[99]

In *E. coli* K-12 RW193 the virtually identical uptake rates of the ^{55}Fe-label of ^{55}Fe-ferrichrome and the ^{3}H-label of chromic ^{3}H-desferri-ferrichrome were observed suggesting transport of ferrichrome as an intact siderophore. The fact that kinetically inert Λ-*cis* chromic desferri-ferrichrome isomer was taken up also demonstrated that the Λ-*cis* coordination isomer, which is the naturally occurring isomer in all ferrichrome-type siderophores, is the one that is transported. From transport studies in *Salmonella typhimurium* LT-2 *enb-7* it was inferred that besides the enterobactin uptake route additional mechanisms for hydroxamate siderophores were operative. While ferrichrome seemed to be taken up as an intact chelate molecule, ferrioxamine B appeared to only deliver its iron to the cell, since the ^{3}H-label of ^{3}H-ferrioxamine B was not found to be enriched in the cells. Neither cis- nor trans-geometrical coordination isomers of chromic ferrioxamine B were accepted by the uptake system which was taken as a proof for a reductive removal of iron during uptake. Transport of ferrioxamines in *Salmonella* obviously proceeds by rapid dissociation of iron from the ligand molecule while ferrichrome was processed more slowly. These very early transport studies in genera of the Enterobacteriaceae indicated quite clearly that different siderophore receptors are present and to a certain extent that also different mechanisms may be operative. A comparison of iron uptake in *E. coli* using different hydroxamate siderophores revealed that ferrichromes (ferrichrome, ferricrocin, ferrichrysin) were all taken up at comparable rates, whereas ferrichrome A was not taken up.[100] This investigation also showed that coprogen and Fe-rhodotorulate were taken up in *E. coli* confirming the function of an earlier identified FhuE receptor protein which specifically recognizes coprogen and Fe-rhodotorulate.[101] Saturation kinetics with ferricrocin showed: K_m = 0.15, V_{max} = 43.5 pmol/min mg. These values are close to those reported for ferrichrome.[104] Ferric rhodotorulate uptake gave values of K_m = 1.4 μM, V_{max} = 22.9 pmol/min mg. Ferrioxamine B showed a moderate transport activity, whereas the cyclic ferrioxamine E was devoid of any transport activity in *E. coli*.[100]

Following the fate of ferrichrome during uptake in *E. coli* indicated that the ligand was modified and excreted into the medium, disproving the idea that the ferrichrome ligand may be used for another round of transport.[103] Evidence was presented that ferrichrome is acetylated at the N–OH group giving rise to a N–O–CO–CH$_3$ residue which is assumed to lower the binding affinity for iron. In a cell-free system it was subsequently shown that reduction of the ferrichrome complex is a prerequisite for this modification.[103] While this mechanism of ferrichrome iron utilization argues against the entrance of the intact ferrichrome molecule into the cytoplasm, there is evidence that ^{3}H-ferrichrome is actively transported in inner membrane vesicles of *E. coli* K-12.[104] Valinomycin was a powerful inhibitor of transport while nigericin had a slight effect, if any, suggesting that ferrichrome transport may be more dependent on the electrical potential than on the transmembrane hydrogen ion gradient. Since ferrichrome is a neutral molecule, it was suggested that it probably permeates the plasma membrane by an alkali metal symport mechanism. Some indication for binding of cations by ferrichrome has been reported, although direct proof for the biological involvement of such a mechanism has not been given so far[105]. However, in the light of the high iron to ligand ratio observed during ferrichrome transport in *E. coli*, it seems questionable that ferrichrome enters the cytoplasm as an intact complex.

While fungal hydroxamates like ferrichromes are exogenous siderophores for *E. coli*, the ability to produce hydroxamate siderophores in the family of Enterobacteriaceae is not uncommon. For example aerobactin (Figure 7), a dihydroxamate siderophore, was first detected in *Aerobacter aerogenes* (=*Enterobacter aerogenes*).[106] It was subsequently detected in plasmid ColV-harboring *E. coli* strains and it was also shown to be chromosomally encoded in *Shigella* strains.[107,108] The genetics of plasmid and chromosomally encoded aerobactin biosynthesis will be treated comprehensively elsewhere in this book. Other enterobacterial genera, i.e., *Enterobacter agglomerans* (=*Erwinia herbicola*) have been shown to produce the trihydroxamate siderophore ferrioxamine E and several minor ferrioxamine

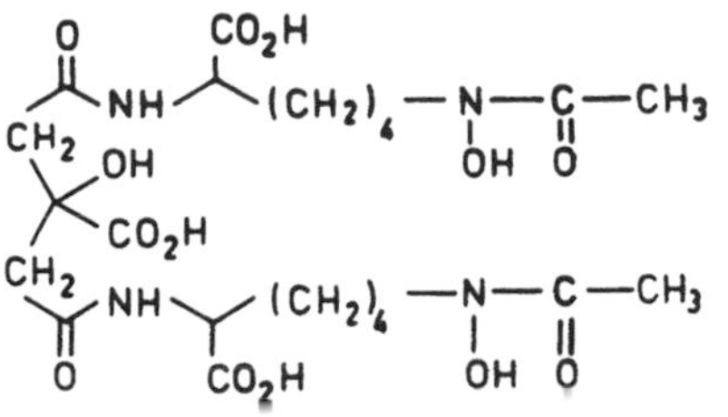

FIGURE 7.   Structure of aerobactin.

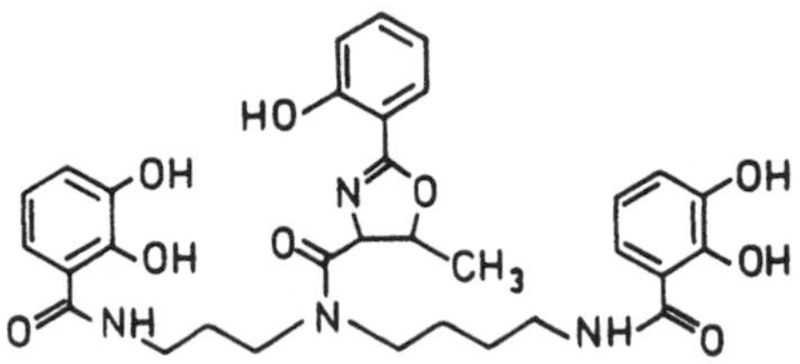

FIGURE 8.   Structure of parabactin.

products.[32] Recently, *Hafnia alvei* also proved to be a ferrioxamine producer.[70] There are indications that hydroxamate siderophores may occur in the genus *Salmonella* (*S. stanleyville* 207/81) although a structure has not yet been reported.[109] While in most enterobacteria siderophores and/or outer membrane receptors have been detected, a hemolytic strain of *Serratia marcescens* has been shown to possess a mechanistically novel iron transport system (SFU) which is obviously not dependent on outer membrane proteins or any of the known siderophores.[110]

PARACOCCUS

Paracoccus denitrificans excretes a hexacoordinate catecholamide iron chelator, parabactin (Figure 8).[111,112] The formation constant with iron(III) at alkaline pH is $K_f = 10^{48}$ and the coordination about the metal center is Λ. *Paracoccus* is a member of the Gram-negative bacteria for which an outer membrane is a characteristic feature. Growth in low-iron medium resulted in the production of parabactin, and the expression of five iron regulated outer membrane proteins (M_r 85,000, 83,000, 78,000, 74,000, 72,000).[113] Growth of iron-deprived cells in ferric citrate (20 μM) medium repressed parabactin production and iron deprivation-induced membrane protein production, but led instead to production of a M_r 23,000 outer membrane protein. While the high molecular weight outer membrane proteins are suggested to be involved in the high affinity siderophore transport, the 23,000 Da proteins are suggested to be involved in a non-siderophore mediated, low affinity iron uptake pathway. The mechanism by which *Paracoccus denitrificans* utilizes parabactin has been examined using ^{55}Fe-parabactin, the Ga(III)complex of parabactin, ^{3}H ferric parabactin, enantiopara-bactin, parabactin A as well as a large number of homologues.[114] While the Fe-complex was rapidly taken up by *P. denitrificans*, the Ga-complex was not, suggesting a reductive removal of the metal during transport. Moreover, the data suggested that *P. denitrificans* exhibits outer membrane stereospecificity as shown by the uptake of L-parabactin and the exclusion of D-parabactin. The L-parabactin adopts a Λ-configuration about the metal center whereas D-parabactin adopts a Δ-configuration. Both enantioparabactin (= D-parabactin) and L-parabactin A form Δ chelates with ferric iron, and both were found to be unable to supply iron to *P. dentrificans* in a highly specific manner. The overall sensitivity of the ferric parabactin-mediated iron-uptake system in *P. denitrificans* was also evaluated by examining

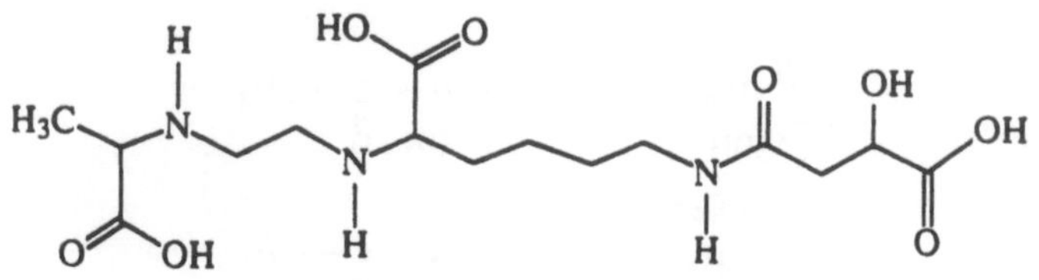

FIGURE 9.   Structure of rhizobactin.

the abilities of various parabactin analogues to reverse EDTA-induced iron starvation. When present in relatively high amounts most of the analogues tested were able to stimulate growth. However, parabactin, homoparabactin and norparabactin proved to be the most effective siderophores. While uptake of iron from parabactins possessing an intact oxazoline ring showed significant stereospecificity, iron acquisition from the A forms of ferric D- and L-parabactin, in which the oxazoline ring is hydrolyzed to the open chain threonyl structure, nonstereospecific.[115] The kinetics of uptake of the natural siderophore, L-parabactin, revealed biphasic kinetics by Lineweaver-Burke analysis. At concentrations of 1 to 10 μM (K_m = 3.9 μM, V_{max} = 494 pmol Fe/min mg) a low affinity transport system is active, whereas to 0.1 to 1 μM (K_m = 0.24, V_{max} = 108 pmol Fe/min mg) a high affinity system is operative.[116] The K_m of the high affinity uptake is comparable with the binding affinity which has been reported for the purified ferric parabactin receptor of the outer membrane. Several other catecholamide siderophores, i.e., L-homoparabactin, L-agrobactin as well as L-vibriobactin, all possessing an intact oxazoline ring derived from L-threonine and showing a Λ-coordination about the iron center, were taken up by *P. denitrificans* with high affinity. Agrobactin has been shown to be biologically active not only in *Agrobacterium tumefaciens* but also in *P. dentrificans*.[117] Moreover, while natural agrobactin and parabactin, possessing a Λ coordination around the metal center failed to show growth response in *E. coli*, their open forms (agrobactin A and parabactin A), possessing a Δ coordination about the metal center, were found to be active.[118]

RHIZOBIA

The genera *Azorhizobium, Bradyrhizobium* and *Rhizobium* are involved in symbiotic nitrogen fixation and are collectively referred to as *Rhizobia*. Since nitrogenase, ferredoxin, hydrogenase and cytochromes contain iron, there is an absolute need for *Rhizobia* to acquire iron both from the plant host as symbionts and as free-living microorganisms in the soil. Therefore, the requirements for iron of *Rhizobia* are probably quite different depending on their environment. There are now several reports on the occurrence of siderophores in *Rhizobia*.[119-124] However, only a few of the screened *Rhizobia* strains showed siderophore production. The majority of strains studied were devoid of any iron chelating agents. For instance, in a search for siderophore production using five different siderophore assays only 1 of 20 strains of *B. japonicum* produced citric acid as a siderophore.[125] Another report revealed that 19 out of 52 strains of *R. phaseoli* showed siderophore-like activity in a bioassay.[126] Moreover, only a few reports are available on siderophore utilization which was assessed either by growth assays or by transport studies employing radiolabeled ferric siderophores.[119,121,124,126] These results revealed that (1) multiple siderophore receptors for different siderophore types exist, i.e., for rhizobactin (Figure 9), ferrichrome, ferrioxamine, pseudobactin, anthranilic acid and citric acid, and (2) ferric citrate complexes seem to be used by all *Rhizobia* and are probably a major source of iron both for the plant host as well as for the *Rhizobia* in the nodules. It remains to be established whether hydroxamate, catecholate or rhizobactin siderophores of *Rhizobia* are actually required for the symbiotic phase of life or whether they are only required for competition and survival in the soil. One

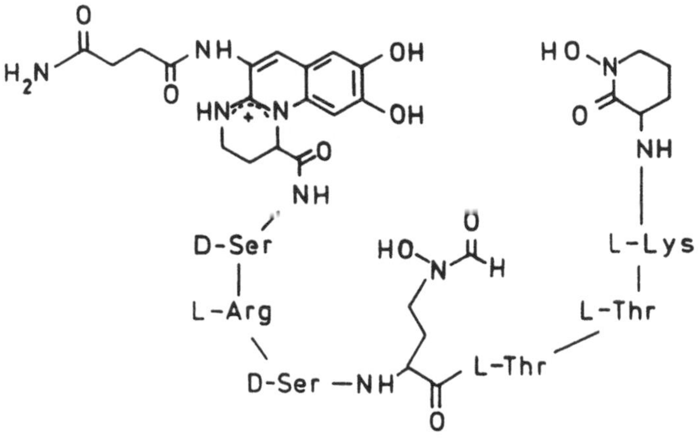

FIGURE 10.   Structure of pseudobactin.

has also to consider the possibility that rhizobial production of siderophores other than ferric citrate, and with higher stability constants, may be deleterious to the plants.

PSEUDOMONAS AND AZOTOBACTER

The genus *Pseudomonas* comprises a group of aerobic, non-fermentative, Gram-negative bacteria which are ubiquitous in soil and water. The fluorescent *Pseudomonads*, e.g., *P. fluorescens*, *P. aeruginosa*, *P. putida*, and *P. syringae* have been shown to produce the characteristic yellow-green, water-soluble fluorescent siderophores named pyoverdins or pseudobactins (Figure 10).[127-132] Typically these siderophores contain three bidentate ligands: the catecholate type ligand dihydroxyquinoline (chromophore), linked to a peptide chain containing a β-hydroxyaspartyl or a $^\delta N$-hydroxyornithyl residue and at the C-terminal end a cyclic δ-N-hydroxyornithine residue. The amino acids in the peptide chain consist of 7 to 10 amino acids with different chiral configurations. The alternating D- and L-amino acids obviously prevent proteolytic degradation.

The diversity in pyoverdin peptide structure has raised the question of strain specificity. Meyer was the first to consider this aspect in detail by investigating iron uptake mediated by pyoverdins in a number of fluorescent Pseudomonads, i.e., *P. aeruginosa*, *P. fluorescens*, *P. putida*, *P. chloroaphis*, and the phytopathogenic *P. tolaasii*.[133] Strain specificity of pyoverdin-mediated iron utilization was systematically studied in whole cells by cross-feeding and iron uptake studies and also by binding to isolated outer membranes. As expected, iron incorporation from the homologous pyoverdins (strain specific pyoverdin) was observed in all cases. The experiments involving heterologous pyoverdins revealed two types of response: (1) four strains, *P. putida* ATCC 12633, *P. tolaasii* NCPPB 2192, *P. fluorescens* W, and *P. fluorescens* ATCC 17400 did not incorporate iron from any of the heterologous pyoverdins, (2) three strains, *P. aeruginosa* PAO, *P. chlororaphis* ATCC 9446, and *P. fluorescens* ATCC 13525 did not incorporate iron from pyoverdins of the first group mentioned above, but recognized all pyoverdins within the second group. In each case, however, the homologous pyoverdins function as the best siderophore. These results clearly showed that there is indeed a strain specific recognition by membrane located receptors in *Pseudomonas* strains which may be the basis for competition in the rhizosphere. The amount of iron taken up was in the range of 0.5 to 1.0 nmol/mg after an incubation period of 30 min. When pyoverdins were added *in vitro* to isolated outer membranes originating from different *Pseudomonas* strains, the most efficient binding was always obtained in the homologous system which nicely corresponded to the specificity of iron transport.[133] The results obtained with iron transport measurements were also confirmed by growth promoting tests. Analyzing the effects of homologous and heterologous pyoverdins on the growth of one strain, it was observed that the best stimulation of growth was usually obtained with the homologous pyoverdin. However, in some cases pyoverdins were found to be inhibitory to some strains and in some other cases stimulation was found to be superior with heterologous pyoverdins, suggesting that the results of growth stimulation were not as clear as those obtained with the transport of radiolabeled pyoverdins. Some discrepancies of the growth stimulating experiments may be explained in part by interligand exchange under prolonged incubation periods when production of the homologous siderophore exceeds that of the added heterologous siderophores. In general, however, from the results presented in this paper it seems clear that the observed specificity of pyoverdin-mediated iron uptake relies on the differences in recognition capacity and binding efficiency of the outer membrane receptors.

Strain specific utilization of pseudobactins (= pyoverdins) has also been demonstrated in the plant-beneficial *Pseudomonas* strains *P. putida* WCS358 and *P. fluorescens* WCS 374.[134] Under iron-limited conditions *Pseudomonas putida* WCS358 produces a siderophore, pseudobactin 358, whereas *P. fluorescens* WCS374 produces pseudobactin 374, both of which seem to be essential for their plant growth stimulating ability. However, a difference in specificity of siderophore utilization was observed. While strain WCS358 was able to grow with the siderophores from both strains, strain WSC374 was unable to grow with the siderophore from strain WCS358. Transport experiments using both types of siderophores in both strains confirmed the narrow specificity of strain WCS374 and the broad specificity of strain WCS358.

The kinetic data of pseudobactin 358-mediated iron uptake in strain WCS358 showed $K_m = 0.23$ μM and $V_{max} = 0.14$ nmol/min mg dry weight. Transport of pseudobactin 358 was sensitive to sodium azide, 2,4-dinitrophenol, carbonyl cyanide *m*-chlorophenylhydrazone, indicating an energy requiring transport process. Phosphate bond energy seemed not to be involved in the transport of siderophores from *Pseudomonas*, since arsenate was without any effect. Nigericin, which reduces the proton gradient (Δ pH) over the cytoplasmic membrane by exchange of K^+ ions for H^+ ions did not significantly influence iron uptake. Contrary to that, valinomycin, a potassium ionophore which is known to reduce the electrochemical potential ($\Delta\Psi$) across the cytoplasmic membrane, reduced iron uptake mediated by pseudobactins. From these results the authors concluded that for pseudobactin-mediated iron uptake the proton motive force, and in particular, the electrochemical part, is most important.

Members of the genus *Azotobacter* are free-living nigrogen-fixing soil bacteria. Under iron-deficient conditions, *Azotobacter vinelandii* synthesizes large amounts of a yellow green fluorescent water soluble compound, azotobactin, which shows a striking similarity with the pyoverdins.[131] *Azotobacter chroococcum* which is the species being most frequently isolated from soil has been shown to produce a pyoverdin-type siderophore.[135] Other species, i.e., *A. macrocytogenes* and *A. paspali* grown in iron-limited medium have also been shown to produce fluorescent compounds similar to the pyoverdins, although their structures have not been elucidated so far.[135] Siderophore mediated uptake of iron has been investigated in *Azotobacter vinelandii*.[136,137] Uptake appeared biphasic, with both an initial rapid and an ensuing slower uptake, both being energy dependent. Uptake was inhibited by the addition of NaCl, KCl, LiCl, and $MgCl_2$ to the incubation medium, suggesting a more sophisticated interaction between azotobactin and its cognate membrane receptor.

FURTHER GRAM-NEGATIVE BACTERIA

The genus *Azospirillum* comprises Gram-negative, nitrogen-fixing bacteria which are associated with roots of grasses and are believed to enhance growth. *A. lipoferum* D-2 produces catechol-type siderophores containing 2,3-dihydroxybenzoic acid, 3,5-dihydroxybenzoic acid and salicylate as phenolic compounds and leucine and lysine as amino acids. Siderophores from *Azospirillum* have been named spirillobactins, although a definite structure was not reported by these authors.[138] Further siderophores containing catecholates combined with one or two amino acid residues have been described recently. Thus, chrysobactin, a 2,3-dihydroxybenzoic acid linked to the N^α-terminus of D-lysyl-L-serine has been shown to occur in *Erwinia chrysanthemi*.[139] Other examples are the amonabactins, isolated from *Aeromonas*, containing a catecholate-type structure and either L-tryptophane or L-phenyl alanine.[140] Spirillobactin, isolated from *A. brasiliense* has been shown to function as an iron-carrier in the producing species and also in *A. lipoferum*, but not in *A. amazonense*.[141] The kinetic data of transport determined in *A. brasiliense* showed $K_m = 0.23$, $V_{max} = 0.27$ nmol/min mg. Transport was mediated by an active process which was inhibited by CCCP and 2,4-DNP and no uptake occurred at 0 to 4°C. *Legionella* species which occur naturally in a water environment and include the pathogen *Legionella pneumophila* do not synthesize the common chemical types of siderophores, although iron is required as a trace element for growth in chemically defined media.[142]

Several other Gram-negative bacteria within different genera have been reported to either produce siderophores or express iron repressible outer membrane proteins. Vibriobactin, isolated from *Vibrio cholerae*,[143] possesses a structure containing an oxazoline ring and thus resembles parabactin, agrobactin and mycobactin (mycobactin family). Although no transport data are available, the biological activity in the producing strain, *V. cholerae* Lou15, has been demonstrated.[143] *Vibrio anguillarum*, a fish pathogen, has been shown to produce anguibactin, containing a mixed catecholate/hydroxamate siderophore, derived from 2,3-dihydroxybenzoic acid, cysteine (thazoline) and *N*-hydroxyhistamine.[144] Many isolates of this bacterium possess a plasmid-mediated iron uptake system which has been shown to be strongly correlated to virulence.[145] Hydroxamate production has also been observed in the Gram-negative *Aquaspirillum magnetotacticum* which contains the magnetosomes consisting of the iron oxide magnetite (2% of dry weight).[146] Surprisingly, this hydroxamate production occurred at high (20 μ*M*) and not at low (5 μ*M*) iron concentrations. The isolated but not purified compound stimulated growth of the enterobactin-deficient *Salmonella typhimurium* enb-7. Moreover, three outer membrane proteins ranging from 72,000 to 85,000 Da were coordinately produced at iron concentrations conducive to hydroxamate production. *Haemophilus parainfluenzae* produced at least four iron-repressible outer membrane proteins and a variety of iron-repressible outer membrane proteins from five other *H. parainfluenzae* strains have been reported, suggesting the existence of a high-affinity iron uptake system.[147] Iron-repressible outer membrane proteins have also been observed in *Pasteurella haemolytica* and *Bordetella pertussis*.[148,149]

Although the expression of iron-regulated outer membrane proteins may indicate the involvement of siderophores, other iron sources may be used. In *Haemophilus influenzae*, for example, the expression of outer membrane proteins, the utilization of siderophones and the utilization of heme iron have been observed, suggesting that multiple sources of iron can be used.[150] Haptoglobin forms a stable association with hemoglobin which suppresses acquisition of hemoglobin by certain bacteria. However, several hemolytic bacteria are able to obtain iron from hemoglobin-haptoglobin complexes which may enhance their survival in blood.[151] *Vibrio vulnificus* was unable to grow in deferrated medium without an additional iron source, but was able to grow with the addition of a hemoglobin-haptoglobin complex.[152,153] In contrast, heme-binding serum proteins did not provide essential iron to *Neisseria* species.[154] Both, *N. meningitidis* and *N. gonorrhoeae* can acquire iron from transferrin or lactoferrin without the involvement of siderophores.[155-160] *H. parainfluenzae*, *P. haemolytica*, and *B. pertussis* are found primarily on mucosal surfaces, where lactoferrin is a major iron-sequestering glycoprotein. Although these Gram-negative bacteria express iron-regulated outer membrane proteins, they do not appear to secrete siderophores.

TABLE 2
Kinetic Data of Siderophore Iron Transport in Fungi

Organism	Siderophore	K_m (μM)	V_{max} (pmol/min mg)	Ref.
Neurospora crassa	Ferricrocin	5	200	190
	Ferrirubin	0.05	—	191
	Chloroacetylferrichrysin	8	200	202
	Coprogen	5	200	191
	p-Azidobenzoylcoprogen B	6	200	201
Penicillium parvum	Ferrichrome	7	20	177
Aspergillus quadricinctus	Ferrichromes	5	100	211
Aspergillus fumigatus	Ferrichromes	5	20—30	211
Aspergillus melleus	Ferrichrysin	5	100	211
	Ferrichrome, ferricrocin	5	50	211
Rhodotorula pilimanae	Fe-rhodotorulate	6.8	600	175
	Enantio-Fe-rhodotorulate	4.8	360	175
Stemphylium botyrosum	Coprogen B	2.8	—	223
	Dimerum acid	2.2	—	223

The kinetic data of bacterial siderophore transport have been compiled for comparison purposes in Table 1. Moreover, special mutant strains are presented in Table 2 which may be used to detect bacterial and fungal siderophores in soil or other materials according to the bioassay techniques described by Szaniszlo,[11] or may serve as a means for further differentiation of certain bacterial groups as described by Reissbrodt and Rabsch.[71]

FUNGI

USTILAGO

The role of ferrichrome (Figure 11) as a ferric ionophore in the smut fungus *Ustilago sphaerogena* has been studied by Emery.[161] In this investigation it was shown for the first time that the ligand desferri-ferrichrome functions as a shuttle to bring iron into the cell. Other trihydroxamate siderophores like the ferrioxamines, ferrichrome A, ferrichrome A trimethylester as well as the monohydroxamates benzohydroxamic acid and fusarinine showed no iron transport activity in this fungus. Also nonhydroxamates, such as EDTA and citrate were ineffective in transporting iron into the cells. After ingress of ferrichrome and egress of the ligand, addition of iron to the medium reforms ferrichrome which then again enters the cells, confirming the view that desferri-ferrichrome can be used repeatedly for iron transport by the producing organism. Experiments using the ^{14}C-labeled ligand, desferri-ferrichrome, confirmed the fact that the iron-free ligand is not taken up by the cells. Ferrichrome transport in fungi is not linked to enzymatic modification or destruction of the ligand as has been reported in *E. coli* for the transport of ferrichrome and for the transport of enterobactin. Ferrichrome mediated iron transport in *Ustilago sphaerogena* was shown to be sensitive to respiratory inhibitors, such as azide, cyanide, and anaerobiosis.[161] However, ferrichrome taken up by the cells could not be released by these inhibitors, suggesting that transport but not retention is energy dependent. The maximum rate of uptake occurred at about pH 7 and fell to zero below pH 5 or above pH 8. The optimal temperature was 30°C with complete inhibition above 50°C. The specificity of transport with respect to the metal revealed that besides ferric ions, aluminum and gallium were also transported, both of which form stable chelates with desferri-ferrichrome. The copper complex was, however, biologically ineffective.

Double-label experiments with both the iron and ligand labeled have revealed two basic mechanisms for iron uptake from ferrichrome and ferrichrome A.[162] Whereas ferrichrome

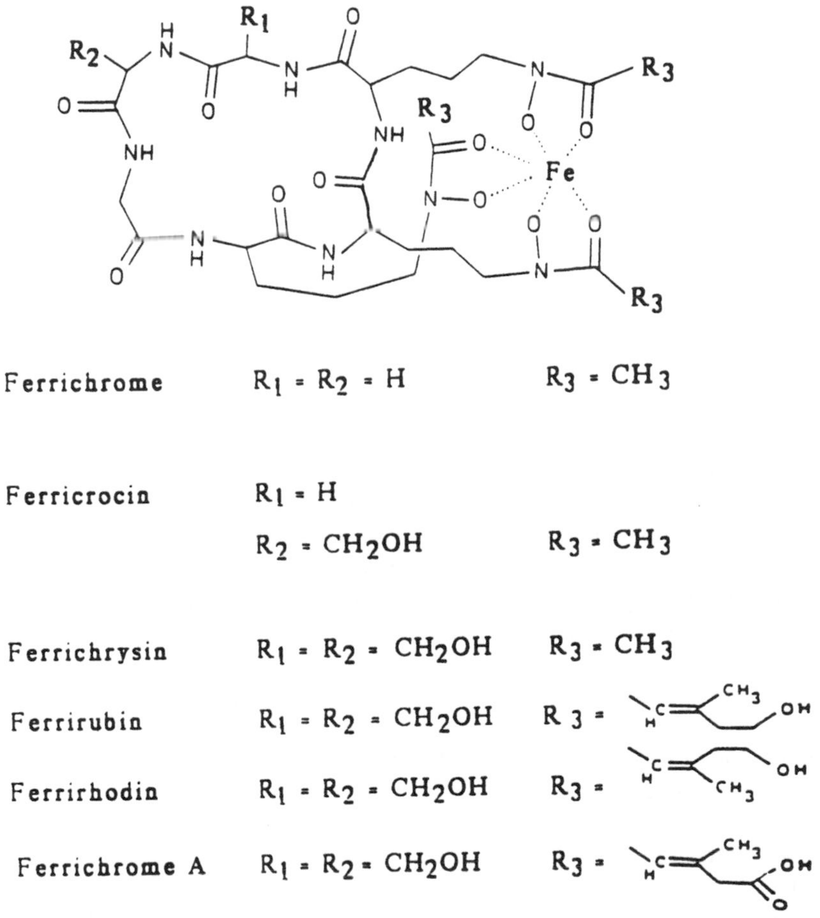

FIGURE 11. Structures of ferrichrome-type siderophores.

is taken up inside the cell to deliver iron, ferrichrome A binds to the cell membrane where iron is removed from the ligand. Uptake of ^{51}Cr(III) desferri-ferrichrome A in *Ustilago spaerogena* was nearly identical to that of Fe(III) ferrichrome A uptake. However, release of free ligand could not be observed, confirming that the chromium complex remained stable and the metal could not be dissociated from the complex by cellular reductases. On the other hand unlabeled ferrichrome A effectively competed with the chromium complex during uptake, indicating competition for the same receptor. When ^{67}Ga(III)desferri-ferrichrome was supplied to the cells of *U. spaerogena*, ^{67}Ga uptake was similar to that of ^{59}Fe from ferrichrome A. However, the results with ^{14}C-labeled ligand were different for the gallium and iron chelates. Whereas the ^{14}C-labeled ligand of ferrichrome A reappeared in the medium, the ligand from the Ga(III)desferriferrichrome A could not be released, since gallium has no stable 2 + oxidation state.

Ferrichrome A transport in *U. sphaerogena* has been followed by EPR spectroscopy.[163] Disappearance of the EPR signal was taken as a proof that reductive removal of iron takes place at the membrane during uptake of ferrichrome A. The EPR data confirmed the results with radiolabeled ferrichrome. Iron chelated to citrate was taken up by an energy-dependent process in *U. sphaerogena*.[164] However, the citrate was not taken up with the metal, indicating that citrate has some of the characteristics of ferrichrome A iron uptake in iron-deficient cells. The rate of reduction of iron from iron-citrate as determined from EPR measurements occurred at the same rate as iron transport. Moreover, like iron uptake from ferrichrome A,

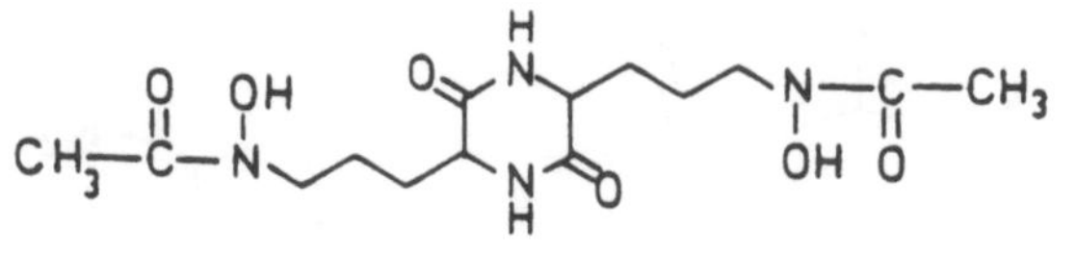

FIGURE 12.   Structure of rhodotorulic acid.

iron uptake from iron citrate is completely inhibited by dipyridyl, a ferrous chelator, confirming the view that ferrichrome A and Fe-citrate deliver their iron to the cell by a reductive step. It may be assumed that this kind of siderophore mediated iron transport may occur in a wide variety of fungi which do not produce any siderophore but still have the capacity to use iron bound to different siderophores.

The occurrence of ferrichrome and ferrichrome A is a characteristic trait of the *Ustilaginaceae*.[165] Several other related genera of the family of *Ustilaginaceae*, such as *Graphiola, Protomyces,* and *Tilletaria* have been reported to produce ferrichrome and ferrichrome A. From a strain of *Neovossia indica,* a smut fungus of the order Tilletiales, ferrichrome, ferrichrome C, and a novel tetraglycylferrichrome have been isolated.[166] Tetraglycylferrichrome revealed transport properties nearly identical to ferrichrome, indicating that enlargement of the peptide ring by one additional glycine does not seriously influence the transport properties of ferrichrome. Interestingly, the formerly named *U. violacea* (Pers.) Rouss, has been replaced in the genus *Microbotryum* because of morphological characteristics and also because of its production of rhodotorulic acid.[167] The regulation of ferrichrome and ferrichrome A biosyntheses has been shown to be related to the available iron content of the medium and to the oxidative metabolism of the fungus.[164] Ferrichrome A, the more effective chelator, was preferentially synthesized under more extreme conditions of iron stress, but completely repressed when the cells were supplied with sufficient iron. In contrast, biosynthesis of ferrichrome was strongly but not completely repressed by iron. Mutants of *Ustilago maydis,* defective in the biosynthesis of ferrichrome and ferrichrome A have been obtained.[168] Class I mutants no longer produced ferrichrome while retaining the ability to produce ferrichrome A. Class II mutants were defective in the production of both ferrichrome and ferrichrome A. Genetic and biochemical analysis suggested that class II mutants were defective in the ability to hydroxylate L-ornithine to δ-N-hydroxyornithine, the first step in the biosynthesis of these siderophores. Class II mutants could be complemented by DNA clones (cosmids) containing a dominant selectable marker for hygromycin B resistance.

RHODOTORULA

Compared to filamentous fungi, iron transport in yeast cells is easier to perform, as pipetting is possible at all growth stages. Strains of *Rhodotorula* produce rhodotorulic acid (Figure 12), a dihydroxamate type siderophore, composed of two N-acetyl-N-hydroxyornithine residues connected head-to-head in a diketopiperazine ring.[169,170] The Fe(III), Al(III) and Cr(III) complexes of rhodotorulic acid (RA) have been isolated and characterized.[171] The stability constant (log β_3 = 31.1) of the ferric complex of RA has been determined potentiometrically and spectrophotometrically.[172] The iron complex exists as a dimer of formulation Fe_2RA_3 at pH values from 4 to 11. Below pH 4 the dimer dissociates into the monomeric cationic species $FeRA^+$ in which both of the hydroxamate groups of RA are coordinated to a single ferric iron. Analysis of CD spectra of Fe_2RA_3 and comparison with ferrichrome A clearly showed that Fe_2RA_3 has a Δ-cis absolute configuration around both metal centers which corresponds to the Δ absolute configuration of other diketopiperazine containing siderophores of the coprogen family.[173] Transport studies with FeRA have revealed that the ferric rhodotorulate complex is taken up in *Rhdotorula pilimanae* in a so-called taxi mechanism,[174] in which the ligand functions as an iron taxi which brings the iron to the cell without penetration of the cell membrane. The mechanism is very similar to that proposed for ferrichrome A in *Ustilago*. However, in *Rhodotorula* iron transfer to the cell is obviously not linked to a reductive removal of iron, as shown by the use of Ga-rhodotorulate possessing no stable 2 + oxidation state.[175] The process is one of active transport, since it exhibits many features, including pH (6.5) and temperature (37°C) optima with pronounced sensitivity to azide, cyanide, 2,4-dinitrophenol and iodacetamide. *Rhodotorula pilimanae* does not accept ferrioxamines or ferrichrome A, but citrate was as effective as rhodotorulic acid.

Studies on the stereospecificity of siderophore-mediated iron uptake in *R. pilimanae* with *enantio*-rhodotorulic acid and isomers of chromic rhodotorulate have shown that the organism discriminates preferentially between optical (Δ/Λ) rather than between geometrical (cis/trans) arrangements of the ligands.[176] Thus, confirming the view that stereoselective recognition of optical isomers takes place during iron uptake mediated by RA and that this recognition primarily involves the right-handed Δ coordination of the metal center. Stereoselective transport of siderophores in *Penicillum parvum*, using *enantio*-ferrichrome, had shown earlier that in this case the left-handed Λ-coordination propeller of the metal center was the preferred coordination.[177] Using synthetic analogues of RA it was shown that the diketopiperazine ring of RA may be replaced by simple chains of *n*-methylene groups without losing the property of transporting iron to the cells.[176] By decreasing the number of bridging methylene groups from 6 to 3 a decrease of transport was observed, suggesting an optimum in the iron-iron distances of the analogues. When the terminal methyl residues at both ends of RA were replaced by isopropyl residues an increase in uptake was observed. Summarizing the results with synthetic analogues of RA it was suggested that three features acount for the observed differences in uptake: (1) the terminal residues, (2) the iron-iron distances and (3) the chirality at the metal center and its effect on the position of the terminal groups.

NEUROSPORA CRASSA

Neurospora crassa is a fast growing, saprophytic, filamentous fungus, known as the red bread mold. It is often isolated from cereals but also occurs in the soil. While the anamorph was formerly classified into the *Moniliana*, it has been recently introduced into a new genus named *Chrysonilia*.[178] Numerous mutants of *N. crassa* have been compiled by Perkins et al.,[179] and are deposited at the Fungal Genetics Stock Center, Dept. of Microbiology, University of Kansas Medical Center, U.S. Since ornithine is a constituent of all fungal siderophores, a mutant blocked in the biosynthesis of ornithine fails to produce siderophores.[180] This has indeed been confirmed by using the triple mutant *N. crassa (arg-5 ota aga)*.[181,182] A characteristic feature of *N. crassa* is its ability to produce siderophores from two different siderophore classes: coprogen (Figure 13) and ferricrocin.[183,184]

Coprogen transport has been studied using the triple mutant *N. crassa (arg-5 ota aga)* which does not produce siderophores when grown in the absence of L-ornithine.[181] These studies revealed that besides coprogen and ferricrocin citrate also served as an iron transport agent. Iron uptake mediated by citrate, however, differed markedly in its uptake characteristics from those mediated by coprogen or ferrichromes, suggesting the presence of a low-affinity iron transport system. Members of the ferrioxamine class, such as ferrioxamine B and ferrioxamine D_1 were devoid of any transport properties in *N. crassa*. This apparently is true for most but not all fungi. It had been shown earlier that ferricrocin was found in mycelia as well as in conidia, but not in the culture filtrate of *N. crassa*.[185] Thus in low-iron medium (0.1 to 1 μM) the synthesized desferri-ferricrocin remains intracellular, whereas desferri-coprogen is excreted into the incubation medium. Using Mössbauer spectroscopy it was possible to demonstrate that after uptake of coprogen in *N. crassa*, iron is slowly removed from the complex but that intracellular ferricrocin is a relatively stable pool, functioning as the main iron storage compound in mycelia.[186,187] Moreover, ferricrocin was the predominant siderophore in conidia.[188,189] Ferritin-like structures were absent in mycelia

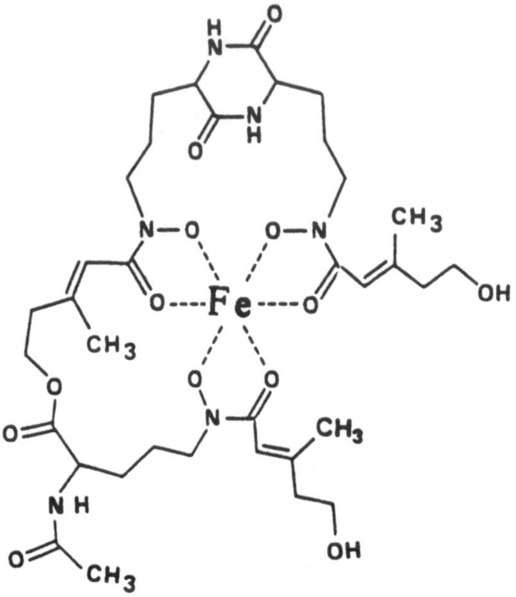

FIGURE 13.   Structure of coprogen.

and conidia of *N. crassa*. These findings clearly confirmed that ferricrocin is synthesized in mycelia mainly for iron storage purposes to be finally included in conidiospores during sporulation. The presence of ferricrocin in conidia allows rapid germination as has been shown in detail by Horowitz et al.[183]

Early transport studies with ferrichrome-type siderophores in the siderophore-free mutant, *N. crassa* (*arg-5 ota aga*), revealed that except for ferrichrome A and ferrirubin all other ferrichrome-type siderophores were taken up.[190] Reevaluation of the transport kinetic data showed K_m values in the range of 1 to 5 μM for ferrichrome, ferricrocin and ferrichrysin.[191] Transport of ^{14}C-labeled coprogen was competitvely inhibited by ferrichrome-type compounds (ferrichrome K_i = 1 μM, ferricrocin K_i = 3.0 μM, ferrirubin K_i = 0.07 μM) which led to the conclusion that in *N. crassa* a common siderophore transport system for ferrichromes and coprogen is present.[191] However, from the differing structural features of the ferrichromes and coprogen it was inferred that the transport systems must possess different recognition sites.

A further important aspect of siderophore iron transport in *N. crassa* is the stereospecificity of ferrichrome uptake. The first evidence for stereospecific ferrichrome uptake came from studies with *enantio*-ferrichrome.[177,192] These investigations clearly showed that enantiomeric ferrichromes (enantio-ferrichrome, enantio-ferricrocin) synthesized from D-ornithine instead from L-ornithine are not recognized by the fungal transport systems. Since both the peptide backbone as well as the metal center adopt opposite configurations in enantio-ferrichrome, a clear assignment as to which part of the molecule is responsible for stereospecific recognition could not be made. Several lines of evidence, however, point to the importance of the metal center for recognition of siderophores in fungi. Thus, small changes of the N-acyl residues surrounding the metal center significantly changed their transport properties.[193] This was found to be the case when the acetyl residues were successively replaced by trans-anhydromevalonyl residues in the ferrichrome/ferrirubin series or when the anhydromevalonyl residues of coprogen were successfully replaced by acetyl residues in the coprogen/neocoprogen series.[193,194] These results clearly demonstrated that the metal center and its surrounding functionalities are of primary importance during the uptake process. A similar conclusion was drawn from the results obtained with retrohydroxamate ferrichrome and desmethyl-retrohydroxamate ferrichrome. The methyl groups of the acetylhydroxamic acid residues obviously have a significant function during the initial recognition process of uptake. Further proof for the importance of the metal center for siderophore recognition came from studies with des(diserylglycyl)ferrirhodin, DDF.[193] DDF is a ferrirhodin derivative lacking the cyclic ser-ser-gly insert of the cyclic peptide ring but which contains an intact Λ-cis absolute configuration of the metal center.[196] DDF was bound to the binding sites of the transport system in N. crassa but transport of iron into the cells was no longer possible.[193] Summarizing these results, it appears that there is considerable evidence for the metal center to be the most important domain for the initial interaction with the putative binding sites in fungal siderophore transport systems.

Transport of ferrichromes and coprogens in N. crassa has been shown to be sensitive towards a variety of respiratory poisons and uncouplers, i.e., potassium cyanide, sodium azide, dinitrophenol, indicating that siderophore transport in fungi is an energy consuming process. N. crassa and probably most other fungi possess an active plasma membrane ATPase, hydrolyzing ATP by simultaneous extrusion of protons which results in the formation of a membrane potential of about 100 to 200 mV inside negative.[197] Several agents affecting ATP production inhibit siderophore uptake. Arsenate also causes a sudden change of siderophore uptake.[198] Both, intracellular decrease of ATP as well as depolarization of the membrane potential affect siderophore transport in N. crassa, as both events are intimately connected to one another. However, there is evidence that the membrane potential is the actual driving force for siderophore transport in fungi.[199] Membrane depolarization was achieved by adding high amounts of glucose (1 mM) to glucose starved cells of N. crassa. The development of a high-affinity glucose transport system allows a sudden entrance of glucose by a proton-symport mechanism which in turn depolarizes the membrane. Under these conditions siderophore transport was inhibited transiently but resumed after some minutes due to restoration of the membrane potential. When ATPase was simultaneously inhibited by diethylstilbestrol or dicyclohexylcarbodiimide a complete inhibition of siderophore transport occurred. The presence of ATPase inhibitors alone had no effect on siderophore uptake, confirming the view that membrane depolarization is the driving force for siderophore transport across the plasma membrane. Further support for the requirement of a membrane potential for fungal siderophore transport came from studies with inverted plasma membrane vesicles.[200] However, our knowledge on siderophore transport in fungi is still in a preliminary state; so far no membrane protein has been identified which is involved in siderophore transport. Recent SDS gel electrophoretic studies with membrane proteins isolated from N. crassa, grown in iron-deficient and -sufficient medium, revealed no difference in the plasma membrane protein pattern, excluding the existence of overproduced iron-regulated plasma membrane proteins. As an alternative explanation, constitutive transport proteins have been postulated which are expressed in the presence of conidial siderophore iron stores.[189] Attempts to identify membrane proteins involved in coprogen transport have been made with the use of a photoaffinity label, p-azidobenzoylcoprogen B, which upon illumination reacts covalently with membrane components. In the dark p-azidobenzoylcoprogen B was taken up by the transport system of N. crassa (arg-5 ota aga) in a concentration-dependent manner (K_m = 6 μM, V_{max} = 0.2 nmol/min mg) and was also a competitive inhibitor of coprogen uptake, K_i = 5 μM. Photolysis of cells with near-ultraviolet light during transport in the presence of the photoaffinity label resulted in approximately 50% inhibition of both coprogen and ferrichrysin uptake. Uptake of the photolabel itself was also inhibited upon illumination. However, due to a variety of difficulties (unspecific binding, position and orientation of the label, low yield of labeled protein) identification of transport proteins has not yet been achieved. One derivative, chloroacetyl-ferrichrysin, proved to be an unexpectedly potent reversible inhibitor (K_i = 0.4 μM) of both ferrichrysin and coprogen uptake[202] similar to the natural siderophore, ferrirubin. This probably opens new perspectives by focusing on reactive labels linked to the N-acyl residues, surrounding the metal center and not to the backbone of the molecule, as has been done with coprogen B. The N-acyl structure has been shown to be of fundamental importance for recognition and binding of siderophores.[193] The study of the importance of the N-acyl residues of siderophores was supported by the isolation of minor compounds from other fungi. Thus, ferrichromes have been isolated which contained heterogenous N-acyl residues called asperchromes, as they originated from *Aspergillus ochraceous*.[203,204] Natural derivatives of coprogen have been isolated from *Epicoccum purpurascens* and *Curvularia lunata* which contained one or two N-acetyl residues instead of N-trans-anhydromevalonyl residues. These were named triornicin, isotriornicin, neocoprogen I and neocoprogen II.[205-207] Both asperchromes and neocoprogens were valuable tools for determining molecular recognition of siderophores by membrane located transport systems in *N. crassa*.[193] Various novel compounds of the coprogen family have recently been isolated from the genus *Alternaria* which contains plant-pathogenic species. Thus, *Alternaria alternata* produces several dimethyl coprogens whereas *Alternaria longipes* produces hydroxycoprogens.[208,209] However, the transport properties of these compound have not yet been investigated.

ASPERGILLUS AND PENICILLIUM

Most species of the fungal genera *Aspergillus* and *Penicillium* are vegetative (anamorph) forms of *Ascomycetes*. Because the sexual stages (teleomorph forms) are unknown they are classified as Deuteromycetes or Fungi Imperfecti. Strains of *Aspergillus* and *Penicillium* are widespread on earth and seem to be especially adapted to saprophytic life. A variety of strains are pathogenic to plants and animals. However, adaptation to the host is never exclusive, i.e., all pathogenic strains can survive as saprophytes on earth or other substrates. Rapid acquisition of iron by siderophores and transport systems seems to play a crucial role in all these fungi. In a first systematic search for siderophores in the *Aspergillales*, ferrichrome, ferricrocin, ferrichrysin, ferrirubin, ferrirhodin and coprogen were reported.[210] Ferrichrome is produced by *A. niger* and *A. quadricinctus*. Ferricrocin is produced by *A. fumigatus*, *A. nidulans*, *A. humicola*, *A. versicolor* and *A. viridi-nutans*, while *A. melleus*, *A. terreus*, and *A. oryzae* produce mainly ferrichrysin. *Aspergillus* strains showed a marked specificity with respect to the species own siderophores.[211] However, siderophores of the ferrichrome family were taken up by young mycelia of all strains tested, irrespective of the main ferrichrome-type siderophore these strains produce in low-iron cultures. Thus, *A. quadricinctus*, a ferrichrome producer, *A. fumigatus*, a ferricrocin producer, and *A. melleus*, a ferrichrysin producer, were equally effective in utilizing ferrichrome, ferricrocin and ferrirubin, and to a lesser degree ferrichrysin. Transport kinetic data revealed $K_m = 5$ μM and $V_{max} = 50$ to 100 pmol/min mg for all examined ferrichromes. The sensitivity towards sodium azide clearly demonstrated that siderophore iron uptake in Aspergilli is an active and energy-requiring transport process. Uptake of iron from coprogen and ferrioxamines, however, could not be observed, indicating that siderophore transport system(s) in Aspergilli are particularly designed to utilize ferrichrome-type siderophores.

Aspergillus ochraceus has been shown to produce ferrichrysin, ferrirubin and, in minor quantities, a variety of structurally different ferrichrome-type siderophores which have been named asperchromes.[203] The asperchromes possess a common cyclic -orn-orn-orn-ser-ser-gly- peptide moiety, like ferrichrysin, but differ in the ornithyl-⁸N-acyl residues. Besides siderophores of the ferrichrome family, *Aspergillus* strains produce triacetylfusigen, also named N,N',N'' triacetylfusarinine C (Figure 14), a cyclic triester composed of three molecules of N^α-acetyl-N^δ-(cis-5-hydroxy-3-methylpent-2-enoyl)-N^δ-hydroxy-L-ornithine.[212-214] Triacetylfusarinine has also been detected in *Mycelia sterilia* EP-76 (previously incorrectly identified as *Penicillium* sp.).[215,216] Iron uptake studies in this fungus have revealed that *M. sterilia* EP-76 possesses at least three distinctive pathways of iron uptake: (1) a true sider-

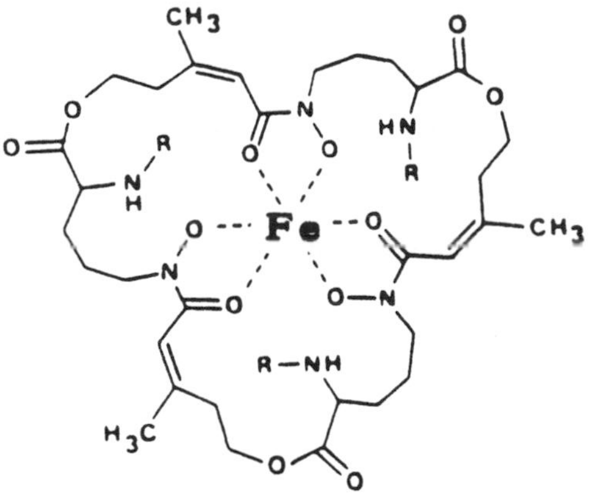

R • CO–CH₃

FIGURE 14. Structure of *N,N,',N''*-triacetylfusarinine
C (= triacetylfusigen).

ophore system utilizing ferric-triacetylfusarinine and involving cellular hydrolysis of the
chelate for iron release; (2) a ferrichrome transport system; and (3) an exchange mechanism
in which iron from other complexes is incorporated into extracellular triacetylfusarinine.
Stereochemical aspects of ferric triacetylfusarinine transport in *M. sterilia* have been examined using chromic complexes of triacetylfusarinine.[217] These studies gave evidence that
the predominant Δ-coordination isomer in solution is not the one which is biologically active
in iron transport but rather the Λ-coordination isomer is taken up. The rapid uptake of Λ-
cis-Fe-triacetylfusarinine from a Δ-*cis*-Fe-triacetylfusarinine solution by cells of *Mycelia
sterilia* EP-76 was explained by a very rapid equilibration of the two isomers. This equilibration is possible because of the d^5 electronic configuration of iron, which makes the
ferric ligand complexes kinetically labile with respect to isomerization and ligand exchange.

Strains of *Penicillium* have also been reported to produce a variety of siderophores of
the ferrichrome family. Contrary to the Aspergilli, some strains of *Penicillium* are able to
produce coprogen in addition to ferrichromes.[210] It is interesting to note that strains of
Aspergillus and *Penicillium* produce specific cellular siderophores in addition to the well-
known siderophores of the culture medium.[185] Thus, the cellular siderophore of *Aspergillus
nidulans* is ferricrocin, that of *Penicillium chrysogenum* is coprogen. Early investigations
on the stereospecific uptake of siderophores in strains of *Aspergillus* and *Penicillium* indicated
that the mirror images, the *enantio*-ferrichromes, were not recognized by the ferrichrome
transport system.[177,192] Siderophore recognition and transport obviously depend on two structural features of the siderophore, the chirality of the coordination geometry around the metal
center and the structure and chirality of the ligand backbone. Since both features are reversed
in the *enantio*-ferrichromes a definite assignment of one of these characters seems difficult.
However, transport data with *Neurospora crassa* indicated that the chirality around the metal
center is crucial for recognition, while the structure of the ligand backbone is possibly
required for the function of subsequent transport across the fungal membrane.[193]

FURTHER FILAMENTOUS FUNGI

Siderophores of the fusarinine-type have been isolated from *Fusarium roseum* and related
species, all of which occur as monomers, dimers, and linear trimers.[218] The cyclic trimer
of cis-fusarinine has been named fusigen.[219] The acetylated form, *N,N',N''*-triacetylfusarine
C (= tracetylfusigen) is a common siderophore in the genus *Aspergillus*.[212] *Fusarium di-*

merum and *Myrothecium* sp. have been shown to produce dimerum acid and coprogen B which are both biosynthetically related.[220] While dimerum acid represents a dihydroxamate, coprogen B is a trihydroxamate. The transport properties of fusarinines, dimerum acid and coprogen have been studied in *Gliocladium virens* which produces predominantly the monohydroxamates (cis- and trans-fusarinines) and the dihydroxamate (dimerum acid) and minor quantities of coprogen B, coprogen and ferricrocin.[221,222] These studies showed that in this fungus the mono- and dihydroxamates and also ferricrocin behaved as good iron transport agents, whereas the coprogens behaved as weak iron transporting agents.[222] *Stemphylium botyrosum* f. sp *lycopersici*, the causal agent of leaf spot and foliage blight of tomato, has been shown to produce dimerum acid, coprogen B and an unidentified monohydroxamate as the major siderophores.[223] Kinetic studies of initial uptake vs. siderophore concentration exhibited saturation kinetics with apparent $K_M = 2.8$ μM for coprogen B and $K_m = 2.2$ μM for dimerum acid. Fe-rhodotorulate, which is structurally related to dimerum acid and ferrichrome, functioned also as siderophore in this fungus. Ferrioxamines, however, were ineffective siderophores. Dimerum acid and coprogen B were also isolated from the plant pathogen, *Verticillium dahliae,* which is a soil-borne pathogen of numerous crop plants.[224,225] It has been reported that the severity of disease in calcareous soils is reduced by addition of iron salts, suggesting that the fungal pathogens strongly compete for iron. Ferrichrome-type siderophores have been isolated from dermatophyte fungi. Thus, *Microsporum gypseum* produces ferricrocin and an unidentified compound,[226] while *Trichophyton mentagrophytes* produces ferrichrome.[227] Ferrirhodin has been identified as the principal siderophore of *Botrytis cinerea* which is a common plant pathogen causing infections in all kinds of plant surfaces, crops, fruits, grapes and seedlings.[228] A similar siderophore pattern has been observed in *Verucobotrys,* confirming a chemotaxonomical relationship between the two genera, *Botrytis* and *Verucobotrys.*[229] Transport studies in *Botrytis cinerea* showed the following order of uptake: ferrichrysin (100%), ferrirubin (57%), ferrirhodin (45%) and coprogen (6%), indicating that the chemically more stable ferrichrome-type siderophores are the preferred iron carriers in this fungus, which may be of relevance as *Botrytis* is known to rapidly acidify its incubation medium. Cellulose-degrading fungi, such as *Trichoderma viride,* have recently been shown to produce dimerum acid and fusarinine.[230] Most interestingly, ectomycorrhizal fungi have been shown to produce hydroxamate siderophores by the use of the *Arthrobacter*-JG9 bioassay.[14] Some of the ectomycorrhizal fungi from *Ericaceae* have recently been shown to produce hydroxamate siderophores and their structure elucidation is in progress.[231]

In several fungal genera, i.e., *Phycomyces, Mucor* and *Rhizopus*, hydroxamate siderophores have not been detected.[230] However, it has been reported that desferrioxamine (Desferal) treatment of dialysis patients, increased the risk of severe and fatal forms of mucormycosis, from which *Rhizopus rhizopodiformis* could be isolated.[232,233] This may indicate that *Rhizopus* can either utilize iron via ferrioxamine B or that desferrioxamine B can alleviate the fungistatic effect of serum proteins. Current studies, however, suggest that strains of *Rhizopus* can produce a siderophore which, however, is not a hydroxamate.[229]

Geotrichum candidum is also unable to produce any hydroxamate-type siderophore, but is able to take up ferric and ferrous iron and also accepts iron from several fungal siderophores, such as ferrichrome, Fe-rodotorulate, coprogen and even ferrioxamine B.[234,235] Interestingly, the ferrioxamine uptake route seems to be independent from the ferrichrome uptake route, as inhibition was uncompetitive.

YEASTS

The common yeast *Saccharomyces cerevisiae* is unable to synthesize hydroxamate type siderophores although application of the chrome azurol S assay to glucose grown cultures gave positive results due to the fermentative production of carboxylic acids. However, *S.*

cerevisiae can take up iron from various hydroxamate type siderophores, like ferricrocin, ferrioxamine B, Fe-rhodotorulic acid and from ferric citrate.[237,238] The question as to whether or not a reduction step is required during ferrioxamine B uptake has been studied in greater detail. Uptake of iron from ferrioxamine B has been shown to be dependent on a ferri-reductase only when the concentration is high. In this case ferrozine functions as an inhibitor of iron uptake. At low extracellular concentration (7 μM), uptake appeared to be essentially nonreductive, as iron uptake was ferrozineinsensitive and did not occur in heme-deficient cells which lacked an inducible ferri-reductase.[238] From these results the authors concluded that in *S. cerevisiae* two mechanisms contribute to the uptake of ferrioxamine B: one with high affinity, by which the siderophore is internalized and another, with lower affinity, in which iron is dissociated from the ligand prior to uptake. A mutant of *S. cerevisiae* (Fre⁻) has recently been isolated which lacked both the ferric reductase activity and the ability to take up ferric iron from the medium, suggesting that the ferric reductase is involved in iron uptake.[239]

Candida albicans is an imperfect, dimorphic yeast. It is often an associate of human beings and other warm-blooded animals and it is frequently encountered as an opportunistic pathogen. The incidence of both superficial and invasive candidiosis has increased markedly over the last few decades. The genus *Candida* includes diploid white asporogenous yeasts and probably includes a number of species that are believed to be imperfect forms of ascomycete species.[29] Since *C. albicans* possesses all the enzymes for glycolysis, the hexose monophosphate shunt, and the tricarboxylic acid cycle, it behaves metabolically like other fungi. Siderophores have not yet been isolated from species of *Candida albicans*, although there are reports on color formation when iron was added to low-iron culture supernatants.[240-242] It may well be that *Candida*, like *Saccharomyces*, satifies the need for iron by reductive mechanisms and also by transport systems for exogenous hydroxamate siderophores. The human pathogenic yeast, *Histoplasma capsulatum*, has been shown to produce fusarinines, dimerum acid, and coprogen B which might enhance fungal virulence in infected tissue, although this has still to be demonstrated.[243] Yeasts of the genus *Cryptococcus* have revealed different results concerning the production of siderophores. Since *Cryptococcus neoformans* is a human pathogen, iron acquisition in this organism is of considerable importance. Although spent low-iron medium contained neither hydroxamates nor organic acids or any other iron chelator, exogenous ferrioxamine B stimulated growth, suggesting that siderophore uptake mechanisms still exist, while the biosynthetic capacity of siderophores has possibly been lost.[244] *Cryptococcus melibiosum* has been reported to produce ferrichrome C.[170] These different findings obviously indicate that the genus *Cryptococcus* is not uniform. Recently, the genus *Myxozyma* has been separated from the genus *Cryptococcus* because of several distinct traits.[245]

CONCLUDING REMARKS

A great variety of microorganisms have been shown to produce siderophores and it is to be expected that new siderophore structures will be elucidated in the near future. It is interesting to note that there is a considerable variation in the kinds of iron-binding ligands, all of which seem to have their cognate receptors or recognition sites within membrane located siderophore transport systems. Regulation of siderophore biosynthesis and iron transport are only partly understood but production of siderophores, iron-deprivation or cross-feeding by siderophores are essential events in microecology. Of special interest is the involvement of siderophores in natural extreme or low-iron environments, i.e., in calcareous soils, at the plant root, in plant, animal or human tissues, which may explain beneficial and deleterious effects as well as virulence of microbial pathogens in the host. While the ability to take up iron from different siderophores is a highly conserved property, the ability to synthesize siderophore ligands has obviously been lost in some microbial genera during evolution. Transport specificity is observed in several siderophore classes containing characteristic structural elements. Stereospecificity is found in those siderophores which contain chiral elements. Moreover, different mechanisms of transport seem to exist, some of which involve the uptake of the intact chelate, while others seem to react with the cell membrane and deliver only the iron atom to the cells. In all cases, however, a specific interaction with siderophore transport systems located in the membrane is required. The kinetic data of several well-studied fungi have been compiled in Table 3.

Finally, a list of siderophores and producing organisms is presented in Tables 4 and 5 which allows a comparison of organisms on a chemosystematic level and also provides a source of organisms for the production of the various siderophores.

TABLE 3
Indicator Strains for the Detection of Siderophores[a]

Indicator strains	Siderophores detected	Siderophores not detected	Ref.
Salmonella typhimurium TA2700	Enterobactin (2,3-DHBS)₂ (2,3-DHBS)₃	2,3-DHBA Aerobactin Ferrichrome(s) Coprogen Ferrioxamines	77
E. coli LG1522	Aerobactin Fe-rhodotorulate	Enterobactin 2,3-DHBA (2,3-DHBS)₂ (2,3-DHBS)₃ Ferrichrome Coprogen Ferrioxamines	255
Arthrobacter flavescens JG9	Ferrichrome(s) Coprogen Ferrioxamines Fe-rhodotorulate	Aerobactin Enterobactin 2,3-DHBA (2,3-DHBS)₂ (2,3-DHBS)₃	11
Salmonella stanleyville GRR32	Ferrichrome Coprogen Ferrioxamines (2,3-DHBS)₂ (2,3-DHBS)₃	Aerobactin Enterobactin 2,3-DHBA	76
Salmonella stanleyville GRR17	Ferrichrome(s) Ferrioxamines Enterobactin (2,3-DHBS)₂ (2,3-DHBS)₃	Coprogen Aerobactin 2,3-DHBA	76
Salmonella typhimurium SR1001	2,3-DHBA	Enterobactin (2,3-DHBS)₂ (2,3-DHBS)₃ Aerobactin Ferrichrome(s) Coprogen Ferrioxamines	76
Escherichia coli P8	Enterobactin Coprogen	Ferrichrome(s)	102
Escherichia coli H1774	Enterobactin Ferrichrome(s)	Coprogen	101
Escherichia coli H1876	Ferrichrome Coprogen	Enterobactin	81
Erwinia herbicola FM13	Enterobactin Ferrichrome Coprogen	Ferrioxamines	32, 33
Morganella morganii SBK3	Colibactin (unidentified)	Enterobactin (2,3-DHBS)₂ (2,3-DHBS)₃ Aerobactin Ferrichrome(s) Coprogen Ferrioxamines	74

Note: 2,3-DHBA = 2,3-dihydroxybenzoic acid, (2,3-DHBS)₂ = linear dimer of 2,3-dihydroxybenzoylserine, (2,3-DHBS)₃ = linear trimer of 2,3-dihydroxybenzoylserine.

[a] Modified siderophore pattern analysis according to Rabsch and Reissbrodt.[71,76]

TABLE 4
Bacterial Siderophores and Producing Organisms

Siderophore	Producing organism	Ref.
Aerobactin	*Aerobacter aerogenes* (=*Enterobacter aerogenes*)	106
	Escherichia coli ColV	107
	Shigella flexneri	108
	Shigella boydii	109
	Salmonella spp.	246
	Citrobacter, Proteus, Morganella, Serratia	246
	Yersinia intermedia, Y. fredriksenii, Y, kristensii	72
Agrobactin	*Agrobacterium tumefaciens*	117, 118
Amonabactin	*Aeromonas hydrophila*	140
Anguibactin	*Vibrio anguillarum*	144, 145
Arthrobactin	*Arthrobacter pascens*	4
Azotobactin	*Azotobacter vinelandii*	247
Azotochelin	*Azotobacter vinelandii*	248
Chrysobactin	*Erwinia chrysanthemi*	139
Enterobactin (=Enterochelin)	*Escherichia coli, Salmonella typhimurium,*	67, 68
	Shigella spp., *Klebsiella* spp., *Enterobacter* spp.	249
Exochelin MS	*Mycobacterium smegmatis*	43
Exochelin MB	*Mycobacterium bovis*	43
Ferribactin	*Pseudomonas fluorescens*	29
Ferrioxamine (A_1, A_2, B, C, D_1, D_2, E, F, G, H)	*Streptomyces pilosus, S. griseus, S. griseoflavus, S. olivaceus, S. aureofaciens, S. galilaeus, S. lavendulae*	19—25
Ferrioxamine B	*Arthrobacter simplex*	29
Ferrioxamine E, D_2, B	*Erwinia herbicola*	32
Ferrioxamine E	*Pseudomonas stutzeri*	30
Ferrioxamine E	*Chromobacterium violaceum*	29
Ferrioxamine G	*Hafnia alvei*	70
Mycobactin A-T	*Mycobacterium* (different species)	250
Nocobactin(NA, NB)	*Nocardia asteroides, N. brasiliensis*	251
Parabactin	*Paracoccus denitrificans*	111, 112
Parabactin A	*Paracoccus denitrificans*	111, 112
Pyoverdin(e) Pa	*Pseudomonas aeruginosa*	128
Pyoverdin Pf	*Pseudomonas fluorescens*	131
Pyoverdin 358	*Pseudomonas putida* WCS358	134
Pseudobactin	*Pseudomonas putida* B10	132
Pseudobactin 7SR1	*Pseudomonas fluorescens* ATCC 13525	129
Pyochelin	*Pseudomonas aeruginosa*	252
Rhizobactin	*Rhizobium meliloti*	119
Schizokinen	*Bacillus megaterium*	52
Schizokinen	*Anabaena* spp.	62, 63
Spirillobactin	*Azospirillum brasiliense*	141
Staphyloferrin	*Staphylococcus aureus*	63
Vibriobactin	*Vibrio cholerae*	143

TABLE 5
Fungal Siderophores and Producing Organisms

Siderophore	Producing organism	Ref.
Ferrichrome	*Aspergillus niger, A. quadricinctus*	210, 211
	Penicillium parvum	177
	Ustilago sphaerogena, U. maydis	161, 165, 168
	Neovossia indica	166
	Trichophyton mentagrophytes	227
Ferrichrome A	*Ustilago sphaerogena, U. maydis*	161, 165, 168

TABLE 5 (continued)
Fungal Siderophores and Producing Organisms

Siderophore	Producing organism	Ref.
Ferrichrome C	*Cryptococcus melibiosum* (= *Myxozyma*)	170, 245
	Neurospora crassa	183
	Neovossia indica	166
	Aspergillus oryzae, A. duricaulis	212
Tetraglycylferrichrome	*Neovossia indica*	166
Asperchrome (A, B_1, B_2, B_3, C, D_1, D_2, D_3, E)	*Aspergillus ochraceus*	203
Malonichrome	*Fusarium roseum*	253
Ferricrocin	*Aspergillus* spp., *A. fumigatus*-group	212
	A. viridi-nutans	212
	Neurospora crassa	183
	Microsporum canis	226
Ferrichrysin	*Aspergillus melleus, A. ochraceus*	210, 203
Ferrirubin	*Penicillium variabile, P. rugulosum*	200
	Aspergillus ochraceus	203
	Paecilomyces varioti	210
Ferrirhodin	*Aspergillus nidulans, A. versicolor*	210
	Botrytis cinerea	228
	Verucobotrys	229
Des(diserylglycyl)-ferrirhodin (DDF)	*Aspergillus ochraceus*	196
Fusarinine (A, B)	*Fusarium roseum*	218
Fusarinine C (= Fusigen)	*Fusarium cubense, F. roseum*	219
	Aspergillus fumigatus	219
	Giberella fujikuroi	219
	Penicillium chrysogenum	219
	Paecilomyces varioti	219
Triacetylfusarinine C (= Triacetylfusigen)	*Aspergillus fumigatus*-group	212, 213
	A. deflectus	214
	Mycelia sterilia	215, 217
Dimerum acid	*Fusarium dimerum*	220
	Stemphylium botyrosum	223
	Verticillium dahliae	224
	Gliocladium virens	221
Coprogen	*Neurospora crassa*	184
	Penicillium camemberti, P. chrysogenum	210
	P. citrinum, P. notatum, P patulum	210
	P. urticae	210
	Curvularia lunata	207
	Epicoccum purpurascens	205
Coprogen B	*Fusarium* sp. *F. dimerum*	220
	Myrothecium striatisporium, M. verrucaria	220
	Gliocladium virens	221
	Stemphylium botyrosum	223
	Verticillium dahliae	225
	Histoplasma capsulatum	243
Triornicin	*Epicoccum purpurascens*	205
Isotriornicin	*Epicoccum purpurascens*	206
Neocoprogen I (= Isotriornicin)	*Curvularia lunata*	207
Neocoprogen II	*Curvularia lunata*	207
Dimethylcoprogen	*Alternaria alternata*	208
Hydroxycoprogen	*Alternaria longipes*	209
Rhodotorulic acid	*Rhodotorula pilimanae, R. glutinis, R. rubra*	170
	Leucosporidium spp.	170
	Rhodosporidium spp.	170
	Sporidiobolus spp.	170

TABLE 5 (continued)
Fungal Siderophores and Producing Organisms

Siderophore	Producing organism	Ref.
	Sporobolomyces spp.	170
	Microbotryum violaceum	167
	M. major	167
	Sphacelotheca spp.	254

REFERENCES

1. **Hesseltine, C. W., Pidacks, C., Whithill, A. R., Bohonos, N., Hutchings, B. L., and Williams, J. H.,** Coprogen, a new growth factor for coprophilic fungi, *J. Am. Chem. Soc.,* 74, 1362, 1952.
2. **Neilands, J. B.,** A cristalline organo iron compound from a rust fungus *Ustilago sphaerogena, J. Am. Chem. Soc.,* 74, 4846, 1952.
3. **Lochhead, A. G., Burton, M. O., and Thexton, R. H.,** A bacterial growth factor synthesized by a soil bacterium, *Nature,* 170, 282, 1952.
4. **Linke, W. D., Crueger, A., and Diekmann, H.,** *Arch. Microbiol.,* 85, 44, 1972.
5. **Neilands, J. B.,** Some aspects of microbial iron metabolism, *Bacteriol. Rev.,* 21, 101, 1957.
6. **Demain, A. L. and Hendlin, D.,** Iron transport compounds as growth stimulators for *Microbacterium* sp., *J. Gen. Microbiol.,* 21, 72, 1959.
7. **Hendlin, D. and Demain, A. L.,** An absolute requirement for iron transport factors by *Microbacterium lacticum* 8181, *Nature,* 184, 1894, 1959.
8. **Alexanian, S., Diekmann, H., and Zähner, H.,** Stoffwechselprodukte von Mikroorganismen. 94. Mitteilung. Vergleich der Wirkung von Sideraminen als Wuchsstoffe und als Antagonisten der Sideramine, *Arch. Microbiol.,* 82, 55, 1972.
9. **Burnham, B. F. and Neilands, J. B.,** Studies on the metabolic function of the ferrichrome compounds, *J. Biol. Chem.,* 236, 554, 1961.
10. **Powell, P., Szaniszlo, P. J., and Reid, C. C. P.,** Confirmation of occurrence of hydroxamate siderophores in soil by a novel *Escherichia coli* bioassay, *Appl. Environ. Microbiol.,* 46, 1080, 1983.
11. **Crowley, D. E., Reid, C. C. P., and Szaniszlo, P. J.,** Microbial siderophores as iron sources for plants, in *Iron Transport in Microbes, Plants and Animals,* Winkelmann, G., van der Helm, D., and Neilands, J. B., Eds., VCH, Weinheim, 1987, 375.
12. **Lockhead, A. G.,** Soil bacteria and growth promoting substances, *Bacteriol. Rev.,* 22, 145, 1958.
13. **Lochhead, A. G. and Burton, M. O.,** Incidence in soil of bacteria requiring vitamin B_{12} and the terregens factor, *Soil Sci.,* 82, 237, 1956.
14. **Szaniszlo, P. J., Powell, P. E., Reid, C. P. P. Reid, and Cline, G. R.,** Production of hydroxamate siderophore iron chelators by ectomycorrhizal fungi, *Mycologia,* 72, 1158, 1981.
15. **Emery, T., Emery, L., and Olsen, R. K.,** Retrohydroxamate ferrichrome, a biomimetic analogue of ferrichrome, *Biochem. Biophys. Res. Commun.,* 119, 1191, 1984.
16. **Shanzer, A., Libmann, J., Lazar, R., Tor, Y., and Emery, T.,** Synthetic ferrichrome analogues with growth promotion activity for *Arthrobacter flavescens, Biochem. Biophys. Res. Commun.,* 157, 389, 1988.
17. **Shanzer, A., Libmann, J., Lazar, R., and Tor, Y.,** Receptor mapping with artificial siderophores, *Pure Appl. Chem.,* 61, 1529, 1989.
18. **Bickel, H., Gäumann, E., Keller-Schierlein, W., Prelog, V., Vischer, E., Wettstein, A., and Zähner, H.,** Über eisenhaltige Wachstumsfaktoren, die Sideramine, und ihre Antagonisten, die eisenhaltigen Antibiotika Sideromycine, *Experientia,* 16, 129, 1960.
19. **Bickel, H., Bosshardt, R., Gäumann, E., Reusser, P., Vischer, E., Voser, W., Wettstein, A., and Zähner, H.,** Stoffwechselprodukte von Actinomyceten, 26. Mitt. Über die Isolierung und Charakterisierung der Ferrioxamine A-F, neuer Wuchstoffe der Sideramin-Gruppe, *Helv. Chim. Acta,* 43, 2118, 1960.
20. **Keller-Schierlein, W., Mertens, P., Prelog, V., and Walser, A.,** Die Ferrioxamine A_1, A_2 und D_2, *Helv. Chim. Acta,* 48, 710, 1965.
21. **Bickel, H., Hall, G. E., Keller-Schierlein, W., Prelog, V., Vischer, E., and Wettstein, A.,** Über die Konstitution von Ferrioxamin B, *Helv. Chim. Acta,* 43, 2129, 1960.
22. **Bickel, H., Keberle, H., and Vischer, E.,** Stoffwechselprodukte von Mikroorganismen, 43. Mitt. Zur Kenntnis von Ferrioxamin B, *Helv. Chim. Acta,* 46, 1385, 1963.

23. **Keller-Schierlein, W. and Prelog, V.**, Über das Ferrioxamin E. Ein Beitrag zur Konstitution des Nocardamins, *Helv. Chim. Acta,* 44, 1981, 1961.

24. **Keller-Schierlein, W. and Prelog, V.**, Ferrioxamin G, *Helv. Chim. Acta,* 45, 590, 1962.

25. **Adapa, S., Huber, P., and Keller-Schierlein, W.**, 179. Stoffwechselprodukte von Mikroorganismen, 216. Mitteilung. Isolierung, Strukturaufklärung und Synthese von Ferrioxamin H, *Helv. Chim. Acta,* 65, 1818, 1982.

26. **Bickel, H., Gäumann, E., Nussberger, G., Reusser, P., Vischer, E., Voser, W., Wettstein, A., and Zähner, H.**, Über die Isolierung und Charakterisierung der ferrimycine A_1 and A_2, neuer Antibiotika der Sideromycin-Gruppe, *Helv. Chim. Acta,* 43, 2105, 1960.

27. **Keller-Schierlein, W., Huber, P., and Kawaguchi, H.**, Chemistry of Danomycin, an iron containing antibiotic, in *Natural Products and Drug Development,* Alfred Benzo Symposium 20. Krogsgaard-Larsen, P., Christensen, S. B., and Kofod, H., Eds., Munksgaard, Copenhagen, 1984, 213.

28. **Zähner, H., Hütter, R., and Bachmann, E.**, Stoffwechselprodukte von Actinomyceten. 23. Mitteilung. Zur Kenntnis der Sideromycinwirkung, *Arch. Microbiol.,* 36, 325, 1960.

29. **Müller, A. and Zähner, H.**, Stoffwechselprodukte von Mikroorganismen. Ferrioxamine aus Eubacteriales, *Arch. Microbiol.,* 62, 257, 1968.

30. **Meyer, J. M. and Abdallah, M. A.**, The siderochromes of nonfluorescent *Pseudomonas:* production of nocardamine by *Pseudomonas stutzeri, J. Gen. Microbiol.,* 118, 125, 1980.

31. **Yang, C. and Leong, J.**, Production of desferrioxamines B and E from a ferroverdin-producing *Streptomyces* species, *J. Bacteriol.,* 149, 381, 1982.

32. **Berner, I., Konetschny-Rapp, S., Jung, G., and Winkelmann, G.**, Characterization of ferrioxamine E as the principal siderophore of *Erwinia herbicola (Enterobacter agglomerans), Biol. Met.,* 1, 51, 1988.

33. **Berner, I. and Winkelmann, G.**, Ferrioxamine transport and the identification of the ferrioxamine E receptor protein (FoxA) in *Erwinia herbicola, Biol. Met.,* 2, 197, 1990.

34. **Müller, G. and Raymond, K.**, Specificity and mechanism of ferrioxamine-mediated iron transport in *Streptomyces pilosus, J. Bacteriol.,* 160, 304, 1984.

35. **Müller, G., Matzanke, B. F., and Raymond, K. N.**, Iron transport in *Streptomyces pilosus* mediated by ferrichrome siderophores, rhodotorulic acid, and enantio-rhodotorulic acid, *J. Bacteriol.,* 160, 313, 1984.

36. **Leong, J. and Raymond, K. N.**, Coordination isomers of biological iron transport compounds. IV. Geometrical isomers of chromic desferriferrioxamine B, *J. Am. Chem. Soc.,* 97, 293, 1975.

37. **Ratledge, C. and Hall, M. J.**, Influence of metal ions on the formation of mycobactin and salicylic acid in *Mycobacterium smegmatis* grown in static culture, *J. Bacteriol.,* 108, 312, 1971.

38. **Snow, G. A.**, Mycobactins: iron chelating growth factors from mycobacteria, *Bacteriol. Rev.,* 34, 99, 1970.

39. **Ratledge, C.**, Nutrition, growth and metabolism, in *The Biology of the Mycobacteria,* Vol. 1, Raledge, C. and Stanford, J. L., Eds., Academic Press, London, 1982, 185.

40. **Hall, R. M. and Ratledge, C.**, A simple method for the production of mycobactin, the lipid-soluble siderophore, from mycobacteria, *FEMS Microbiol. Lett.,* 15, 133, 1982.

41. **Hall, R. M.**, Mycobactins, how to obtain them and how to employ them as chemotaxonomic characters for the mycobacteria and related organisms, *Actinomycetes,* 19, 92, 1986.

42. **Macham, L. P., Ratledge, C., and Nocton, J. C.**, Extracellular iron acquisition by Mycobacteria: role of the exochelins and evidence against the participation of mycobactin, *Infect. Immun.,* 12, 1242, 1975.

43. **Macham, L. P., Stephenson, M. C., and Ratledge, C.**, Iron transport in *Mycobacterium smegmatis* and the isolation, purification and function of exochelin MS, *J. Gen. Microbiol.,* 101, 41, 1977.

44. **Stephenson, M. C. and Ratledge, C.**, Iron transport in *Mycobacterium smegmatis:* uptake of iron from ferriexochelin, *J. Gen. Microbiol.,* 101, 193, 1979.

45. **Messenger, A. J. M. and Ratledge, C.**, Iron transport in *Mycbacterium smegmatis:* uptake of iron from ferric citrate, *J. Bacteriol.,* 149, 131, 1982.

46. **Hall, R. M., Sritharan, M., Messenger, A. J. M., and Ratledge, C.**, Iron transport in *Mycobacterium smegmatis:* occurrence of iron-regulated envelope proteins as potential receptors for iron uptake, *J. Gen. Microbiol.,* 133, 2107, 1987.

47. **Ratledge, C. and Marshall, B. J.**, Iron transport in *Mycobacterium smegmatis:* the role of mycobactin, *Biochim. Biophys. Acta,* 279, 58, 1972.

48. **Ratledge, C. Patel, P. V., and Mundy, J.**, Iron transport in *Mycobacterium smegmatis:* location of mycobactin by electron microscopy, *J. Gen. Microbiol.,* 128, 1559, 1982.

49. **Ratledge, C., Macham, L. P., Brown, K. A., and Marshall, B. J.**, Iron transport in *Mycobacterium smegmatis:* a restricted role for salicylic acid in the extracellular environment, *Biochim. Biophys. Acta,* 372, 39, 1974.

50. **Messenger, A. J. M., Hall, R. M., and Ratledge, C.**, Iron uptake process in *Mycobacterium raccae* R877R, a mycobacterium lacking mycobactin, *J. Gen. Microbiol.,* 132, 845, 1986.

51. **Byers, B. R., Powell, M. V., and Lankford, C. E.**, Iron-chelating hydroxamic acid (schizokinen) active in initiation of cell division in *Bacillus megaterium, J. Bacteriol.,* 93, 286, 1967.

52. **Mullis, K. B., Pollack, J. R., and Neilands, J. B.,** Structure of schizokinen, an iron-transport compound from *Bacillus megaterium, Biochemistry,* 10, 4894, 1971.

53. **Ito, T. and Neilands, J. B.,** Products of "low-iron fermentation" with *Bacillus subtilis;* isolation, characterization and synthesis of 2,3-dihydroxybenzoylglycine, *J. Am. Chem. Soc.,* 80, 4645, 1958.

54. **Davis, W. B. and Byers, B. R.,** Active transport of iron in *Bacillus megaterium:* role of secondary hydroxamic acids, *J. Bacteriol.,* 107, 491, 1971.

55. **Haydon, A. H., Davis, W. B., Arceneaux, J. E. L., and Byers, B. R.,** Hydroxamate recognition during iron transport from hydroxamate-iron chelates, *J. Bacteriol.,* 115, 912, 1973.

56. **Arceneaux, J. E. L., Davies, W. B., Downer, D. N., Haydon, A. H., and Byers, B. R.,** Fate of labeled hydroxamates during iron transport from hydroxamate-iron chelates, *J. Bacteriol.,* 115, 919, 1973.

57. **Ashwell, J. E. Haydon, A. H., Truner, H. R., Dawkins, C. A., Arceneaux, J. E. L., and Byers, B. R.,** Specificity of siderophore receptors in membrane vesicles of *Bacillus megaterium, J. Bacteriol.,* 130, 173, 1977.

58. **Arceneaux, J. E. L. and Byers, B. R.,** Ferrisiderophore reductase activity in *Bacillus megaterium, J. Bacteriol.,* 141, 715, 1980.

59. **Peters, W. J. and Warren, R. A. J.,** The mechanism of iron uptake in *Bacillus subtilis, Can. J. Microbiol.,* 16, 1285, 1970.

60. **Armstrong, J. E. and Van Baalen, C.,.** Iron transport in microalgae: the isolation and biological activity of a hydroxamate siderophore from blue-green alga *Agmenellum quadruplicatum, J. Gen. Microbiol.,* 111, 253, 1979.

61. **Simpson, F. B. and Neilands, J. B.,** Siderochromes in cyanophyceae: isolation and characterization of schizokinen from *Anabaena* sp., *J. Phycol.,* 12, 44, 1976.

62. **Lammers, P. J. and Sanders-Loehr, J.,** Active transport of ferric schizokinen in *Anabaena* sp., *J. Bacteriol.,* 151, 288, 1982.

63. **Meiwes, J., Fiedler, H.-P., Haag, H., Zähner, H., Konetschny-Rapp, S., and Jung, G.,** Isolation and characterization of staphyloferrin A, a compound with siderophore activity from *Staphylococcus hyicus* DSM 20459, *FEMS Microbiol. Lett.,* 67, 201, 1990.

64. **Konetschny-Rapp, S., Jung, G., Meiwes, J., and Zähner, H.,** Staphyloferrin A: a structurally new siderophore form staphylococci. *Eur. J. Biochem.,* 191, 65, 1990.

65. **Archibald, F.,** *Lactobacillus plantarum,* an organism not requiring iron, *FEMS Microbiol. Lett.,* 19, 29, 1983.

66. **Strachan, R. C., Aranha, H., Lodge, J. S., Arceneaux, J. E. L., and Byers, B. R.,** Teflon chemostat for studies of trace metal metabolism in *Streptococcus mutans* and other bacteria, *Appl. Environm. Microbiol.,* 43, 257, 1982.

67. **O'Brien, I. G. and Gibson, F.,** The structure of enterochelin and related 2,3-dihydroxy-N-benzoylserine conjugates from *Escherichia coli, Biochim. Biophys. Acta,* 215, 393, 1970.

68. **Pollack, J. R. and Neilands, J. B.,** Enterobactin, an iron transport compound from *Salmonella typhimurium, Biochem. Biophys. Res. Commun.,* 38, 989, 1970.

69. **Perry, R. D. and San Clementes, C. L.,** Siderophore synthesis in *Klebsiella pneumoniae* and *Shigella sonnei* during iron deficiency, *J. Bacteriol.,* 140, 1129, 1979.

70. **Reissbrodt, R., Rabsch, W. J., Chapeaurouge, A., Jung, G., and Winkelmann, G.,** Isolation and identification of ferrioxamines G and E as siderophores in *Hafnia alvei, Biol. Met.,* 3, 54, 1990.

71. **Reissbrodt, R. and Rabsch, W.,** Further differentiation of *Enterobacteriacea* by means of siderophore-pattern analysis, *Zbl. Bakt. Hyg. A,* 268, 306, 1988.

72. **Heesemann, J.,** Chromosomal-encoded siderophores are required for mouse virulence of enteropathogenic *Yersinia* species, *FEMS Microbiol. Lett.,* 48, 229, 1987.

73. **Martinez, J. L., Cercendo, E., Baquero, F., Perez-Dias, J. C., and Delgado-Ibarren, A.,** Incidence of aerobactin production in gram-negative hospital isolates, *FEMS Microbiol. Lett.,* 43, 351, 1987.

74. **Rabsch, W.,** personal communication, 1990.

75. **Pollack, J. R., Ames, B. N., and Neilands, J. B.,** Iron transport in *Salmonella typhimurium* mutants blocked in the biosynthesis of enterobactin, *J. Bacteriol.,* 104, 635, 1970.

76. **Rabsch, W., Tkacik, J., Lindemann, W., Mikula, J., and Reissbrodt, R.,** Different systems for iron supply of *Salmonella typhimurium* and *Escherichia coli* wild strains as a tool for a typing scheme, *Zentralbl. Bacteriol. Int. J. Med. Microbiol.,* in press.

77. **Luckey, M., Pollack, J. R., Wayne, R., Ames, B. N., and Neilands, J. B.,** Iron uptake in *Salmonella typhimurium:* utilization of exogenous siderochromes as iron carriers, *J. Bacteriol.,* 111, 731, 1972.

78. **Rabsch, W. and Reissbrodt, R.,** Investigations of *Salmonella* strains from different clinical-epidemiological origin with phenolate and hydroxamate (aerobactin)-siderophore bioassays, *J. Hyg. Epidemiol. Microbiol. Immunol. (Praha),* 32, 353, 1988.

79. **McDougal, S. and Neilands, J. B.,** Plasmid- and chromosome-coded aerobactin synthesis in enteric bacteria: insertion sequences flank operon in plasmid-mediated systems, *J. Bacteriol.,* 159, 300, 1984.

80. **Langmann, L., Young, I. G., Frost, G. E., Rosenberg, H., and Gibson, F.,** Enterochelin system of iron transport in *Escherichia coli:* mutations affecting ferric-enterochelin esterase, *J. Bacteriol.,* 112, 1142, 1972.

81. **Hantke, K.,** 2,3-Dihydroxybenzoylserine—a siderophore in *Escherichia coli, FEBS Lett.,* 67, 5, 1990.

82. **Nikaido, H. and Rosenberg, E. Y.,** Cir and Fiu proteins in the outer membrane of *Escherichia coli* catalyze transport of monomeric catechols: study with β-lactam antibiotics containing catechol and analogous groups, *J. Bacteriol.,* 172, 1361, 1990.

83. **Curtis, N. A. C., Eisenstadt, R. L., East, S. J., Cornford, R. J., Walker, L. A., and White, J.,** Iron-regulated outer membrane proteins of *Escherichia coli* K-12 and mechanism of action of catechol-substituted cephalosporins, *Antimicrob. Agents Chemother.,* 32, 1897, 1988.

84. **Heidinger, S., Braun, V., Pecoraro, V. L., and Raymond, K. N.,** Iron supply to *Escherichia coli* by synthetic analogs of enterochelin, *J. Bacteriol.,* 153, 109, 1983.

85. **Ecker, D. J., Matzanke, B. F., and Raymond, K. N.,** Recognition and transport of ferric enterobactin in *Escherichia coli, J. Bacteriol.,* 167, 666, 1986.

86. **Lee, C.-W., Ecker, D. J., and Raymond, K. N.,** The pH-dependent reduction of ferric enterobactin probed by electrochemical methods and its implications for microbial iron transport, *J. Am. Chem. Soc.,* 49, 6920, 1985.

87. **Cass, M. E., Garret, T. M., and Raymond, K. N.,** The salicylate mode of bonding in protonated ferric enterobactin analogues, *J. Am. Chem. Soc.,* 111, 1677, 1989.

88. **Gutermann, S.,** Colicin B: mode of action and inhibition by enterochelin, *J. Bacteriol.,* 114, 1217, 1973.

89. **Hantke, K. and Braun, V.,** A function common to iron-enterochelin transport and action of colicins B, I, V in *Escherichia coli, FEBS Lett.,* 59, 277, 1975.

90. **Gutermann, S. and Dann, L.,** Excretion of enterochelin by *exbA* and *exbB* mutants of *Escherichia coli, J. Bacteriol.,* 114, 1225, 1973.

91. **Hollifield, C. W. and Neilands, J. B.,** Ferric enterobactin transport system in *Escherichia coli* K-12, Extraction, assay, and specificity of the outer membrane receptor, *Biochemistry,* 17, 1922, 1978.

92. **Jalal, M. A. F. and van der Helm, D.,** Purification and crystallization of ferric enterobactin receptor protein, FepA, from the outer membranes of *E. coli* UT5600/pBB2, *FEBS Lett.,* 243, 366, 1989.

93. **Neilands, J. B., Erickson, T. J., and Rastetter, W. H.,** Stereospecificity of the ferric enterobactin receptor of *Escherichia coli, J. Biol. Chem.,* 256, 3831, 1981.

94. **Venuti, M. C., Rastetter, W. H., and Neilands, J. B.,** 1,3,5-Tris(N,N',N''-2,3-dihydroxyben-zoyl)aminomethylbenzene, a synthetic iron chelator related to enterobactin, *J. Med. Chem.,* 22, 123, 1979.

95. **Ecker, D. J., Loomis, L. D., Cass, M. E., and Raymond, K. N.,** Substituted complexes of enterobactin and synthetic analogues as probes of the ferric-enterobactin receptor in *Escherichia coli, J. Am. Chem. Soc.,* 110, 2457, 1988.

96. **Pugsley, A. and Reeves, P.,** Uptake of ferrienterochelin by *Escherichia coli:* energy-dependent stage of uptake, *J. Bacteriol.,* 130, 26, 1977.

97. **Lodge, J. S. and Emery, T.,** Anaerobic iron uptake by *Escherichia coli, J. Bacteriol.,* 801, 1984.

98. **Hantke, K.,** Ferrous iron transport mutants in *Escherichia coli* K-12, *FEMS Microbiol. Lett.,* 44, 53, 1987.

99. **Leong, J. and Neilands, J. B.,** Mechanisms of siderophore iron transport in enteric bacteria, *J. Bacteriol.,* 126, 823, 1976.

100. **Matzanke, B. F., Müller, G. I., and Raymond, K. N.,** Hydroxamate siderophore mediated iron uptake in *E. coli:* stereospecific recognition of ferric rhodotorulic acid, *Biochem. Biophys. Res. Commun.,* 121, 922, 1984.

101. **Hantke, K.,** Identification of an iron uptake system specific for coprogen and rhodotorulic acid in *Escherichia coli, Mol. Gen. Genet.,* 191, 301, 1983.

102. **Hartmann, A. and Braun, V.,** Iron transport in *Escherichia coli:* uptake and modification of ferrichrome, *J. Bacteriol.,* 143, 246, 1980.

103. **Schneider, R., Hartmann, A., and Braun, V.,** Transport of the iron ionophore ferrichrome in *Escherichia coli* K-12 and *Salmonella typhimurium* L T2, *FEMS Microbiol. Lett.,* 11, 115, 1981.

104. **Negrin, R. and Neilands, J. B.,** Ferrichrome transport in inner membrane vesicles of *Escherichia coli* K12, *J. Biol. Chem.,* 253, 2339, 1978.

105. **Hider, R. C., Drake, A. F., Kuroda, R., and Neilands, J. R.,** Symport transport of ferrichrome-type siderophores, *Naturwissenschaften,* 67, 136, 1990.

106. **Gibson, F. and Magrath, D.,** The isolation and characterization of a hydroxamic acid (aerobactin) formed by *Aerobacter aerogenes* 62-1, *Biochim. Biophys. Acta,* 192, 175, 1969.

107. **Williams, P.,** Novel iron uptake system specified by ColV plasmids: an important component in the virulence of invasive strains of *Escherichia coli, Infect. Immun.,* 29, 411, 1980.

108. **Lawlor, K. M. and Payne, S. M.,** Aerobactin genes in *Shigella* spp., *J. Bacteriol.,* 160, 266, 1984.

109. **Rabsch, W., Paul, P., and Reissbrodt, R.,** A new hydroxamate siderophore for iron supply of *Salmonella, Acta Microbiol. Hung.,* 34, 85, 1987.

110. **Zimmermann, L., Angerer, A., and Braun V.,** Mechanistically novel iron(III) transport system in *Serratia marcescens, J. Bacteriol.,* 171, 238, 1989.

111. **Tait, G. T.,** The identification and biosynthesis of siderochromes formed by *Micrococcus denitrificans, Biochem. J.,* 146, 191, 1975.

112. **Peterson, T. and Neilands, J. B.,** Revised structure of a catecholamide spermidine siderophore from *Paracoccus denitrificans, Tetrahedron Lett.,* 50, 4805, 1979.

113. **Wee, S., Hardesty, S., Madiraju, M. V. V. S., and Wilkinson, B. J.,** Iron-regulated outer membrane proteins and non-siderophore-mediated iron acquisition by *Paracoccus denitrificans, FEMS Microbiol. Lett.,* 51, 33, 1988.

114. **Bergeron, R. J., Weimar, W. R., and Dionis, J. B.,** Demonstration of ferric L-parabactin-binding activity in the outer membrane of *Paracoccus denitrificans, J. Bacteriol.,* 170, 3711, 1988.

115. **Bergeron, R. J., Dionis, J. B., Elliot, G. T., and Kline, S. J.,** Mechanism and stereospecificity of the parabactin-mediated iron transport system in *Paracoccus denitrificans, J. Biol. Chem.,* 260, 7936, 1985.

116. **Bergeron, R. J. and Weimar, W. R.,** Kinetics of iron acquisition from ferric siderophores by *Paracoccus denitrificans, J. Bacteriol.,* 172, 2650, 1990.

117. **Ong, S. A., Peterson, T., and Neilands, J. B.,** Agrobactin, a siderophore from *Agrobacterium tumefaciens, J. Biol. Chem.,* 254, 1860, 1979.

118. **Neilands, J. B., Peterson, T., and Ong, S. A.,** High affinity iron transport in microorganisms. Iron(III) coordination compounds of the siderophores agrobactin and parabactin, in *Inorganic Chemistry in Biology and Medicine,* Martell, A., Ed., American Chemical Society, Washington, D.C., 1980, 263.

119. **Smith, M. J., Shoolery, J. N., Schwyn, B., Holden, I., and Neilands, J. B.,** Rhizobactin, a structurally novel siderophore from *Rhizobium meliloti, J. Am. Chem. Soc.,* 107, 1739, 1985.

120. **Modi, M., Shah, K. S., and Modi, V. V.,** Isolation and characterization of catechol-like siderophores from cowpea *Rhizobium* RA-1, *Arch. Microbiol.,* 141, 156, 1985.

121. **Rioux, C. R., Jordan, D. C., and Rattray, J. B. M.,** Iron requirement of *Rhizobium leguminosarum* and secretion of anthranilic acid during growth on iron-deficient medium, *Arch. Biochem. Biophys.,* 248, 175, 1986.

122. **Nambiar, P. T. C. and Sivaramakrishnan, S.,** Detection and assay of siderophores in cowepea rhizobia (*Bradyrhizobium*) using radioactive Fe (^{59}Fe), *Lett. Appl. Microbiol.,* 4, 37, 1987.

123. **Patel, N. H., Chakraborty, R. N., and Desai, S. B.,** Isolation and partial characterization of phenolate siderophore from *Rhizobium leguminosarum* IARI 102, *FEMS Microbiol. Lett.,* 56, 131, 1987.

124. **Skorupska, A., Derylo, M., and Lorkiewicz, Z.,** Siderophore containing 2,3-dihydroxybenzoic acid and threonine formed by *Rhizobium trifolii, Acta Biochim. Pol.,* 35, 119, 1988.

125. **Bosch, I., Meidl, E. J., Hoult, M., Plessner, O., and Guerinot, M. L,** Iron uptake and metabolism in the Bradyrhizobium/soybean symbiosis, in *Nitrogen Fixation: Hundred Years After,* Bothe, H., de Bruijn, F. J., and Newton, W. E., Eds., Gustav Fischer, New York, 1988, 652.

126. **Carillo-Castaneda, G. and Peralta, J. R. V.,** Siderophore-like activities in *Rhizobium phaseoli, J. Plant. Nutr.,* 11, 935, 1988.

127. **Meyer, J.-M. and Abdallah, M. A.,** The fluorescent pigment of *Pseudomonas fluorescens:* biosynthesis, purification and physicochemical properties, *J. Gen. Microbiol.,* 107, 319, 1978.

128. **Wendenbaum, S., Demange, P., Dell, A., Meyer, J.-M., and Abdallah, M. A.,** The structure of pyoverdine Pa, the siderophore of *Pseudomonas aeruginosa, Tetrahedron Lett.,* 24, 4877, 1983.

129. **Yang, C.-C. and Leong, J.,** Structure of pseudobactin 7SR1, a siderophore from plant-deleterious *Pseudomonas, Biochemistry,* 23, 3534, 1984.

130. **Briscot, G., Taraz, K., and Budzikiewiz, H.,** Pyoverdin-type siderophores from *Pseudomonas aeruginosa, Liebigs Ann. Chem.,* 1989, 367, 1989.

131. **Demange, P., Wendenbaum, S., Linget, C., Mertz, C., Cung, M. T., Dell, A., and Abdallah, M. A.,** Bacterial siderophores: structure and NMR assignment of pyoverdins Pa, siderophores of *Pseudomonas aeruginosa* ATCC 15692. *Biol. Met.,* 3, 155, 1990.

132. **Teintze, M., Hossain, M. B., Barnes, C. L., Leong, J., and van der Helm, D.,** Structure of ferric pseudobactin, a siderophore from a plant growth promoting *Pseudomonas, Biochemistry,* 20, 6446, 1981.

133. **Hohnadel, D. and Meyer, J.-M.,** Specificity of Pyoverdine-mediated iron uptake among fluorescent *Pseudomonas* strains, *J. Bacteriol.,* 170, 4865, 1988.

134. **DeWeger, L. A., van Arendonk, J. J. C. M., Recourt, K., van der Hofstad, G. A. J. M., Weisbeek, P. J., and Lugtenberg, B.,** Siderophore-mediated uptake of Fe^{3+} by the plant growth-stimulating *Pseudomonas putida* strain WCS358 and by the other rhizosphere microorganisms, *J. Bacteriol.,* 170, 4693, 1988.

135. **Page, W. J.,** Iron-dependent production of hydroxamate by sodium-dependent *Azotobacter chroococcum, Appl. Environ. Microbiol.,* 53, 1418, 1987.

136. **Knosp, O., von Tigerstrom, M., and Page, W.,** Siderophore-mediated uptake of iron in *Azotobacter vinelandii, J. Bacteriol.,* 159, 341, 1984.

137. **Page, W. J. and Huyer, M.**, Derepression of *Aztobacter vinelandii* siderophore system, using iron-containing minerals to limit iron repletion, *J. Bacteriol.*, 158, 496, 1984.

138. **Saxena, B., Modi, M., and Modi, V. V.**, Isolation and characterization of siderophores from *Azospirillum lipoferum* D-2, *J. Gen. Microbiol.*, 132, 2219, 1986.

139. **Persmark, M., Expert, D., and Neilands, J. B.**, Isolation, characterization, and synthesis of chrysobactin, a compound with siderophore activity from *Erwinia chrysanthemi*, *J. Biol. Chem.*, 264, 3187, 1989.

140. **Bargouthi, S., Young, R., Arceneaux, J. E. L., and Byers, B. R.**, Physiological control of amonabactin biosynthesis in *Aeromonas hydrophila*, *Biol. Met.*, 2, 155, 1989.

141. **Bachhawat, A. and Gosh, S.**, Iron transport in *Azospirillum brasiliense:* role of the siderophore spirillobactin, *J. Gen. Microbiol.*, 133, 1759, 1987.

142. **Reeves, M. W., Pine, L., Neilands, J. B., and Balows, A.**, Absence of siderophore activity in *Legionella* species grown in iron-deficient media, *J. Bacteriol.*, 154, 324, 1983.

143. **Griffith, L. G., Sigel, S. P., Payne, S. M., and Neilands, J. B.**, Vibriobactin, a siderophore from *Vibrio cholerae*, *J. Biol. Chem.*, 259, 383, 1984.

144. **Jalal, M. A. F., Hossain, M. B., van der Helm, D., Sanders-Loehr, J., Actis, L. A., and Crosa, J. H.**, Structure of anguibactin, a unique plasmid-related bacterial siderophore from the fish pathogen *Vibrio anguillarum*, *J. Am. Chem. Soc.*, 111, 292, 1989.

145. **Crosa, J. H.**, Genetics and molecular biology of siderophore-mediated iron transport in bacteria, *Microbiol. Rev.*, 53, 517, 1989.

146. **Paoletti, L. C. and Blakemore, R. P.**, Hydroxamate production by *Aquaspirillum magnetotacticum*, *J. Bacteriol.*, 167, 73, 1986.

147. **Morton, D. J. and Williams, P.**, Characterization of the outer-membrane proteins of *Haemophilus parainfluenzae* expressed under iron-sufficient and iron-restricted conditions, *J. Gen. Microbiol.*, 135, 445, 1989.

148. **Deneer, H. G. and Potter, A. A.**, Iron-repressible outer membrane proteins of *Pasteurella haemolytica*, *J. Gen. Microbiol.*, 135, 435, 1989.

149. **Redhead, K., Hill, T., and Chart, H.**, Interaction of lactoferrin and transferrins with the outer membrane of *Bordetalla pertussis*, *J. Gen. Microbiol.*, 133, 891, 1987.

150. **Williams, P, Morton, D. J., Towner, K. J., Stevenson, P., and Griffith, E.**, Utilization of enterobactin and other exogenous iron sources by *Haemophilus influenzae*, *H. parainfluenzae* and *H. paraphrophilus.*, *J. Gen. Microbiol.*, 136, 2343, 1990.

151. **Francis, R. T., Jr., Booth, J. W., and Becker, R. R.**, Uptake of iron from hemoglobin and hemoglobin-haptoglobin complex by hemolytic bacteria, *J. Biochem.*, 14, 767, 1985.

152. **Zakaria-Meehan, Massad, G., Simpson, L. M., Travis, J. C., and Oliver, J. D.**, Ability of *Vibrio vulnificus* to obtain iron from hemoglobin-haptoglobin complexes, *Infect. Immun.*, 56, 275, 1988.

153. **Helms, S. D., Oliver, J. D., and Travis, J. C.**, Role of heme compounds and haptoglobin in *Vibrio vulnificus* pathogenicity, *Infect. Immun.*, 45, 345, 1984.

154. **Dyer, D. W., West, E. P., and Sparling, P. F.**, Effects of serum carrier proteins on the growth of pathogenic Neisseriae with heme-bound iron, *Infect. Immun.*, 55, 2171, 1987.

155. **Simonson, C., Brener, D., and Devoe, I. W.**, Expression of a high-affinity mechanism for acquisition of transferrin iron by *Neisseria meningitidis*, *Infect. Immun.*, 36, 107, 1982.

156. **West, S. E. H. and Sparling, P. F.**, Response of *Neisseria gonorrhoeae* to iron limitation: alterations in expression of membrane proteins without apparent siderophore production, *Infect. Immun.*, 47, 388, 1985.

157. **Archibald, F. S. and DeVoe, I. W.**, Removal of iron from human transferrin by *Neisseria meningitidis*, *FEMS Microbiol. Lett.*, 6, 159, 1979.

158. **Archibald, F. S. and De Voe, I. W.**, Iron acquisition by *Neisseria meningitidis in vitro*, *Infect. Immun.*, 27, 322, 1980.

159. **Mickelsen, P. A. and Sparling, P. F.**, Ability of *Neisseria gonorrhoeae, Neisseria meningitidis* and commensal *Neisseriae* species to obtain iron from transferrin and iron compounds, *Infect. Immun.*, 33, 555, 1981.

160. **Mickelsen, P. A., Blackman, E., and Sparling, P. F.**, Ability of *Neisseria gonorrhoeae, Neisseria meningitidis* and commensal *Neisseria* species to obtain iron from lactoferrin, *Infect. Immun.*, 35, 915, 1982.

161. **Emery, T.**, Role of ferrichrome as a ferric ionophore in *Ustilago sphaerogena*, *Biochemistry*, 10, 1483, 1971.

162. **Ecker, D. J. and Emery, T.**, Iron uptake from ferrichrome A and iron citrate in *Ustilago sphaerogena*, *J. Bacteriol.*, 155, 616, 1983.

163. **Ecker, D. J., Lancaster, J. R., and Emery, T.**, Siderophore iron transport followed by electron paramagnetic resonance spectroscopy, *J. Biol. Chem.*, 257, 8623, 1982.

164. **Ecker, D. J., Passavant, C. W., and Emery, T.**, Role of two siderophores in *Ustilago sphaerogena*. Regulation of biosynthesis and uptake mechanisms, *Biochim. Biophys. Acta*, 720, 242, 1982.

165. **Deml, G. and Oberwinkler, F.**, Studies in Heterobasidiomycetes, Part 22. A survey on siderophore formation in low-iron cultured anther smuts of *Caryophyllaceae*, *Zbl. Bakt. Hyg. I. Abt. Orig. C.*, 3, 475, 1982.

166. **Deml, G., Voges, K., Jung, G., and Winkelmann, G.**, Tetraglycylferrichrome — The first heptapeptide ferrichrome, *FEBS Lett.*, 173, 53, 1984.

167. **Deml, G. and Oberwinkler, F.**, Studies in Heterobasidiomycetes. Part 24. On *Ustilago violacea* (Pers.) Rouss. from *Saponaria officinalis* L., *Phytopathol. Z.*, 104, 345, 1982.

168. **Wang, J., Budde, A. D., and Leong, S. A.**, Analysis of ferrichrome biosynthesis in the phytopathogenic fungus *Ustilago maydis*: cloning of an ornithine-N^5-oxygenase gene, *J. Bacteriol.*, 171, 2811, 1989.

169. **Atkin, C. L. and Neilands, J. B.**, Rhodotorulic acid, a diketopiperazine dihydroxamic acid with growth factor activity. I. Isolation and characterization, *Biochemistry*, 7, 3734, 1968.

170. **Atkin, C. L. Neilands, J. B., and Phaff, H. J.**, Rhodotorulic acid from species of *Leucosporidium*, *Rhodosporidium*, *Rhodotorula*, *Sporidiobolus* and *Sporobolomyces*, and a new alanine-containing ferrichrome from *Cryptococcus melibiosum*, *J. Bacteriol.*, 103, 722, 1970.

171. **Carrano, C. J. and Raymond, K. N.**, Coordination chemistry of microbial iron transport compounds. 10. Characterization of the complexes of rhodotorulic acid, a dihydroxamate siderophore, *J. Am. Chem. Soc.*, 100, 5371, 1978.

172. **Carrano, C. J., Cooper, S. R., and Raymond, K. N.**, Coordination chemistry of microbial iron transport compounds. 11. Solution equilibria and electrochemistry of ferric rhodotorulate complexes, *J. Am. Chem. Soc.*, 101, 599, 1979.

173. **Hossain, M. B., Jalal, M. A. F., Benson, B. A., Barnes, C. L., and van der Helm, D.**, Structure and conformation of two coprogen-type siderophores: neocoprogen I and neocoprogen II, *J. Am. Chem. Soc.*, 109, 4948, 1987.

174. **Carrano, C. J. and Raymond, K. N.**, Coordination chemistry of microbial iron transport compounds: rhodotorulic acid and iron uptake in *Rhodotorula pilimanae*, *J. Bacteriol.*, 136, 69, 1978.

175. **Müller, G., Isowa, Y., and Raymond, K. N.**, Stereospecificity of siderophore-mediated iron uptake in *Rhodotorula pilimanae* as probed by enantiorhodotorulic acid isomers of chromic rhodotorulate, *J. Biol. Chem.*, 260, 13921, 1985.

176. **Müller, G., Barclay, S. J., and Raymond, K. N.**, The mechanism and specificity of iron transport in *Rhodotorula pilimanae* probed by synthetic analogs of rhodotorulic acid, *J. Biol. Chem.*, 260, 13916, 1985.

177. **Winkelmann, G. and Braun, V.**, Stereoselective recognition of ferrichrome by fungi and bacteria, *FEMS Microbiol. Lett.*, 11, 237, 1981.

178. **Arx, J. A.**, On *Monilia sitophila* and some families of Ascomycetes, *Sydowia*, 34, 13, 1981.

179. **Perkins, D. D., Radford, A., Newmeyer, D., and Björkman, M.**, Chromosomal loci of *Neurospora crassa*, *Microbiol. Rev.*, 46, 426, 1982.

180. **Davies, R. H.**, Utilization of exogenous and endogenous ornithine by *Neurospora crassa*, *J. Bacteriol.*, 96, 389, 1968.

181. **Winkelmann, G. and Zähner, H.**, Stoffwechselprodukte von Mikroorganismen. 115. Mitteilung. Eisenaufnahme bei *Neurospora crassa*. I. Zur Spezifität des Eisentransportes, *Arch. Microbiol.*, 88, 49, 1973.

182. **Winkelmann, G.**, Surface iron polymers and hydroxy acids. A model of iron supply in sideramine-free fungi, *Arch. Microbiol.*, 121, 43, 1979.

183. **Horowitz, N. H., Charlang, G., Horn, G., and Williams, N. P.**, Isolation and identification of the conidial growth factor of *Neurospora crassa*, *J. Bacteriol.*, 127, 135, 1976.

184. **Keller-Schierlein, W. and Diekmann, H.**, Stoffwechselprodukte von Mikroorganismen. 85. Mitteilung. Zur Konstitution des Coprogens, *Helv. Chim. Acta*, 53, 2035, 1970.

185. **Charlang, G., Bradford, N., Horowitz, N. H., and Horowitz, R. M.**, Cellular and extracellular siderophores of *Aspergillus nidulans* and *Penicillium chrysogenum*, *Mol. Cell Biol.*, 1, 94, 1981.

186. **Matzanke, B. F., Bill, E., Müller, G. I., Trautwein, A. X., and Winkelmann, G.**, Metabolic utilization of ^{57}Fe-labeled coprogen in *Neurospora crassa*. An in vivo Mössbauer study, *Eur. J. Biochem.*, 162, 643, 1987.

187. **Matzanke, B. F., Bill, E., Trautwein, A. X., and Winkelmann, G.**, Ferricrocin functions as the main intracellular iron-storage compound in mycelia of *Neurospora crassa*, *Biol. Met.*, 1, 18, 1988.

188. **Matzanke, B. F., Bill, E., Trautwein, A. X., and Winkelmann, G.**, Role of siderophores in iron storage in spores of *Neurospora crassa* and *Aspergillus ochraceus*, *J. Bacteriol.*, 169, 5873, 1987.

189. **Huschka, H. and Winkelmann, G.**, Iron limitation and its effect on membrane proteins and siderophore transport in *Neurospora crassa*, *Biol. Met.*, 2, 108, 1989.

190. **Winkelmann, G.**, Metabolic products of microorganisms. 132. Uptake of iron by *Neurospora crassa*. III. Iron transport studies with ferrichrome-type compounds, *Arch. Microbiol.*, 98, 39, 1974.

191. **Huschka, H., Naegeli, H. U., Leuenberger-Ryf, H., Keller-Schierlein, W., and Winkelmann, G.**, Evidence for a common siderophore transport system but different siderophore receptors in *Neurospora crassa*, *J. Bacteriol.*, 162, 715, 1985.

192. **Winkelmann, G.**, Evidence for stereospecific uptake of iron chelates in fungi, *FEBS Lett.*, 97, 43, 1979.

193. **Huschka, H., Jalal, M. A. F., van der Helm, D., and Winkelmann, G.,** Molecular recognition of siderophores in fungi: role of iron-surrounding N-acyl residues and the peptide backbone during membrane transport in *Neurospora crassa, J. Bacteriol.,* 167, 1020, 1986.

194. **Winkelmann, G., Berner, I., and Huschka, H.,** Structure-activity relationship of siderophores in fungi, *J. Plant. Nutr.,* 11, 883, 1988.

195. **Emery, T., Emery, L., and Olson, R. K.,** Retrohydroxamate ferrichrome, a biomimetic analogue of ferrichrome, *Biochim. Biophys. Res. Commun.,* 119, 1191, 1984.

196. **Jalal, M. A. F., Galles, J. L., and van der Helm, D.,** Structure of des(diserylglycyl)ferrirhodin, DDF, a novel siderophore from *Aspergillus ochraceous, J. Org Chem.,* 50, 5642, 1985.

197. **Bowman, B. J. and Bowman, E. J.,** H$^+$-ATPases from mitochondria, plasma membranes, and vacuoles of fungal cells, *J. Membrane Biol.,* 94, 83, 1986.

198. **Müller, G. and Winkelmann, G.,** Arsenate causes an immediate loss of bound siderophores from cytoplasmic membranes of fungi, *FEMS Microbiol. Lett.,* 9, 149, 1980.

199. **Huschka, H., Müller, G., and Winkelmann, G.,** The membrane potential is the driving force for siderophore iron transport in fungi, *FEMS Microbiol. Lett.,* 20, 125, 1983.

200. **Winkelmann, G. and Huschka, H.,** Molecular recognition and transport of siderophores in fungi, in *Iron Transport in Microbes, Plants and Animals,* Winkelmann, G., van der Helm, D., and Neilands, J. B., Eds., VCH Verlagsgesellschaft, Weinheim, 1987, 317.

201. **Bailey, T. C., Kime-Hunt, E. M., Carrano, C. J., Huschka, H., and Winkelmann, G.,** A photoaffinity label for the siderophore-mediated iron transport system in *Neorospora crassa, Biochim. Biophys. Acta,* 883, 299, 1986.

202. **Carrano, C. J., Bailey, C. T., and Bonadies, J. A.,** Transport properties of N-acyl derivatives of the coprogen and ferrichrysin classes of siderophores in *Neurospora crassa, Arch. Microbiol.,* 146, 41, 1986.

203. **Jalal, M. A. F., Mocharla, R., Barnes, C. L., Hossain, M. B., Powell, D. R., Eng-Wilmot, D. L., Grayson, S. L., Benson, B. A., and van der Helm, D.,** Extracellular siderophores from *Aspergillus ochraceus, J. Bacteriol.,* 158, 683, 1984.

204. **Jalal, M. A. F., Mocharla, R., and van der Helm, D.,** Separation of ferrichromes and other hydroxamate siderophores of fungal origin by reversed-phase chromatography, *J. Chromatogr.,* 301, 247, 1984.

205. **Frederick, C. B., Bentley, M. D., and Shive, W.,** Structure of triornicin, a new siderophore, *Biochemistry,* 20, 2436, 1981.

206. **Frederick, C. B., Bentley, M. D., and Shive, W.,** The structure of the fungal siderophore, isotriornicin, *Biochem. Biophys. Res. Commun.,* 105, 133, 1982.

207. **Hossain, M. B., Jalal, M. A. F., Benson, B. A., Barnes, C. L., and van der Helm, D.,** Structure and conformation of two coprogen-type siderophores: neocoprogen I and neocoprogen II, *J. Am. Chem. Soc.,* 109, 4948, 1987.

208. **Jalal, M.A. F., Love, S. K., and van der Helm, D.,** N$^\alpha$-Dimethylcoprogens: three novel trihydroxamate siderophores from pathogenic fungi, *Biol. Met.,* 1, 4, 1988.

209. **Jalal, M. A. F. and van der Helm, D.,** Siderophores of highly phytopathogenic, *Alternaria longipes, Biol. Met.,* 2, 11, 1989.

210. **Zähner, H., Keller-Schierlein, W., Hütter, R., Hess-Leisinger, K., and Deer, A.,** Stoffwechselprodukte von Mikroorganismen. 40. Mitteilung. Sideramine aus Aspergillaceen, *Arch. Microbiol.,* 45, 119, 1963.

211. **Wiebe, C. and Winkelmann, G.,** Kinetic studies on the specificity of chelate iron uptake in *Aspergillus, J. Bacteriol.,* 123, 837, 1975.

212. **Diekmann, H. and Krezdorn, E.,** Stoffwechselprodukte von Mikroorganismen. 150. Ferricrocin, Triacetylfusigen und andere Sideramine aus Pilzen der Gattung Aspergillus, Gruppe Fumigatus, *Arch. Microbiol.,* 106, 191, 1975.

213. **Hossain, M. B., Eng-Wilmot, D. L., Loghry, R. A., and van der Helm, D.,** Circular dichroism, crystal structure, and absolute configuration of the siderophore ferric N,N',N''-triacetylfusarinine $FeC_{39}H_{57}N_6O_{15}$, *J. Am. Chem. Soc.,* 102, 5766, 1980.

214. **Anke, H.,** Metabolic products of microorganisms. 163. Desferritriacetylfusigen, an antibiotic from *Aspergillus deflectus, J. Antibiot.,* 30, 125, 1977.

215. **Adjimani, J. P. and Emery, T.,** Iron uptake in *Mycelia sterilia* EP76, *J. Bacteriol.,* 169, 3664, 1987.

216. **Emery, T.,** Fungal ornithine esterases: relationship to iron transport, *Biochemistry,* 15, 2723, 1976.

217. **Adjimani, J. P. and Emery, T.,** Stereochemical aspects of iron transport in *Mycelia sterilia* EP-76, *J. Bacteriol.,* 170, 1377, 1988.

218. **Emery, T.,** Isolation, characterization, and properties of fusarinine, a δ-hydroxamic acid derivative of ornithine, *Biochemistry,* 4, 1410, 1965.

219. **Diekmann, H.,** Stoffwechselprodukte von Mikroorganismen, 56. Mitteilung. Fusigen — ein neues Sideramin aus Pilzen, *Arch. Microbiol.,* 58, 1, 1967.

220. **Diekmann, H.,** Stoffwechselprodukte von Mikroorganismen. 81. Mitteilung. Vorkommen und Strukturen von Coprogen B und Dimerumsäure, *Arch. Microbiol.,* 73, 65, 1970.

221. **Jalal, M. A. F., Love, S. K., and van der Helm, D.,** Siderophore mediated iron(III) uptake in *Gliocladium virens*. I. Properties of cis-fusarinine, trans-fusarinine, dimerum acid, and their ferric complexes, *J. Inorg. Biochem.*, 28, 417, 1986.

222. **Jalal, M. A. F., Love, S. K., and van der Helm, D.,** Siderophore mediated iron(III) uptake in *Gliocladium virens*. II. Role of ferric mono- and dihydroxamates as iron transport agents, *J. Inorg. Biochem.*, 29, 259, 1987.

223. **Manulis, S., Kashman, Y., and Barash, I.,** Identification of siderophores and siderophore-mediated uptake of iron in *Stemphylium botryosum, Phytochemistry*, 26, 1317, 1987.

224. **Harrington, G. J. and Neilands, J. B.,** Isolation and characterization of dimerum acid from *Verticillium dahliae, J. Plant. Nutr.*, 5, 675, 1982.

225. **Barash, I., Zion, R., Krikun, J., and Nachmias, A.,** Effect of iron status on Verticillium wilt disease and on in vitro production of siderophores by *Verticillium dahliae, J. Plant Nutr.*, 11, 893, 1988.

226. **Bentley, M. D., Aderegg, R. J., Szaniszlo, P., and Davenport, R. F.,** Isolation and identification of the principal siderophore of the dermatophyte *Microsporum gypseum, Biochemistry*, 25, 1455, 1986.

227. **Barash, I.,** personal communication, 1989.

228. **Winkelmann, G.,** Isolation and identification of the principal siderophore of the plant pathogenic fungus *Botrytis cinerea, Biol. Met.*, 1, 90, 1988.

229. **Winkelmann, G.,** unpublished data, 1989.

230. **Winkelmann, G.,** Siderophore production in *Trichoderma viride, Biol. Met.*, in press.,

231. **Schuler, R. and Haselwandter, K.,** Hydroxamate production by ericoid mycorrhizal fungi, *J. Plant Nutr.*, 11, 907, 1988.

232. **Boelaert, J. R., van Roost, G. F., Vergauwe, P. L., Verbanck, J. J., DeVroey, C., and Segaert, M. F.,** The role of desferrioxamine in dialysis-associated mucormycosis: report on three cases and review of the literature, *Clin. Nephrol.*, 29, 261, 1988.

233. **Van Cutsem and Boelaert, J. R.,** Effects of deferoxamine, feroxamine and iron on experimental mucormycosis (zygomycosis), *Kidney Int.*, 36, 1061, 1989.

234. **Mor, H., Pasternak, M., and Barash, I.,** Uptake of iron by *Geotrichum candidum*, a non-siderophore producer, *Biol. Met.*, 1, 99, 1988.

235. **Mor, H. and Barash, I.,** Characterization of siderophore-mediated iron transport in *Geotrichum candidum*, a non-siderophore producer, *Biol. Met.*, 2, 209, 1990.

236. **Neilands, J. B., Konopka, K., Schwyn, B., Coy, M., Francis, R. T., Paw, B., and Bagg, A.,** Comparative biochemistry of microbial iron assimilation, in *Iron Transport in Microbes, Plants and Animals*, Winkelmann, G., van der Helm, D., and Neilands, J. B., Eds., VCH, Weinheim, 1987, chap. 1.

237. **Lesuisse, E., Raguzzi, F., and Crichton, R. R.,** Iron uptake by the yeast *Saccharomyces cerevisiae*: involvement of a reduction step, *J. Gen. Microbiol.*, 133, 3229, 1987.

238. **Lesuisse, E. and Labbé, P.,** Reductive and non-reductive mechanisms of iron assimilation by the yeast *Saccharomyces cerevisiae, J. Gen. Microbiol.*, 135, 257, 1989.

239. **Dancis, A., Klausner, R. D., Hinnebusch, A. G., and Barriocanal, J. G.,** Genetic evidence that ferric reductase is required for iron uptake in *Saccharomyces cerevisiae, Mol. Cell. Biol.*, 10, 2294, 1990.

240. **Holzberg, M. and Artis, W.,** Hydroxamate siderophore production by opportunistic and systemic fungal pathogens, *Infect. Immun.*, 40, 1134, 1983.

241. **Ismail, A., Bedell, G. W., and Lupan, D. M.,** Siderophore production by the pathogenic yeast, *Candida albicans, Biochem. Biophys. Res. Commun.*, 132, 1160, 1985.

242. **Ismail, A., Bedell, G. W., and Lupan, D. M.,** Effect of temperature on siderophore production by *Candida albicans, Biochem. Biophys. Res. Commun.*, 132, 1160, 1985.

243. **Burt, W.,** Identification of coprogen B and its breakdown products from *Histoplasma capsulatum, Infect. Immun.*, 35, 990, 1982.

244. **Jacobson, E. S. and Petro, M. J.,** Extracellular iron chelation in *Cryptococcus neoformans, J. Med. Vet. Mycol.*, 25, 415, 1987.

245. **van der Walt, J. P., Weijman, A. C. M., and von Arx, J. A.,** The anamorphic yeast genus *Myxozyma* gen. nov. Sydowia, *Ann. Mycol. Ser. II.*, 34, 191, 1981.

246. **Payne, S.,** Iron and virulence in the family *Enterobacteriaceae, CRC Crit. Rev. Microbiol.*, 16, 81, 1981.

247. **Fukasawa, K., Goto, M., Sasaki, K., and Hirata, Y.,** Structure of a yellow-green peptide produced by iron-deficient *Azotobacter vinelandii* strain O, *Tetrahedron Lett.*, 28, 5359, 1972.

248. **Corbin, J. L. and Bulen, W. A.,** The isolation and identification of 2,3-dihydroxybenzoic acid and 2-N, 6-N-di(2,3-dihydroxybenzoyl)-L-lysine formed by iron-deficient *Azotobacter vinelandii, Biochemistry*, 8, 757, 1969.

249. **Earhart, C.,** Ferrienterobactin transport in *Escherichia coli*, in *Iron Transport in Microbes, Plants and Animals*, Winkelmann, G., van der Helm, D., and Neilands, J. B., Eds., VCH Verlagsgesellschaft, Weinheim, 1987, 67.

250. **Snow, G. A.,** Mycobactins: iron-chelating growth factors from bacteria, *Bacteriol. Rev.*, 34, 99, 1970.

251. **Ratledge, C. and Patel, P. V.,** The isolation, properties and taxonomic relevance of lipid-soluble, iron-binding compounds (the nocobactins) from *Nocardia, J. Gen. Microbiol.,* 93, 141, 1976.

252. **Cox, C. D., Rinehart, K. L., Moore, M. L., and Cook, J. C.,** Pyochelin: novel structure of an iron-chelating growth promoter for *Pseudomonas aeruginosa, Proc. Natl. Acad. Sci. U.S.A.,* 78, 4256, 1981.

253. **Emery, T.,** Malonichrome, a new iron chelate from *Fusarium roseum, Biochim. Biophys. Acta,* 629, 382, 1980.

254. **Deml, G.,** Studies in Heterobasidiomycetes. Part 34. A survey on siderophore formation in low-iron cultured smuts from the floral parts of *Polygonaceae, Syst. Appl. Microbiol.,* 6, 23, 1985.

255. **Carbonetti, N. H. and Williams, P. H.,** A cluster of five genes specifying the aerobactin iron uptake system of plasmid ColV-K30, *Infect. Immun.,* 46, 7, 1984.

GENETICS OF BACTERIAL IRON TRANSPORT

Volkmar Braun and Klaus Hantke

INTRODUCTION

Bacteria must frequently adapt to drastic changes in their environment. For example, *Escherichia coli* released from the gut can survive for some period of time in soil or water. Adaptation includes induction of transport systems for the uptake of available substrates, and the synthesis of enzymes required for substrate metabolism. Such adaptation occurs by regulation of genes, usually at the level of transcription initiation frequency. At the same time, synthesis of gene products which are no longer required is switched off.

The transport mechanism for any substrate into bacteria is not known at the atomic level. Only such detailed knowledge enables us to understand the molecular events that take place when a polar substrate crosses the hydrophobic layer of a biological membrane with the aid of a protein. No X-ray structure is available for any transport protein isolated from a membrane, as has been elucidated for enzymes, enzyme substrate complexes, and enzyme substrate analog complexes in the transition state during catalysis. Nevertheless, a wealth of knowledge has been assembled from biochemical and, above all, from genetic studies. Mutants are used to define a transport system because they most clearly demonstrate the specificity of a transport system for a certain substrate, or a group of similar substrates. By genetic studies, the number of transport genes and the polypeptides they encode can be deduced. This is achieved by cloning the transport genes on plasmids. Chromosomal DNA is fragmented using restriction endonucleases and the resulting fragments inserted into plasmid vector DNA. Plasmids containing transport genes are selected by complementation of chromosomal transport mutants. Subclones are then constructed to reduce the size of the insert DNA so that it consists almost exclusively of the transport region being studied. It is often convenient to cleave the DNA fragment encoding several transport genes into fragments which code for only one or two genes. This can be done using combinations of different plasmids which are compatible in one cell, and using a host strain in which the entire chromosomally encoded transport region has been deleted. Various combinations of DNA fragments on the plasmids are then tested for their ability to restore the entire transport system. The cloned DNA is then sequenced, the encoded polypeptides determined from sequence and biochemical analyses, and the proteins are subsequently localized in subfractions of the bacterial cell.

Nucleotide sequences provide valuable information on the overall physico-chemical properties of proteins and guide experiments aimed at subcellular localization of proteins. Usually, transport proteins occur in numbers (100 molecules per cell) that are too low to be detected by standard staining procedures after electrophoretic separation on polyacrylamide gels run in the presence of sodium dodecyl sulfate (SDS-PAGE). The proteins have to be either labeled radioactively, or oversynthesized by subcloning the genes of interest on expression vectors. Detection of the labeled transport proteins among the many membrane proteins is often difficult. Frequently, comparison of transport-positive wild type cells with transport-negative mutants (lack transport components) identifies the proteins in question. Oversynthesis also has its problems because it is often detrimental to the host cells because they cannot tolerate greatly increased amounts of hydrophobic proteins in their membranes. Therefore, overexpression of transport genes is preferably performed under controlled conditions. The transport genes are cloned downstream an operator sequence whose transcription is inhibited by a repressor protein. Cells are maintained and grown under repressive conditions. When the culture has reached a certain density, expression of the transport genes is induced by inactivation of the repressor. This is achieved by adding a corepressor, for example IPTG (isopropyl β-D-thiogalactoside) to inactivate the frequently used *lac* repressor, or by raising the temperature from 30° to 42°C in case of temperature-sensitive repressors such as the cI_{857} repressor of phage lambda. The best expression system to identify a membrane protein was recently developed from phage T7.[1] In this system, the structural gene of a transport protein is cloned behind the strong promoter of the phage's gene *10* contained on a multicopy vector (pT7-5,pT7-6). The T7 RNA polymerase exclusively recognizes the gene *10* promoter so that only the transport gene is transcribed. The *E. coli* RNA polymerase is inhibited by the addition of rifampicin which does not inhibit the T7 polymerase. The T7 polymerase is cloned on a second plasmid (pGP1-2) and its expression is under control of the lambda p_L promoter which is regulated by the cI_{857} repressor. Synthesis of T7 polymerase is induced at 42°C which in turn transcribes the transport gene. Addition of ^{35}S-methionine exclusively labels the cloned transport protein which can then be identified by autoradiography after SDS-PAGE.

A second system more suitable to produce sufficient amounts of a transport protein and thus enable it to be isolated from the membrane is also based on phage T7.[2] The desired gene is cloned downstream of the gene *10* promoter and transcribed by T7 polymerase. The polymerase gene is contained as a single copy in the bacterial chromosome under the control of the IPTG-inducible *lacUV5* promoter. To avoid transcription by the few polymerase molecules which escape repression under uninduced conditions, cells are used which synthesize the T7 lysozyme which is capable of inhibiting the polymerase unless it is overproduced. If the protein under study is toxic, cells start to die after they have synthesized sufficient amounts of protein for isolation.

Mutations inserted at defined sites identify essential amino acids involved in transport function of a protein. The properties of defined mutants also allow one to deduce models for the folding of a particular protein within a membrane, which then are tested and refined with the help of site-specific mutations. Recently, a method was developed which employs genetically constructed fusions between a periplasmic enzyme of *E. coli* (alkaline phosphatase), lacking the leader peptide, and N-terminal portions of a membrane protein.[3] The latter portion directed alkaline phosphatase across the cytoplasmic membrane provided the fusion sites were in loops of the membrane protein exposed to the periplasm. Data obtained by combining the above methods provide a rather detailed picture of the primary structure, active sites and the folding of transport proteins within a membrane which are instrumental for the interpretation of X-ray structures once they are available. Interestingly, the power of the genetic methods is so high that with genetic data alone a transport mechanism can be outlined in broad terms without any biochemical studies. Comparing genetic data of a newly studied transport system with another better characterized transport system frequently provides clues for a possible transport mechanism which can then be pursued experimentally. These few examples of presently used genetic methods illustrate the potential of genetics to approach technically difficult problems, and to obtain information about transport systems more rapidly than with any other techniques.

TRANSPORT OF SUBSTRATES INTO *E. COLI*

The uptake of substrates into the best studied organism, *E. coli*, will be outlined in general terms to contrast it with the very special features of iron transport which will be discussed in detail below.

UPTAKE ACROSS THE OUTER MEMBRANE

Gram-negative bacteria like *E. coli* are surrounded by two membranes, the outer membrane and the cytoplasmic membrane.[4-6] Hydrophilic substrates not larger than 600 to 700

Da diffuse through water-filled pores of the outer membrane formed by the most abundant proteins in this membrane.[7] However, for some substrates stereochemical recognition takes place between the substrate and an outer membrane protein. This has been demonstrated for maltodextrins which are recognized by the LamB protein,[8] and for nucleosides which interact with the Tsx protein.[9-12] The PhoE protein forms a more efficient pore than do the porins[6] both for inorganic and organic phosphate but does not seem to specifically recognize phosphate but rather it displays a broad specificity for anions.[13] Before uptake studies had been performed, genetic evidence pointed to the role of these proteins in transport of certain substrates. Synthesis of these proteins is regulated at the transcriptional level by maltodextrins, nucleosides and phosphates, respectively. Maltodextrins (the intracellular regulatory compound is maltotriose) convert a protein (MalT) to an activator, nucleosides inactivate two repressor proteins (DeoR, CytR), and phosphate starvation induces a complex regulatory network resulting in PhoE synthesis. These proteins facilitate the diffusion of the substrates across the outer membrane. The smaller homologs of the maltodextrins, phosphates, and the nucleosides, can also pass through the porins so that the specific proteins are not absolutely required. They increase the rate of diffusion, and they are essential for the uptake of the larger homologs across the outer membrane into the periplasmic space. As will be discussed later, uptake of iron across the outer membrane follows an entirely different mechanism. Receptor proteins are essential components and transport is regulated.

TRANSPORT ACROSS THE CYTOPLASMIC MEMBRANE

Actual transport against a concentration gradient, and stereochemical recognition between the substrate and the transport proteins occurs in the cytoplasmic membrane. Energy required for active transport is provided in the form of an electrochemical potential across the cytoplasmic membrane by electron transport chains located therein, or by ATP hydrolysis through the membrane-bound ATPase. Certain substrates, such as lactose, are transported across the cytoplasmic membrane of *E. coli* by a single protein, the Lac Permease, by a process driven by the electrochemical potential.[14]

Periplasmic Binding-Protein-Dependent Transport (PBT)

The present knowledge of binding-protein-dependent transport will be briefly outlined as iron transport seems to be analogous to this mechanism.

Transport of many amino acids, peptides, certain sugars and anions follows a so-called binding protein-dependent transport mechanism in which proteins in the periplasmic space (located between the outer and the cytoplasmic membrane) are involved.[14,15] The binding proteins recognize certain substrates and deliver them to the integral membrane proteins of the cytoplasmic membrane. They are essential constituents of these transport systems. The periplasmic binding proteins can be released by an osmotic shock treatment involving plasmolysis of cells in 15% sucrose (which counterbalances the internal osmotic pressure) in the presence of EDTA (releasing Mg^{2+} ions supposed to stabilize the outer membrane). Upon rapid dilution of the plasmolysed cells into a low-salt Tris/Mg^{2+} buffer, periplasmic proteins are released from the cells. Alternatively, cells are converted into spheroplasts by a similar procedure but with the inclusion of lysozyme to degrade the murein (peptidoglycan) layer. Such treated cells show greatly reduced transport rates which can be restored by adding back the binding protein in the presence of Ca^{2+} (increases the permeability of the outer membrane).[14] Substrate bound to the binding protein rather than the free substrate is accepted by the cytoplasmic transmembrane protein(s). Usually the periplasmic proteins are synthesized in large excess with respect to the membrane proteins, and are the best characterized transport proteins. The structure of several (for arabinose, ribose, galactose, sulfate) has been resolved to the atomic scale by X-ray analysis. They exhibit similar conformations composed of two globular domains forming a cleft at the substrate binding site, linked by a flexible hinge.[16]

The very hydrophobic integral cytoplasmic membrane proteins accept substrates from the binding proteins. Usually two such proteins are found in a single system which exhibit sequence similarities, for example HisQ and HisM of histidine, MalF and MalG of maltose, PstA and PstC of phosphate, and OppB and OppC of the peptide transport system.[15] But there are also exceptions to this rule. The high affinity arabinose transport system contains only one hydrophobic protein of the usual size (34,000 Da),[17] and variations also occur in iron transport systems (see later).

Characteristic of the PBT system is the involvement of a polar but nevertheless membrane-bound protein which displays sequences also found in nucleotide binding proteins. In fact, it has long been known that ATP either directly or indirectly serves as an energy source for PBT. ATP binding, but not ATP hydrolysis, has been demonstrated for the MalK, HisP and OppD proteins.[18] There is evidence that these proteins are bound to the inside of the cytoplasmic membrane.[19]

The PBT system is a high affinity transport system (K_M 1 μM and below) which concentrates substrates inside the cell against a very large gradient (in the order of 10^5).

IRON (III) TRANSPORT INTO *E. COLI*

Iron in the Fe^{3+} form in aerobic conditions and physiological pH (around 7) is extremely insoluble and only scarcely available for bacteria under all natural conditions. As a response to iron-starvation, bacteria synthesize elaborate iron supply systems which are composed of low molecular weight Fe^{3+}-complexing compounds, termed siderophores, and Fe^{3+}-siderophore transport systems. For *E. coli* five iron(III) transport systems have been characterized.[20-23] They use the siderophores aerobactin, citrate and enterochelin (synonymous with enterobactin), synthesized by *E. coli*, the siderophores ferrichrome (and the closely related ferricrocin and ferrichrysin), coprogen, and rhodotorulic acid synthesized by various fungi, and ferrioxamine B formed by *Arthrobacter simplex*. The five systems are defined by five specific outer membrane receptors for aerobactin, ferrichrome (ferricrocin, ferrichrysin), Fe^{3+}-citrate, Fe^{3+}-enterochelin, and one common receptor for Fe^{3+}-coprogen, Fe^{3+}-rhodotorulic acid, and Fe^{3+}-ferrioxamine B. Fe^{3+} ligated by the hydroxamate siderophores (aerobactin, ferrichrome, coprogen, rhodotorulic acid, ferrioxamine B) is transported across the cytoplasmic membrane by the same transport components, while Fe^{3+}-citrate and Fe^{3+}-enterochelin use different transport proteins. It seems that Fe^{3+} ligated by different siderophores is transported across the two membranes of *E. coli* by basically the same mechanism although different proteins are involved. In the following these transport systems will be discussed.

IRON(III)-HYDROXAMATE TRANSPORT INTO *E. COLI*

Transport of Fe^{3+} via siderophores of the hydroxamate type are described together because they share most of the transport proteins (Figure 1). One of the characteristic features of the Fe^{3+} transport systems is the absolute requirement for outer membrane receptor proteins. These proteins can barely be visualized by staining after SDS-PAGE of isolated outer membranes from cells grown in high iron media, indicating a concentration of roughly 5000 receptor protein molecules per cell. However, they become major outer membrane proteins under iron-limiting growth conditions, adding up to about 100,000 molecules per cell (see next Chapter). The receptors bind Fe^{3+}-siderophores with apparent Michaelis-Menten constants (K_M) below 0.1 μM. For example, a K_M value of 0.06 μM was estimated for the binding of ferrichrome to the FhuA receptor.[24] A possible mechanism for the transport of Fe^{3+}-siderophores across the outer membrane is proposed in a later chapter. The receptors are specific for their siderophores as outlined in Figure 1. No cross-reactivity has been observed. Mutants or natural variants lacking a receptor are unable to take up Fe^{3+} via a

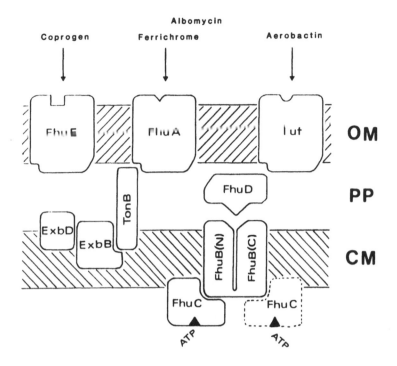

FIGURE 1. Model of the iron(III) hydroxamate transport systems of *E. coli*. The outer membrane (OM) receptor proteins FhuA, FhuE and Iut bind the different iron(III) siderophores which are transported across the OM through the activity of the TonB protein which itself is influenced by the ExbB and ExbD proteins and the energized state of the cytoplasmic membrane. All the iron(III) hydroxamate compounds require the FhuD Protein in the periplasmic space (PP), and the FhuB and FhuC proteins in the cytoplasmic membrane (CM). ATP denotes a potential ATP binding site at the FhuC protein as deduced from the nucleotide sequence. The stoichiometry of the transport proteins has not been determined. FhuB(N) and FhuB(C) indicates that the FhuB protein has double the size of transmembrane proteins usually found in periplasmic-binding protein-dependent transport systems and contains an internal homology suggesting that both halves are derived from a common ancestor gene by duplication and subsequent fusion. (Designed by W. Köster.)

related siderophore. Such mutants are easy to obtain because most receptors serve multiple functions in that they are also binding sites for bacteriophages and bacteriocins (colicins). Mutants resistant to the phages and colicins have usually lost the receptor, and in turn the ability to take up the Fe^{3+} siderophore. A convenient way to isolate mutants in this system is to search for clones which are spontaneously resistant to the antibiotic albomycin. Albomycin shares structural features with ferrichrome, and is taken up by the same transport system. Mutations in the genes encoding proteins for ferrichrome uptake (Figure 1) were isolated using albomycin, and the mutations were complemented with chromosomal DNA fragments. Four genes, termed *fhuA*, *fhuB*, *fhuC*, and *fhuD*, were found by this approach.[25] Their order, organization in an operon, transcription polarity on the circular *E. coli* chromosome was determined, the products of the genes were identified and localized in the cellular compartments.[25,27] The genes were sequenced,[26,28-30] and the molecular weights of the polypeptides deduced from the nucleotide sequences (Figure 2) agreed with the sizes determined by SDS-PAGE. There is no information available on the transmembrane arrangement of the FhuA, FhuB, and FhuC polypeptides. FhuA (and the other receptors) is a polar protein as is common for outer membrane proteins for which data on conformation and arrangement in the membrane have been obtained.[31-33] Fusion proteins consisting of N-

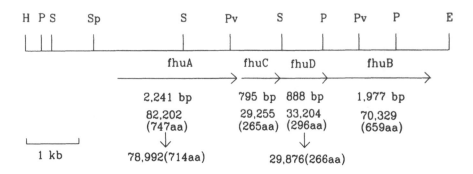

FIGURE 2.   Arrangement of the *fhu* genes on the chromosome of *E. coli* K-12. The genes bear the same names as the proteins they encode (Figure 1). The horizontal arrow indicates the transcription polarity. The size of the genes is given in base pairs (bp) and the molecular weights of the proteins and the number of amino acids (aa) is presented below. The FhuA and FhuD proteins are synthesized as precursors containing N-terminal signal sequences which are released upon transport of the proteins across the cytoplasmic membrane. The cleavage sites of the following restriction endonucleases are included: H, *Hind*III; E, *Eco*RI; S, *Sal*I; Pv, *Pvu*II; P, *Pst*I; Sp, *Sph*I.

terminal FhuA portions, ranging from 11 to 90% of the mature protein, and a truncated alkaline phosphatase, lacking the signal sequence and 5 amino acids of the mature protein, were exported across the cytoplasmic membrane but not properly inserted into the outer membrane.[34] Another genetic construct, in which the C-terminal 24 amino acids of FhuA were replaced by 3 random amino acids, was for the most part proteolytically degraded but sufficient amounts were incorporated into the outer membrane to confer cells nearly wild type sensitivity to phages T5 and T1, three orders of magnitude less sensitivity to colicin M, and seven orders less to phage φ80 (Figure 1): cells were resistant to albomycin and unable to take up ferrichrome because they did not bind to the altered receptor.[35] Mutations exerting no extensive conformational alterations in the FhuA protein will eventually lead to a model on the arrangement of FhuA in the outer membrane, and on binding-epitopes for the various ligands. Since all the ligands approach FhuA from the outside of the cell, defined localized mutations will be useful for mapping FhuA sites exposed on the cell surface.

After the Fe^{3+}-siderophores have bound to the outer membrane receptors they are translocated into the periplasmic space and from there across the cytoplasmic membrane into the cytoplasm. The Fe^{3+}-hydroxamates use a common system defined by the periplasmic protein FhuD,[27] the very hydrophobic protein FhuB in the cytoplasmic membrane,[27] and the membrane-associated but rather polar FhuC protein.[25,27] The latter contains two domains which is typical of nucleotide-binding proteins.[36]

Ferrichrome is modified after Fe^{3+} and has been released by reduction to Fe^{2+}.[37,38] Modification probably occurs at one of the N-hydroxyl groups involved in binding of Fe^{3+} since the modified compound transported iron 100 times less efficiently. Modified ferrichrome contains one acetyl group so that acetylation seems to cause inactivation. Modified ferrichrome is released from the cells. This could be a mechanism by which cells rid themselves of a potentially dangerous iron ligand inside the cell.

Albomycin is also modified after transport into the cell. The antibiotically active portion is cleaved from the iron carrier and remains inside the cell.[39] The iron carrier is degraded by peptidase N in *E. coli* and by peptidase N and peptidase A in *Salmonella typhimurium*.[40] Degradation is necessary for antibiotic activity since peptidase N-negative mutants of *E. coli* are albomycin resistant, and in fact albomycin serves as a siderophore and supports growth of mutants on iron-limited medium.

Iron transport via **aerobactin** gained wide interest when it was shown that this transport system contributed to the virulence of human and animal *E. coli* isolates.[41-46] The genes

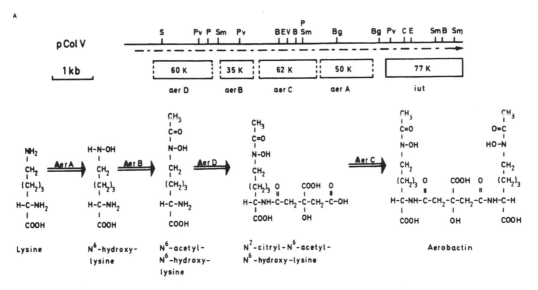

FIGURE 3.   Genes designated *aer*A (*iucD*), *aer*B (*iucB*), *aer*C (*iucC*), and *aer*D (*iucA*) that encode the enzymes for biosynthesis of aerobactin. This operon also codes for the outer membrane receptor protein Iut but the additional transport proteins are encoded in the *fhu* operon (Figures 1 and 2). The reactions catalyzed by the enzymes are indicated. Cleavage sites of restriction endonucleases are marked B, *Bam*HI; Bg, *Bgl*II; C, *Cla*I; E, *Eco*RI; EV, *Eco*RV; P, *Pst*I; Pv, *Pvu*I; S, *Sal*I; Sm, *Sma*I; pColV, plasmid pColV.

encoding the aerobactin biosynthetic enzymes[47,48] and the outer membrane receptor protein (Figure 3) are encoded on large conjugative plasmids, usually pColV plasmids, or on the chromosome.[42,43] The *iut* receptor gene has been sequenced and found to encode a protein consisting of 700 amino acids which is synthesized as a precursor containing a 25 residue signal sequence.[49] Interestingly, the size of this receptor seems to vary in closely related species. Molecular weights of 74,000, 76,000, and 85,000, based on the electrophoretic mobility on SDS-PAGE, have been reported for the receptor in *E. coli*, *Klebsiella edwardsii*, and *Enterobacter cloacae*, respectively.[49] Aerobactin is apparently reused as an iron carrier and not inactivated like ferrichrome and enterochelin.[50] Transport assays show that recycling aerobactin transports on average three iron ions.

The **FhuE receptor** serves as a binding site for Fe^{3+}-**coprogen, Fe^{3+}-ferrioxamine B,** and Fe^{3+}-**rhodotorulic acid.** All three iron-siderophores can deliver iron to *E. coli* but the transport rate is highest with coprogen.[51,52] The *fhuE* gene has been sequenced and found to encode a protein consisting of 729 amino acids including an unusually long signal sequence of 36 residues.[51] Processing of the FhuE precursor to the mature protein is rather slow and further delayed in C-terminally truncated FhuE derivatives.[50] A genetically constructed FhuE derivative, in which 29 residues at the C-terminal end were replaced by five random residues, was exported across the cytoplasmic membrane but no longer inserted into the outer membrane.[51] An equivalent construct of FhuA had the 23 C-terminal amino acids replaced by three random residues.[35] Most of this protein remained in the periplasm, and was degraded by trypsin added to spheroplasts. This was also observed with the FhuE derivative. However, a small portion of the FhuA derivative was properly inserted into the outer membrane since cells expressed some of the FhuA-related phage and colicin sensitivities.

It has been reported that patients treated with ferrioxamine B (Desferal®, the methane sulfonate derivative) for iron overload symptoms are susceptible to infections by *Yersinia enterocolitica*.[54,55] To examine whether *Y. enterocolitica* contains a FhuE-like receptor for the uptake of Fe^{3+}-ferrioxamine, B, *Y. enterocolitica* chromosomal DNA fragments were cloned into an *E. coli fhuE* mutant to test for restoration of FhuE activity. A 3.2 kb fragment enabled the *fhuE* mutant to grow on coprogen, and ferrioxamine B and D as sole iron sources,

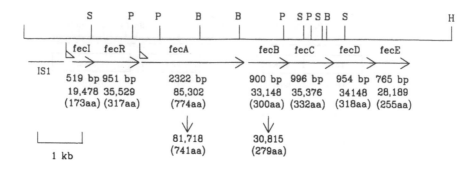

FIGURE 4.   The arrangement of the *fec* genes on the chromosome of *E. coli* K-12 and *E. coli* B. In addition to the published structural genes encoding transport components,[61] the genes *fecI* and *fecR* coding for two regulatory proteins[63,66] have been included. Upstream of the *fec* operon is located an insertion element (IS1) which may be responsible for the difficulties in mapping and cloning this region.[62,65] For further explanation see Figure 2.

but also on ferrimycin which is an antibiotic for *Yersinia*. A ferrimycin-resistant mutant of *Y. enterocolitica* became sensitive when transformed with the cloned DNA fragment. Apparently, *Y. enterocolitica* contains a FhuE equivalent receptor which is also active in *E. coli*.[55]

IRON(III)-DICITRATE TRANSPORT INTO *E. COLI*

The citrate-inducible iron transport system is of particular interest with regard to its regulation. It was detected in *E. coli* mutants which could not synthesize their own siderophore but were able to grow on low-iron media supplemented with citrate.[56] An apparent K_m of 0.2 μM and a V_{max} of 66 patoms Fe^{3+} per mg dry weight per min was determined which is in the range found for other *E. coli* iron transport systems. The system was also defined by 2 classes of mutants, one class was cotransducible with *argF* using phage P1, and the other class was not.[57] The *argF*-cotransducible mutations lack an outer membrane protein[59] which is inducible when cells are grown on a citrate medium.[58] The genes, designated *fec* (Figure 4), have been cloned[60] and sequenced.[61] They form an operon of five genes. The proteins were identified after cloning into expression vectors, and they have been localized subcellularly. The FecA protein represents the outer membrane receptor protein. The FecB protein was found in the periplasmic space, the FecC,D proteins are very hydrophobic and form integral cytoplasmic membrane proteins. The polar FecE protein is also found in the cytoplasmic membrane and contains two regions of extensive sequence homology with nucleotide-binding proteins.[61] Since the five genes are located close together within a DNA region of 6 kb, and phage P1 transduces 80 kb, it is not understood why only *fecA* was cotransduced with *argF*.[57,62] The gene order in this region of the chromosome is *proAB argF fecABCDE lac*.[62] This region contains a number of insertion elements which can cause deletions and inversions.[64] This may also be the reason why DNA fragments containing the entire *fecA* gene could not be cloned from *E. coli* K-12 making us turn to *E. coli* B from which *fecA* cloning was possible.[60] Comparison of the nucleotide sequences of *fecA* and *fecB* of *E. coli* K-12 and B revealed ten single nucleotide replacements which resulted in six conservative amino acid exchanges (Ala/Thr, Thr/Met, Leu/Val).

For induction of the Fe^{3+}-dicitrate transport system, citrate and a small amount of iron are required in the growth medium. Larger amounts of iron suppress the system since it is also under the Fur repressor control (see next chapter). When the iron in the medium is trapped by addition of deferri-ferrichrome, the Fe^{3+}-dicitrate transport system is not induced.[65] In addition to citrate fluorocitrate and phosphocitrate are strong inducers although the latter transport iron very poorly. The intracellular citrate concentration can be increased tenfold without affecting induction. These data suggest that the inducer does not have to enter the cytoplasm, although induction increases the cytoplasmic transcription initiation frequency.[62] In fact, transport-deficient mutants, except those mutated in *fecA* and *tonB*, are just as well inducible as transport-proficient cells.[62,65] It is concluded that Fe^{3+}-dicitrate is the inducer and that it has only to be transported across the outer membrane through the action of the FecA and TonB proteins. In the Periplasm it binds to a transmembrane signaling protein, which in the substrate-protein complex configuration, leads to transcription initiation by either directly binding through a region exposed in the cytoplasm to the operator upstream of *fecA*, or by interacting with a soluble cytoplasmic protein which acts as an activator protein or repressor. An activator could bind to the transmembrane protein and be released when this is occupied by Fe^{3+}-dicitrate, and in turn bind to the operator. The reverse could take place in case of a repressor.

Recently, we succeeded in cloning and sequencing the region upstream of *fec*.[63] Two open reading frames were found, designated *fecI* and *fecR*, which encode proteins consisting of 173 and 317 amino acids (Figure 4). Cells transformed with a multicopy plasmid carrying *fecI* strongly induced the transport system in the absence of Fe^{3+}-dicitrate.[66] This suggests that the *fecI* gene product functions as an activator which when overproduced induces the transport system. Transformants carrying only *fecR* on a multicopy plasmid were not induced, and could not be induced by Fe^{3+}-dicitrate. Cells containing *fecI* and *fecR* on plasmids exhibited an Fe^{3+}-dicitrate regulated *fec* gene expression. A simple model predicts FecR to be the transmembrane signaling protein to which FecI binds in the uninduced state. When FecR is missing, FecI does not need the inducer to bind to the operator and to induce transcription.

IRON(III)ENTEROCHELIN TRANSPORT INTO *E. COLI*

Enterochelin (enterobactin) is the endogenous siderophore of *E. coli* and for many enterobacteria. It is a cyclic trimer of 2,3-dihydroxybenzoylserine (DBS). The genes for its synthesis and for the uptake system are clustered in the 13 min region of the *E. coli* chromosome (Figure 5).

Enterochelin Biosynthesis

The compound 2,3-dihydroxybenzoic acid (DHB) is formed from chorismic acid which is the common precursor for aromatic amino acids, *p*-aminobenzoic acid, *p*-hydroxybenzoic acid and menaquinone[67] (Figure 5). The gene products of the *entABC* genes catalyze the formation of dihydroxybenzoic acid and those of *entDEF* fuse L-serine to DHB and cyclize the product to enterochelin. These genes have been cloned and some of them have been sequenced. However, since there are many genes in the cluster, and genes for synthesis and uptake are intermingled (Figure 5) the system is very complex and some uncertainties remain. Published data has been summarized in Table 1 and only some of the new findings will be commented on in this review.

Some confusion existed as to the identities of *entA*, the gene for 2,3-dihydro-2,3-dihydroxybenzoate dehydrogenase and *entC*, the gene for isochorismate synthetase, since the genetically characterized *entC401* mutation turned out to be in *entA*. This was shown by complementation of *entC401* with the cloned *entA* gene.[73] The amino acid sequence of EntA (2,3-dihydro-2,3-dihydroxybenzoate dehydrogenase) exhibited clear relationships to dehydrogenases and showed no similarities to other chorismate-utilizing enzymes as one would expect for a protein with EntC activity (isochorismate synthetase). After sequencing the gene encoding a 44-kDa protein it became evident that this was EntC because of its structural relationship to other chorismate utilizing enzymes such as TrpE (component of the anthranilate synthetase) and PabB (component of the *p*-aminobenzoate synthetase). The analysis of an insertion mutation, constructed *in vitro* in the cloned gene and recombined

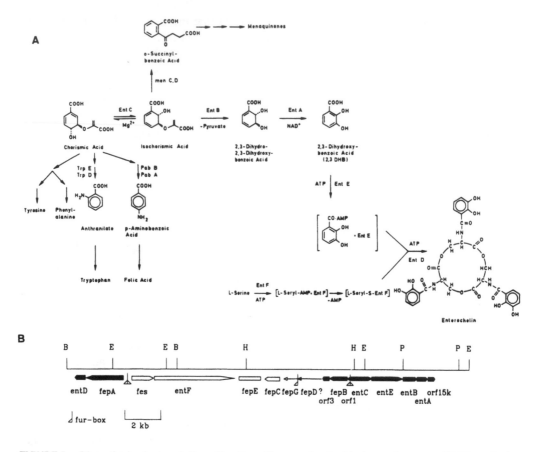

FIGURE 5. Biosynthesis of enterochelin and location of the genes involved in the *ent-fep* operon. (A) Biosynthesis of enterochelin and other aromatic compounds derived from chorismic acid. (B) Order of the enterochelin biosynthesis and transport genes between 13 and 14 min on the genetic map of *E. coli*. ♦ genes sequenced, ◊ gene products identified by polyacrylamide gel electrophoresis; the direction of *fepE* transcription is not clear. Restriction sites: B (*Bam*HI); E (*Eco*RI); H (*Hind*III); P (*Pvu*II).

TABLE 1
Enterochelin Synthesis and Uptake

Gene	Protein (mol. wt.)	Local	Function	Ref.
entA	26,249	cyt.	2,3-Dihydro-2,3-dihydroxybenzoate dehydrogenase	68, 69
entB	32,558	cyt.	2,3-Dihydro-2,3-dihydroxybenzoate synthase	68, 69
entC	42,917	cyt.	Isochorismate synthetase	70
entD	23,579	CM	Enterochelin synthetase	71
entE	58,299	cyt.	2,3-Dihydroxybenzoate-AMP ligase	72, 73
entF	ca. 160,000	cyt.	Serine-AMP ligase	74
entG	—	—	Activity seems to reside in EntB	69, 79
fepA	79,908	OM	Receptor for enterochelin and DBS$_n$	75
fepB	34,255	PPL	Possibly binding protein for enterochelin	76
fepC	ca. 31,500	CM	Enterochelin and DBS PBT	77
fepD	?	CM	Enterochelin and DBS PBT	77
fepE	ca. 43,000	CM	Enterochelin and DBS PBT	77
fepG	?	CM	Enterochelin and DBS PBT	76
fes	ca. 44,000	cyt.	Enterochelin esterase	78

into the chromosomal locus *entC*, exhibited strong polar effects on the following genes[70] and disclosed the reason why possibly *entC401* was used to define *entC:* it was one of the rare mutants *only* affected in the isochorismate synthetase activity (EntC) while mutations in *entC*, by the polarity effect, were also affected in other enzymes (Figure 5). However, the properties of the *entC401* mutant also show that EntA is necessary for the isochorismate synthetase activity of EntC. Possibly these proteins are only active or stable in the cell as a multimeric enzyme complex.

The gene *entG* was defined in a Mu-lysogen as being unable to make enterochelin from dihydroxybenzoate.[79] The genes *entA* and *entB* were also affected in this mutant. *In vitro* enzymic complementation assays showed the presence of EntD, EntE and EntF activities in the mutant strain, which led to the definition of *entG* as a fourth enzyme for the enterochelin synthetase complex.[79] Complementation studies with plasmids and DNA sequencing failed to detect a gene product for *entG* between *entB* and *entA*.[68,69] However, for full expression of the EntG$^+$ complementing activity the C-terminal end of EntB was necessary.[73] From these observations it is assumed that there is an interaction between the catechol synthesizing enzyme complex and the enterochelin synthetase,[69] which may be necessary for the enzymatic activity of the latter enzyme complex.

After synthesis the siderophore has to be excreted into the medium. Very little is known about this process. It is assumed that immediately after completion the siderophore is secreted possibly with the help of a membrane associated protein of the biosynthetic complex. A good candidate for this activity in the enterochelin system is the EntD protein, which was localized at the inner side of the cytoplasmic membrane.[71]

Iron(III) Enterochelin Transport

The uptake of iron-enterochelin requires the TonB-dependent outer membrane receptor FepA, and the products of five genes for transport across the cytoplasmic membrane. Only the sequences of the *fepA* gene[75] and the periplasmic protein *fepB*[76] are known. After cloning, four *fepB* gene products were identified with apparent molecular weights between 31,500 and 36,500.[77] In the amino acid sequence derived from the DNA sequence three possible leader peptidase cleavage sites were found.[76] It is unknown if the processed FepB forms have different activities in the transport process.

A short open reading frame, ORF1 (68 amino acid residues), was identified between the promoter region with a Fur box and the start of *fepB* (Figure 5).[76] The function of ORF1 is not clear. Downstream of *fepB* another open reading frame, ORF3, codes for a protein of 174 amino acid residues. A Tn5 insertion in the cloned gene when placed into the chromosome by recombination surprisingly yielded no phenotype.[76] Another Fur box downstream of ORF3 seems to regulate the expression of *fepG* and *fepC*.[76] The gene products of *fepD* and *fepG* have not been visualized on polyacrylamide gels. This is not uncommon since it has always been difficult in the PBT systems to identify the very hydrophobic cytoplasmic membrane components.[26] They are made in very low amounts and show an unusual behavior when subjected to SDS-polyacrylamide gel electrophoresis. Unusual for PBT systems is the description of five genes, not counting the ORFs, involved in transport through the cytoplasmic membrane while other systems have only three or four genes.

The gene *fes* encodes the B subunit of the enterochelin esterase and is also located in the *ent fep* gene cluster.[80] No mutation in the A subunit is known up to now. *fes* Mutants are normal in biosynthesis and uptake of enterochelin but on iron deficient medium they are unable to grow in contrast to their parent strain which can grow on iron deficient medium.[80] This clearly demonstrates that Fes has a function in iron-enterochelin uptake. Depending on conditions selected some iron may be removed from the complex. This was shown using enterochelin analogs which lacked the fragile ester bond and which could not be hydrolyzed by Fes. Some of these analogs were still able to supply cells with sufficient iron indicating that cleavage of the ring was not absolutely necessary for the removal of iron from the chelate complex.[81]

The 15k protein (Figure 5), which has been sequenced,[69] clearly belongs to the *entC* operon. No function could be assigned to this protein since a chromosomal mutant displayed no phenotype.[69]

Iron(III) Dihydroxybenzoylserine Transport

Although *fepA* mutants are unable to transport iron-enterochelin, they are still able to transport iron with the degradation product of enterochelin, 2,3-dihydroxybenzoylserine. This indicated that there are other ways for its iron complex to enter the periplasmic space. It was shown that mainly FepA and Fiu, and to a minor extent also Cir, are involved in the uptake of iron-2,3-dihydroxybenzoylserine.[82] Cir and Fiu are two iron-regulated outer membrane receptors, Cir for colicins I and V, Fiu for colicins G,H[83] and E492.[82a] Their natural substrates, presumably catechol-siderophores, are unknown. DBS-iron is only taken up across the cytoplasmic membrane via the Fep binding-protein-dependent transport system.[82]

It is surprising that dihydroxybenzoate, the precursor for enterochelin, is mainly taken up through the outer membrane via Cir and Fiu and only to a minor extent via FepA. In the presence of strong iron chelators (e.g., 0.2 mM 2,2'dipyridyl) growth promotion is only observed when the cells are able to make enterochelin from dihydroxybenzoate. It was shown that MECAM is mainly taken up via FepA but other transport routes seemed to exist.[81] Examination with defined mutants in *fepA, cir,* and *fiu* showed that Cir was the second receptor for MECAM while Fiu had only a weak activity.[82a] The catechol myxochelin A [N,N'-bis-(2,3-dihydroxybenzoyl)-lysinol][84] was found to be taken up via FepA, Cir, and Fiu in this order of preference.[82a]

The gene encoding the Cir protein has been sequenced.[85,86] Sequence comparison with those of other siderophore receptors showed that FepA is the closest relative to Cir in the receptor protein family. This was not unexpected since both proteins recognize catechol-siderophores. The *fiu* gene has not yet been sequenced.

The first hint for a catechol specificity by Cir and Fiu came from studies with catechol-substituted cephalosporins. It was found that catechol-substituted cephalosporins were taken up in a TonB-dependent manner.[87] Unexpectedly, the uptake of these cephalosporins was *not* dependent on FepA but rather on Cir and Fiu. Astonishingly, the receptors Cir and Fiu tolerated a broad spectrum of chemically modified catechol-β-lactams, indicating that these receptor proteins have a relatively low substrate specificity.[87]

UNUSUAL Fe^{3+} TRANSPORT SYSTEM OF *SERRATIA MARCESCENS*

A cloned chromosomal DNA fragment of *S. marcescens* conferred an *E. coli* mutant, unable to produce enterochelin, the ability to grow on iron-limited medium.[88] The astonishing property of this iron transport system is its independence of the TonB function. In addition, no outer membrane receptor protein and no siderophore has been related to this transport system. The DNA fragment was sequenced and found to contain three open reading frames encoding three proteins (Figure 6). All three proteins have been identified by SDS-PAGE. SfuA is found in the periplasmic space. SfuB, as deduced from the nucleotide sequence, must be a very hydrophobic protein and can only be detected by *sfuB* gene expression in an *in vitro* transcription/translation system. Upon heating it does not enter the gel, and changes its electrophoretic mobility relative to standard proteins under different conditions. These properties are also found with the very hydrophobic FhuB,[26,27] and FecC,D proteins.[61] SfuC is polar but nevertheless membrane-associated and has two consensus sequences found in nucleotide binding proteins.[89] This system displays all the characteristics of the other iron transport systems regarding transport across the cytoplasmic membrane. Yet it is lacking a transport protein in the outer membrane. Since transport occurs independent of TonB, it is unlikely that a chromosomally encoded protein of *E. coli* substitutes the receptor. Growth on iron-limited media is slow, and transport rates were rather low. The latter measurements

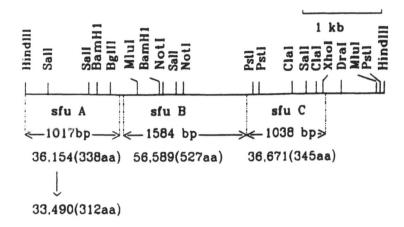

FIGURE 6. Arrangement of the iron(III) transport genes of *Seratia marcescens*. The genes are transcribed from left to right.

suffered from the lack of information about the iron(III) ligand recognized by the transport system. It is possible that this incomplete transport system, with respect to the other iron transport systems, exists as such in *S. marcescens* but it may also consist of additional components. However, the interesting observation remains that it suffices to supply iron to *E. coli* in contrast to the other Fe^{3+} transport systems which are inactive in the absence of an outer membrane receptor or TonB activity.

TRANSPORT OF Fe^{3+}-SIDEROPHORES ACROSS THE OUTER MEMBRANE

The receptor proteins in the outer membrane are the distinctive features of the Fe^{3+}-siderophore transport systems. There may be two reasons for cell surface receptors. The Fe^{3+}-siderophores are too large to diffuse with sufficient rates through the water-filled channels formed by the porin proteins (the latter exclude molecules larger than 600 Da). The Fe^{3+}-siderophores are larger with the exception of Fe^{3+}-dicitrate (443 Da). In addition, the iron-free siderophores exhibit a very low affinity for the transport system and diffuse out of the cells. After being loaded with iron, they are fixed to the receptor once they contact the cell surface. By this means iron-siderophores are extracted from the medium and concentrated at the cell surface, from where they are taken up into the cell by an active process. Alternatively, they would diffuse in and out of the porin channels until they encounter a transport protein in the periplasmic space. Fixation at surface receptors and active transport across the outer membrane accelerates the overall uptake rate considerably.

Receptor proteins are synthesized in response to iron limitation in such amounts that they become major outer membrane proteins with copy numbers between 10,000 and 100,000 per cell. In the iron-repressed state they amount to about 1000 to 5000 molecules (see next chapter). The high concentration and the limited variety of proteins in the outer membrane facilitated the detection of the Fe^{3+}-siderophore receptors. In fact, the FhuA receptor protein (formerly named TonA protein) was the second membrane protein of *E. coli* (after the murein-lipoprotein[90]) to be isolated in a pure and active form.[91] Interest in this protein originally focused on its properties as a phage and colicin receptor (Figure 1) until in a search for *tonB* mutants, using albomycin for selection, *fhuA* mutants were isolated. Most of the mutants were resistant to phages T5, T1, and colicin M,[92] suggesting FhuA to be involved in albomycin uptake. Inhibition of killing by colicin M through ferrichrome supported the notion of a common receptor. The same conclusion was reached by competition experiments between ferrichrome and phage φ80.[93] Iron-regulated outer membrane proteins were also observed in *Salmonella typhimurium*.[94] Following these initial observations the receptor proteins for Fe^{3+}-dicitrate, Fe^{3+}-enterochelin,[95] and Fe^{3+}-aerobactin were identified.[96]

The Fe^{3+}-siderophores not only bind to the receptors of metabolizing (energized) but also to nonmetabolizing (unenergized) cells. For example, ferrichrome inhibited binding of phage T5 to energy-deprived cells[97] and to the isolated FhuA protein. This test can easily be performed in a quantitative way because T5 binding triggers release of phage DNA which in turn inactivates the phage so that the surviving phage can be counted by the classical plaque assay. Interestingly, ferrichrome inhibited T5 binding to energized cells only poorly.[97] At a 1 μM ferrichrome concentration, inhibition of T5 binding to unenergized cells was 90% whereas a 100-fold higher concentration of ferrichrome inhibited T5 binding to energized cells less than 40%. T5 binding inhibition of a *tonB* mutant was more pronounced as binding was nearly 100% abolished by 0.1 μM ferrichrome, regardless of the energy state of the cell. The same results were obtained using colicin M in place of ferrichrome as the T5 competitor. As long as the cells were unenergized, or were *tonB*, colicin M remained at the receptor and inhibited T5 binding.[98] These results led us to propose that the FhuA receptor has two functional states. In the unenergized conformation ferrichrome is able to inhibit T5 binding, and colicin M stays at the receptor but is not transported across the outer membrane. This could be shown by so-called trypsin rescue experiments which are based on trypsin degradation of colicins as long as they are bound to the receptors and have not been internalized.[98] In the energized state colicin M is taken up by the cells, and can no longer be degraded by trypsin.

Inhibition of T5 binding by ferrichrome could not be explained by lack of ferrichrome uptake into unenergized cells or *tonB* mutants since binding of phages T1 and φ80 was inhibited in energized *tonB* wild type cells. The latter two phages have the interesting properties of testing the metabolic state of the cells they are going to infect. They bind irreversibly only to energized *tonB* wild type cells which are able to support multiplication of the phages. Energization can proceed via the electron transport chains, or by ATP hydrolysis through the membrane-bound H^+-ATPase.[99] T1 host range mutants exist which can infect *tonB* mutants indicating that DNA uptake does not require the TonB function.[99] It appears that DNA release from the wild type T1 and φ80 phage heads is only triggered by the energized FhuA conformation.

The requirement of an energized state for receptor function is always linked with the requirement for the TonB function. This led us to suggest that FhuA receptor activity is regulated by the energized state of the cell, and that the TonB protein serves as a coupling device between the cytoplasmic membrane and the outer membrane receptors.[99] There are no metabolites known outside the cytoplasmic membrane which could regulate outer membrane receptor activity. Recently, the TonB protein was localized in the cytoplasmic membrane.[100,101] The rather polar amino acid sequence, deduced from the nucleotide sequence,[102] contains an accumulation of hydrophobic amino acids at the N-terminal end which may anchor the TonB protein in the cytoplasmic membrane, with the rest extending into the periplasmic space. In fact, in spheroplasts the TonB protein has been degraded by trypsin and proteinase K, indicating exposure to the periplasmic space.[101,103]

Recently, genetic evidence for an interaction between the TonB protein and the FhuA protein has been obtained. Mutations in the *fhuA* gene, which abolished or reduced TonB-dependent FhuA activities, were suppressed by mutations in the *tonB* gene.[104] Point mutations were introduced by site-directed mutagenesis into a region of *fhuA* which exhibited strong homology to all, and only to TonB-dependent receptors and colicins[104-108] (Figure 7). This consensus sequence was designated "TonB box" to indicate its involvement in coupling to the TonB protein. In all proteins is the TonB box located close to the N-terminal end. The TonB box of the FhuA protein reads Asp Thr Ile Thr Val. Replacement of Ile with Pro abolished growth on ferrichrome and conferred resistance to albomycin. A *tonB* suppressor mutation, resulting in a TonB protein with an amino acid exchange of Gln to Lys at position 165 of the open reading frame (position 160 of the actual translation product[101,102]), introduced

			Asp	Thr	Ile	Thr	Val				
FhuA	(6-15)	Glu	Asp	Thr	Ile	Thr	Val	Thr	Ala	Ala	Pro
FhuE	(5-14)	Glu	Glu	Thr	Val	Ile	Val	Glu	Gly	Ser	Ala
IutA	(5-14)	Asp	Glu	Thr	Phe	Val	Val	Ser	Ala	Asn	Arg
FecA	(22-31)	Gly	Phe	Thr	Leu	Ser	Val	Asp	Ala	Ser	Leu
FepA	(11-20)	Asp	Asp	Thr	Ile	Val	Val	Thr	Ala	Ala	Glu
Cir	(5-14)	Gly	Glu	Thr	Met	Val	Val	Thr	Ala	Ser	Ser
BtuB	(5-14)	Pro	Asp	Thr	Leu	Val	Val	Thr	Ala	Asn	Arg
Colicin B	(16-25)	Gly	Asp	Thr	Met	Val	Val	Trp	Pro	Ser	Thr
Colicin D	(16-25)	Gly	His	Ser	Met	Val	Val	Trp	Pro	Ser	Thr
Colicin M	(1-10)	Met	Glu	Thr	Leu	Thr	Val	His	Ala	Pro	Ser
Colicins Ia/Ib	(22-31)	His	Glu	Ile	Met	Ala	Val	Asp	Ile	Tyr	Asn

FIGURE 7. Consensus sequence in TonB-dependent outer membrane receptor proteins and in colicins, termed TonB box. The numbers indicate the positions of the amino acid residues in the mature proteins. The most highly conserved residues are underlined. Colicins are toxic proteins synthesized by certain strains of *E. coli* and kill only *E. coli* cells or very close relatives. The uptake of the colicins listed depends on receptor proteins, on TonB, ExbB and ExbD, and on cellular energy. Transport of colicins across the outer membrane displays many features of the transport of iron(III) siderophores and of vitamin B_{12}.

on a plasmid, restored growth on ferrichrome, sensitivity to albomycin and colicin M. A Gln to Leu exchange at the same position only increased colicin M sensitivity tenfold. Replacement of Val by Asp in FhuA resulted in an inactive FhuA protein that regained sensitivity to colicin M when combined with the TonB mutants. Cells remained albomycin resistant and unable to use ferrichrome. Asp could be replaced with glycine without loss of any FhuA activity. Furthermore, replacement of Asp with Ala in colicin B (Figure 7) did not reduce the activity in contrast to a Pro replacement which abolished activity.[109] In 9 out of 11 sequences the first residue of the consensus pentapeptide is occupied by an acidic amino acid. FecA with Phe and colicin D with His at this position are the exceptions but naturally occurring active variants. It seems that the function of the TonB box is not determined by amino acid side chains occurring at certain positions but rather by the conformation of the entire region which is dependent on the interaction of all the residues. Computer-assisted analysis of conformation changes revealed an increase in the flexibility of the FhuE protein when Val (3rd position in Figure 7) was replaced with Leu, Gln, Arg, and Pro. Fe^{3+}-coprogen transport rates decreased in the same order.[51] Therefore, a more rigid structure seems to conform better with the proposed binding of the TonB box region to the TonB protein. Suppression of a mutation in one protein by a mutation in another protein suggests a physical and functional interaction between the two proteins. The impaired interaction caused by a mutation in one protein can be restored by an appropriate mutation in the other protein. At present, this is the only evidence for a direct interaction between receptors, colicins and the TonB protein.

The first spontaneous *tonB* suppressor mutations were identified in a mutant that could bind but not transport vitamin B_{12}.[89] The mutant, named *btuB451*, expressed a BtuB outer membrane receptor protein for vitamin B12 in which Leu of the TonB box was replaced by Pro. Phenotypic revertants were able to grow on a medium supplemented with 5 n*M* vitamin B12. The methionine auxotrophic strain (*metE*) used in these studies required vitamin B_{12} for methionine biosynthesis. Growing cells were not reverted in *btuB* but mutated in *tonB*. Both *tonB* suppressor mutations (Gln/Leu, Gln/Lys) restored uptake of vitamin B_{12} equally well. However, in a strain mutated in the *exb* locus only *tonB* (Gln/Leu) transformants grew with similar rates to BtuB wild type cells while *tonB* (Gln/Lys) transformants formed only very small colonies after several days of incubation.[103]

The term *exb* originally defined mutations which conferred insensitivity to colicin B due to the excretion of an inhibitor which was shown to be enterochelin.[111] Iron uptake of *exb* mutants was reduced to a different extent for various siderophores.[112] Iron limitation led to an increased synthesis of enterochelin which inhibited binding of colicin B by competing for the same receptor (FepA) (see chapters on specificity of iron transport in bacteria and fungi and on pyoverdins and pseudobactins.) In this respect *exb* mutants resemble *tonB* mutants. In fact, *exb*-related functions were always *tonB*-related functions. The *exb* locus was cloned and sequenced.[113] It encoded two genes, designated *exbB* and *exbD*, which expressed two proteins composed of 244 and 141 amino acids. The available chromosomal mutations in the *exb* locus affected both genes, as revealed by complementation analysis using plasmids carrying *exbB*, or *exbD*, or both genes.[113] *exb* Mutants were resistant to albomycin and deficient in ferrichrome transport. They exhibited 10- to 1000-fold reduced sensitivity to the TonB-dependent colicins B, D, and M.[113] Overproduction of the TonB protein in *exb* mutants increased colicin sensitivity to near *exb* wild type level, depending on the colicin used.[103] This shows partial compensation of the ExbB,D functions by overproduction of TonB. If the ExbB,D proteins contribute to the activity of the TonB protein, an increase in the TonB population may supply enough active TonB molecules so that ExbB,D are partially dispensable. In fact, ExbB affects TonB. TonB activity and the TonB protein are unstable. The activity of TonB rapidly declined after cessation of TonB synthesis.[114,115] Radioactively labeled TonB protein exhibited a half-life of about 10 min.[101,103] Degradation of TonB was inhibited when ExbB was overproduced at the same time.[103] TonB and ExbB had to be overproduced to be detectable on an autoradiogram after SDS-PAGE. Expression of chromosomally encoded *exbB* was not sufficient to prevent the decline of overproduced TonB protein. ExbB also prevented degradation of ExbD and delayed proteolysis by trypsin added to spheroplasts. ExbB failed to prevent proteolysis of TonB, and ExbD did not affect TonB stability. Trypsin degraded ExbB to defined products suggesting the protected regions to be embedded in the cytoplasmic membrane.

Two lines of evidence were obtained for the involvement of the ExbB,D proteins in transport across the outer membrane, although the proteins were primarily found in the cytoplasmic membrane fraction.[113] $^{55}Fe^{3+}$-labeled ferrichrome was released by a surplus of unlabeled ferrichrome from *exb* mutants suggesting that most of the ferrichrome was still bound to the overproduced FhuA receptor and had not crossed the outer membrane. In addition, the requirement of ExbB,D for colicin M sensitivity could be overcome by rendering the outer membrane temporarily permeable through osmotic shock treatment.[103] The same procedure made *tonB* mutants sensitive to colicin M.[98] A model consistent with these data proposes a trimer consisting of TonB, ExbB, and ExbD in the cytoplasmic membrane. ExbB physically stabilizes TonB and thus increases TonB activity. The model does not exclude the possibility that ExbB,D chemically modify TonB and in this way contribute to TonB activity and stability. However, the high amounts of ExbB required to stabilize TonB argue against a catalytic process and are more in favor of a stoichiometric relation.

The ExbB and ExbD proteins revealed 25% identity and more than 70% similarity (conservative amino acid replacements) with the amino acid sequences of the TolQ and TolR proteins, respectively.[113] The latter proteins are encoded by the *tolQ* and *tolR* genes to which the *tolA* and *tolB* genes are closely linked.[116,117] Group A colicins (A, E1, E2, E3, K, N), in contrast to the TonB-dependent group B colicins (B, D, M, I, V) require the *tolQRAB*-encoded functions for uptake into cells (except E1 uptake which is *tolB*-independent). Since *tol* mutants are sensitive to 2.5% cholate, an outer membrane alteration was suspected to be the cause of resistance to the A group colicins. These colicins have very different target sites and modes of action so that the common denominator affected by *tol* mutations must

```
PROTEIN   RESIDUE                          CONSERVED SEQUENCE

FhuC        30    L S L T F P A G K V T G L I G H N G S G K S T L - L K M L G R H
FecE        21    V S L S L P I G K I T A L I G P N G C G K S T L - L N C F S R L
SfuC   (Sm) 22    I D L Q V A A G S R T A I V G P S G S G K T T L - L R I I A G F
BtuD        19    L S G E V R A G E I L H L V G P N G A G K S T L - L A R M A G M

HisP        25    V S L Q A R A G D V I S I I G S S G S G K S T F - L R C I N F L
MalK        22    I N L D I H E G E F V V F V G P S G C G K S T L - L R M I A G L
PstB        29    I N L D I A K N Q V T A F I G P S G C G K S T L - L R T F N K M
OppD1  (Ec) 43    V T L R L Y E G E T L G V V G E S G C G K S T F - A R A I I G L
OppD2  (Ec) 42    L N F S L R A G E T I G I V G E S G S G K S Q I - A F A L M G L
OppD   (St) 40    L N F T L R A G E T L G I V G E S G S G K S Q S R L R - L M G L
RbsA   (N)  23    A A L N V Y P G R V M A L V G E N G A G K S T M - M K V L T G I

FhuC        141   L S G G E R Q R A W I A M L V A Q D S - R - - - - - - C L L L D E P T S A L D I A H Q V D V L S L V H R
FecE        139   L S G G Q R Q R A F L A M V L A Q N T P - - - - - - - V V L L D E P T T Y L D I N H Q V D L M R L M G E
SfuC   (Sm) 136   L S G G Q Q R V A L A R A L S Q Q - P R - - - - - - L M L L D E P F S A L D T G L R A A T R K A V A E
BtuD        127   L S G G E W Q R V R L A A V V L Q I T P Q A N P A G Q L L L L D E P M N S L D V A Q Q S A L D K I L S A

HisP        154   L S G G Q Q Q R V S I A R A L - A M E P D - - - - - - V L L F D E P T S A L D P E L V G E V L R I M Q Q
MalK        134   L S G G Q Q Q R V A I G R T L V A - E P S - - - - - - V F L L D E P L S N L D A A L R V Q M R I E I S R
PstB        152   L S G G Q Q Q R L C I A R G I - A I R P E - - - - - - V L L L D E P C S A L D P I S T G R I E E L I T E
OppD1  (Ec) 165   F S G G Q C Q R I G I A R A L - I L E P K - - - - - - L I I C D E P V S A L D V S I Q A Q V V N L L Q Q
OppD2  (Ec) 169   F S G G M R Q R V M I A M A L L - C R P K - - - - - - L L I A D E P L S N L D A A L R V Q M R I E I S R
OppD   (St) 167   F S G G M R Q R V M I A M A L L - C R P K - - - - - - L L I A D E P T T A L D V T V Q A Q I M T L L N E
RbsA   (N)  144   L S I G D Q Q M V E I A K V L S F - E S K - - - - - - V I I M D E P T D A L T D T E T E S L F R V I R E
```

Proteins were from *E. coli* unless otherwise indicated (9). (Ec) *Escherichia coli*, (St) *Salmonella typhimurium*, (N) amino terminal half to the polypeptide; Sm, *Serratia marcescens*.

FIGURE 8.   Segments of polar transport proteins in the cytoplasmic membrane displaying sequence homologies to nucleotide-binding proteins.

reside in a common uptake step. In a sense the Tol functions are the pendant of the TonB, ExbB and ExbD functions. Sequence homology of ExbB with TolQ, and of ExbD with TolR prompted an investigation as to whether Tol and Exb functions can replace each other. The only TonB-dependent function which apparently was completely ExbBD-independent was the irreversible adsorption of phages T1 and φ80. The plating efficiency of these phages was also not altered on *tol* mutants. However, double mutants in the *tolQ* and *exbB* genes were completely phage resistant[118] and were also resistant to colicins E1, E2, and K tested.[119] These results indicate that division into group A and group B colicins, based on their uptake routes, is still valid because single mutations in the *tol* or the *tonB exbBD* genes confer strongly reduced sensitivities for either group A or group B colicins. However, double mutants in *tolQ* and *exbB* genes are completely resistant to group A and group B colicins demonstrating that *tolQ* and *exbB* encoded functions can partially replace each other in single mutants. The structural similarities of TolQ and ExbB, maintained after a divergent evolution from a common ancestor, still preserved features of a once common function. The mutual replacement is not perfect but still very pronounced.

TRANSPORT OF IRON(III) SIDEROPHORES ACROSS THE CYTOPLASMIC MEMBRANE

The iron transport systems consist of a periplasmic protein, one or two very hydrophobic proteins in the cytoplasmic membrane, and a polar membrane-bound protein containing two regions which are characteristic for nucleotide binding proteins (Figure 8). These properties are also found in periplasmic binding protein-dependent transport systems (see chapter on Structures, Coordination Chemistry and Functions of Microbial Iron Chelates.) Binding to a periplasmic protein was only demonstrated for Fe^{3+}-dicitrate to FecB with an estimated affinity (K_D) of 0.7 μM.[120] The periplasmic FepB, FhuD, FebP and SfuA proteins are required for transport. In contrast, most of the BtuE protein of the mechanistically similar vitamin B_{12} transport system can be released by osmotic shock treatment but is not required for vitamin B_{12} transport.[121] The BtuE protein, as deduced from the nucleotide sequence, does not contain a signal sequence which is required for export into the periplasmic space.

Therefore, BtuE most likely is not a periplasmic binding protein. A vitamin B_{12}-binding protein was isolated, but not correlated to any of the sequenced *btu* transport genes.[122]

The specificity of the ferrichrome transport system is not very narrow. Addition of one succinate residue to ferricrocin did not alter the transport rate despite conversion of a neutral molecule to a negatively charged derivative.[123] Ferrichrysinyl disuccinate transported iron with more than half the rate of ferrichrome.[124] The octahedral ambient of Fe^{3+}, which is probably primarily recognized by the transport proteins, is the same in these ferrichrome derivatives. Also enantio-ferrichrome was transported by *E. coli* with 37% of the ferrichrome rate.[125] The variety of Fe^{3+}-hydroxamates accepted by the FhuBCD proteins is astonishing (Figure 1) although they are transported with different rates (ferrichrome, aerobactin, and coprogen much faster than ferrioxamine B and rhodotorulic acid). Fe^{3+} is mostly surrounded by N–O and C=O groups but the configuration differs, and in aerobactin one N–O C=O pair is replaced with OH C–O.[126]

Energization studies of the Fe^{3+}-siderophore transport systems are hampered by the energy-consuming transport across two membranes which may have different energy requirements. Using *hemA* and *uncA* mutants devoid of cytochromes and lacking H^+-ATPase activity, respectively, energization of ferrichrome transport could either be powered by membrane-bound electron transport chains or by ATP hydrolysis. Arsenate at the low concentration of 10 μM inhibited ferrichrome transport into the *hemA* mutant. Much higher concentrations of cyanide (4 mM) were required to inhibit ferrichrome transport into the *uncA* mutant.[127] Glutamine transport, measured in parallel, was inhibited to a similar extent by arsenate but less by cyanide than ferrichrome transport. It is possible that membrane energization via electron transport chains mainly affects uptake through the outer membrane as has been found for the mechanistically very similar uptake of vitamin B_{12}.[122] There is no evidence that transport across the outer membrane is coupled with transport across the cytoplasmic membrane. Rather, it seems that these are two completely independent events catalyzed by different proteins. However, the periplasmic proteins could be involved in the release of the iron-siderophores from the receptors.

RECEPTOR-DEPENDENT IRON UPTAKE FROM TRANSFERRINS

Transferrin in the serum is only partially saturated (30%) by iron. Even traces of iron are avidly bound by transferrin leaving no free iron available for bacteria. Lactoferrin fulfills a similar function in secretory fluids. Iron deprivation is an important factor determining the unspecific defense of the host against infections.[128] Pathogenic bacteria have solved the problem of iron limitation in different ways. Some produce siderophores to sequester iron while others use still unknown mechanisms to fulfill their iron requirements.

One very interesting system has been elucidated during the last years in the highly adapted pathogens *Neisseria gonorrhoeae*, *N. meningitidis*, and *Haemophilus* species. These organisms, with the exception of some species of *Haemophilus*,[130] do *not* seem to produce siderophores.[129] It was found that transferrin, and sometimes also lactoferrin, were able to supply the pathogenic *Neisseria* with iron.[131,132] Direct contact between the iron carrier protein and the bacterial cell surface is necessary for efficient iron supply.[133] Using transferrin conjugated to peroxidase it was possible to identify a highly specific human transferrin receptor in *N. meningitidis*.[134] Transferrin and lactoferrin receptors were demonstrated in *N. meningitidis*,[135] *N. gonorrhoeae*,[136] *H. influenzae*,[137] and *Bordetella pertussis*.[138] Iron uptake studies with lactoferrin and transferrin in *N. gonorrhoeae* indicate that the uptake systems are probably efficient enough to adequately supply the organism with iron in the human body.[139] A very impressing fact is the high specificity of these iron regulated receptors for either human transferrin or human lactoferrin. The receptors identified to date have molecular weights in the range between 70,000 and 105,000 Da which is in the same order of magnitude

TABLE 2
Examples of Iron Regulated Envelope Proteins in Bacteria

Organism	Size of proteins in kilodaltons	Ref.
Gram-Negative Bacteria		
Escherichia coli	83, 81, 80, 79, 76, 74.5,[a] 74	20
Shigella flexneri	83, 81, 74	144
Shigella species	83—72	145
Salmonella typhimurium	83, 81, 74	146
Salmonella typhi	83, 78, 69	147
Klebsiella aerogenes	83, 78,75,73,70,69	148
Yersinia species	240, 190, 89, 81, 79, 70, 68, 27.5	149
Proteus species	69—83	150
Erwinia chrysanthemi	78, 82, 88	151
Vibrio cholerae	220, 77, 76, 75, 73, 62	152
Vibrio anguillarum	(86) 79,[a] 79	153,154
Vibrio vulnificus	81, 74	155
Aeromonas salmonicida	86, 79	156
Haemophilus species	86, 76, 74, 73, 70	157, 130
Haemophilus pleuropneumoniae	105, 76	158
Pasteurella multicoda	84	159
Neisseria gonorrhoeae	97, 88, 80, 37, 22	160
Campylobacter jejuni	82, 76, 74	161
Pseudomonas species	80, 75, 14	162, 163
Azotobacter vinelandii	93, 85, 81, 77	164
Azotobacter chroococcum	76—70	165
Azomonas macrocytogenes	74, 70	166
Azospirillum brasilense	87, 83, 78, 72	167
Aquaspirillum magnetotacticum	58, 55, (72, 76, 85)[b]	168
Paracoccus denitrificans	84—76	169
Synechococcus PCC7942	92, 50, 48, 35	170
Anacystis nidulans		171
Gram-Positive Bacteria		
Mycobacterium smegmatis	180, 84, 29, 25	172

[a] Proteins determined by plasmid.
[b] Produced at relatively high iron concentrations (20 μM which also allowed synthesis of hydroxamate).

as most of the siderophore receptors (see Table 2). A high specificity for the transferrins of their human or porcine hosts has also been found in *H. influenzae* and *H. haemolyticus* (human origin) and *H. pleuropneumoniae* and *H. parasuis* (porcine origin).[130] It is possible, but not proven, that these iron uptake systems determine the host specificity of these bacteria.

Only the uptake system of *B. pertussis* seems to be less specific as lactoferrin can compete for transferrin in binding assays.[138] Further experiments, in particular cloning and sequencing the genes involved, will be necessary for a better characterization of these interesting receptor proteins and the whole iron uptake system.

IRON REGULATION

In the 1960s, the first reports on iron regulation in *E. coli* dealt with the biosynthesis of 2,3-dihydroxybenzoylserine and its repression by iron. It was speculated that dihydroxybenzoylserine may have a regulatory function by removing iron from a repressor. This would

TABLE 3
Iron Regulated Virulence Factors and Other Systems Not Involved in Iron Transport

Organism	Regulated system	Ref.
Toxins		
Escherichia coli	Colicin V	173
Escherichia coli	Hemolysin	174
Escherichia coli	Shiga-like toxin I	175, 176
Shigella dysenteriae	Shiga toxin	177
Serratia marcescens	Hemolysin	178
Vibrio cholerae	Hemolysin	179
Pseudomonas aeruginosa	Exotoxin A	180
Pseudomonas aeruginosa	Hemagglutinin	181
Pseudomonas aeruginosa	Elastase	181
Listeria monocytogenes	Hemolysin	182
Corynebacterium diphtheriae	Diphtheria toxin	183
Other Systems		
Vibrio parahaemolyticus	Swarmer cell differentiation	184
Vibrio harveyi	Bioluminescence	185
Azotobacter vinelandii	Induction of competence	186
Anacystis nidulans	Induction of competence	187

lead to a derepression of the biosynthetic operon.[140] Young and Gibson reported that iron alone was sufficient to repress the synthesis of enzymes involved in the conversion of chorismic acid to dihydroxybenzoic acid.[141] During these studies they also detected enterochelin, the cyclic trimer of dihydroxybenzoylserine.

Nearly 10 years later the SDS polyacrylamide gel electrophoresis technique of Laemmli[142] modified by Lugtenberg[143] was applied to the separation of outer membrane proteins of *E. coli*. It was realized that a certain group of high molecular weight proteins was only produced under low iron growth conditions.[95] These proteins were later found to be part of siderophore transport systems. In the meantime, the outer membranes of nearly any Gram-negative bacterium examined revealed iron regulated outer membrane proteins, often in the molecular weight range of 70,000 to 80,000 Da. A partial listing of studied strains is given in Table 2. It is assumed that most of these proteins are siderophore receptors as has been shown for *E. coli*. The molecular weights of the iron-regulated proteins given were determined by SDS-polyacrylamide gel electrophoresis. In many cases it may be that other strains from the same species may have a different pattern of iron-regulated outer membrane proteins. Caution is also necessary in interpreting proteins produced under low iron stress, as some proteins are produced by several "stress conditions".[160]

Interestingly, there are also genes which are either unrelated or only indirectly related to iron transport, which are regulated by iron (Table 3). One group of general interest is a variety of bacterial toxins which are only produced under low-iron conditions. Not only toxins of medical importance such as some hemolysins but also the production of the long known colicin V[173] are iron regulated. With respect to hemolysins, this may have a relation to iron uptake since a large amount of iron-containing hemoglobin is liberated when an erythrocyte is lysed. However, only in some strains of *E. coli* is the hemolysin iron regulated.[174] Expression of the luminescence genes in *Vibrio harveyi* is also regulated by iron,[185] however in other luminescent bacteria the regulation seems to be different. In all these cases it is assumed, and sometimes it has been shown, that bacteria are deprived of iron inside their hosts. It has been suggested that low iron is used as an environmental signal by bacteria

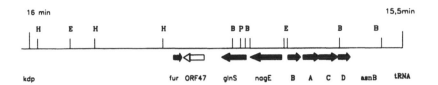

FIGURE 9. Organization of genes in the neighborhood of *fur* on the *E. coli* chromosome.
♦ genes sequenced, ◊ ORF47, product of an unknown gene. Restriction sites: B (*Bgl*II); E
(*Eco*RI); H (*Hind*III); P (*Pvu*I).

to indicate that they are in their host.[188,189] The swarming of *Vibrio parahemolyticus* is
dependent on the physical hinderance of the polar flagellum, and iron limitation.[184] In this
case, low iron may signal overcrowding, and the differentiation to swarming cells may be
beneficial for the organism.

MOLECULAR MECHANISM OF IRON-REGULATION IN *E. COLI*
Fur Protein as a Repressor

Operon fusions to Mud1(Ap*lac*) were used to study regulation of the genes involved
with iron uptake in *E. coli* and to isolate constitutive mutants[190] which showed high expression
of the transport genes irrespective of the iron concentration in the medium (*fur* mutants;
*f*erric *u*ptake *r*egulation). The operon fusions were also helpful in cloning and characterizing
the *fur* gene, which was mapped near 15.5 min on the genetic map of *E. coli*.[191] From the
restriction map of this region, *fur* can now be placed in the neighborhood of *glnS* (Figure
9).[191-193] The cloned gene was sequenced[194] and found to code for a 16.795 kDa protein
which was very rich in histidine (8%).

The Fur protein was purified by metal-ion-affinity chromatography on zinc iminodiace-
tate agarose,[195] or zinc chelate Sepharose.[196] The protein is strongly bound to these supports
possibly because of its high histidine content. Fur is eluted from these columns with a
gradient of histidine. DNA-binding assays were used to study the specific interactions of
the purified repressor protein with DNA from the promoter region of the *aer (iuc)* operon.[197,198] The DNA binding regions of the Fur protein were identified in the promoter
regions of *aer*,[198] *fur*,[199] and *cir* by DNase I footprinting.[200] Fur binding sites contained a
19bp consensus sequence which is found in the promoter region of all genes negatively
regulated by iron (Figure 10). Most of the regulated genes which have been sequenced are
part of ferric siderophore uptake systems, and some are involved in the biosynthesis of
aerobactin or enterochelin. *sodA* is the gene for the manganese containing superoxide dismutase which has been shown to be iron regulated.[210] *sltA* codes the Shiga-like toxin I and
is one representative of a whole group of bacterial toxins which are known to be regulated
by iron (Table 3). Interestingly, there is a second toxin, Shiga-like toxin II, which has about
57% homology to toxin I but is *not* iron regulated and has no Fur binding site in the
corresponding promoter region.[203] The regulation of toxin II is not understood but since
there are strains carrying both toxins on two phages, the second toxin may be expressed
under conditions which not necessarily involve iron deficiency.

Also in other organisms, not closely related to *E. coli*, very similar sequences are
observed in the promoter regions of iron-regulated genes (Figure 11).

To date there have been no studies to quantitate the binding affinity of Fur to different
binding sites. Only by a statistical approach, calculating the deviation from the consensus
sequence, a ranking of the different putative Fur binding sites has been attempted.[200]

For the specific binding of Fur to DNA the presence of one of the following divalent
heavy metals Mn^{2+}, Fe^{2+}, Co^{2+}, Cu^{2+}, Cd^{2+}, and partially Zn^{2+}, is essential.[198] In most
experiments Mn^{2+} was used instead of Fe^{2+}, since the latter is readily oxidized under aerobic
conditions. It is interesting that Mn^{2+} also under certain *in vivo* conditions may bind to Fur.

Gene	Nucleotide No.[a]	Sequence	Reference
in *E.coli*			
aerA (*iucA*)	349	G A T A A T G A G A A T C A T T A T T	198,201
aerA (*iucA*)	369	C A T A A T T G T T A T T A T T T A	198,201
fhuA	383	C T T T A T A A T A A T C A T T C T C	202
fepA	415	A T T A T T G A T A A C T A T T T G C	202a
fepA	427	T A T A T T G A T A A T A T T A T T G	202a
fhuE	62	T A C A A A C A A A A T T A T T C G C	52
fhuE	40	G C G T A T A T T T C T C A T T T G C	52
fecA	144	G A A A A T A T T C T T A T T T C G	63
fecA	138	T G T A A G G A A A A T A A T T C T T	63
cir	2	T G G A T T G A T A A T T G T T A T C	85,86
cir	8	G A T A A T T G T T A T C G T T T G C	85,86
tonB	271	G A A T A T G A T T G C T A T T T G C	102
exbB	506	G A G A A C G A C T A T C A A T T C G	113
exbC	1118	A G C A A C G G C A A T C G G C C T C	113
fepB	-13	G A A A A T G A G A A G C A T T A T T	70,76
entC	7	A T A A A T G A T A A T C A T T A T T	70
fes	391	G C A A A T G C A A A T A G T T A T C	202a
fes	415	T A T T A T C A A T A T A T T T C T G	202a
fur	186	T A T A A T G A T A C G C A T T A T C	194
fecI	1170	T G T A A T G A T A A C C A T T C T C	63
fecI	1176	G A T A A C C A T T C T C A T A T T A	63
sltA	204	G A A T A T G A T T A T C A T T T T C	203,175,
sodA	48	G A T A A T C A T T T T C A A T A T C	204
consensus		G A T A A T G A T A A T C A T T A T C	
Serratia marcescens			
shlB	320	G A T T G T C A T A A T T T C C C C C	205
sfuA	54	T T T A A T A C G A A T C G T T T T C	89
Vibrio cholerae			
hly		A A T A A T A T G A A T A T C A G T A	206
Corynebacterium diphtheriae			
tox	-56	T A T A A T T A G G N_9C C T A A T T A T T	207
Anacystis nidulans			
isiA	-51	A C T T A TTG A G A A T T A T T G T A	208
isiA	-61	C T T A A TAT C A A C T T A T T G A G	208
Synechococcus sp.			
irpA	-212	T A A A A A T G A N_9T C A T T T T T A	209

a) Nucleotide numbering according to references cited.

FIGURE 10.   Sequences of proved or assumed Fur binding sites.

It was found that with low Mg^{2+} medium the iron regulated gene *fhuF* was repressed by the addition of Mn^{2+}.[211] A new gene, *ird*, and its operon fusion with Mud1 has been described and is positively regulated by Fur and Fe^{2+}. Also an induction of this fusion was observed after addition of Mn^{2+}.[211] On low Mg^{2+}/high Mn^{2+} medium about 50% of the selected manganese-resistant colonies were mutated in *fur*. This was interpreted to indicate that Mn^{2+} binding to Fur leads to a repression of all iron uptake systems which thus starves the cells for iron and results in growth inhibition.[211]

It is assumed that the Fur protein, as with other DNA-binding proteins, binds as a dimer or multimer to its DNA recognition sequence. Proof for this hypothesis was obtained by the isolation of negatively complementing mutations in the cloned *fur* gene. Large amounts of these plasmid encoded mutated Fur proteins in a *fur*+ host led to a derepression of Fur-regulated genes. This is explained by the formation of mixed oligomers which were inactive *in vivo* and *in vitro*.[196]

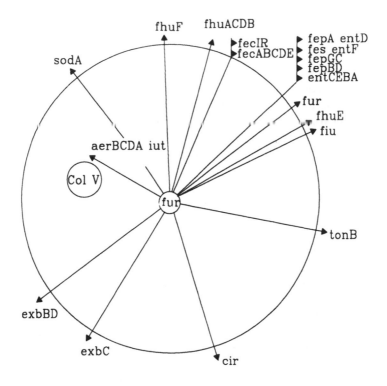

FIGURE 11. Iron regulated genes in *E. coli*.

FUR AS A POSITIVE REGULATOR?

As previously mentioned, Fur with iron as cofactor seems not only to act as a repressor but may act directly or indirectly as an activator in the case of *ird* Fur-Fe^{2+}.[211] Similarly, it has been observed that *sodB*, the gene for the iron-dependent superoxide dismutase, is not expressed in a *fur* mutant but is expressed normally in the same strain after transformation with a fur$^+$ plasmid.[212] It is concluded that Fur either directly or indirectly controls SodB biosynthesis. The *sodB* mutation leads to increased secretion of enterochelin.

Many spontaneous Fur mutants isolated by the Mn^{2+} selection procedure are unable to grow on succinate, fumarate, and acetate.[211] In this case Fur directly, or indirectly, seems to activate these catabolic pathways. It is interesting that the *fur* gene is regulated by Fur itself (Figure 10), and by the catabolite-activator protein (CAP)[199] which leads to high levels of Fur when cells are grown on succinate, fumarate, or acetate.

OUTLOOK

Employment of genetics combined with identification of the gene products and their subcellular localization has resulted in a basic understanding of the way iron(III) is transported into Gram-negative bacterial cells. This goal was achieved in a very short time. Conceptually new and therefore most interesting is the existence of transport systems across the outer membrane which shed a new light on the structure and function of this cell compartment. Translocation of the iron(III) siderophores across the outer membrane is certainly not a pure diffusion process but rather a regulated, energy-consuming step. How energy is provided and how the activity of the outer membrane receptors is regulated is the most interesting aspect for future research. The components of the outer membrane transport system have been roughly defined. They consist of the receptor proteins with specificities for siderophore structural groups, and the TonB, ExbB, ExbD, TolQ, TolR, and TolA proteins which are common for all the iron(III) transport systems. Since the receptors are constituents of the outer membrane and the other components reside, as far as it is known, in the cytoplasmic membrane, an explanation of the transport mechanism has to take into account the functional and physical interaction between proteins in two adjacent membranes. This *inter*-membrane interaction of transport components and the energy dependence makes the development of a transport-active cell-free membrane vesicle system difficult. Such a system will ultimately be required to characterize the transport mechanism in molecular terms. In the meantime the molecular biological approach revealed a broad compositional outline of the transport systems and how they may function. The rewarding result of these studies was the recognition of a great similarity between the various iron transport systems in one organism (*E. coli*), with the perspective that very similar iron transport mechanisms seem to exist in other Gram-negative bacteria, that transport across the cytoplasmic membrane presumably follows the periplasmic-binding-protein-dependent transport mechanism, and that all of these features are shared with the vitamin B_{12} transport.

The application of this knowledge may result in the designing of improved iron chelators for treatment of iron overload diseases which do not support microbial growth and increase susceptibility to microbial infections. Antibiotics which enter cells via iron transport system show a substantially reduced minimal inhibitory concentration and thus less toxic side effects. Such compounds are, for example, the albomycins,[40] the ferrimycins,[213] the catechol cephalosporins,[87,214] and a rifamycin derivative.[215] Since siderophores also form complexes with aluminum(III) they may serve as leads for designing ligands to cure the increasing intoxication by this metal.

REFERENCES

1. **Tabor, S. and Richardson, C. C.,** A bacteriophage T7 RNA polymerase/promoter system for controlled exclusive expression of specific genes, *Proc. Natl. Acad. Sci. U.S.A.*, 82, 1074, 1985.
2. **Studier, F. W. and Moffatt, B. A.,** Use of bacteriophage T7 RNA polymerase to direct selective high-level expression of cloned genes, *J. Mol. Biol.*, 189, 113, 1986.
3. **Manoil, C. and Beckwith, J.,** A genetic approach to analyzing membrane protein topology, *Science*, 233, 1403, 1986.
4. **Braun, V. and Hantke, K.,** Bacterial cell surface receptors, in *Organization of Prokaryotic Cell Membranes*, Vol. II, Ghosh, B. K., Ed., CRC Press, Boca Raton, FL, 1981, 1.
5. **Braun, V., Fischer, E., Hantke, K., Heller, K., and Rotering, H.,** Functional aspects of gram-negative cell surfaces, in *Subcellular Biochemistry*, Vol. 11, Roodyn, D. B., Ed., Plenum Press, New York, 1985, 103.
6. **Lugtenberg, B. and van Alphen, L.,** Molecular architecture and functioning of the outer membrane of *Escherichia coli* and other gram-negative bacteria, *Biochim. Biophys. Acta*, 737, 51, 1983.
7. **Nikaido, H. and Vaara, M.,** Outer membrane, in *Escherichia coli and Salmonella typhimurium, Cellular and Molecular Biology*, Vol. 1, Neidhardt, F. C., Ed., American Society for Microbiology, Washington, D.C., 1987, 7.
8. **Ferenci, T.,** Selectivity in solute transport: binding sites and channel structure in maltoporin and other bacterial sugar transport proteins, *BioEssays*, 10, 3, 1989.
9. **Hantke, K.,** Phage T6-colicin K receptor and nucleoside transport in *Escherichia coli*, *FEBS Lett.*, 70, 109, 1976.
10. **Krieger-Brauer, H. J. and Braun, V.,** Functions related to the receptor protein specified by the *tsx* gene of *Escherichia coli*, *Arch. Microbiol.*, 124, 233, 1980.
11. **Maier, C., Bremer, E., Schmid, A., and Benz, R.,** Pore-forming activity of the Tsx protein from the outer membrane of *Escherichia coli*: demonstration of a nucleoside-specific binding site, *J. Biol. Chem.*, 263, 2493, 1988.
12. **Benz, R., Schmid, A., Maier, C., and Bremer, E.,** Characterization of the nucleoside-binding site inside the Tsx channel of *Escherichia coli* outer membrane, *Eur. J. Biochem.*, 176, 699, 1988.
13. **Benz, R. and Bauer, K.,** Permeation of hydrophilic molecules through the outer membrane of gram-negative bacteria, *Eur. J. Biochem.*, 176, 1, 1988.

14. **Hengge, R. and Boos, W.**, Maltose and lactose transport in *Escherichia coli:* examples of two different types of concentrative transport systems, *Biochim. Biophys. Acta*, 737, 443, 1983.

15. **Ames, G. F.-L.**, Bacterial periplasmic transport systems: structure, mechnism, and evolution, *Annu. Rev. Biochem.*, 55, 397, 1986.

16. **Quiochio, F. A.**, Molecular features and basic understanding of protein-carbohydrate interaction. The arabinose binding protein sugar complex, *Curr. Top. Microbiol.*, 139, 135, 1988.

17. **Scripture, J. B., Voelker, C., Miller, S., O'Donneil, R. T., Polgar, L., Rade, J., Horazdovsky, B. F., and Hogg, R. W.**, High affinity L-arabinose transport operon, *J. Mol. Biol.*, 197, 37, 1987.

18. **Higgins, C. F., Gallagher, M. P., Mimmaok, M. L., and Pearce, S. R.**, A family of closely related ATP-binding subunits from prokaryotic and eukaryotic cells, *Bio Essays*, 8, 111, 1988.

19. **Gallagher, M. P., Pearce, S. R., and Higgins, C. F.**, Identification and localization of the membrane-associated, ATP-binding subunits of the oligopeptide permease of *Salmonella typhimurium*, *Eur. J. Biochem.*, 180, 133, 1989.

20. **Braun, V.**, The iron-transport systems of *Escherichia coli*, in *The Enzymes of Biological Membranes*, Vol. 3, Martonosi, A. N., Ed., Plenum Press, New York, 1985.

21. **Braun, V.**, The unusual features of the iron transport systems of *Escherichia coli*, *TIBS*, 10, 75, 1985.

22. **Braun, V., Hantke, K., Eick-Helmerich, K., Köster, W., Preler, U., Sauer, M., Schäffer, S., Schöffler, H., Staudenmaier, H., and Zimmermann, L.**, Iron transport systems in *Escherichia coli*, in *Iron Transport in Microbes, Plants and Animals*, Winkelmann, G., van der Helm, D., and Neilands, J. B., VCH Verlagsgesellschaft, Weinheim, 1987, 35.

23. **Braun, V. and Winkelmann, G.**, Microbial iron transport. Structure and function of siderophores, in *Progress in Clinical Biochemistry and Medicine*, Vol. 5, Springer-Verlag, Heidelberg, 1987, 69.

24. **Wookey, P., Hussein, S., and Braun, V.**, Functions in outer and inner membranes of *Escherichia coli* for ferrichrome transport, *J. Bacteriol.*, 146, 1158, 1981.

25. **Fecker, L. and Braun, V.**, Cloning and expression of the *fhu* genes involved in iron(III)-hydroxamate uptake by *Escherichia coli*, *J. Bacteriol.*, 156, 1301, 1983.

26. **Köster, W. and Braun, V.**, Iron hydroxamate transport of *Escherichia coli:* nucleotide sequence of the *fhuB* gene and identification of the protein, *Mol. Gen. Genet.*, 204, 435, 1986.

27. **Köster, W. and Braun, V.**, Iron-hydroxamate transport into *Escherichia coli* K12: localization of FhuD in the periplasm and of FhuB in the cytoplasmic membrane, *Mol. Gen. Genet.*, 217, 233, 1989.

28. **Burkhardt, R. and Braun, V.**, Nucleotide sequence of the *fhuC* and *fhuD* genes involved in iron(III)hydroxamate transport: domains in FhuC homologous to ATP-binding proteins, *Mol. Gen. Genet.*, 209, 49, 1987.

29. **Coulton, J. W., Mason, P., and Allatt, D. D.**, *fhuC* and *fhuD* genes for iron (III)-ferrichrome transport into *Escherichia coli* K-12, *J. Bacteriol.*, 169, 3844, 1987.

30. **Coulton, J. W., Reid, G. K., and Campana, A.**, Export of hybrid proteins FhuA'-'LacZ and FhuA'-'PhoA to the cell envelope of *Escherichia coli* K-12, *J. Bacteriol.*, 170, 2267, 1988.

31. **Rosenbusch, J.**, Three-dimensional structure of membrane proteins, in *Bacterial Outer Membranes as Model Systems*, Inouye, M., Ed., John Wiley & Sons, New York, 1987, 141.

32. **Vogel, H. and Jahnig, F.**, Models for the structure of outer-membrane proteins of *Escherichia coli* derived from raman spectroscopy and prediction methods, *J. Mol. Biol.*, 190, 191, 1986.

33. **Jab, B. K.**, Molecular design of PhoE porin and its functional consequences, *J. Mol. Biol.*, 205, 407, 1989.

34. **Günter, K. and Braun, V.**, Probing FhuA'-'PhoA fusion proteins for the study of FhuA export into the cell envelope of *Escherichia coli* K12, *Mol. Gen. Genet.*, 215, 69, 1988.

35. **Schultz, G., Ullrich, F., Heller, K. J., and Braun, V.**, Export and activity of hybrid FhuA'-'Iut receptor proteins and of truncated FhuA' proteins of the outer membrane of *Escherichia coli*, *Mol. Gen. Genet.*, 216, 230, 1989.

36. **Duncan, T. M., Parsonage, D., and Senior, A. E.**, Structure of the nucleotide-binding domain in the β subunit of *Escherichia coli* F_1-ATPase, *Eur. Biochem. Soc.*, 208, 1, 1986.

37. **Hartmann, A. and Braun, V.**, Iron transport in *Escherichia coli:* uptake and modification of ferrichrome, *J. Bacteriol.*, 143, 246, 1980.

38. **Schneider, R., Hartmann, A., and Braun, V.**, Transport of the iron ionophore ferrichrome in *Escherichia coli* K-12 and *Salmonella typhimurium* LT2, *FEMS Microbiol. Lett.*, 11, 115, 1981.

39. **Hartmann, A., Fiedler, H.-P., and Braun, V.**, Uptake and conversion of the antibiotic albomycin by *Escherichia coli* K-12, *Eur. J. Biochem.*, 99, 517, 1979.

40. **Braun, V., Günthner, K., Hantke, K., and Zimmermann, L.**, Intracellular activation of albomycin in *Escherichia coli* and *Salmonella typhimurium*, *J. Bacteriol.*, 156, 308, 1983.

41. **Williams, P. H.**, Novel iron uptake system specified by ColV plasmids: an important component in the virulence of invasive strains of *Escherichia coli*, *Infect. Immun.*, 26, 925, 1979.

42. **Crosa, J. H.**, The relationship of plasmid-mediated iron transport and bacterial virulence, *Annu. Rev. Microbiol.*, 38, 69, 1984.

43. **Payne, S. M.,** Iron and virulence in the family Enterobacteriaceae, *CRC Crit. Rev. Microbiol.*, 16, 81, 1988.

44. **Orskov, C., Svanborg-Eden, K., and Orskov, F.,** Aerobactin production of serotyped *Escherichia coli* from urinary tract infections, *Med. Microbiol. Immunol.*, 177, 9, 1988.

45. **Wittig, W., Prager, P., Tietze, E., Seltmann, G., and Tschäpe, H.,** Aerobactin-positive *Escherichia coli* as causative agents of extra-intestinal infections among animals, *Arch. Exp. Vet. Med. Leipzig*, 42, 221, 1988.

46. **Johnson, J. R., Moseley, S. L., Roberts, P. L., and Stamm, W. E.,** Aerobactin and other virulence factor genes among strains of *Escherichia coli* causing urosepsis: association with patient characteristics, *Infect. Immun.*, 56, 405, 1988.

47. **Gross, R., Engelbrecht, F., and Braun, V.,** Identification of the genes and their polypeptide products responsible for aerobactin synthesis by pColV plasmids, *Mol. Gen. Genet.*, 201, 204, 1985.

48. **deLorenzo, V. and Neilands, J. B.,** Characterization of *iucA* and *iucC* genes of the aerobactin system of plasmid pColV-K30 in *Escherichia coli*, *J. Bacteriol.*, 167, 350, 1986.

49. **Krone, W. J. A., Steghuis, F., Koningstein, G., van Doorn, C., Roosendaal, B., deGraaf, F. K., and Oudega, B.,** Characterization of the pColV-K30 encoded cloacin DF13/aerobactin outer membrane receptor protein of *Escherichia coli*; isolation and purification of the protein and analysis of its nucleotide sequence and primary structure, *Microbiol. Lett.*, 26, 153, 1985.

50. **Braun, V., Brazel-Faisst, C., and Schneider, R.,** Growth stimulation of *Escherichia coli* in serum by iron(III) aerobactin. Recycling of aerobactin, *FEMS Microbiol. Lett.*, 21, 99, 1984.

51. **Sauer, M., Hantke, K., and Braun, V.,** Sequence of the *fhuE* outer membrane receptor gene of *Escherichia coli* K-12 and properties of mutants, *Mol. Microbiol.*, 4, 427, 1990.

52. **Sauer, M., Hantke, K., and Braun, V.,** Ferric-coprogen receptor FhuE of *Escherichia coli*: processing and sequence common to all TonB-dependent outer membrane receptor proteins, *J. Bacteriol.*, 169, 2044, 1987.

53. **Melby, K., Slohrdahl, S., Guttenberg, T. J., and Nordbo, S. A.,** Septicaemia due to *Yersinia enterocolitica* after oral overdoses of iron, *Br. Med. J.*, 285, 467, 1982.

54. **Brock, J. H. and Ng, J.,** The effect of desferrioxamine on the growth of *Staphylococcus aureus, Yersinia enterocolitica* and *Streptococcus faecalis* in human serum: uptake of desferrioxamine-bound iron, *Microbiol. Lett.*, 20, 439, 1983.

55. **Bäumler, A. and Hantke, K.,** unpublished results.

56. **Forst, G. E. and Rosenberg, H.,** The inducible citrate-dependent iron transport system in *Escherichia coli* K12, *Biochim. Biophys. Acta*, 330, 90, 1973.

57. **Woodrow, G. C., Langman, L., Young, I. G., and Gibson, F.,** Mutations affecting the citrate-dependent iron uptake system in *Escherichia coli*, *J. Bacteriol.*, 133, 1524, 1978.

58. **Hancock, R. E. W., Hantke, K., and Braun, V.,** Iron transport in *Escherichia coli* K-12: involvement of the colicin B receptor and of a citrate-inducible protein, *J. Bacteriol.*, 127, 1370, 1976.

59. **Wagegg, W. and Braun, V.,** Ferric citrate transport in *Escherichia coli* requires outer membrane receptor protein FecA, *J. Bacteriol.*, 145, 156, 1981.

60. **Pressler, U., Staudenmaier, H., Zimmermann, L., and Braun, V.,** Genetics of the iron dicitrate transport system of *Escherichia coli*, *J. Bacteriol.*, 170, 2716, 1988.

61. **Staudenmaier, J., Van hove, B., Yaraghi, Z., and Braun, V.,** Nucleotide sequences of the *fecBCDE* genes and locations of the proteins suggest a periplasmic-binding-protein-dependent transport mechanism for iron(III)dicitrate in *Escherichia coli*, *J. Bacteriol.*, 171, 2626, 1989.

62. **Zimmermann, L., Hantke, K., and Braun, V.,** Exogenous induction of the iron dicitrate transport system of *Escherichia coli* K-12, *J. Bacteriol.*, 159, 271, 1984.

63. **Staudenmaier, H. and Braun, V.,** unpublished results.

64. **Knott, V., Blake, D. J., and Brownlee, G. G.,** Completion of the detailed restriction map of the *E. coli* genome by the isolation of overlapping cosmid clones, *Nucl. Acid. Res.*, 17, 5901, 1989.

65. **Hussein, S., Hantke, K., and Braun, V.,** Citrate-dependent iron transport system in *Escherichia coli* K-12, *Eur. J. Biochem.*, 117, 431, 1981.

66. **Van hove, B. and Braun, V.,** unpublished results.

67. **Weische, A., Johanni, M., and Leistner, E.,** Biosynthesis of o-succinylbenzoic acid I: cell free synthesis of o-succinylbenzoic acid from isochorismic acid in enzyme preparations from vitamin K producing bacteria, *Arch. Biochem. Biophys.*, 256, 212, 1987.

68. **Liu, J., Duncan, K., and Walsh, C. T.,** Nucleotide sequence of a cluster of *Escherichia coli* enterobactin biosynthesis genes: identification of *entA* and purification of its product 2,3-dihydro-2,3-dihydroxybenzoate dehydrogenase, *J. Bacteriol.*, 171, 791, 1989.

69. **Nahlik, M. S., Brickman, T. J., Ozenberger, B. A., and McIntosh, M. A.,** Nucleotide sequence and transcriptional organisation of the *Escherichia coli* enterobactin biosynthesis cistrons *entB* and *entA*, *J. Bacteriol.*, 171, 784, 1989.

70. **Ozenberger, B. A., Brickman, T. J., and McIntosh, M. A.**, Nucleotide sequence of *Escherichia coli* isochorismate synthetase and evolutionary relationship of isochorismate synthetase and other chorismate-utilizing enzymes, *J. Bacteriol.*, 171, 775, 1989.

71. **Armstrong, S. K., Pettis, G. S., Forester, L. J., and McIntosh, M. A.**, The *Escherichia coli* enterobactin biosynthesis gene, *entD*: nucleotide sequence and membrane localization of its products, *Mol. Microbiol.*, 3, 757, 1989.

72. **Pickett, C. L., Hayes, L. D., and Earhart, C. F.**, Molecular cloning of the *Escherichia coli* K-12 *entACGBE* genes, *FEMS Microbiol. Lett.*, 24, 77, 1984.

73. **Nahlik, M. S., Fleming, P. T., and McIntosh, M. A.**, Cluster of genes controlling synthesis and activation of 2,3-dihydroxybenzoic acid in production of enterobactin in *Escherichia coli*, *J. Bacteriol.*, 169, 4163, 1987.

74. **Pettis, G. S. and McIntosh, M. A.**, Molecular characterization of the *Escherichia coli* enterobactin cistron *entF* and coupled expression of *entF* and the *fes* gene, *J. Bacteriol.*, 169, 4154, 1989.

75. **Lundrigan, M. D. and Kadner, R. J.**, Nucleotide sequence of the gene for the ferrienterochelin receptor FepA in *Escherichia coli*, *J. Biol. Chem.*, 261, 10797, 1986.

76. **Elkins, M. F. and Earhart, C. F.**, Nucleotide sequence and regulation of the *Escherichia coli* gene for ferrienterobactin transport protein *fepB*, *J. Bacteriol.*, 171, 5443, 1989.

77. **Ozenberger, E. A., Nahlik, M. S., and McIntosh, M. A.**, Genetic organization of multiple *fep* genes encoding ferric enterobactin transport functions in *Escherichia coli*, *J. Bacteriol.*, 169, 3638, 1987.

78. **Fleming, T. P., Nahlik, M. S., Neilands, J. B., and McIntosh, M. A.**, Physical and genetic characterization of cloned enterobactin genomic sequences from *Escherichia coli* K-12, *Gene*, 34, 47, 1985.

79. **Woodrow, G. C., Young, I. G., and Gibson, F.**, Mu-induced polarity in the *Escherichia coli* K-12 *ent* gene cluster: evidence for a gene (*entG*) involved in the biosynthesis of enterochelin, *J. Bacteriol.*, 124, 1, 1975.

80. **Langmann, L., Young, I. G., Frost, G. E., Rosenberg, H., and Gibson, F.**, Enterochelin system of iron transport in *Escherichia coli*: mutations affecting ferric-enterochelin esterase, *J. Bacteriol.*, 112, 1142, 1972.

81. **Heidinger, S., Braun, V., Pecoraro, V. L., and Raymond, K. N.**, Iron supply to *Escherichia coli* by synthetic analogs of enterochelin, *J. Bacteriol.*, 153, 109, 1983.

82. **Hantke, K.**, Dihydroxybenzoylserine—a siderophore for *Escherichia coli*, *FEMS Microbiol. Lett.*, 67, 5, 1990.

82a. **Hantke, K.**, unpublished results.

83. **Hantke, K.**, Identification of an iron uptake system specific for coprogen and rhodotorulic acid in *Escherichia coli* K12, *Mol. Gen. Genet.*, 191, 301, 1983.

84. **Kunze, B., Bedorf, N., Kohl, W., Höfle, G., and Reichenbach, H.**, Myxochelin A, a new iron-chelating compound from *Angiococcus disciformis* (Myxobacterales) production, isolation, physico-chemical and biological properties, *J. Antibiot.*, 42, 14, 1989.

85. **Nau, C. D. and Konisky, J.**, Evolutionary relationship between the *tonB*-dependent outer membrane transport proteins: nucleotide and amino acid sequences of the *Escherichia coli* colicin I receptor gene, *J. Bacteriol.*, 171, 1041, 1989.

86. **Nau, C. D. and Konisky, J.**, Evolutionary relationship between the tonB-dependent outer membrane transport proteins: nucleotide and amino acid sequences of the *Escherichia coli* colicin I receptor gene. Author's corrections, *J. Bacteriol.*, 171, 4530, 1989.

87. **Curtis, N. A. C., Eisenstadt, R. L., East, S. J., Cornford, R. J., Walker, L. A., and White, A. J.**, Iron-regulated outer membrane proteins of *Escherichia coli* K-12 and mechanism of action of catechol-substituted cephalosporins, *Antimicrob. Agents Chemother.*, 32, 1879, 1988.

88. **Zimmermann, L., Angerer, A., and Braun, V.**, Mechanistically novel iron(III) transport system in *Serratia marcescens*, *J. Bacteriol.*, 171, 238, 1989.

89. **Angerer, A., Gaisser, S., and Braun, V.**, Nucleotide sequences of the *sfuA*, *sfuB* and *sfuC* genes of *Serratia marcescens* suggest a periplasmic binding protein-dependent transport mechanism, *J. Bacteriol.*, 172, 572, 1990.

90. **Braun, V. and Rehn, K.**, Chemical characterization, spatial distribution and function of a lipoprotein (murein-lipoprotein) of the *E. coli* cell wall, *Eur. J. Biochem.*, 10, 426, 1969.

91. **Braun, V., Schaller, K., and Wolff, H.**, A common receptor protein for phage T5 and colicin M in the outer membrane of *Escherichia coli*, *Biochim. Biophys. Acta*, 328, 87, 1973.

92. **Hantke, K. and Braun, V.**, Membrane receptor-dependent iron transport in *Escherichia coli*, *FEBS Lett.*, 49, 301, 1975.

93. **Wayne, R. and Neilands, J. B.**, Evidence for a common binding site for ferrichrome compounds and bacteriophage ϕ80 in the cell envelope of *Escherichia coli*, *J. Bacteriol.*, 121, 497, 1975.

94. **Bennett, R. L. and Rothfield, L. J.**, Genetic and physiological regulation of intrinsic proteins of the outer membrane of *Salmonella typhimurium*, *J. Bacteriol.*, 127, 498, 1976.

95. **Braun, V., Hancock, R. E. W., Hantke, K., and Hartmann, A.,** Functional organization of the outer membrane of *Escherichia coli*: phage and colicin receptors as components of iron uptake systems, *J. Supramol. Struct.*, 5, 37, 1976.

96. **Bindereif, A., Braun, V., and Hantke, K.,** The cloacin receptor of ColV-bearing *Escherichia coli* is part of the Fe^{3+}-aerobactin transport system, *J. Bacteriol.*, 150, 1472, 1982.

97. **Hantke, K. and Braun, V.,** Functional interaction of the *tonA/tonB* receptor system in *Escherichia coli*, *J. Bacteriol.*, 135, 190, 1978.

98. **Braun, V., Frenz, S., Hantke, K., and Schaller, K.,** Penetration of colicin M into cells of *Escherichia coli*, *J. Bacteriol.*, 142, 162, 1980.

99. **Hancock, R. E. W. and Braun, V.,** Nature of the energy requirement for the irreversible adsorption of bacteriophage T1 and φ80 to *Escherichia coli*, *J. Bacteriol.*, 125, 409, 1976.

100. **Plastow, G. S. and Holland, J. B.,** Identification of the *Escherichia coli* inner membrane peptide specified by a lambda *tonB* transducing bacteriophage, *Biochem. Biophys. Res. Commun.*, 90, 1007, 1979.

101. **Postle, K. and Skare, J. T.,** *Escherichia coli* TonB protein is exported from the cytoplasm without proteolytic cleavage of its aminoterminus, *J. Biol. Chem.*, 263, 11000, 1988.

102. **Postle, K. and Good, R. F.,** DNA sequence of the *Escherichia coli tonB* gene, *Proc. Natl. Acad. Sci. U.S.A.*, 80, 5235, 1983.

103. **Fischer, E., Günter, K., and Braun, V.,** Involvement of ExbB and TonB in transport across the outer membrane of *Escherichia coli*: phenotypic complementation of *exb* mutants by overexpressed *tonB* and physical stabilization of TonB by ExbB, *J. Bacteriol.*, 171, 5127, 1989.

104. **Schöffler, H. and Braun, V.,** Transport across the outer membrane of *Escherichia coli* K12 via the FhuA receptor is regulated by the TonB protein of the cytoplasmic membrane, *Mol. Gen. Genet.*, 217, 378, 1989.

105. **Köck, J., Ölschläger, T., Kamp, R. M., and Braun, V.,** Primary structure of colicin M, an inhibitor of murein biosynthesis, *J. Bacteriol.*, 169, 3358, 1987.

106. **Schramm, E., Ölschläger, T., Tröger, W., and Braun, V.,** Sequence, expression and localization of the immunity protein for colicin B, *Mol. Gen. Genet.*, 211, 176, 1988.

107. **Mankovich, J. A., Lai, P. H., Gokul, N., and Konisky, J.,** DNA and amino acid sequence analysis of structural and immunity genes of colicin Ia and Ib, *J. Bacteriol.*, 168, 228, 1986.

108. **Roos, U., Harkness, R. E., and Braun, V.,** Assembly of colicin genes from a few DNA fragments. Nucleotide sequence of colicin D, *Mol. Microbiol.*, 3, 891, 1989.

109. **Mende, J. and Braun, V.,** Import-defective colicin B derivatives mutated in the tonB box, *Mol. Microbiol.*, 4, 1523, 1990.

110. **Heller, K., Kadner, R. J., and Günter, K.,** Suppression of the *btuB451* mutation by mutations in the *ton*B gene suggests a direct interaction between TonB and TonB-dependent receptor proteins in the outer membrane of *Escherichia coli*, *Gene*, 64, 147, 1988.

111. **Gutermann, S. and Dann, L.,** Excretion of enterochelin by *exbA* and *exbB* mutants of *Escherichia coli*, *J. Bacteriol.*, 114, 1225, 1973.

112. **Hantke, K. and Zimmermann, L.,** The importance of the *exbB* gene for vitamin B_{12} and ferric iron transport, *FEMS Microbiol.*, 12, 31, 1981.

113. **Eick-Helmerich, K. and Braun, V.,** Import of biopolymers into *Escherichia coli*: nucleotide sequences of the *exbB* and *exbD* genes are homologous to those of the *tolQ* and *tolR* genes, respectively, *J. Bacteriol.*, 171, 5117, 1989.

114. **Bassford, P. J., Schnaitman, C. A., and Kadner, R. J.,** Functional stability of the *bfe* and *tonB* gene products in *Escherichia coli*, *J. Bacteriol.*, 130, 750, 1977.

115. **Kadner, R. K. and McElhaney, G.,** Outer membrane-dependent transport systems in *Escherichia coli*: turnover of TonB function, *J. Bacteriol.*, 134, 1020, 1978.

116. **Sun, T.-P. and Webster, R. E.,** *fii*, a bacterial locus required for filamentous phage infection and its relation to colicin-tolerant *tolA* and *tolB*, *J. Bacteriol.*, 165, 107, 1986.

117. **Sun, T.-P. and Webster, R. E.,** Nucleotide sequence of a gene cluster involved in entry of E colicins and single-stranded DNA of infecting filamentous bacteriophages into *Escherichia coli*, *J. Bacteriol.*, 169, 2667, 1987.

118. **Braun, V.,** The structurally related *exbB* and *tolQ* genes are interchangeable in conferring *tonB*-dependent colicin, phage and albomycin sensitivity, *J. Bacteriol.*, 171, 6387, 1989.

119. **Braun, V. and Hantke, K.,** unpublished results.

120. **Unger, G. and Braun, V.,** unpublished results.

121. **Rioux, C. R. and Kadner, R. J.,** Vitamin B_{12} transport in *Escherichia coli* K12 does not require the *btuE* gene of the *btuCED* operon, *Mol. Gen. Genet.*, 217, 301, 1989.

122. **Bradbeer, F., Kenley, J. S., DiMasi, D. R., and Leighton, M.,** Transport of vitamin B_{12} in *Escherichia coli*. Corrinoid specificities of the periplasmic B_{12}-binding protein and of energy-dependent B_{12} transport, *J. Biol. Chem.*, 125, 1032, 1978.

123. **Coulton, J. W., Naegeli, H.-U., and Braun, V.,** Iron supply of *Escherichia coli* with polymer-bound ferricrocin, *Eur. J. Biochem.*, 99, 39, 1979.

124. **Schneider, R., Hartmann, A., and Braun, V.,** Transport of the iron ionophore ferrichrome in *Escherichia coli* K-12 and *Salmonella typhimurium* LT2, *FEMS Microbiol. Lett.*, 11, 115, 1981.
125. **Winkelmann, G. and Braun, V.,** Stereoselective recognition of ferrichrome by fungi and bacteria, *FEMS Microbiol. Lett.*, 11, 237, 1981.
126. **Winkelmann, G.,** Iron complex products (siderophores), *Biotechnology*, 4, 215, 1986.
127. **Eberspächer, B. and Braun, V.,** The involvement of cytochromes in the uptake of ferrichrome by *Escherichia coli* K-12, *FEMS Microbiol. Lett.* 7, 61, 1980.
128. **Weinberg, E. D.,** Iron withholding: a defense against infection and neoplasia, *Physiol. Rev.*, 64, 65, 1984.
129. **West, S. E. H. and Sparling, P. F.,** Response of *Neisseria gonorrhoeae* to iron limitation: alterations in expression of outer membrane proteins without apparent siderophore production, *Infect. Immun.*, 47, 338, 1985.
130. **Morton, D. J. and Williams, P.,** Utilization of transferrin-bound iron by *Haemophilus* species of human and porcine origins, *FEMS Microbiol. Lett.*, 65, 123, 1989.
131. **Archibald, F. S. and DeVoe, J. W.,** Removal of iron from human transferrin by *Neisseria meningitidis*, *FEMS Microbiol. Lett.*, 6, 159, 1979.
132. **Mickelsen, P. A., Blackman, E., and Sparling, P. F.,** Ability of *Neisseria gonorrhoeae*, *Neisseria meningitidis* and commensal *Neisseria* species to obtain iron from lactoferrin, *Infect. Immun.*, 35, 915, 1982.
133. **Simson, C., Bremer, D., and DeVoe, J. W.,** Expression of a high-affinity mechanisms for acquisition of transferrin iron by *Neisseria meningitidis*, *Infect. Immun.*, 36, 107, 1982.
134. **Schryvers, A. B. and Morris, L. J.,** Identification and characterization of the transferrin receptor from *Neisseria meningitidis*, *Mol. Microbiol.*, 2, 281, 1988.
135. **Schryvers, A. B. and Morris, L. J.,** Identification and characterization of the human lactoferrin-binding protein from *Neisseria meningitidis*, *Infect. Immun.*, 56, 1144, 1988.
136. **Lee, T. C. and Schryvers, A. B.,** Specificity of the lactoferrin and transferrin receptors in *Neisseria gonorrhoeae*, *Mol. Microbiol.*, 2, 827, 1988.
137. **Schryvers, A. B.,** Characterization of the human transferrin and lactoferrin receptors in *Haemophilus influenzae*, *Mol. Microbiol.*, 2, 467, 1988.
138. **Redhead, K., Hill, T., and Char, H.,** Interaction of lactoferrin and transferrins with the outer membrane of *Bordetella pertussis*, *J. Gen. Microbiol.*, 133, 891, 1987.
139. **McKenna, W. R., Mickelsen, P. A., Sparling, P. F., and Dyer, D. W.,** Iron uptake from lactoferrin and transferrin by *Neisseria gonorrhoeae*, *Infect. Immun.*, 56, 785, 1988.
140. **Brot, N. and Goodwin, J.,** Regulation of 2,3-dihydroxybenzoylserine synthetase by iron, *J. Biol. Chem.*, 243, 510, 1968.
141. **Young, I. G. and Gibson, F.,** Regulation of the enzymes involved in the biosynthesis of 2,3-dihydroxybenzoic acid in *Aerobacter aerogenes* and *Escherichia coli*, *Biochim. Biophys. Acta*, 177, 401, 1969.
142. **Laemmli, U. K.,** Cleavage of the structural proteins during the assembly of the head of bacteriophage T4, *Nature*, 227, 680, 1970.
143. **Lugtenberg, B., Meijers, J., Peters, R., van der Hoek, P., and van Alphen, L.,** Electrophoretic resolution of the 'major outer membrane protein' of *Escherichia coli* K-12 into four bands, *FEBS Lett.*, 58, 254, 1975.
144. **Payne, S. M., Niesel, D. W., Peixotto, S. S., and Lawlor, K. M.,** Expression of hydroxamate and phenolate siderophores by *Shigella flexneri*, *J. Bacteriol.*, 155, 949, 1983.
145. **Payne, S. M.,** Iron and virulence in *Shigella*, *Mol. Microbiol.*, 3, 1301, 1989.
146. **Bennet, R. L. and Rothfield, L. I.,** Genetic and physiological regulation of intrinsic proteins of the outer membrane of *Salmonella typhimurium*, *J. Bacteriol.*, 127, 498, 1976.
147. **Fernandez-Beros, M. E., Gonzalez, C., McIntosh, M. A., and Cabello, F. C.,** Immune response to the iron-deprivation-induced proteins of *Salmonella typhi* in typhoid fever, *Infect. Immun.*, 57, 1271, 1989.
148. **Williams, P., Brown, M. R. W., and Lambert, P. A.,** Effect of iron deprivation on the production of siderophores and outer membrane proteins in *Klebsiella aerogenes*, *J. Gen. Microbiol.*, 130, 2357, 1984.
149. **Carniel, E., Mazigh, D., and Mollaret, H. H.,** Correlation between the presence of two high molecular weight iron-regulated proteins and the virulence of *Yersiniae*, *Contr. Microbiol. Immunol.*, 9, 259, 1987.
150. **Shand, G. H., Anwar, H., Kadurugamuwa, J., Brown, M. R. W., Silverman, S. H., and Melling, J.,** In vivo evidence that bacteria in urinary tract infection grow under iron-restricted conditions, *Infect. Immun.*, 48, 35, 1985.
151. **Enard, C., Diolez, A., and Expert, D.,** Systemic virulence of *Erwinia chrysanthemi* 3937 requires a functional iron assimilation system, *J. Bacteriol.*, 170, 2419, 1988.
152. **Sigel, S. P. and Payne, S. M.,** Effect of iron limitation on growth, siderophore production, and expression of outer membrane proteins of *Vibrio cholerae*, *J. Bacteriol.*, 150, 148, 1982.
153. **Crosa, J. H. and Hodges, L. L.,** Outer membrane proteins induced under conditions of iron limitation in the marine fish pathogen *Vibrio anguillarum*, *Infect. Immun.*, 31, 223, 1981.

154. **Actis, L. A., Tomalsky, M. E., Farrell, D. H., and Crosa, J. H.,** Genetic and molecular characterization of essential components of the *Vibrio anguillarum* plasmid-mediated iron-transport system, *J. Biol. Chem.,* 263, 2853, 1988.

155. **Wright, A. C., Simpson, L. M., Richardson, K., Maneval, D. R., Jr., Oliver, J. D., and Morris, J. G., Jr.,** Siderophore production and outer membrane proteins of selected *Vibrio vulnificus* strains under conditions of iron limitation, *FEMS Microbiol. Lett.,* 35, 255, 1986.

156. **Chart, H. and Trust, T. J.,** Acquisition of iron by *Aeromonas salmonicida, J. Bacteriol.,* 156, 758, 1983.

157. **Williams, P. and Brown, M. R. W.,** Influence of iron restriction on growth and the expression of outer membrane proteins by *Haemophilus influenzae* and *H. parainfluenzae, FEMS Microbiol. Lett.,* 33, 153, 1986.

158. **Deneer, H. G. and Potter, A. A.,** Effect of iron restriction on the outer membrane proteins of *Actinobacillus (Haemophilus) pleuropneumoniae, Infect. Immun.,* 57, 789, 1989.

159. **Ikeda, J. S. and Hirsh, D. C.,** Antigenically related iron-regulated outer membrane proteins produced by different somatic serotypes of *Pasteurella multicoda, Infect. Immun.,* 56, 2499, 1988.

160. **Keevil, C. W., Davies, D. B., Spillane, B. J., and Mahentiralingam, E.,** Influence of iron-limited and replete continuous culture on the physiology and virulence of *Neisseria gonorrhoeae, J. Gen. Microbiol.,* 135, 851, 1989.

161. **Field, L. H., Headley, V. L., Payne, S. M., and Berry, L. J.,** Influence of iron on growth, morphology, outer membrane protein composition, and synthesis of siderophores in *Campylobacter jejuni, Infect. Immun.,* 54, 126, 1986.

162. **Sokol, P. A. and Woods, D. E.,** Demonstration of an iron-siderophore-binding protein in the outer membrane of *Pseudomonas aeruginosa, Infect. Immun.,* 40, 665, 1983.

163. **Meyer, J. M., Mock, M., and Abdallah, M. A.,** Effect of iron on the protein composition of the outer membrane of fluorescent pseudomonads, *FEMS Microbiol. Lett.,* 5, 395, 1979.

164. **Page, W. J. and Huyer, M.,** Derepression of the *Azotobacter vinelandii* siderophore system using iron-containing minerals to limit iron repletion, *J. Bacteriol.,* 158, 496, 1984.

165. **Page, W. J.,** Iron-dependent production of hydroxamate by sodium-dependent *Azotobacter chroococcum, Appl. Environ. Microbiol.,* 53, 1418, 1987.

166. **Collinson, S. K. and Page, W. J.,** Production of outer-membrane proteins and an extracellular fluorescent compound by iron limited *Azomonas macrocytogenes, J. Gen. Microbiol.,* 135, 1229, 1989.

167. **Bachhawat, A. K. and Ghosh, S.,** Isolation and characterization of the outer membrane proteins of *Azospirillum brasilense, J. Gen. Microbiol.,* 133, 1751, 1987.

168. **Paoletti, L. C. and Blakemore, R. P.,** Hydroxamate production by *Aquaspirillum magnetotacticum, J. Bacteriol.,* 167, 73, 1986.

169. **Hoe, M., Wilkinson, B. J., and Hindahl, M. S.,** Outer membrane proteins induced upon iron deprivation of *Paracoccus denitrificans, Biochim. Biophys. Acta,* 813, 338, 1985.

170. **Scanlan, D. J., Mann, N. H., and Carr, N. G.,** Effect of iron and other nutrient limitations on the pattern of outer membrane proteins in the cyanobacterium *Synechococcus PCC7942, Arch. Microbiol.,* 152, 224, 1989.

171. **Guikema, J. A. and Sherman, L. A.,** Influence of iron deprivation on the membrane composition of *Anacystis nidulans, Plant Physiol.,* 74, 90, 1984.

172. **Hall, R. M., Sritharan, M., Messenger, A. J., and Ratledge, C.,** Iron transport in *Mycobacterium smegmatis*: occurrence of iron-regulated envelope proteins as potential receptors for iron uptake, *J. Gen. Microbiol.,* 133, 2107, 1987.

173. **Chehade, H. and Braun, V.,** Iron-regulated synthesis and uptake of colicin V, *FEMS Microbiol. Lett.,* 52, 177, 1988.

174. **Lebek, G. and Gruenig, H.,** Relation between the hemolytic property and iron metabolism in *Escherichia coli, Infect. Immun.,* 50, 682, 1985.

175. **Calderwood, S. B., Auclair, F., Donohue-Rolfe, A., Keusch, G. T., and Mekalanos, J. J.,** Nucleotide sequence of the Shiga-like toxin genes of *Escherichia coli, Proc. Natl. Acad. Sci. U.S.A.,* 84, 4364, 1987.

176. **DeGrandis, S., Ginsberg, J., Toone, M., Climie, S., Friesen, J., and Brunton, J.,** Nucleotide sequence and promoter mapping of the *Escherichia coli* Shiga-like toxin operon of bacteriophage H-19B, *J. Bacteriol.,* 169, 4313, 1987.

177. **Strockbine, N. A., Jackson, M. P., Sung, L. M., Holmes, R. K., and O'Brien, A. D.,** Cloning and sequencing of the gene for Shiga toxin from *Shigella dysenteriae* Type 1, *J. Bacteriol.,* 170, 1116, 1988.

178. **Poole, K. and Braun, V.,** Iron regulation of *Serratia marcescens* hemolysin gene expression, *Infect. Immun.,* 56, 2967, 1988.

179. **Stoebner, J. A. and Payne, S. M.,** Iron-regulated hemolysin production and utilization of heme and hemoglobin by *Vibrio cholerae, Infect. Immun.,* 56, 2891, 1988.

180. **Lory, S.,** Effect of iron on accumulation of exotoxin A-specific mRNA in *Pseudomonas aeruginosa, J. Bacteriol.,* 168, 1451, 1986.

181. **Bjorn, M. J., Sokol, P. A., and Iglewski, B. H.**, Influence of iron on yields of extracellular products in *Pseudomonas aeruginosa* cultures, *J. Bacteriol.*, 138, 193, 1979.

182. **Cowart, R. E. and Foster, B. G.**, The role of iron in the production of hemolysin by *Listeria monocytogenes*, *Curr. Microbiol.*, 6, 287, 1981.

183. **Fourel, G., Phalipon, A., and Kaczorek, M.**, Evidence for direct regulation of diphtheria toxin gene transcription by Fe^{2+}-dependent DNA-binding repressor, DtoxR, in *Corynebacterium diphtheriae*, *Infect. Immun.*, 57, 3221, 1989.

184. **McCarter, L. and Silverman, M.**, Iron regulation of swarmer cell differentiation of *Vibrio parahaemolyticus*, *J. Bacteriol.*, 171, 731, 1989.

185. **Makemson, J. C. and Hastings, J. W.**, Iron represses bioluminescens and affects catabolite repression of luminescence in *Vibrio harveyi*, *Curr. Microbiol.*, 7, 181, 1982.

186. **Page, W. J. and von Tigerstrom, M.**, Induction of transformation competence in *Azotobacter vinelandii* iron-limited cultures, *Can. J. Microbiol.*, 24, 1590, 1978.

187. **Golden, S. S. and Sherman, L. A.**, Optimal conditions for genetic transformation of the cyanobacterium *Anacystis nidulans* R2, *J. Bacteriol.*, 158, 3642, 1984.

188. **Braun, V., Fischer, E., Hantke, K., and Rotering, H.**, Iron as a signal in bacterial infections, in *Molecular Basis of Viral and Microbial Pathogenesis*, 38. Colloquium Mosbach, Springer-Verlag, Heidelberg, 1987, 151.

189. **Weinberg, E. D.**, Iron withholding: a defense against infection and neoplasia, *Physiol. Rev.*, 64, 65, 1984.

190. **Hantke, K.**, Regulation of ferric iron transport in *Escherichia coli* K 12: isolation of a constitutive mutant, *Mol. Gen. Genet.*, 182, 288, 1981.

191. **Hantke, K.**, Cloning of the repressor protein gene of iron-regulated systems in *Escherichia coli*, *Mol. Gen. Genet.*, 197, 337, 1984.

192. **Yamao, F., Inokuchi, H., Cheung, A., Ozeki, H., and Söll, D.**, *E. coli* glutaminyl-tRNA synthetase I. Isolation and DNA sequence of the *glnS* gene, *J. Biol. Chem.*, 257, 11639, 1982.

193. **Plumbridge, J.**, Organisation of the *Escherichia coli* chromosome between *glnS* and *glnU,V*, *Mol. Gen. Genet.*, 209, 618, 1987.

194. **Schäffer, S., Hantke, K., and Braun, V.**, Nucleotide sequence of the regulatory gene *fur*, *Mol. Gen. Genet.*, 200, 110, 1985.

195. **Wee, S., Neilands, J. B., Bittner, M. L., Hemming, B. C., Haymore, B. L., and Seetharam, R.**, Expression, isolation and properties of Fur (ferric uptake regulation) protein of *Escherichia coli* K 12, *Biol. Met.*, 1, 62, 1988.

196. **Braun, V., Schäffer, S., Hantke, K., and Tröger, W.**, Regulation of gene expression by iron, in *The Molecular Basis of Bacterial Metabolism*, 41, Colloquium Moosbach, Springer-Verlag, Heidelberg, 1990, 164.

197. **Bagg, A. and Neilands, J. B.**, Ferric uptake regulation protein acts as a repressor, employing iron(II) as a cofactor to bind the operator of an iron transport operon in *Escherichia coli*, *Biochemistry*, 26, 5471, 1987.

198. **deLorenzo, V., Wee, S., Herrero, M., and Neilands, J. B.**, Operator sequences of the aerobactin operon of plasmid ColV-K30 binding the ferric uptake regulation (*fur*) repressor, *J. Bacteriol.*, 169, 2624, 1987.

199. **deLorenzo, V., Herrero, M., Giovannini, F., and Neilands, J. B.**, Fur (ferric uptake regulation) protein and CAP (catabolite-activator protein) modulate transcription of *fur* gene in *Escherichia coli*, *Eur. J. Biochem.*, 173, 537, 1988.

200. **Griggs, D. W. and Konisky, J.**, Mechanism for iron-regulated transcription of the *Escherichia coli cir* gene: metal-dependent binding of Fur protein to the promoters, *J. Bacteriol.*, 171, 1048, 1989.

201. **Bindereif, A. and Neilands, J. B.**, Promoter mapping and transcriptional regulation of the iron assimilation system of plasmid ColV-K30 in *Escherichia coli* K-12, *J. Bacteriol.*, 162, 1039, 1985.

202. **Coulton, J. W., Mason, P., Cameron, D. R., Carmel, G., Jean, R., and Rode, H. N.**, Protein fusions of β-galactosidase to the ferrichrome-iron receptor of *Escherichia coli* K 12, *J. Bacteriol.*, 165, 181, 1986.

203. **Jackson, M. P., Neill, R. J., O'Brien, A. D., Holmes, R. K., and Newland, J. W.**, Nucleotide sequence analysis and comparison of the structural genes for Shiga-like toxin I and Shiga-like toxin II encoded by bacteriophages from *Escherichia coli* 933, *FEMS Microbiol. Lett.*, 44, 109, 1987.

204. **Takeda, Y. and Avila, H.**, Structure and gene expression of the *E. coli* Mn-superoxide dismutase gene, *Nucl. Acids Res.*, 14, 4577, 1986.

205. **Poole, K., Schiebel, E., and Braun, V.**, Molecular characterization of the hemolysin determinant of *Serratia marcescens*, *J. Bacteriol.*, 170, 3177, 1988.

206. **Alm, R. A.**, Molecular Characterization of the Haemolysin Determinant of *Vibrio cholerae* O1, Ph.D. thesis, Univ. of Adelaide, Adelaide, 1989.

207. **Tai, S. S. and Holmes, R. K.**, Iron regulation of the cloned diphtheria toxin promoter in *Escherichia coli*, *Infect. Immun.*, 56, 2430, 1988.

208. **Laudenbach, D. E. and Straus, N. A.**, Characterization of a cyanobacterial stress-induced gene similar to *psbC*, *J. Bacteriol.*, 170, 5018, 1988.

209. **Reddy, K. J., Bullerjahn, G. S., Sherman, D. M., and Sherman, L. A.,** Cloning, nucleotide sequence, and mutagenesis of a gene (*irpA*) involved in iron-deficient growth of the cyanobacterium *Synechococcus* sp. strain PCC7942, *J. Bacteriol.*, 170, 4466, 1988.

210. **Schiavone, J. R. and Hassan, H. M.,** The role of redox in the regulation of manganese-containing superoxide dismutase biosynthesis in *Escherichia coli*, *J. Biol. Chem.*, 263, 4269, 1988.

211. **Hantke, K.,** Selection procedure for deregulated iron transport mutants (*fur*) in *Escherichia coli* K 12: *fur* not only affects iron metabolism, *Mol. Gen. Genet.*, 210, 135, 1987.

212. **Niederhoffer, E. C., Naranjo, C. M., and Fee, J. A.,** Relationship of the superoxide dismutase genes, *sodA* and *sodB*, to the iron uptake (*fur*) regulon in *Escherichia coli* K-12, in *Metal Ion Homeostasis: Molecular Biology and Chemistry*, Alan R. Liss, New York, 1989, 149.

213. **Zähner, H., Diddens, H., Keller-Schierlein, W., and Nägeli, H. U.,** Some experiments with semisynthetic sideromycins, *Jpn. J. Antibiot.*, 30(Suppl.), 201, 1977.

214. **Watanabe, N.-A., Nagasu, T., Katsu, K., and Kitoh, K.,** E-0702, a new cephalosporin, is incorporated into *Escherichia coli* cells via the *tonB*-dependent iron transport system, *Antimicrob. Agents Chemother.*, 31, 497, 1987.

215. **Pugsley, A. P., Zimmermann, W., and Wehrli, W.,** Highly efficient uptake of a rifamycin derivative via the FhuA-TonB-dependent uptake route in *Escherichia coli*, *J. Gen. Microbiol.*, 133, 3505, 1987.

PYOVERDINS AND PSEUDOBACTINS

Mohamed A. Abdallah

INTRODUCTION

Siderophores are small molecules of low molecular weight (*circa* 400 to 2000) which are biosynthesized and excreted in large amounts by microorganisms when grown in iron-deficient conditions. These substances bind firmly iron(III) and facilitate its transport into the cells.[1,2]

The Pseudomonads are widespread bacteria which are classified into five groups of genetic homology.[3] The fluorescent *Pseudomonas*, which belong to the first homology group, are characterized by the production of yellow-green, water-soluble, fluorescent compounds, called pyoverdins.

Although the occurrence of these compounds in the cultures of the fluorescent *Pseudomonas* was used as a taxonomic character for many years[3] and although the peptidic nature of these molecules was suspected, the structure and the biological role of pyoverdins remained obscure for a long time. The first reports describing an efficient purification procedure, some physicochemical properties, as well as the iron-transport activity of pyoverdins, concerned a pyoverdin excreted by a strain of *P. fluorescens*.[4,5] This revived the interest of several research groups and led to the only example of X-ray structure elucidation of a pyoverdin, namely, pseudobactin, excreted by *Pseudomonas* B10 which was shown to be a plant growth-promoting factor.[6]

The structure of pseudobactin presented some similarities with that of a yellow-green, water-soluble, fluorescent compound excreted by *Azotobacter vinelandii* and described earlier.[7] Both substances are constituted with very different linear peptide chains bound to closely related chromophores derived from 2,3-diamino-6,7-dihydroxyquinoline.[6-9] The structure of pseudobactin A, a second siderophore of *Pseudomonas* B10, and possessing a dehydro form of the chromophore of pseudobactin, was deduced from this latter by NMR spectroscopy.[10]

Due to the difficulties to obtain crystalline pyoverdins, other classical physical and chemical methods were used in order to elucidate their structures. NMR was applied to one of the plethoric pyoverdins excreted by *P. fluorescens* ATCC 13525,[11] but was insufficient to solve the corresponding structure in spite of the many useful data provided in that study. The combination of Fast Atom Bombardment Mass Spectrometry (FAB-MS) and high resolution NMR appeared to be more successful for the structure elucidation of the pyoverdins of *P. aeruginosa* ATCC 15692.[12,13] Subsequently it has been applied to a number of pyoverdins from several strains of different fluorescent *Pseudomonas*,[14,15] and finally to azotobactin itself,[16] revising the structure of the fluorescent compound excreted by *A. vinelandii* which had been reported many years ago.[7] Meanwhile, other structures were reported using mainly chemical methods (pseudobactin 7SR1[17] excreted by a strain of plant-deleterious *Pseudomonas*) or NOESY NMR (pseudobactin A214,[18] occurring from the cultures of a bean-deleterious strain of *Pseudomonas*). The structure elucidation of pseudobactin 358[19] excreted by a strain of *P. putida* WC358 was tentatively performed using essentially Edman degradation, but this structure has not been confirmed yet. The use of FAB-MS again was more effective and allowed the structure elucidation of the pyoverdins excreted by a strain of *P. fluorescens*[20] and the revision of the structure of pyoverdin Pa,[21] which was independently confirmed using more elaborate NMR sequencing methods.[22]

In this report we will focus on the structural characteristics and the physicochemical properties of pyoverdins and pseudobactins. We will first comment on the culture conditions and describe the latest improvements in the purification procedures of pyoverdins. The biological and genetic aspects of pyoverdins and pseudobactins are reviewed in the chapter by Weisbeek and Marugg.

GENERAL PROPERTIES OF PYOVERDINS AND PSEUDOBACTINS

STRAINS AND CULTURE MEDIUM

Pseudomonas strains are grown in aerobic conditions. The culture medium generally used[4] has the following composition per liter: $K_2H PO_4$, 6 g; KH_2PO_4, 3 g; $(NH_4)_2 SO_4$, 1 g; $MgSO_4$, $7H_2O$, 0.2 g; succinic acid, 4 g. It is adjusted to pH 7.0 before sterilization. In some instances, in order to ensure the best production of pyoverdins the culture medium is first deferrated using either Chelex 100 or 8-hydroxyquinoline. The predetermined amount of a ferric chloride solution is then added to the culture flasks before sterilization. The flasks are carefully prewashed with hydrochloric acid and rinsed with distilled water. The conditions used to produce azotobactin are different and have been reported elsewhere.[16]

ISOLATION AND PURIFICATION OF THE PYOVERDINS

The bacteria are grown aerobically at 25°C in conical flasks, each containing 0.5 l of culture medium and subject to mechanical agitation. After 48 h, the culture medium is centrifuged and the pyoverdins extracted from the supernatant.
Two methods can be used:

Complexation-Decomplexation Method

This is based on the ability of the pyoverdin-iron(III) or pyoverdin-aluminum(III) complexes to be extracted into organic solvents such as phenol/methylene chloride mixtures. It is a modification of the procedure earlier described[4] for the purification of pyoverdin Pf. After 48 h, the cultures are centrifuged (20,000 × *g* for 30 min at 4°C), and the pH of the supernatant adjusted to 5.0 by careful addition of formic acid and, when necessary, extracted with ethyl acetate in order to remove several by-products produced by the bacteria.[23]

The aqueous phase is treated with a 2 *M* iron chloride or aluminum chloride solution, the precipitate of inorganic phosphates removed by centrifugation, and the supernatant concentrated under reduced pressure to 10% its initial volume, saturated by addition of sodium chloride and extracted with a mixture of methylene chloride-phenol (1:1 v/w) (3 × 200 ml/l of concentrated supernatant). The organic phase containing the metal complexes is separated from the aqueous phase after centrifugation (4000 rpm, 10 min, room temperature), then treated with equal volumes of diethyl ether and water. The aqueous phase is washed with ether, evaporated under reduced pressure, and lyophilized.

The crude pyoverdin-iron(III) (or aluminum[III]) complexes are chromatographed on a suitable ion-exchange Sephadex column (CM C-25 or DEAE A-25) eluted with pyridine-acetic acid buffer at pH 5.0. The fractions are monitored by spectrophotometry, at 403 nm, pooled, and lyophilized.

The decomplexation of the pyoverdins-Fe(III) (or pyoverdins-Al[III]) is performed according to an earlier described procedure,[4] with slight modifications. The pH of each fraction is adjusted to 3.0 by addition of a 10% solution of acetic acid. Two volumes of a 5% solution of 8-hydroxyquinoline in methylene chloride are added and the biphasic mixture transferred into a stoppered flask and vigorously shaken for 30 min on a shaker. The aqueous phase is separated from the organic phase in a separatory funnel, and its pH readjusted to 3.0 before a new treatment with 8-hydroxyquinoline. Four treatments are necessary to remove the metal from the pyoverdins. Finally the aqueous phase is washed three times with methylene chloride in order to remove the excess of 8-hydroxyquinoline. After evaporation under reduced pressure, each pyoverdin thus obtained is chromatographed on a suitable Sephadex ion-exchange column as above, eluted first isocratically with pyridine-acetic acid buffer at pH 5.0, then with a linear gradient of the same buffer (0.1 to 2 M).

This procedure cannot be used when the metal complexes are not soluble in the phenol/methylene chloride phase. A hydrophobic chromatography method using octadecylsilane as a stationary phase has been developed[16] and appears to be the method of choice to separate the crude pyoverdins from the inorganic salts present in the culture media.[22,24-26]

Hydrophobic Chromatography Method

After centrifugation, the bacterial supernatant is filtered through a 0.45-μm membrane (Millipore), adjusted to pH 4.0, and applied to a column of octadecylsilane (Lichroprep RP 18, 40 to 63 μm, Merck, Darmstadt).[15,16,22,24-26] The pyoverdins are eluted with a 1:1 mixture of acetonitrile/0.05 M pyridine-acetic acid buffer pH 5.0, and chromatographed on a suitable Sephadex ion-exchange column (CM C-25 or DEAE A-25), first eluted isocratically with pyridine-acetic acid buffer pH 5.0, then with a linear gradient of the same buffer (0.05 to 2 M). The fractions are monitored by spectrophotometry at 380 nm, pooled, evaporated, treated with three to five equivalents of iron chloride, chromatographed first on an octadecylsilane column and then on a suitable Sephadex ion-exchange column eluted as above. Each fraction can give rise to several pyoverdin-Fe(III) complexes at this stage, corresponding to as many pyoverdins present in the culture.[25] In addition, each fraction isolated from this chromatographic step has to be purified by preparative HPLC, since it can be often constituted with several closely related pyoverdins-Fe(III) complexes. This is crucial for any further structural, physicochemical, or biochemical determination involving pyoverdins.

The complexes isolated after preparative HPLC are deferrated using 8-hydroxyquinoline as above, and the free ligands purified by ion-exchange column chromatography as above.

CRITERIA OF PURITY

The criteria used to check the purity of these compounds and of their complexes are film electrophoresis and HPLC. In film electrophoresis, which is generally performed at pH 5.0, each pyoverdin (or its iron complex) is positively (or negatively) charged.[4,6,11-16,22,24-26] On excitation at 350 nm, the free ligands present a yellowish fluorescence and the aluminum complexes a blue one. Iron complexes are nonfluorescent. Very good electrophoretic analyses are obtained on cellulose acetate sheets at 300 V, during 30 min, in 0.1 M pyridine-acetic acid, pH 5.0, buffer.

HPLC analyses of the free ligands require the pretreatment of the columns by EDTA solutions to remove any trace of metallic cations which are a main source of artifacts for these types of compounds.[27] The eluent system which contains EDTA and octylsulfonic acid allows a very good and reproducible separation of pyoverdins by ion-pair liquid chromatography in the presence of octylsulfonic acid.[22,28]

BIOSYNTHESIS OF THE PYOVERDINS

Few or many pyoverdins can be isolated from fluorescent *Pseudomonas* supernatants depending on the strain investigated. Some of them are really biosynthesized by the bacteria, but can give rise to closely related pyoverdins as the pH of the medium increases during the culture. HPLC analysis of bacterial supernatants as a function of time of culture has been performed in the case of *P. aeruginosa* ATCC 15692.[22] It has shown that pyoverdin Pa and pyoverdin Pa B are the major siderophores excreted by *P. aeruginosa*. After 40 h culture, measurable amounts of pyoverdin Pa A were observed. The appearance of this latter coincides with the decrease of pyoverdin Pa. This is due to the hydrolysis of pyoverdin Pa into pyoverdin Pa A. This hydrolysis is faster as the pH becomes more alkaline.

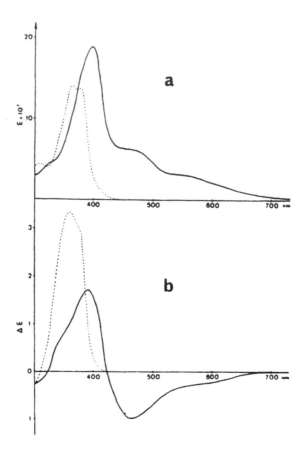

FIGURE 1.   Absorption spectra (a) and circular dichroism spectra (b) of pyoverdins ($\cdots$) and their iron (III) complex (———) at pH 5.0

SPECTROPHOTOMETRIC CHARACTERISTICS (SEE FIGURE 1)

The absorption characteristics of pyoverdins in the 350- to 700-nm range reflect the spectral properties of their common fluorescent chromophore. The absorption spectra of the pyoverdins-Fe(III) complexes are not pH dependent between pH 3 and 10. They exhibit a maximum at 403 nm ($\epsilon = 1.9 \times 10^4 \, M^{-1} \cdot cm^{-1}$) and two shoulders at 460 nm ($\epsilon = 6.4 \times 10^3 \, M^{-1} \cdot cm^{-1}$) and 540 nm ($\epsilon = 3.0 \times 10^3 \, M^{-1} \cdot cm^{-1}$). The spectra of the aluminum(III) and gallium(III) complexes present only the maximum at 403 nm ($\epsilon = 1.9 \times 10^4 \, M^{-1} \cdot cm^{-1}$). Those of the free ligands are pH sensitive and show isosbestic points at 373 nm. At pH 4.2, they present two maxima at 365 nm ($\epsilon = 1.4 \times 10^4 \, M^{-1} \cdot cm^{-1}$) and 380 nm ($\epsilon = 1.4 \times 10^4 \, M^{-1} \cdot cm^{-1}$).

The circular dichroism spectra of the iron complexes are not affected by pH variations. They show a positive Cotton effect at 400 nm ($\Delta \epsilon = +1.9$) and two negative Cotton effects at 460 nm ($\Delta \epsilon = -1.13$) and 540 nm ($\Delta \epsilon = -0.3$). The circular dichroism spectra of the free pyoverdins are pH sensitive. At pH 4.0, they present a positive Cotton effect at 365 nm ($\Delta \epsilon = +3.33$) and at 380 nm ($\Delta \epsilon = +2.67$).[6,14,15,22,24-26]

Due to the extra imidazolone ring present in the chromophore of azotobactin, the spectrophotometrical characteristics of this siderophore are different from those of classical pyoverdins: at pH 5.0, the free ligand shows a maximum at 380 nm ($\epsilon = 2.35 \times 10^4 \, M^{-1} \cdot cm^{-1}$) and a shoulder at 366 nm ($\epsilon = 1.96 \times 10^4 \, M^{-1} \cdot cm^{-1}$), whereas the iron complex has a maximum at 412 nm ($\epsilon = 2.30 \times 10^4 \, M^{-1} \cdot cm^{-1}$) and two shoulders at 450 nm ($\epsilon = 1.0 \times 10^4 \ M^{-1}\cdot\text{cm}^{-1}$) and at 550 nm ($\epsilon = 2.0 \times 10^3 \ M^{-1}\cdot\text{cm}^{-1}$). The shape of these spectra, however, is different from that of the spectra of pyoverdins, confirming the differences in the structure of the chromophore.[14-16]

Acid hydrolysis of pyoverdins can cause some changes in the spectrophotometric characteristics of their chromophore.

In mild acid conditions (6 M, HCl, 90°C, 30 min), when the acyl substituent present on the nitrogen atom at C-3 is not a keto acid, the chromophore does not show important spectral variations even after loss of the acyl moiety. In stronger hydrolysis conditions (6 M HCl, 110°C, 24 h), the acyl substituent is removed and replaced by a hydroxyl group. This causes a hypsochromic shift of 10 nm, and the trishydroxylated chromophore obtained presents two maxima at 354 nm ($\epsilon = 1.7 \times 10^4 \ M^{-1}\cdot\text{cm}^{-1}$) and 368 nm ($\epsilon = 1.6 \times 10^4 \ M^{-1}\cdot\text{cm}^{-1}$).[14,29]

When the acyl substituent is a keto acid like α-ketoglutaric acid, there is an important change in the color of the solution of pyoverdin; the initially pale yellow solution becomes deep orange-red. At pH 4.0 the absorption spectra of the chromophoric compounds isolated present a maximum at 410 nm ($\epsilon = 2.2 \times 10^4 \ M^{-1}\cdot\text{cm}^{-1}$) which shifts to 460 nm ($\epsilon = 2.95 \times 10^4 \ M^{-1}\cdot\text{cm}^{-1}$) at pH 6.7 with an isosbestic point at 426 nm ($\epsilon = 1.65 \times 10^4 \ M^{-1}\cdot\text{cm}^{-1}$). These changes are due to the formation of a fused pyrazinone ring on the chromophore.[22]

FAB-MS CHARACTERISTICS (SEE FIGURE 2)

FAB-MS appears to be a method of choice to obtain very useful informations in the structure elucidation of pyoverdins. FAB-MS gives the molecular mass of the ligands and of their metal complexes, and establishes the 1:1 stoichiometry of the complexes indicating the presence of three bidentate chelating groups per ligand. For many pyoverdins, FAB-MS gave the complete sequence of the peptide chain, with both C- and N-terminal fragments observable.

In N-terminal fragmentation, the classical type of fragments **1**, **2**, and **3** were not observed.[30] This is not surprising since in pyoverdins, due to the permanent positive charge of the chromophore, these fragments would be twice positively charged. Instead, amide ions **4**, aldehydic ions **5**, and Cα fragment ions **6** are observed and interpreted. The fragments of 44 and 45 mass units below the amide ions **4** have been assigned to ions **7** and **8**.[14-16,22,24-26]

In the C-terminal fragmentation, after loss of the chromophore, the C-terminal fragments of pyoverdins are identical to those generally observed in classical peptides. The only particularity is the cleavage across the chromophore which seems to be a very general feature for pyoverdins.[12,14,15,22,24-26] This yields a chromophoric fragment **9** and a peptidic fragment **10**.

More structural data can be obtained using FAB-MS, namely, the differences observed between the pyoverdins produced by a given strain of *Pseudomonas*. These pyoverdins possess in common the same C-terminal ion which occurs from the pyoverdin characteristic cleavage across the saturated ring of the chromophore. This also indicates that the mass differences between these pyoverdins are due to a modification occurring in the acyl group bound to the amine at position C-3 of the chromophore.[13-15,20-22,24-26]

NMR CHARACTERISTICS (SEE FIGURE 3)

NMR is also a very powerful method used for the structure determination of pyoverdins. They all show a common part in the aromatic region, characteristic of their common fluorescent chromophore. Their ¹H NMR spectra are characterized by the presence at low field of three aromatic protons occurring as singlets at 7.90, 7.10, and 6.90 ppm assigned, respectively, to protons H-4, H-5, and H-8 of the chromophore, and a broad singlet at *circa*

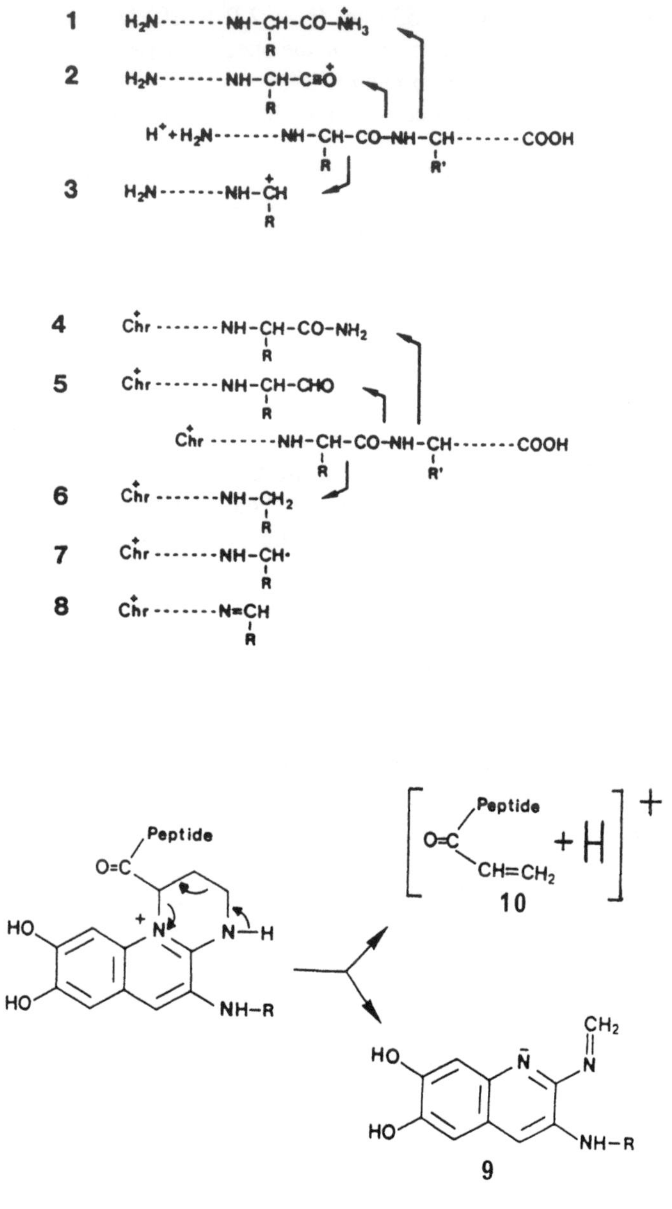

FIGURE 2.   FAB-MS fragments obtained with classical peptides and with pyoverdins.

5.70 ppm, corresponding to proton H-11 on the asymmetric carbon in the vicinity of the quinoline ring. These signals are slightly shifted when the chromophore contains an extra imidazolone ring as in azotobactin,[16] and their chemical shifts are pH dependent.

In the ^{13}C NMR spectra of pyoverdins, this chromophore is characterized by the presence of nine signals between 100 and 160 ppm: three tertiary and six quaternary carbon atoms. These signals have been assigned using [1]H - [13]C, [15]N - [13]C, as well as long-range coupling values.[15,16,22] The complete assignment of the other signals of pyoverdins in their ^{1}H or ^{13}C spectra has been generally performed using two-dimensional techniques such as COSY and heteronuclear correlations.[14-16,22,24-26] Moreover, sequencing of the peptide has been performed by NMR in H$_2$O using a combination of more elaborate two-dimensional techniques such as TOCSY, HOHAHA, and ROESY when the sequence was not completely readily obtainable by FAB-MS, as in the case of the pyoverdins of *P. aeruginosa* ATCC 15692.[22]

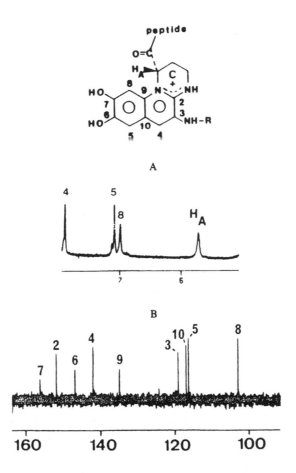

FIGURE 3.   The characteristic signals due to the chromophore
part in the ^{1}H NMR (A) and ^{13}C NMR (B) spectra of pyoverdins.

CONFIGURATION OF THE AMINO ACIDS AND OF THE CHROMOPHORE IN PYOVERDINS (SEE FIGURE 4)

The amino acids present in pyoverdins can occur in either configuration. Their stereo-chemical analysis is performed after derivatization of a pyoverdin total acid hydrolyzate (HCl or HI), followed by gas chromatography or GC-MS on a capillary L-Chirasil-Val capillary column. The derivatives are N-perfluoroacyl O-alkyl esters, generally N-penta-fluoropropionyl O-methyl esters.[31] The same experiment performed on purified hydrolytic fragments of pyoverdins, allows the complete stereochemical assignment of the peptide chain, when the same amino acids occur in both configurations in the molecule.

The stereochemistry of the chromophore of all the pyoverdins so far investigated was established by comparison of their circular dichroism spectra with the spectrum of pseu-dobactin B10, the structure of which was determined by X-ray crystallography.[6] This con-figuration has always been found to be (S), even for azotobactin, in spite of the different shape of its circular dichroism spectrum.[14,16]

PHYSICAL CHEMISTRY OF PYOVERDINS AND PSEUDOBACTINS

The protonation constants of the free ligands, the binding constants of their iron(III) complexes, as well as the spectra of the different species in solution have been measured combining spectrophotometry and potentiometry. The stability constant, relative to the equi-librium

$$L^{4-} + Fe(III) \rightleftarrows LFe^-$$

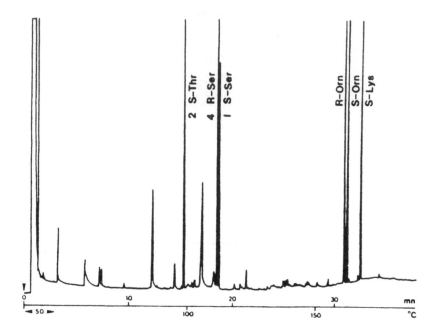

FIGURE 4. Example of stereochemical analysis of the amino acids of pyoverdins by gas chromatography using L-Chirasil-Val stationary phase.

for pyoverdin Pa A was found to be log K_1 = 30.8 ± 0.9. In this respect pyoverdin Pa A is closer to ferrichromes than to enterobactin. At physiological pH, a neutral species LHFe was observed, with a constant log K_2 = 43 ± 1 corresponding to the equilibrium

$$L^{4-} + H^+ + (Fe(III) \rightleftarrows LHFe$$

The redox potential values were determined by cyclic voltammetry and found to be −0.490 V/NHE for pyoverdin Pa A −Fe(III) complex suggesting that iron can be released to the cells by a reductive process.[15]

CONCLUSION

When fluorescent *Pseudomonas* are cultivated in iron-deficient conditions, they excrete several yellow-green, water-soluble fluorescent compounds called pyoverdins. These substances are chromopeptides possessing a chromophore derived from 2,3-diamino-6,7-dihydroxyquinoline bound to a peptidic moiety identical for all the pyoverdins occurring from a given strain and differing generally from strain to strain. The chromophore has the same configuration (S) for all the pyoverdins investigated so far. For a given strain, the pyoverdins differ only in the acyl substituent bound to the amino group on C-3 of the chromophore, which can in many cases derive from succinic acid, malic acid, and α-ketoglutaric acid.

The diversity in the structure of the pyoverdins shows that there are large structural differences between them in the shape of their peptidic moiety: it can be linear, partly cyclic, or completely cyclic.

Many linear pyoverdins have been reported: the structure of pseudobactin (see Figure 5), excreted by *Pseudomonas* B10, was elucidated by X-ray crystallography.[6,32] The structure of its dihydro analogue, pseudobactin A (see Figure 6), was deduced by NMR spectroscopy.[10] Pseudobactin A214 (see Figure 7) is linear but has its C terminal free,[18] in contrast to pseudobactin B10. Its structure was established using classical NOESY NMR. Pseudobactin

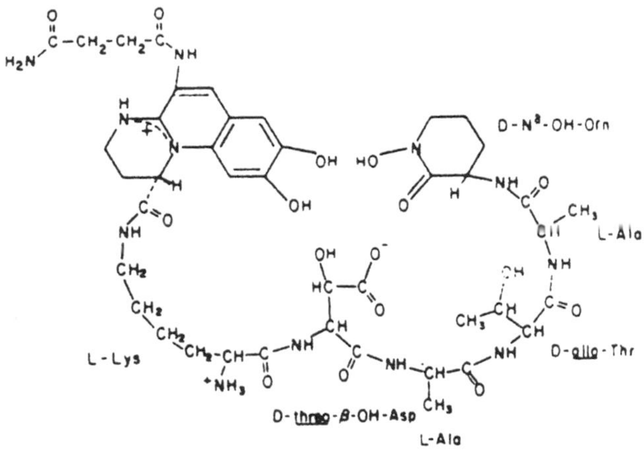

FIGURE 5. The structure of pseudobactin.

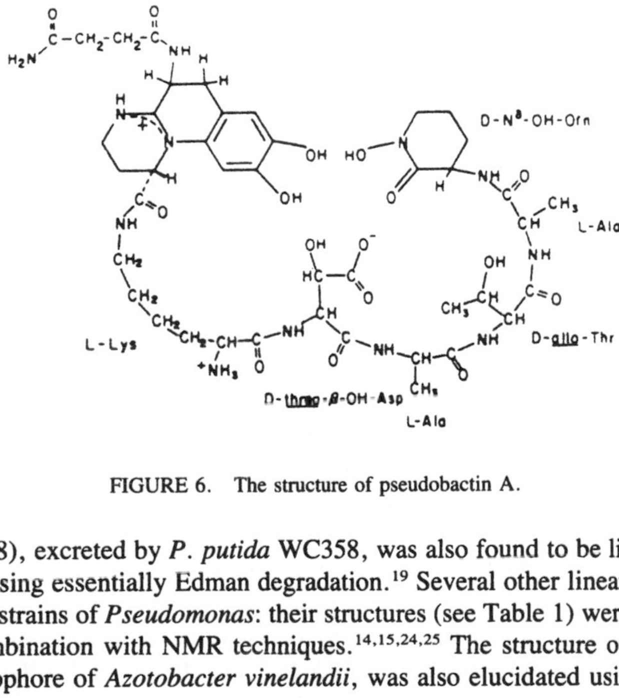

FIGURE 6. The structure of pseudobactin A.

358 (see Figure 8), excreted by *P. putida* WC358, was also found to be linear. Its structure was elucidated using essentially Edman degradation.[19] Several other linear pyoverdins were found in several strains of *Pseudomonas*: their structures (see Table 1) were elucidated using FAB-MS in combination with NMR techniques.[14,15,24,25] The structure of azotobactin (see Figure 9), siderophore of *Azotobacter vinelandii*, was also elucidated using the same techniques.[16]

Partly cyclic pyoverdins were reported.[20] They have their N terminal blocked by the chromophore and their C terminal belongs to a cyclic pseudodepsipeptide (see Figure 10). Pyoverdin Pa was initially thought to be linear[12-14] (see Figure 11A). However, its FAB-MS did not show the complete fragmentation pattern given by the other linear pyoverdins. Its structure was revised and found to be, in fact, partly cyclic (see Figure 11B) independently by FAB-MS on the dansylated peptides obtained by partial hydrolysis[21] and by NMR sequencing.[22] A totally cyclic structure established using essentially chemical methods was reported for pseudobactin 7SR1 (see Figure 12), where the chromophore is bound to a serine of the cyclic peptide via an ester linkage.[17]

The diversity of these structures illustrates the complexity of the problem of the structure elucidation of the pyoverdins. In the absence of crystals FAB-MS is a method of choice for sequencing the peptide moiety of the siderophores. However, in few cases such as pyoverdin

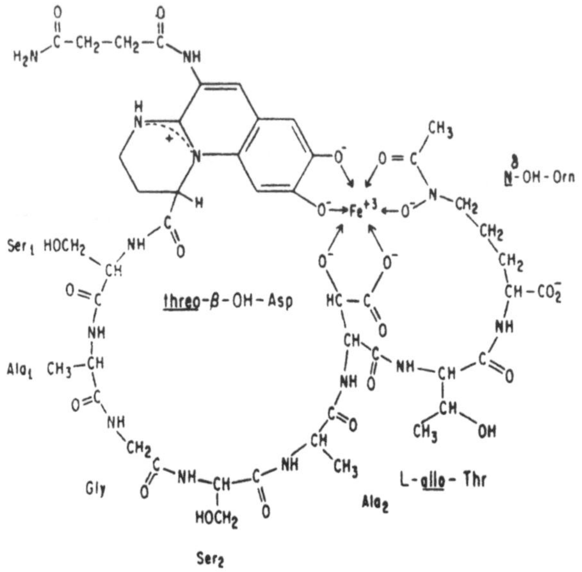

FIGURE 7. The structure of pseudobactin A214.

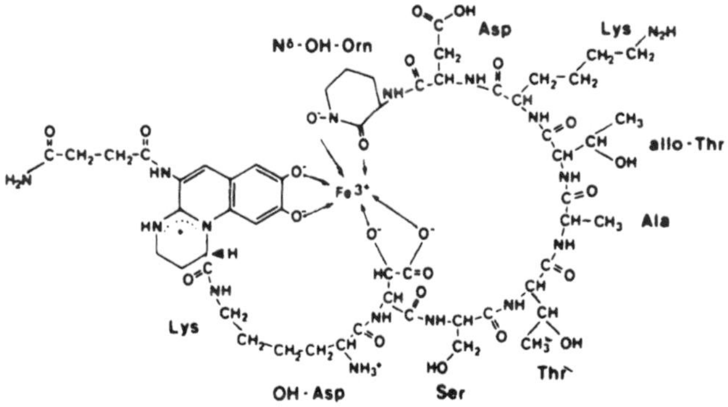

FIGURE 8. The structure of pseudobactin 358.

Pa where FAB-MS does not give the complete sequence, an additional NMR study in H_2O using a combination of two-dimensional methods such as 2D-TOCSY NMR and 2D-ROESY NMR is essential for the unambiguous sequencing of the peptide moiety, after determination of the sequential correlations between the CHα and the NH protons.

A general structure for all the pyoverdins (and azotobactin) is represented in Figure 13 where the acyl moiety can eventually be extended to other diacids.[26]

All these pyoverdins and pseudobactins (and azotobactin) so far investigated possess three bidentate groups giving very stable octahedral complexes with metals like Fe(III), Al(III), Ga(III), or Cr(III): the catechol group of the chromophore, one hydroxamate group and one hydroxyaspartic group, or two hydroxamate groups on the peptidic part.

A fluorescent compound excreted by *P. syringae* has recently been reported as possessing two hydroxyaspartic acid moieties and no hydroxyornithine,[32] but no structural confirmation has been yet reported.

TABLE 1
Sequence of Pyoverdins Occurring from Different Fluorescent *Pseudomonas* as Determined by FAB-MS

Bacteria	Structures of pyoverdins
Pseudomonas putida ATCC 12633	Chr-L-Asp-L-Lys-D-OHAsp-(D-L)Ser-L-Thr-D-Ala-D-Glu-(D-L)Ser-L-cOHOrn
P. chlororaphis ATCC9446	Chr-(D-L)Ser-L-Lys-Gly-(D-L)OHOrn-(D-L)Ser-(L-Lys(D-L)OHOrn — H₂O)
P. fluorescens ATCC 13525	Chr-(D-L)Ser-L-Lys-Gly-(D-L)OHOrn-(D-L)Ser-(L-Lys-(D-L)OHOrn — H₂O)
P. putida CFBP 2461	Chr-L-Asp(D-L)-Lys-D-OHAsp-D-Ser-L-Thr-D-Ala(D-L)-Lys-L-Thr-L-cOHOrn
P. fluorescens SB 83	Chr-D-Ala-L-Lys-L-Thr-D-Ser-L-OHOrn-L-cOHOrn
P. tolaasii NCPPB 2192	Chr-D-Ser-L-Lys-L-Ser-D-Ser-L-Thr-D-Ser-L-OHOrn-L-Thr-D-Ser-D-cOHOrn

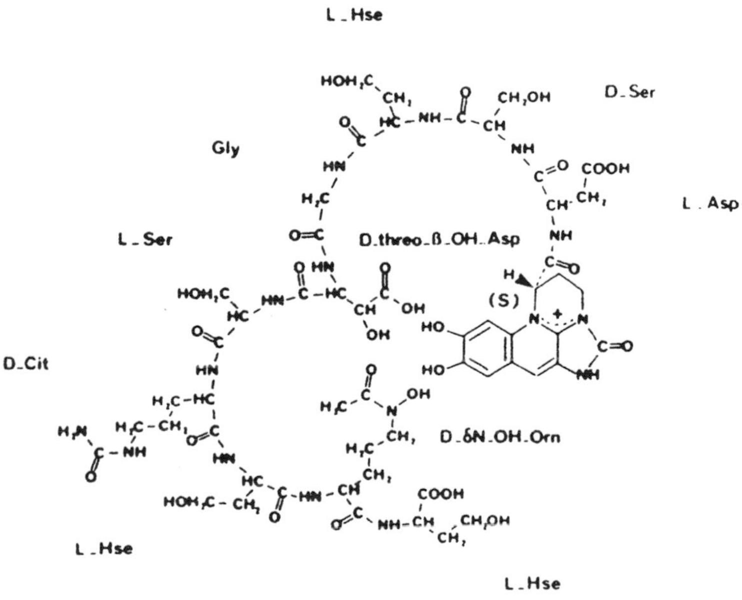

FIGURE 9.   The revised structure of azotobactin.

NOTE ADDED IN PROOF:

The structures of pyoverdins produced by two different strains of fluorescent *Psuedo-monas* have been reported after the manuscript had been sent to the publisher.[34,35]

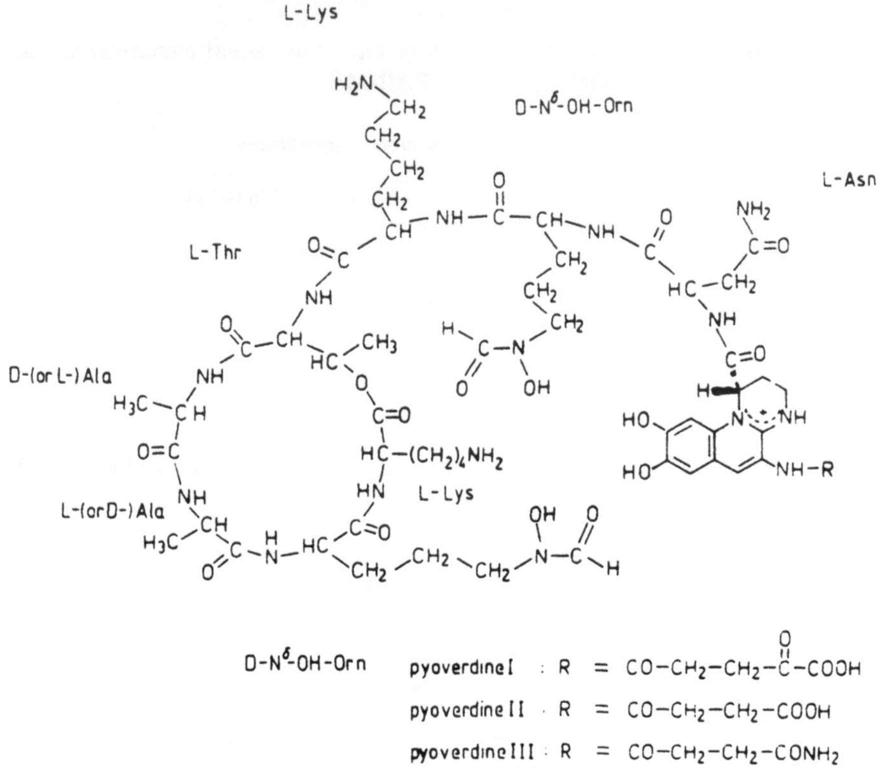

FIGURE 10.   The structure of pyoverdins I, II, and III from a strain of *Pseudomonas fluorescens*.

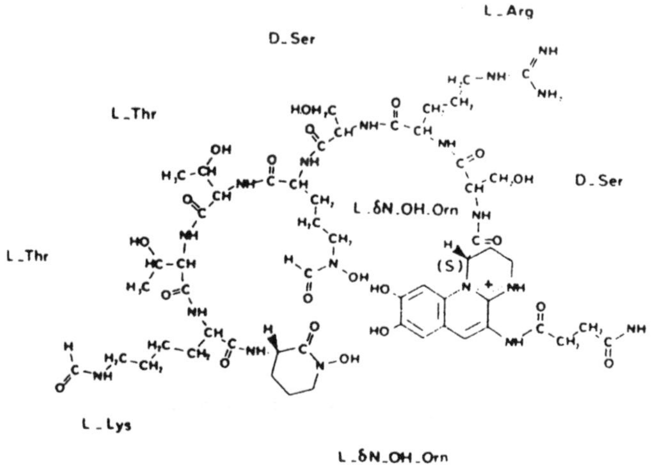

A

FIGURE 11(A).   The former structure of pyoverdin Pa; (B) the revised structure of pyoverdin Pa.

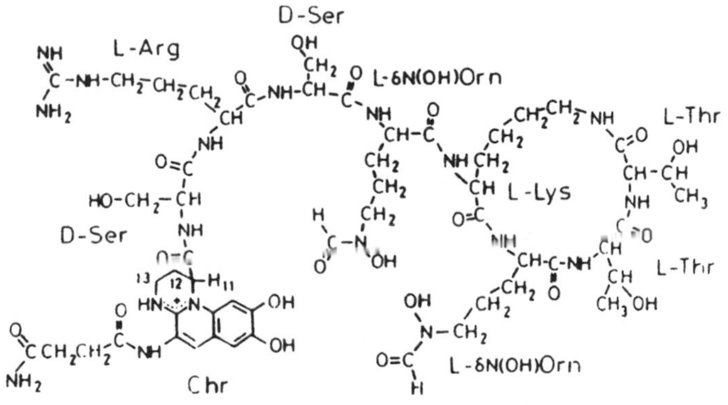

FIGURE 11(B).

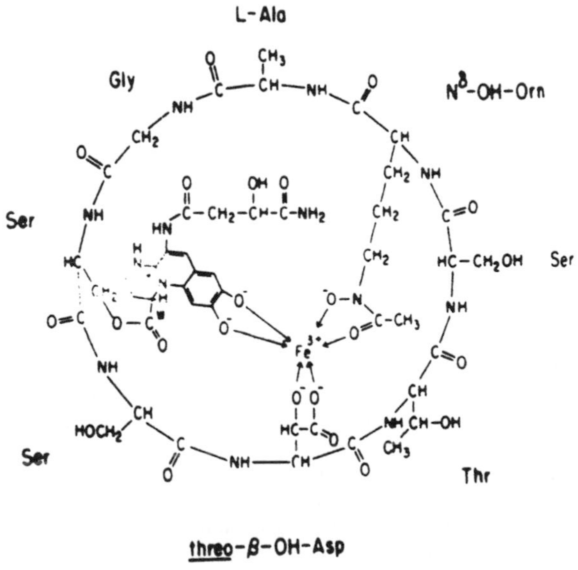

FIGURE 12. The structure of pseudobactin 7SR1.

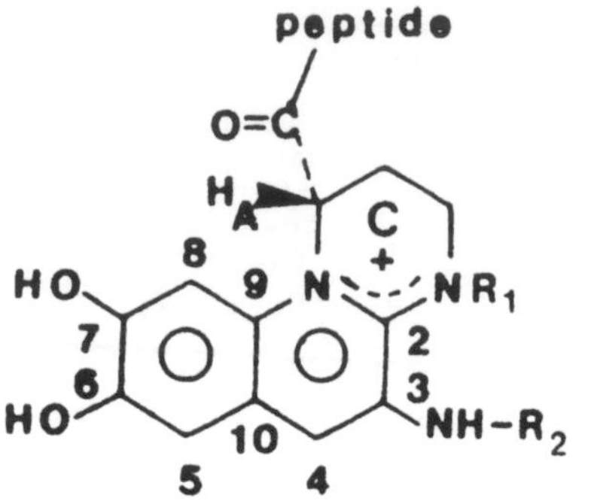

peptide

$R_1 = H \quad R_2 = CO-CH_2-CH_2-CONH_2$
$R_1 = H \quad R_2 = CO-CH_2-CH_2-COOH$
$R_1 = H \quad R_2 = CO-CH_2-CHOH-CONH_2$
$R_1 = H \quad R_2 = CO-CH_2-CHOH-COOH$
$R_1 = H \quad R_2 = CO-CH_2-CH_2-CO-COOH$
$R_1 , R_2 = CO$

FIGURE 13.   General structure of pyoverdins and azotobactin.

REFERENCES

1. **Neilands, J. B.,** Microbial iron compounds, *Annu. Rev. Biochem.*, 50, 715, 1981.
2. **Hider, R. C.,** Siderophore mediated absorption of iron, in *Structure and Bonding 58,* Clarke, M. J. et al., Eds., Springer-Verlag, Berlin, 1984, 25.
3. **Stanier, R. Y., Palleroni, N. J., and Doudoroff, M.,** The aerobic Pseudomonads: a taxonomic study, *J. Gen. Microbiol.*, 43, 159, 1966.
4. **Meyer, J.-M., and Abdallah, M. A.,** The fluorescent pigment of *Pseudomonas fluorescens:* biosynthesis, purification and physicochemical properties, *J. Gen. Microbiol.*, 107, 319, 1966.
5. **Meyer, J.-M. and Hornsperger, J.-M.,** Role of pyoverdin$_{Pf}$, the iron-binding fluorescent pigment of *Pseudomonas fluorescens,* in iron-transport, *J. Gen. Microbiol.*, 107, 329, 1978.
6. **Teintze, M., Hossain, M. B., Barnes, C. L., Leong, J., and van der Helm, D.,** Structure of ferric pseudobactin, a siderophore from a plant growth promoting *Pseudomonas, Biochemistry*, 20, 6446, 1981.
7. **Fukasawa, K., Goto, M., Sasaki, K., Hirata, S., and Sato, S.,** Structure of the yellow-green fluorescent peptide produced by iron-deficient *Azotobacter vinelandii* Strain O, *Tetrahedron*, 28, 5359, 1972.
8. **Corbin, J. L., Karle, I. L., and Karle, J.,** Crystal structure of the chromophore from the fluorescent peptide produced by iron-deficient *Azotobacter vinelandii, J. Chem. Soc. Chem. Commun.*, p. 186, 1970.
9. **Karle, I. L. and Karle, J.,** Structure of the chromophore from the fluorescent peptide produced by iron-deficient *Azotobacter vinelandii, Acta Crystallogr. Sect. B*, 27, 1891, 1971.
10. **Teintze, M. and Leong, J.,** Structure of Pseudobactin A, a second siderophore from plant growth promoting Pseudomonas B10, *Biochemistry*, 20, 6457, 1981.
11. **Philson, S. B. and Lliñas, M.,** Siderochromes from *Pseudomonas fluorescens*. II. Structural homology as revealed by NMR spectroscopy, *J. Biol. Chem.*, 257, 8086, 1982.
12. **Wendenbaum, S., Demange, P., Dell, A., Meyer, J.-M., and Abdallah, M. A.,** The structure of Pyoverdin Pa, the siderophore of *Pseudomonas aeruginosa, Tetrahedron Lett.*, 24, 4877, 1983.
13. **Briskot, G., Taraz, K., and Budzikiewicz, H.,** Siderophore vom Pyoverdin-Typ aus *Pseudomonas aeruginosa, Z. Naturforsch. Teil C*, 41, 497, 1986.
14. **Demange, P., Wendenbaum, S., Bateman, A., Dell, A., and Abdallah, M. A.,** Bacterial siderophores: structure and physicochemical properties of pyoverdins and related compounds, in *Iron Transport in Microbes, Plants and Animals*, Winkelmann, G., van der Helm, D., and Neilands, J. B., Eds., VCH Publishers, Weinheim, West Germany, 1987, 767.
15. **Abdallah, M. A., Albrecht-Gary, A. M., Blanc, S., Dell, A., Demange, P., Linget, C., and MacLeod, J.,** Pyoverdins, a new class of siderophores; structure and physicochemical properties, in 27th Int. Conf. on Coordination Chemistry, Gold Coast, Australia, 1989.

16. **Demange, P., Bateman, A., Dell, A., and Abdallah, M. A.,** Structure of Azotobactin D, a siderophore of *Azotobacter vinelandii* strain D (CCM 289), *Biochemistry,* 27, 2745, 1988.

17. **Yang, C. C. and Leong, J.,** Structure of Pseudobactin 7SR1, a siderophore from a plant-deleterious *Pseudomonas, Biochemistry,* 23, 3534, 1984.

18. **Buyer, J. S., Wright, J. M., and Leong, J.,** Structure of Pseudobactin A 214, a siderophore from a bean-deleterious *Pseudomonas, Biochemistry,* 25, 5492, 1986.

19. **van der Hofstadt, G. A. G. M., Marugg, J. D., Verjans, G. M. G. M., and Weisbeek, P. J.,** Characterization and structural analysis of the siderophore produced by the PGPR *Pseudomonas putida* strain WCS358, in *Iron, Siderophores and Plant Diseases,* Swinburne, T. R., Ed., Plenum Press, New York, 1986, 71.

20. **Poppe, K., Taraz, K., and Budzikiewicz, H.,** Pyoverdin type siderophores from *Pseudomonas fluorescens, Tetrahedron,* 43, 2261, 1987.

21. **Briskot, G., Taraz, K., and Budzikiewicz, H.,** Pyoverdin-type siderophores from *Pseudomonas aeruginosa, Liebigs Ann. Chem.,* p. 375, 1989.

22. **Demange, P., Wendenbaum, S., Linget, C., Mertz, C., Cung, M. T., Dell, A., and Abdallah, M. A.,** Bacterial siderophores: Structure and NMR assignment of Pyoverdins Pa, siderophores of *Pseudomonas aeruginosa* ATCC 15692, *Biol. Metals,* 3, 155, 1990

23. **Itoh, S., Honda, H., Tomita, F., and Suzuki, T.,** Rhamnolipids produced by *Pseudomonas aeruginosa* grown on n-paraffin (mixture of C12, C13 and C14 fractions), *J. Antibiot.,* 24, 856, 1971.

24. **Demange, P., Bateman, A., MacLeod, J. K., Dell, A., Piémont, Y., and Abdallah, M. A.,** Bacterial siderophores: Identification of unusual 3,4,5,6-tetrahydropyrimidine-based amino acids in pyoverdins from two *Pseudomonas fluorescens* strains, *Tetrahedron Lett.,* 31, 7611, 1990.

25. **Demange, P., Bateman, A., Mertz, C., Dell, A., Piémont, Y., and Abdallah, M. A.,** Hores of *Psuedomonas tolaasii* NCPPB 2192 and pyoverdins Pf, siderophores of *Pseudomonas fluorescens* CCM 2798. Identification of an unusual natural amino acid, *Biochemistry,* 29, 11041, 1990.

26. **Linget, C., Azadi, P., MacLeod, J. K., Dell, A., and Abdallah, M. A.,** unpublished work, 1990.

27. **Cramer, S. M., Nathanael, B., and Horvath, C.,** High performance liquid chromatography of Deferoxamine and Ferrioxamine: interference by iron present in the chromatographic system, *J. Chromatogr.,* 295, 405, 1984.

28. **Knox, J. H.,** Ion-exchange and ion-pair chromatography, in *High Performance Liquid Chromatography,* Knox, J. H., Ed., Edinburgh University Press, Edinburgh, 1978, 52.

29. **MacDonald, J. C. and Bishop, G. G.,** Spectral properties of a mixture of fluorescent pigments produced by *Pseudomonas aeruginosa, Biochim. Biophys. Acta,* 800, 11, 1984.

30. **Williams, D. H., Bradley, C. V., Santikarn, S., and Bojesen, G.,** Fast atom bombardment mass spectrometry, *Biochem. J.,* 201, 105, 1982.

31. **Demange, P., Abdallah, M. A., and Frank, H.,** Assignment of the configuration of the amino acids in peptidic siderophores, *J. Chromatogr.,* 438, 291, 1988.

32. **van der Helm, D., Jalal, M. A. F., and Hossain, M. B.,** The crystal structures, conformation and configurations of siderophores, in *Iron Transport in Microbes, Plants and Animals,* Winkelmann, G., van der Helm, D., and Neilands, J. B., Eds., VCH Publishers, Weinheim, West Germany, 1987, 135.

33. **Cody, Y. S. and Gross, D. C.,** Characterization of pyoverdin pss, the fluorescent siderophore of *Pseudomonas syringae* pv syringae, *Appl. Environ. Microbiol.,* 53, 928, 1987.

34. **Mohn, G., Taraz, K., and Budzikiewicz, H.,** New Pyoverdin-type siderophores from *Pseudomonas fluorescens, Z. Naturforsch.,* 45b, 1437, 1990.

35. **Persmark, M., Freyd, T., and Mattiasson, B.,** Purification, characterization and structure of psuedobactin 589A, a siderophore from plant growth promoting *Pseudomonas, Biochemistry,* 29, 7348, 1990.

MOLECULAR GENETICS OF SIDEROPHORE BIOSYNTHESIS IN FLUORESCENT PSEUDOMONADS

Joey D. Marugg and Peter J. Weisbeek

INTRODUCTION

The genus *Pseudomonas* comprises a vast group of bacteria which are found in a variety of natural environments (soils, fresh or sea water), and in many different associations with plants and animals. This ecological diversity is a reflection of their simple nutritional requirements, and their ability to metabolize a wide range of organic compounds. The pseudomonads are typically aerobic, chemoheterotrophic, rod-shaped, polarly flagellated, Gram-negative bacteria that do not sporulate. The G + C content of their DNA is relatively high ranging from 58 to 69 mol %. The genus is divided into five ribosomal RNA homology groups on the basis of RNA and DNA hybridization experiments.[1] The largest group is that of the fluorescent pseudomonads, and includes the species. *P. aeruginosa, P. fluorescens, P. putida,* and *P. syringae.* A distinctive property of the fluorescent pseudomonads, from which their name is derived, is the production and excretion of yellow-green fluorescent, water-soluble, low-molecular-weight compounds. Two different names have been given to these fluorescent compounds: pyoverdin(e)s and pseudobactins. The molecules have a strong affinity for ferric iron, as demonstrated by their iron-binding coefficients varying between 10^{22} and 10^{25} (pH 7), and function as iron(III)-transport agents.[2,3] Because of these properties they belong to the group of microbial products known as **siderophores**.[4]

Many fluorescent *Pseudomonas* spp. are common soil microorganisms and prominent members of the microflora in the rhizosphere of plants. From a number of investigations in recent years it has become clear that some of these fluorescent pseudomonads have considerable potential as biological control agents in agriculture by protecting crop plants against deleterious microorganisms (both bacteria and fungi).[5-7] It has been proposed that the *Pseudomonas* spp. strains, due to the production of fluorescent siderophores, promote plant growth by depriving these deleterious rhizosphere microorganisms of iron, thereby inhibiting their growth.[8,9] Other fluorescent *Pseudomonas* species have an opposite effect; they are plant pathogens.[10]

Some *Pseudomonas* spp., e.g., *P. aeruginosa*, also produce another type of siderophore called pyochelin, which differs considerably in structure and in properties from the pyoverdin/pseudobactin-type siderophores (see below).[11] Pyoverdin as well as pyochelin seem to play a role in the virulence of *P. aeruginosa.* They both promote the removal of iron from transferrin, an iron-binding serum glycoprotein involved in the host defense and highly inhibitory to bacterial growth, and they both stimulate growth in human serum of mutants defective in siderophore production.[12,13] Pyoverdin, however, appears to be more efficient in mediating iron transport in human transferrin or serum.[14] Although pyochelin has been the subject of many investigations in recent years, detailed studies on its biosynthesis are lacking. Pyochelin is a salicylic substituted cysteinyl peptide, and has a relatively low iron-binding coefficient (5×10^5). Recently, it has been shown that pyochelin can be produced chemically in three steps from salicylonitrile, L-cysteine, and L-*N*-methylcysteine.[15] Furthermore, mutants have been isolated that require salicylic acid for pyochelin biosynthesis.[16]

The biosynthetic pathways responsible for the production of pseudobactins/pyoverdins and their regulatory mechanisms still have to be elucidated. Most of our current knowledge on siderophore biosynthesis in fluorescent *Pseudomonas* spp. comes from (genetic) studies on the iron-uptake systems of the rhizosphere-colonizing strain *P. putida* WCS358, and to less extent of another rhizosphere-colonizing strain *Pseudomonas* spp. B10.[17] Both strains are investigated because of their beneficial effects on plants. Their siderophore systems are quite similar in several aspects. The structures of their pseudobactins are very much alike with only minor structural differences (see below), suggesting that their biosynthetic pathways are very similar, too. Furthermore, the many genes involved in pseudobactin biosynthesis have a complex chromosomal organization and are coordinately regulated by iron in both strains. In view of these similarities and from limited evidence obtained with other *Pseudomonas* siderophores we assume that (most of) the features and concepts of siderophore biosynthesis discussed in this chapter are not limited to these two strains, but that they are representative for the fluorescent *Pseudomonas* spp. in general. In this chapter the molecular genetics of pseudobactin biosynthesis in *P. putida* WCS358 will be discussed mainly.

STRUCTURES OF FLUORESCENT SIDEROPHORES

The pyoverdin/pseudobactin siderophores are produced only when the cells are grown under iron-limited conditions, and about 200 to 300 mg/l of siderophore are made.[2] Synthesis closely follows growth and divisions of the cells.[18] The chemical structures of several pyoverdins/pseudobactins have been determined or partially characterized.[2,3,19-27] The complete structure for pseudobactin (Figure 1b), the siderophore from the growth-promoting *Pseudomonas* spp. B10, was determined by single X-ray diffraction and chemical and spectroscopic means.[3] It consists of a linear hexapeptide, which is linked to a fluorescent chromophore, a quinoline derivative. The hydroxamate- and the α-hydroxy acid moieties of the peptide, and the *o*-dihydroxy group of the chromophore form the three iron-chelating groups. The occurrence of both L- and D-amino acids is rather unusual, and their alternate sequence in pseudobactin may be the reason why the compound is not affected by proteolytic enzymes. The structures of other pyoverdins/pseudobactins have been elucidated by using methods like FAB mass spectrometry (e.g., pyoverdin$_{pa}$, Figure 1a), ^{1}H and ^{13}C nuclear magnetic resonance (pseudobactin 7SR1, Figure 1c; pseudobactin A214), or characterized by determination of amino acid composition and sequence.[19-26] The structures of these different siderophores are very similar to pseudobactin: they all possess the same chromophore (except for some minor differences in the acyl residue linked to it), but the (linear or circular) peptide chains, consisting of six to ten amino acids, differ from one strain to the other.

The structure of pseudobactin 358, the siderophore produced by *P. putida* WCS358 has recently been elucidated by combination of ^{1}H and ^{13}C NMR techniques and different amino acid sequencing methods (Figure 1d).[27] In the nonapeptide the α-hydroxyaspartic acid and N^6-hydroxyornithine constitute together with the *o*-hydroxy moiety of the quinoline derivative the three bidentate iron-chelating groups. The structure of pseudobactin 358 resembles the other fluorescent siderophore very much, but is clearly different from, e.g., the pyoverdin produced by *P. putida* ATCC12633.[22]

These structural differences in pyoverdins/pseudobactins are believed to be important in uptake and may play a role in the recognition of these siderophores by bacterial outer membrane receptors.[17,28] Protein receptors for ferric siderophores appear to recognize specifically the geometry of the siderophore backbone surrounding the Fe(III) atom.[29] For pseudobactin 358 it was shown that it could be utilized by only a few of the several hundreds of rhizosphere strains that were tested.[30]

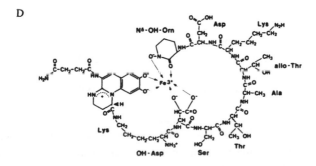

FIGURE 1. Structures of yellow-green fluorescent siderophores, produced by four fluorescent *Pseudomonas* strains. (A) Pyoverdin$_{pa}$; (B) pseudobactin B10; (C) pseudobactin 7SR1; (D) pseudobactin 358.[3,19,20,27]

BIOSYNTHETIC GENES

SIDEROPHORE-DEFECTIVE MUTANTS

Transposon mutagenesis using transposon Tn5 has successfully been applied to *Pseudomonas putida* WCS358. The transposon was introduced into WCS358 with the RP4-derived mobilization system developed by Simon et al.[31] In this system the *Escherichia coli* plasmid pSUP2021, a pBR325 derivative containing the transposon Tn5 and the RP4 origin of transfer replication (Mob-site), is used. In the presence of recipient cells plasmid pSUP2021 is mobilized by the RP4 transfer functions inserted and immobilized in the chromosomal DNA. Since the plasmid is unable to replicate in *Pseudomonas,* recipient cells that have acquired the Tn5-associated antibiotic resistance harbor Tn5 integrated in the chromosome. In a standard mating procedure kanamycin and nalidixic acid-resistant (the latter being required for selection of *Pseudomonas* cells) transconjugants can be found with a frequency of 10^{-6} per recipient cell, whereas the frequency of spontaneous kanamycin-resistant cells is 3×10^{-9}.[32] Auxotrophic mutants are found among the kanamycin and nalidixic acid-resistant transconjugants with a frequency of *circa* 1%, indicating that integration of Tn5 into the chromosome occurs randomly.[32]

A Tn5 mutant collection of WCS358 was used to isolate mutants which were defective in the biosynthesis of the fluorescent siderophore. Fifteen transconjugants yielded 28 mutants that were nonfluorescent (designated as Flu$^-$), unable to grow on iron-limited medium (designated as Sid$^-$), or which exhibited both or intermediate phenotypes.[32] The growth characteristics of the mutants were determined on medium with increasing concentrations of the iron chelator bipyridyl (0.2 to 0.8 mM). The antibiotic activity against an *E. coli* indicator strain was also tested. Wild-type WCS358 inhibits the growth of *E. coli* and a number of other microorganisms.[33] Mutants which do not excrete a siderophore will not be able to bind Fe^{3+} and hence will not act as inhibitors. The phenotypes and characteristics of the different mutants are shown in Table 1.

Six distinct classes of mutants were isolated.[32] Class 1 mutants did not fluoresce but were still able to bind iron very efficiently, and the iron-siderophore complex was taken up normally by the cell. The latter property indicates that at least some alterations in the fluorescent part of the molecule do not influence the binding or uptake process. Class 2 mutants did not fluoresce or sequester iron, meaning that an earlier step in biosynthesis is damaged than in class 1 mutants. The simultaneous disappearance of fluorescence and siderophore activity indicates that the fluorescent group is not synthesized independently from the rest of the molecule and occurs in a later stage of the biosynthesis of the complete molecule (as could already be concluded from the presence of the class 1 mutants). Class 3 was intermediate in phenotype between classes 1 and 2, containing a nonfluorescent mutant with reduced iron binding. The mutants of classes 4 though 6 make either a slightly damaged siderophore or are mutated in a regulation step or in the excretion process. None of the mutants seemed to have a defect in the uptake of the Fe^{3+}-siderophore complex. They all were able to use the siderophore secreted by the wild-type WCS358 under conditions of iron limitation.

The different types of mutants found, together with the probable overall structure of the siderophore molecule, suggest a biosynthetic pathway that starts with the synthesis of the peptide part followed by the synthesis, in several steps, of the fluorescent group.

COMPLEMENTATION OF MUTANTS

To identify genes involved in the siderophore biosynthesis a genomic library was constructed and used in a complementation analysis with the siderophore-defective mutants. The complete chromosome of WCS358 was cloned in fragments in the broad host-range

TABLE 1
Siderophore-Defective Mutants of *P. putida* WCS358

Class	Phenotype[a]	Number	Fluorescent	Max growth conc bipyridyl	In vitro[b] antibiosis
1	Flu⁻Sid⁺	3	−	800 μM	+
2	Flu⁻Sid⁻	18	−	200	−
3	Flu⁻Sid±	1	−	400	+
4	Flu±Sid±	1	±	600	+
5	Flu±Sid⁺	4	±	800	+
6	Flu⁺Sid±	1	+	400	+

[a] Flu⁻: no fluorescence; Flu±: moderate fluorescence; Flu⁺: normal (wt) fluorescence; Sid⁻: growth in presence of max 200 μM bipyridyl; Sid±: growth in presence of 400—600 μM bipyridyl; Sid⁺: growth in presence of max 800 μM bipyridyl. All mutants were grown on KB medium.

[b] Antibiotic activity against an *E. coli* indicator strain; measured by spraying *E. coli* cells on KB agar plates that were previously incubated with spotted *Pseudomonas* WCS358 wild type or mutants.

cosmid pLAFR1.[34] The 21.6-kb vector has a unique *Eco*RI restriction site, encodes resistance against tetracycline, and contains the bacteriophage lambda *cos* site. In the presence of helper cells with the plasmid pRK2013 (which carries the conjugal transfer functions) it can be mobilized to a wide scale of Gram-negative bacteria, including *Pseudomonas* species.[35,36] The resulting genomic library of WCS358 consisted of more than 1000 independent clones. The sizes of the DNA inserts of the obtained cosmids ranged between 17 and 29 kb, with an average of 26 kb.[32]

Each of the 28 WCS358 siderophore-deficient mutants was subsequently conjugated with this genomic library. All non- and slightly fluorescent mutants (class 1 to 5) were screened for complementation of the defect by restoration of the wild-type fluorescence. The fluorescent mutant of class 6 was complemented by restoring the property to grow under iron-limited conditions.[32] Thirteen distinct cosmid clones could be isolated that complemented the mutations of 13 different siderophore-defective mutants. The other 15 mutants could not be complemented with the gene bank used. This may be a reflection of polarity induced by the Tn5 insertions, but one cannot exclude the possibility that the gene bank used is incomplete.[32]

Restriction enzyme analysis, together with the complementation data, allowed us to arrange the complementing cosmids in five groups, designated as A, B, C, D, and E. Each group complemented a specific mutant or a set of mutants.

Group A consisted of six clones, which together complemented nine mutants belonging to class 1 and 2 (one Flu⁻Sid⁺ mutant and eight Flu⁻Sid⁻ mutants, respectively). The Flu⁻Sid⁻ mutants were not only restored in their fluorescence, but also regained the ability to grow under iron limitation. The *Eco*RI restriction enzyme analysis showed that the cosmids had several fragments in common, representing regions of overlap between them. The groups B (one cosmid), C (three), D (one), and E (two) all complemented one particular mutant which belonged to separate classes (6, 4, 1, and 5, respectively; Table 1). *Eco*RI restriction enzyme analysis revealed that the cosmids isolated from clones within group C, and also within group E, contained overlapping DNA stretches. The inserts of the cosmids belonging to different groups (A through E) showed no resemblance with each other and therefore contained DNA coming from different parts of the genome.[32]

These results demonstrate the involvement of at least five gene clusters (with a minimum of nine genes; see below) in the biosynthesis of pseudobactin 358. A similar genetic complexity was also found in the fluorescent *Pseudomonas* spp. strain B10. In this case, chemically or UV light-induced nonfluorescent mutants were individually complemented by conjugation of a pLAFR1 cosmid gene bank and identification of fluorescent transconjugants.[37] Eight distinct cosmid clones were sufficient to complement most of the nonfluorescent mutants. The complementation data obtained for these eight cosmids showed that a minimum of 12 genes arranged in four gene clusters were essential for pseudobactin biosynthesis in B10.[37] In a similar way, at least four complementation groups were shown to be involved in the biosynthesis of a not-further-characterized yellow-green fluorescent pigment (most probably a siderophore) from *P. syringae* pv. *syringae* JL2000.[38]

This large number of genes involved in pseudobactin biosynthesis is not surprising considering the complex molecular structure of pseudobactin(s). The molecule, with a fluorescent hydroxyquinoline moiety and of several unusual amino acids, like D-N^8-hydroxyornithine, requires a great number of specific enzymes for its biosynthesis. It may well be that these special enzymes are organized in multienzyme protein complexes similar to those of the classical peptide antibiotics.[39]

TRANSCRIPTIONAL ORGANIZATION

The genetic and transcriptional organization of the major gene cluster A, which consists of six overlapping cosmid clones complementing nine Flu$^-$Sid$^-$ and Flu$^-$Sid$^+$ mutants disturbed in the biosynthesis of pseudobactin 358, was determined.[40] A restriction map of the cluster is given in Figure 2B. The position of the mutants was used to distinguish several transcriptional units in this cluster. Determination of the transposon integration site in the chromosome of each mutant was done by digestion of the mutant chromosomal DNA with various restriction enzymes, separation of the fragments on agarose gels, followed by blotting onto nitrocellulose filters and hybridization with a [^{32}P]-labeled cosmid clone of cluster A.[40] As probes cosmids pMA1 and pMA3 (Figure 2C) were chosen, since they cover the entire region of cluster A, and complemented all nine mutants.[32] Hybridizations of, e.g., chromosomal *Hind*III and *Xho*I digests in combination with the known restriction maps of group A as well as Tn5 allow accurate mapping of the Tn5 insertions, relative to the flanking *Hind*III and *Xho*I sites.[41] The resulting positions within this cluster for all the Tn5 insertions are given in Figure 2A. A number of other mutants that did not map in cluster A by complementation were also tested by hybridization with the same probes. Two Flu$^-$Sid$^-$ mutants, JM212 and JM217, were found to contain Tn5 insertions inside the region of cluster A (Figure 2A).

This analysis of cluster A showed the presence of at least four transcriptional units.[40] The first transcriptional unit (I) overlaps part of a 3.2-kb *Eco*RI fragment, in which the mutation of Flu$^-$Sid$^+$ mutant JM101 resides and which was shown to be necessary for complementation of JM101. A second, large, transcriptional unit (II) stretches out on the *Eco*RI 13.5-kb fragment, in which the mutations of Flu$^-$Sid$^-$ mutants JM201, JM203, JM204, JM205, JM211, and JM213 are located, and the adjacent fragment of 4.8 kb, which is essential for complementation. At the other end probably also a part of the flanking 3.4-kb fragment contributes to this transcriptional unit. A third transcriptional unit (III) is located on the *Eco*RI fragments of 2.5 and 3.4 kb, since the latter fragment carries the Tn5 insertions of Flu$^-$Sid$^-$ mutants JM209 and JM214, while the first fragment is needed for restoration of the wild-type phenotype. The hybridization results with the Flu$^-$Sid$^-$ mutants JM212 and JM217 demonstrated the presence of an additional transcriptional unit. This fourth transcriptional unit (IV) covers the *Eco*RI 6.1-kb fragment and extends outside cluster A, as mutants JM212 and JM217, both with insertions in this fragment, were not complemented by any of the cosmids of cluster A. Furthermore, these (and the following) results suggest that the rescue of the mutants after introduction of cloned wild-type DNA fragments occurs not very likely via homologous recombination, but has to be the result of true complementation. This implies that conclusions on the possible presence of transcriptional units on such

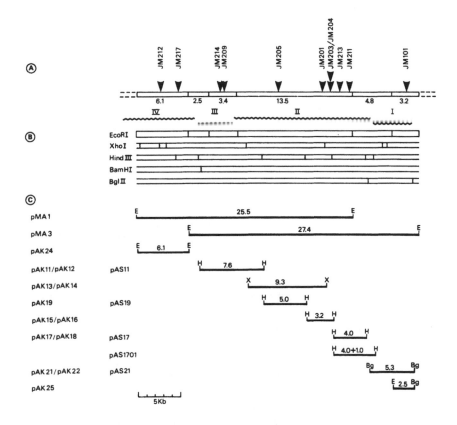

FIGURE 2. (A) Physical map of cluster A. The presence of the transcriptional units I through IV and their approximate limits (represented by wavy lines) were deduced from complementation data and the mapping of the Tn5 insertions of 11 siderophore-defective mutants. These Tn5 integration sites are marked by arrowheads. The sizes of the EcoRI fragments on the map are given in kilobases. (B) Restriction map of cluster A. The map was constructed by combining the mapping data of the cosmids pMA1 and pMA3. (C) Schematic overview of the subclones of cluster A that were constructed. The inserts of the subclones, and of the cosmids pMA1 and pMA3, are represented by the heavy lines. The fragments were cloned into the broad host-range vector pKT240 (first column), and some of them also in the transcription vector pSPT18 (second column). The numbers denote the fragment sizes in kilobases.

fragments are legitimate. The correlated physical and genetic map with the approximate limits of the transcriptional units is indicated in Figure 2A.

To determine the transcriptional organization of this region in more detail, the DNA of cluster A was cloned in smaller fragments and analyzed again for complementation of the mutants. The mobilizable, broad host-range plasmid pKT240 was used as a vector.[42] The cloned fragments, together with their positions on the map, are given in Figure 2C. Each of these plasmid clones was mobilized to its corresponding siderophore-defective mutants. Transconjugants were screened for restoration of yellow-green fluorescence on iron(III)-limited medium as a measure for siderophore biosynthesis.

The plasmids pAK21 and pAK22, with the same 5.3-kb BglII fragment in different orientations, complement mutant JM101. Based on the polarity of Tn5 insertions, this means that a complete transcriptional unit is present on this cloned fragment.[41] The 3.2-kb EcoRI fragment that overlaps the right half of this fragment (Figure 2C) contains the Tn5 insertion in JM101 and is needed for proper complementation of JM101. Therefore a subclone was constructed to check if this region of overlap was sufficient for complementation. The

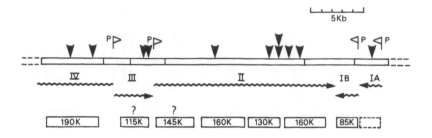

FIGURE 3.   Genetic organization of cluster A. The localization and orientation of the transcriptional units (IA, IB, II, III, IV) are indicated by the wavy lines. Arrowheads on the *Eco*RI restriction map represent Tn5 insertions in the siderophore-defective mutants. Flags (with the letter P) represent promoter regions. The proposed location and the sizes of the polypeptide products encoded by the different subclones are indicated by the closed boxes in the lower part of the figure. The size of the protein represented by the hatched box is unknown.

resulting plasmid pAK25 with the *Bgl*II-*Eco*RI insert of 2.5 kb indeed restores the mutation of JM101, and limits the position of the transcriptional unit (IA) between the borders of this subclone. The *Hin*dIII fragments of 5.0, 3.2 and 4.0 kb, and a 9.3-kb *Xho*I fragment are part of the large transcriptional unit (II) in the middle of cluster A (Figure 2). None of these fragments in pAK13 through pAK19 (Figure 2C) is able to complement any of the mutants JM201, JM203, JM204, JM205, JM211, and JM213, whereas cosmids with inserts which entirely overlap the four subclone fragments (e.g., pMA3) complement each of them. This again shows the presence of a large transcriptional unit on the 13.5-kb *Eco*RI fragment extending within or beyond the 4.0-kb *Hin*dIII fragment at the right end. The plasmids pAK11 and pAK12 both contain the *Hin*dIII 7.6-kb fragment, but in a different orientation. Both complement the mutants JM209 and JM214, which means that the complete transcriptional unit III, which covers the 2.5- and 3.4-kb *Eco*RI fragments (Figure 2), is present within the 7.6-kb *Hin*dIII insert of the plasmids. The *Eco*RI 6.1-kb fragment in pAK24 does not rescue the mutants JM212 and JM217, which confirms the complementation data of pMA1. It demonstrates that this fragment contains only part of the transcriptional unit IV that is disturbed in the two mutants.

Expression of the various subclones (Figure 2C) was studied in *E. coli* minicells.[40] The analysis not only confirmed the presence of the four transcriptional units (IA, II, III, and IV), but also gave information on the directions of several of these transcriptional units, as shown in Figure 3. The analysis also revealed the presence of a fifth transcriptional unit (IB), located between transcriptional units IA and II. The polypeptide products that were encoded by the various plasmids appeared to be relatively large, with apparent molecular weights varying from 85,000 to 190,000.[40] The tentative localization and sizes of the polypeptides are also shown in Figure 3. It is not yet clear whether these transcriptional units constitute large open reading frames or consist of operon-like structures with multiple genes. The proteins which are produced in minicells do not have to be complete products of the biosynthetic pathway, but they may be truncated forms of one or several large proteins. It is therefore not possible to determine precisely the total number of genes on this cluster.

Altogether, the data demonstrate that several genes for biosynthesis of pseudobactin 358 are clustered on a limited region of the *P. putida* WCS358 genome. In addition to that the siderophore biosynthesis also requires the involvement of at least four to five other unlinked gene clusters.[32] A similar scattering of the biosynthetic genes on the chromosome has been observed in *P. aeruginosa* PAO. In this strain mutants defective in biosynthesis of pyoverdin have been mapped genetically. One set of mutations was located at 65 to 70 min on the chromosome, while a second set had a map position at 35 min.[43,44] In general, it seems that in the different fluorescent *Pseudomonas* species the large number of siderophore biosynthetic genes are organized in (sometimes very large) gene clusters dispersed around the genome.

REGULATION

RNA-RNA HYBRIDIZATION

Siderophores are essential elements in the life cycle of the organism only when there is shortage of available iron. Most microorganisms therefore have adapted their genetic capacity for the biosynthesis of siderophores such that the genes are only expressed under iron-deficiency conditions. This is not an absolute rule but the siderophore genes dealt with in this chapter all regulate their expression in an iron-dependent way. Pseudobactin 358 therefore is produced only under iron-limited conditions.[32] Under these conditions certain outer membrane proteins are also expressed specifically.[45]

To address the question of whether this regulation operates at the level of RNA synthesis (or its processing) or at the level of protein synthesis (or processing), the steady-state levels of transcripts of the siderophore genes were determined under different conditions. Hybridization of total cellular RNA with single-stranded RNA probes (with known polarity) was performed. The 7.6-kb *Hind*III and the 5.30kb *Bgl*II fragments from cluster A were cloned into the transcription vector pSPT18, resulting in pAS11 and pAS21, respectively (Figure 2).[40] With this vector, which contains both the SP6 and T7 phage promoters, each strand of a cloned insert can be transcribed into a (radioactive) RNA molecule *in vitro*. This RNA was used as a probe in a Northern hybridization experiment; only RNA with opposite polarity hybridizes to such a probe, and in this way the direction of the transcriptional units on these DNA regions was also confirmed.

Total RNA was isolated from cultures of wild-type WCS358 which were grown under iron(III)-rich and -limited conditions, fractionated on a denaturing agarose gel, blotted to nitrocellulose, and subsequently hybridized separately with each of the [^{32}P]-UTP labeled RNA probes derived from the correct strands of the inserts of pAS11 and pAS21. Hybridization to a specific RNA band was only observed when RNA was isolated from cultures which were grown under iron(III)-limited conditions, and not when RNA was isolated from cultures grown in iron(III)-rich conditions. In both cases the hybridizing bands corresponded with messenger RNAs of about 2.4 to 2.8 kb. These results demonstrate that expression of the genes of these transcriptional units is regulated by iron at the transcriptional level. This effect of iron can be mediated via a system with a regulator protein (activator or repressor) which allows initiation of transcription under iron-limited conditions strictly. This implies that the promoter or operator regions of the respective genes contain specific features that are recognized by such a regulator protein in the absence or presence of iron. Alternatively, the mRNA stability may be influenced by iron in such a way that the rate of mRNA degradation is (rapidly) increased under iron-rich conditions.

CLONING OF IRON-REGULATED PROMOTERS

Since the positions and directions of four transcriptional units within the major biosynthetic gene cluster A have been determined, it was possible to identify the corresponding promoter regions on the DNA (Figure 4). The 7.6-kb *Hind*III fragment contains two promoters, corresponding to transcriptional units II and III (Figure 4). At the other end of cluster A on a 3.2-kb *Eco*RI fragment, promoters of two transcriptional units, IA and IB, were present (Figure 4). Units IA, II, and III encode biosynthetic genes, while unit IB encodes the putative receptor protein for ferric pseudobactin 358 (see below).[40,46]

These promoters were analyzed for their activity and regulation by the construction of transcriptional fusions with the structural genes for β-galactosidase (*lacZ*) and catechol-2,3-dioxygenase (CDO; *xylE*).[64] The 7.6-kb *Hind*III and the 3.2-kb *Eco*RI fragments were cloned

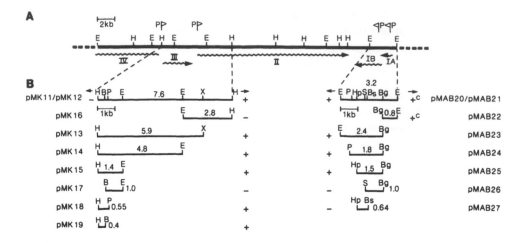

FIGURE 4.   Localization of iron-regulated promoters by subcloning in promoter probe vectors pMP77 and pMP220. (A) Physical map of cluster A, showing *Eco*RI (E) and *Hind*III (H) restriction sites. The orientation and direction of transcriptional units IA, IB, II, III, and IV (wavy lines) are shown. Flags represent possible promoter regions. (B) Overview of the different fragments that were cloned in plasmids pMP77 and pMP220. (Partial) restriction maps of the 7.6-kb *Hind*III fragment (left half) and the 3.2-kb *Eco*RI fragment (right half) are shown. Arrows indicate the direction of the inserts with respect to the *xylE* or *lacZ* genes. X, *Xho*I; Hp, *Hpa*I; Bs, *Bst*EII; S, *Sal*I; B, *Bam*HI; Bg, *Bgl*II; P *Pst*I. −, no promoter activity; +^{c}, constitutive promoter activity; +, iron-regulated promoter activity.

in the broad host-range promoter probe vectors pMP220 and pMP77. Vector pMP220 (IncP; 10.5 kb) encodes a tetracycline resistance gene and contains a promoter less *Escherichia coli lacZ* gene located downstream of several unique sites present in a multilinker.[47] The mobilizable vector pMP77 (IncQ; 12.5 kb) carries resistance genes for chloramphenicol and streptomycin, and contains the promoterless *xylE* gene, from the *xylDEGF* operon downstream of several unique cloning sites.[48] Expression of both the *xylE* and the *lacZ* genes is dependent on the integration of promoter sequences in the proper orientation, upstream of these genes. Both enzymes are normally absent in *Pseudomonas putida* and are easily and quantitatively assayable.[49,50]

Both orientations of the 7.6-kb *Hind*III fragment were cloned in the corresponding site of pMP77. The resulting plasmids pMK11 and pMK12, and control pMP77 were used to transform *E. coli* JM101 cells. When the transformants were grown in cultures under high and low iron conditions and assayed for CDO, no activities at all were measured. This is an indication that *P. putida* promoters are not recognized by *E. coli* RNA polymerase. The fusion plasmids were then mobilized, by use of the pRK2013 helper plasmid, into *P. putida* WCS358 cells. When WCS358(pMK11) transconjugant cells were grown in culture under iron-limited and iron-rich conditions, high CDO activities were measured only when the cells were grown under iron limitation (Table 2). WCS358 transconjugant cells, carrying pMK12 or pMP77, did not express CDO when grown in iron-rich or low-iron cultures (Table 2). The orientation of the promoter activity on the 7.6-kb *Hind*III fragment in pMK11 was in accordance with the direction of transcription as determined previously.[40]

The position of the promoter activity on the DNA was further localized by cloning several subfragments of the 7.6-kb *Hind*III fragment in plasmids pMP77 and pMP220. These fragments and their relative orientation with respect to the reporter gene are shown in Figure 4 (left part). The (specific) activities of the various constructs for CDO and β-galactosidase, determined in WCS358 cell extracts, are presented in Table 2. A 2.8-kb *Eco*RI-*Hind*III fragment located at the right part of the 7.6-kb *Hind*III fragment exhibited no promoter activity (pMK16, Figure 4; Table 2). Iron-regulated CDO activities were found with constructs containing the 5.9-kb *Hind*III-*Xho*I fragment (pMK13), the 4.8-kb (partial) *Hind*III-

TABLE 2
CDO- and β-Galactosidase Activities in Cell Extracts of *P. putida* WCS358 Containing Various Plasmids[a]

Plasmid	CDO activity (mU/mg of protein)		β-Galactosidase activity (U)	
	+200 μM bipyridyl	+100 μM Fe(Cl)₃	+200 μM bipyridyl	+100 μM Fe(Cl)₃
pMP77	0	0		
pMK11	43,000	0		
pMK12	0	0		
pMK13	5,900	0		
pMK14	2,400	0		
pMK15	2,700	0		
pMK16			0	0
pMK17			0	0
pMK18			1,150	15
pMK19			2,950	24

[a] Cultures were grown in KB medium containing bipyridyl or Fe(Cl)₃. β-Galactosidase was assayed by the *o*-nitrophenyl-β-D-galacto-pyranoside assay essentially as described by Miller.[49] Catechol-2,3-dioxygenase (CDO; EC 1.13.11.2) was assayed as described by Nozaki.[50] One milliunit corresponds to the formation of 1 nmol of 2-hydroxymuconic semialdehyde per minute at 24°C.

*Eco*RI fragment (pMK14), or the 1.4-kb *Hin*dIII-*Eco*RI fragment (pMK15). A 10- to 15-fold decrease in enzymatic activity directed by these fragments was observed as compared to the activity given by the intact 7.6-kb *Hin*dIII fragment (Table 2). These differences are not understood, but may be a reflection of differences in transcript stability of the various constructs. The results demonstrated that an iron-regulated promoter activity was retained in the 1.4-kb *Hin*dIII-*Eco*RI fragment. Smaller derivatives of the 1.4-kb fragment were analyzed in vector pMP220. These further deletions revealed that the promoter could be localized within the borders of a 0.42-kb *Hin*dIII-*Bam*HI fragment (Figure 4), as construct pMK19 expressed the *lacZ* gene very efficiently in an iron-regulated fashion (Table 2).

In a similar way the second promoter region, present on the 3.2-kb *Eco*RI fragment (Figure 4; right half), was analyzed for further identification of the promoter(s). This fragment was cloned in both orientations in the *Eco*RI site upstream of the *lacZ* gene in pMP220. Both constructs were mobilized into *P. putida* WCS358, and plasmid pMAB20, with the fragment in the correct orientation, expressed β-galactosidase, but only when the cells were grown under iron-limited conditions (Table 3). Subcloning of this fragment revealed that the enzyme activity depended on the presence of a 2.4-kb *Bgl*II-*Eco*RI fragment (Figure 4; Table 3, pMAB23). This fragment was further reduced in size, and the smallest fragment that could be constructed without loss of β-galactosidase activity was a 1.5-kb *Bgl*II-*Hpa*I fragment as cloned in pMAB25 (Figure 4; Table 3). Two overlapping subfragments, i.e., a 1.0-kb *Bgl*II-*Sal*I and a 0.64-kb *Bst*EII-*Hpa*I fragment (Figure 4; pMAB26 and pMAB27, respectively) did not produce β-galactosidase (Table 3), indicating that promoter activity is disturbed in both fragments.

In *E. coli* a constitutive promoter activity was also measured (pMAB21; Table 3); the subcloning revealed that this activity was directed from the 0.8-kb *Bgl*II-*Eco*RI border fragment (Figure 4; pMAB22). It is very well possible that this is the promoter of a not yet identified gene that is constitutively expressed both in *E. coli* and *Pseudomonas*. This is in contrast with both iron-regulated promoters which were shown not to be active in *E. coli*.

TABLE 3
β-Galactosidase Activities in
Cell extracts of *P. putida*
WCS358 Containing Various
Plasmids[a]

| | β-Galactosidase activity (U) | |
Plasmid	+200 μ*M* bipyridyl	+100 μ*M* Fe(Cl)₃
pMAB20	170	8
pMAB21	230	380
pMAB22	290	390
pMAB23	200	10
pMAB24	180	4
pMAB25	750	15
pMAB26	0	0
pMAB27	0	0

[a] Cultures were grown in KB medium containing bipyridyl or Fe(Cl)₃.

In summary, two siderophore promoters have been identified and localized that are expressed in an iron-dependent way. The relatively high expression levels in iron-limited cells and the apparent absence of transcription in iron-rich cells imply a very stringent control of the transcriptional activity of the promoters studied here. The results also suggest that regulation is mediated via a diffusible factor acting *in trans* that affects the initiation of transcription.

The further investigations were focused on the promoter region on the 0.42-kb *Hind*III-*Bam*HI fragment (derived from the 7.6-kb *Hind*III fragment) to identify the minimal upstream sequences required for iron-regulated promoter activity. The nucleotide sequence of this fragment was determined, and the promoter was localized on a 189-bp *Hae*III-*Bam*HI fragment present within the 0.42-kb *Hind*III-*Bam*HI (Figure 5).[64] The transcription initiation site of this promoter was determined by S1-nuclease mapping within this fragment, which demonstrated that the necessary information for iron-regulated expression is present within 73 bp upstream of the transcription initiation site. A detailed deletion analysis with *Bal*31 revealed that the region from -73 to -67 is required for proper promoter functioning. This region may be involved in recognition or binding of a regulatory protein(s) at the siderophore promoter.

The iron-regulated expression of siderophore biosynthetic genes is complex. Positive regulation, in which an activator protein binds to the promoter under iron-limited conditions, was indicated by the observation that siderophore promoters are not expressed in *E. coli* (and other tested Gram-negative bacteria). The absence of specific activator proteins in these strains is probably the cause for the lack of activity. We recently identified a gene that enabled us to ascertain that activation of transcription is a major regulatory element in the expression of these genes.[65] At the same time there are indications that negative regulation also occurs. At what level this is mediated remains to be determined.

FE³⁺-UPTAKE VIA SPECIFIC RECEPTORS

SIDEROPHORE SPECIFICITY

Upon iron limitation new large outer membrane proteins with apparent molecular weights between 70,000 and 100,0000 are synthesized by fluorescent pseudomonads.[10,45,51,52] The

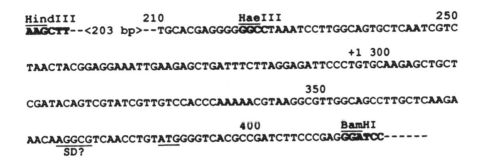

```
HindIII           210              HaeIII                      250
AAGCTT--<203 bp>--TGCACGAGGGGGGCCTAAATCCTTGGCAGTGCTCAATCGTC
                                              +1 300
TAACTACGGAGGAAATTGAAGAGCTGATTTCTTAGGAGATTCCCTGTGCAAGAGCTGCT
                              350
CGATACAGTCGTATCGTTGTCCACCCAAAAACGTAAGGCGTTGGCAGCCTTGCTCAAGA
                         400                    BamHI
AACAAGGCGTCAACCTGTATGGGGTCACGCCGATCTTCCCGAGGGATCC------
      SD?
```

FIGURE 5. Nucleotide sequence of the promoter region present on a 0.42-kb *Hind*III-*Bam*HI fragment. The sequence is numbered from the *Hind*III site. The sequence between nucleotides 7 and 210 is not shown. The transcription initiation site is indicated. The putative Shine-Dalgarno (SD) sequence and start codon are underlined.

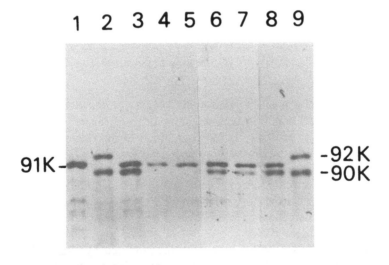

FIGURE 6. Immunoblot analysis of cell envelope preparations run on a 9% sodium dodecyl sulfate-polyacrylamide gel using an antiserum raised against a mixture of 90- and 92-kDa proteins of strain WCS358. Cell envelopes were isolated from cells that were grown under iron limitation. Lanes: 1, WCS374; 2 and 9, WCS358; 3, WCS374(pMR); 4, WCS374(pMR-36); 5, WCS374(pMR-47); 6, WCS374(pAK21); 7, WCS374(pAK22); 8, WCS374(pMA3).

presence of such iron-regulated proteins in the outer membrane suggests that they are involved in the high-affinity siderophore-mediated Fe^{3+}-uptake.[53] A gene encoding an 85,000-MW iron-inducible outer membrane protein was cloned from *Pseudomonas* spp. strain B10, and it was shown that it is the receptor for ferric pseudobactin.[51] In *P. syringae* p.v. *syringae* strain B301D an outer membrane polypeptide of MW 74,000 has been identified which probably serves as the receptor for ferric pyoverdin$_{pss}$.[10]

When strain WCS358 is grown under iron limitation the expression of a number of proteins is induced, among which are two outer membrane proteins with apparent molecular weights of 90,000 and 92,000.[18,45] An antiserum, raised against a mixture of these two proteins, was used in the analysis and characterization of the receptor protein(s) of WCS358. Figure 6 (lanes 2 and 9) shows that indeed only the 90,000- and 92,000-MW proteins of cell envelopes isolated from iron(III)-limited WCS358 cells react in immunoblots with this antiserum. The same antiserum cross-reacts with a protein of 91,000 MW present in cell envelopes from iron(III)-limited cells of *P. fluorescens* strain WCS374 (Figure 6, lane 1), indicating similarities between the membrane proteins of both strains.[45]

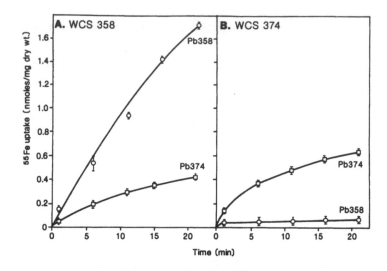

FIGURE 7. $^{55}Fe^{3+}$-uptake mediated by pseudobactin 358 (○) and pseudobactin 374 (□) by cells of strains *P. putida* WCS358 (A) and *P. fluorescens* WCS374 (B) grown under Fe^{3+}-limitation. $^{55}Fe^{3+}$-pseudobactins were used at a final concentration of 1 μM.

The considerable variation in apparent molecular weights for (putative) receptor proteins of fluorescent pseudomonads may well be a reflection of the diversity of the siderophores themselves. This suggests a specificity in the recognition of the siderophore of a certain strain and its cognate receptor protein, a view that is supported by recent experimental evidence.[28,54]

The pseudobactin 358-specific siderophore receptor was identified and characterized using strains *P. putida* WCS358 and *P. fluorescens* WCS374.[46] The amino acid composition of pseudobactin 374, the siderophore of WCS374, has been determined and differs considerably from that of pseudobactin 358.[66]

The specificity in utilization of the pseudobactins produced by strains WCS358 and WCS374 was investigated in $^{55}Fe^{3+}$-uptake experiments using the two purified siderophores.[55] WCS358 cells, which were grown under iron-limited conditions, were able to take up Fe^{3+} from their own Fe^{3+}-siderophore complex as well as from the Fe^{3+}-pseudobactin 374 complex (Figure 7A). WCS374 cells, which were also grown under iron-limited conditions, were only able to incorporate the Fe^{3+} from pseudobactin 374 (Figure 7B). WCS358 and WCS374 were equally efficient in utilizing the iron from the Fe^{3+}-pseudobactin 374 complex, whereas WCS358 was about four times more efficient than WCS374 when it used its own siderophore (Figures 7A and 7B).

A similar conclusion was reached from plating experiments. Both strains were incubated on KB agar plates that contained various concentrations of the purified siderophores. It was found that pseudobactin 358 inhibited the growth of wild-type WCS374 at concentrations of 50 μM and higher, whereas pseudobactin 374 caused no growth inhibition of WCS358 at any of the concentrations used. Also under these conditions ferric pseudobactin 358 cannot be utilized by WCS374, whereas ferric pseudobactin 374 can be used by both strains.

The utilization of a heterologous siderophore by WCS358 is not limited to pseudobactin 374: it has been shown that WCS358 is able to utilize many other siderophores produced by a great number of *Pseudomonas* soil isolates.[30] Some of these siderophores were partly characterized and shown to be different from pseudobactin 358.

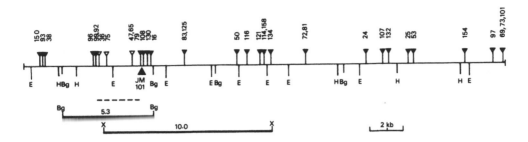

FIGURE 8.   Location of Tn5 insertions on the insert of cosmid pMR. Open triangles refer to the Tn5 insertions that cause a loss of the ability to make WCS374 cells competent for pseudobactin 358 utilization, whereas closed triangles refer to Tn5 insertions that do not affect pseudobactin 358 utilization by WCS374 cells. The broken line limits the DNA region required for utilization of pseudobactin 358. Positions of the 5.3-kb *Bgl*II and the 10.0-kb *Xho*I fragments are indicated by the heavy lines. The location of the genomic Tn5 insertion of the siderophore-defective mutant JM101 is indicated by the large triangle. Restriction enzyme abbreviations: E, *Eco*RI; H, *Hin*dIII; Bg *Bgl*II; X, *Xho*I.

THE GENE ENCODING THE RECEPTOR PROTEIN FOR PSEUDOBACTIN 358

The inability of strain WCS374 to take up iron via pseudobactin 358 was used to identify genes and proteins of WCS358 that are involved in this uptake process. A gene bank containing the complete chromosomal DNA of strain WCS358 was transferred to WCS374, and transconjugant WCS374 cells were tested for whether they had obtained the receptor for ferric pseudobactin 358 and therefore had become competent to use pseudobactin 358.[46] Transconjugants WCS374 colonies were replica-plated to KB agar plates and to KB agar plates supplemented with pseudobactin 358. In this way a single cosmid clone was identified that was able to rescue WCS374 from iron starvation induced by pseudobactin 358. A restriction map of the insert DNA (29.4 kb) of this cosmid pMR was constructed (Figure 8). Southern hybridization experiments, in which cosmid pMR was hybridized with the cosmids of the five siderophore biosynthesis gene clusters (see above), have demonstrated that the pMR insert DNA and the major gene cluster A share a common DNA sequence of about 8.8 kb. This has extended the gene cluster A to approximately 50 kb (Figure 9).[46]

The position of the pseudobactin 358 uptake genes on the pMR insert DNA was analyzed by Tn5 mutagenesis, and by subcloning of the large insert. The analysis of the subclones showed that only plasmids containing a *Bgl*II 5.3-kb fragment in different orientations (pAK21 and pAK22) were able to make WCS374 competent to utilize pseudobactin 358 (Figure 8). The 10.0-kb *Xho*I fragment overlaps most of the *Bgl*II 5.3-kb fragment, but it did not make WCS374 competent to use pseudobactin 358. Therefore the essential DNA had to be present in the more left part of the *Bgl*II fragment (Figure 8). The Tn5 mutagenesis limited the gene to a DNA region of approximately 2.5 kb long within the *Bgl*II 5.3-kb fragment (Figure 8).[46] The positions of the 34 insertions on the pMR insert are shown on the map in Figure 8. Four of the Tn5 insertions (36, 47, 65, and 75) caused a loss of the ability to use pseudobactin 358. These four insertions were all in the *Bgl*II 5.3-kb fragment. All other insertions, including the surrounding insertions 99 and 79, did not affect the pseudobactin 358 utilization of WCS374 cells. The information for pseudobactin 358 utilization is situated within the overlap of cosmids pMR and pMA3 (Figure 9). As expected, cosmid pMA3 also conferred competence for utilization of pseudobactin 358 to WCS374 cells.

The receptor gene is flanked on both sides by biosynthetic genes, as shown by the positions of the Tn5 insertion of several mutants that are defective in the biosynthesis of the siderophore (e.g., JM101 and JM211) (Figure 2). The receptor gene and the biosynthetic genes are on separate transcriptional units. This was demonstrated in a complementation assay using pMR derivatives which carried Tn5 insertions in each of the genes: pMR

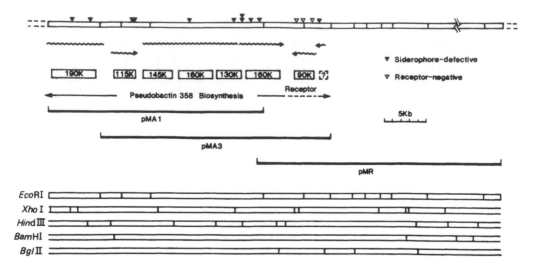

FIGURE 9.   Genetic organization of the gene cluster A involved in siderophore biosynthesis and transport, present on the genome of *P. putida* WCS358. The orientations and directions of transcriptional units (wavy lines), and polypeptide products (closed boxes) are shown. Closed and open triangles refer to the Tn5 insertions of siderophore-defective and receptor-negative mutants, respectively. The restriction map at the bottom of the figure was combined from the data of the overlapping cosmid clones pMA1, pMA3, and pMR (represented by the heavy lines).

derivatives with mutations in the receptor gene were still able to complement the biosynthetic mutant JM101, whereas pMR derivatives which did not complement JM101 still could make strain WCS374 competent for pseudobactin 358 utilization. The previous investigations already demonstrated the presence of two transcriptional units within the 5.3-kb *Bgl*II fragment (IA and IB, Figure 3).[40] One of these transcripts was detected on a Northern blot, and its length measured about 2.4 to 2.8 kb, which fits well with the size of the gene predicted from the Tn5 mapping. A similar organization was observed for the pseudobactin B10-specific receptor and biosynthetic genes.[51]

CHARACTERIZATION OF THE RECEPTOR PROTEIN

In cell envelopes of WCS374 cells that harbor the cosmids pMR or pMA3, and that were grown under iron(II) limitation, a new outer membrane protein was found that reacts with the antiserum mentioned above. It comigrates with the WCS358 protein of 90,000 (Figure 6, lanes 3 and 8). This 90,000-MW protein (and also the 91,000-MW protein) was not observed under iron-rich conditions. This indicates that the mechanism of iron regulation in both strains is very much alike or even identical. The cell envelopes of iron-limited WCS374 cells harboring plasmids pAK21 or pAK22, with the *Bgl*II 5.3-kb fragment of pMR, also contained the 90,000-MW protein (Figure 6, lanes 6 and 7).

Transconjugants of WCS374 harboring the mutagenized pMR derivatives 36 and 47, which are unable to utilize pseudobactin 358 anymore, have also lost this 90,000-MW protein (Figure 6, lanes 4, and 5). This demonstrates that the 90,000-MW protein is involved in the pseudobactin 358-specific iron uptake. Moreover, the position of the 90,000-MW protein in the outer membrane of the cell suggests that it functions as the receptor for ferric pseudobactin 358.

Pseudobactin 358-dependent iron-uptake experiments using WCS374 cells that harbor cosmid pMR or its mutagenized derivatives were used to confirm the previous findings.[46] $^{55}Fe^{3+}$ complexed to pseudobactin 358 and supplied to WCS374(pMR) cells was taken up efficiently, as shown in Figure 10. WCS374 cells without the cosmid did not take up Fe^{3+} from the $^{55}Fe^{3+}$-pseudobactin 358 complex at all (Figure 10). WCS374 cells harboring one of the mutagenized pMR derivatives (36, 47, 65, and 75) were also not able to take up the

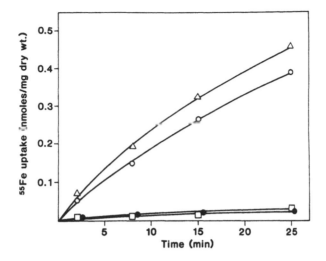

FIGURE 10.   Pseudobactin 358-mediated ^{55}Fe uptake by cells of WCS374 (—●—●—), WCS374(pMR) (—○—○—), WCS374(pAK22) (—△—△—), and WCS374(pMR-36) (—□—□—), grown under iron limitation.

Fe^{3+} from the complex, as shown for pMR-36. WCS374 cells, carrying pAK22 (Figure 10) or pAK21 (data not shown), were as efficient in the uptake of Fe^{3+} from ferric pseudobactin 358 as WCS374(pMR)cells.

Recently, the receptor gene has been sequenced and its product appeared to be a protein of 772 amino acids (86,01 kDa) with a 47-amino acid leader peptide attached in its precursor form. This signal is extremely long for prokaryotes.[67] The receptor protein shares strong homology with four regions of TonB-dependent receptor proteins of *E. coli*, including BtuB, FecA, FepA, FhuA, and IutA.[56-60] This homology suggests the presence of a TonB-like protein in strain WCS358 that is required for Fe(III) transport and interaction of the receptor protein with this protein.

The presence of a single WCS358 protein in the cell envelopes of WCS374 is sufficient to make this strain competent for the utilization of ferric pseudobactin 358. It is likely, however, that other proteins are also involved in the iron-uptake process in *Pseudomonas*. In *E. coli* all siderophore systems, including, e.g., enterobactin and aerobactin, require the participation of several inner membrane-associated proteins in addition to the specific receptors.[61,62] Our results indicate that these not yet identified proteins are either less specific or are very similar in WCS374 and WCS358. They must be able to interact with both WCS374 and WCS358 receptor proteins. The specificity of the uptake process therefore seems mainly to depend on the recognition of the iron-siderophore complex by the 90K outer membrane receptor protein.

The receptor gene from *Pseudomonas* spp. B10 encodes an 85 kDa outer membrane protein.[51] By site-specific exchange mutagenesis mutants were created which did not produce the 85-kDa protein anymore. These mutants were completely unable to take up iron via ferric pseudobactin B10. This shows that strain B10 has only a single receptor protein for uptake of this siderophore. In contrast, when the gene coding for the 90-kDa receptor protein is disrupted in marker-exchange experiments, the mutant cells are still capable of utilizing Fe^{3+} delivered by pseudobactin 358, although this occurs with a largely decreased efficiency. The same mutants are also still able to utilize pseudobactin 374 and other heterologous siderophores, just like the wild-type strain. This suggests that the 90kDa receptor is responsible for the specific iron uptake by pseudobactin 358 only and that WCS358 contains an alternative uptake system which recognizes ferric pseudobactin 358 and heterologous ferric siderophores. An intriguing question is whether the uptake of these other siderophores by WCS358 occurs via a single receptor protein with low specificity, or by a number of more specific receptors. The iron-regulated 92-kDa outer membrane protein may be involved in this broad-specificity uptake system.

SUMMARY AND CONCLUSIONS

This chapter describes the iron uptake systems of the fluorescent pseudomonads, and in particular the efforts that were made to unravel the complex biosynthesis and utilization of pseudobactin 358 in *Pseudomonas putida* strain WCS358. When one bears in mind the complex chemical structure of pseudobactin, it is not hard to envisage that many genes are involved in its synthesis and secretion, the uptake of the ferric pseudobactin, and, finally, the intracellular release of iron. Complementation of siderophore-defective mutants led to the identification of approximately 15 genes distributed over five gene clusters.

It is also clear that the expression of the genes involved in these different steps has to be very efficient and controlled tightly in order to respond quickly and specifically to situations of iron limitation in the environment. The expression of at least part of these genes is iron-dependent and appears to be regulated at the transcriptional level. Strong indications have been obtained demonstrating the possible involvement of a positive regulator that activates transcription at specific promoters under iron stress. Since there are also other indications which point toward negative regulation, a comparison with the siderophore system of *Vibrio anguillarum* is tempting. In this microorganism both positive and negative regulatory factors control siderophore biosynthesis and membrane transport proteins involved in iron translocation.[63]

The ability of fluorescent pseudomonads to take up different pseudobactins is dependent on the presence of specific outer membrane receptor proteins. Although strain WCS358 is capable of utilizing many structurally different pseudobactins, the receptor protein for pseudobactin 358 turned out to be very specific. This indicates that strain WCS358 contains, besides the pseudobactin 358-specific receptor protein, another (or probably more) receptor(s) which recognizes a great number of other similar but distinct pseudobactins. At this moment the identification of these other receptors is under way. The heterologous utilization of pseudobactins by different strains suggests that at least the other components of the membrane transport systems in the fluorescent pseudomonads are very much alike. The pseudobactin specificity, however, seems to be determined by the outer membrane receptor protein(s) exclusively. This specificity of the receptor proteins may vary from strain to strain, but also their number may be variable. It remains to be determined what other (membrane) components are involved in the translocation of ferric pseudobactins and the subsequent release of the bound iron.

REFERENCES

1. **Palleroni, N. J., Kunisawa, R., Contopoulou, R., and Doudoroff, M.,** Nucleic acid homologies in the genus *Pseudomonas, Int. J. Syst. Bacteriol.,* 23, 333, 1973.
2. **Meyer, J. M., Halle, F., Hohnadel, D., Lemanceau, P., and Ratefiarivelo, H.,** Siderophores of *Pseudomonas* — biological properties, in *Iron Transport in Microbes, Plants and Animals,* Winkelmann, G., van der Helm, D., and Neilands, J., Eds., VCH Publishers, Weinheim, West Germany, 1987, 189.
3. **Teintze, M., Hossain, M. B., Barnes, C. L., Leong, J., and van der Helm, D.,** Structure of ferric pseudobactin, a siderophore from a plant growth promoting *Pseudomonas, Biochemistry,* 20, 6446, 1981.
4. **Neilands, J.,** Microbial iron compounds, *Annu. Rev. Biochem.,* 50, 715, 1981.
5. **Burr, T. J. and Caesor, A. J.,** Beneficial plant bacteria, *CRC Crit. Rev. Plant Sci.,* 2, 1, 1984.

6. **Schroth, M. N. and Hancock, J. G.**, Disease-suppressive soils and root-colonizing bacteria, *Science*, 216, 1376, 1982.
7. **Schippers, B., Lugtenberg, B., and Weisbeek, P. J.**, Plant growth control by fluorescent pseudomonads, in *Nonconventional Approaches to Plant Disease Control*, Chet, I., Ed., Wiley-Interscience, New York, 1987, 19.
8. **Kloepper, J. W., Leong, J., Teintze, M., and Schroth, M. M.**, Enhanced plant growth by siderophores produced by plant growth-promoting rhizobacteria, *Nature*, 286, 885, 1980.
9. **Geels, F. P. and Schippers, B.**, Reduction of yield depressions in high frequency potato cropping soil after seed tuber treatments with antagonistic fluorescent *Pseudomonas* spp., *Phytopathol. Z.*, 108, 207, 1983.
10. **Cody, Y. S. and Gross, D. C.**, Outer membrane protein mediating iron uptake via pyoverdin$_{pss}$, the fluorescent siderophore produced by *Pseudomonas syringae* pv. *syringae*, *J. Bacteriol.*, 169, 2207, 1987.
11. **Cox, C. D. and Graham, R.**, Isolation of an iron-binding compound from *Pseudomonas aeruginosa*, *J. Bacteriol.*, 37, 357, 1979.
12. **Cox, C. D. and Adams, P.**, Siderophore activity of pyoverdin for *Pseudomonas aeruginosa*, *J. Bacteriol.*, 48, 130, 1985.
13. **Ankenbauer, R., Sriyosachati, S., and Cox, C. D.**, Effects of siderophores on the growth of *Pseudomonas aeruginosa*, *Infect. Immun.*, 49, 132, 1985.
14. **Sriyosachati, S. and Cox, C. D.**, Siderophore-mediated iron acquisition from transferrin by *Pseudomonas aeruginosa*, *Infect. Immun.*, 52, 885, 1986.
15. **Ankenbauer, R., Toyokuni, T., Staley, A., Rinehart, K. L., Jr., and Cox, D. C.**, Synthesis and biological activity of pyochelin, a siderophore of *Pseudomonas aeruginosa*, *J. Bacteriol.*, 170, 5344, 1988.
16. **Ankenbauer, R. and Cox, C. D.**, Isolation and characterization of *Pseudomonas aeruginosa* mutants requiring salicylic acid for pyochelin biosynthesis, *J. Bacteriol.*, 170, 5364, 1988.
17. **Leong, J.**, Siderophores: their biochemistry and possible role in the biocontrol of plant pathogens, *Annu. Rev. Phytopathol.*, 24, 187, 1986.
18. **Weisbeek, P. J., van der Hofstad, G. A. J. M., Schippers, B., and Marugg, J. D.**, Genetic analysis of the iron-uptake system of two plant growth-promoting *Pseudomonas* strains, in *Iron, Siderophores, and Plant Diseases*, Vol. 117, NATO ASI Series A: Life Sciences, Swinburne, T. R., Ed., Plenum Press, New York, 1986, 299.
19. **Wendenbaum, S., Demange, P., Dell, A., Meyer, J. M., and Abdallah, M. A.**, The structure of pyoverdine$_{pa}$, the siderophore of *Pseudomonas aeruginosa*, *Tetrahedron Lett.*, 24, 4877, 1983.
20. **Yang, C. C. and Leong, J.**, Structure of pseudobactin 7SR1, a siderophore from a plant-deleterious *Pseudomonas* strain, *Biochemistry*, 23, 3534, 1984.
21. **Buyer, J. S., Wright, J. M., and Leong, J.**, Structure of pseudobactin A214, a siderophore from a bean-deleterious *Pseudomonas*, *Biochemistry*, 25, 5492, 1986.
22. **Demange, P., Wendenbaum, S., Bateman, A., Dell, A., and Abdallah, M. A.**, Bacterial siderophores: structure and physicochemical properties of pyoverdins and related compounds, in *Iron Transport in Microbes, Plants and Animals*, Winkelmann, G., van der Helm, D., and Neilands, J., Eds., VCH Publishers, Weinheim, West Germany, 1987, 167.
23. **Philson, S. B. and Llinas, M.**, Siderochromes from *Pseudomonas fluorescens*. I. Isolation and characterization, *J. Biol. Chem.*, 257, 8081, 1982.
24. **Philson, S. B. and Llinas, M.**, Siderochromes from *Pseudomonas fluorescens*. II. Structural homology as revealed by NMR spectroscopy, *J. Biol. Chem.*, 257, 8086, 1982.
25. **Cody, Y. S. and Gross, D. C.**, Characterization of pyoverdin$_{pss}$, the fluorescent siderophore produced by *Pseudomonas syringae* pv. *syringae*, *Appl. Environ. Microbiol.*, 53, 928, 1987.
26. **Torres, L., Perez-Ortin, J. E., Tordera, V., and Beltran, J. P.**, Isolation and characterization of an Fe(III)-chelating compound produced by *Pseudomonas syringae*, *Appl. Environ. Microbiol.*, 52, 157, 1986.
27. **Van der Hofstad, G. A. J. M., Marugg, J. D., Verjans, G. M. G. M., and Weisbeek, P. J.**, Characterization and structural analysis of the siderophore produced by the PGPR *Pseudomonas putida* strain WCS358, in *Iron, Siderophores, and Plant Diseases*, Vol. 117, NATO ASI Series A: Life Sciences, Swinburne, T. R., Ed., Plenum Press, New York, 1986, 71.
28. **Hohnadel, D. and Meyer, J. M.**, Specificity of pyoverdine-mediated iron uptake among fluorescent *Pseudomonas* strains, *J. Bacteriol.*, 170, 4865, 1988.
29. **Huschka, H.-G., Jalal, M. A. F., van der Helm, D., and Winkelmann, G.**, Molecular recognition of siderophores in fungi: role of iron-surrounding N-acyl residues and the peptide backbone during membrane transport in *Neurospora crassa*, *J. Bacteriol.*, 167, 1020, 1986.
30. **Bakker, P. A. H. M.**, Siderophore-Mediated Plant Growth Promotion and Colonization of Roots by Strains of *Pseudomonas* spp., Ph.D. thesis, University of Utrecht, The Netherlands, 1989.
31. **Simon, R., Priefer, U., and Pühler, A.**, A broad host range mobilization system for *in vivo* genetic engineering: transposon mutagenesis in gram-negative bacteria, *Bio/Technology*, 1, 784, 1983.

32. **Marugg, J. D., van Spanje, M., Hoekstra, W. P. M., Schippers, B., and Weisbeek, P. J.,** Isolation and analysis of genes involved in siderophore biosynthesis in plant-growth-stimulating *Pseudomonas putida* WCS358, *J. Bacteriol.,* 164, 563, 1985.

33. **Geels, F. P. and Schippers, B.,** Selection of antagonistic fluorescent *Pseudomonas* spp. and their root colonization and persistence following treatment of seed potatoes, *Phytopathol. Z.,* 108, 193, 1983.

34. **Friedman, A. M., Long, S. R., Brown, S. E., Buikema, W. J., and Ausubel, F. M.,** Construction of a broad host range cosmid vector and its use in the genetic analysis of *Rhizobium* mutants, *Gene,* 18, 289, 1982.

35. **Figurski, D. H. and Helinski, D. R.,** Replication of an origin-containing derivative of plasmid RK2 dependent on a plasmid function provided in trans, *Proc. Natl. Acad. Sci. U.S.A.,* 76, 1648, 1979.

36. **Ditta, G., Stanfield, S., Corbin, D., and Helinski, D. R.,** Broad host range DNA cloning system for gram-negative bacteria: construction of a gene bank of *Rhizobium meliloti, Proc. Natl. Acad. Sci. U.S.A.,* 77, 7347, 1980.

37. **Moores, J. C., Magazin, M. D., Ditta, G. S., and Leong, J.,** Cloning of genes involved in the biosynthesis of pseudobactin, a high-affinity iron transport agent of a plant growth-promoting *Pseudomonas* strain, *J. Bacteriol.,* 157, 53, 1984.

38. **Loper, J. E., Orser, C. S., Panopoulos, N. J., and Schroth, M. N.,** Genetic analysis of fluorescent pigment production in *Pseudomonas syringae* pv. *syringae, J. Gen. Microbiol.,* 130, 1507, 1984.

39. **Kleinkauf, H. and von Döhren, H.,** Biosynthesis of peptide antibiotics, *Annu. Rev. Microbiol.,* 41, 259, 1987.

40. **Marugg, J. D., Nielander, H. B., Horrevoets, A. J. G., van Megen, I., van Genderen, I., and Weisbeek, P. J.,** Genetic organization and transcriptional analysis of a major gene cluster involved in siderophore biosynthesis in *Pseudomonas putida* WCS358, *J. Bacteriol.,* 170, 1812, 1988.

41. **De Bruijn, F. J. and Lupski, J. R.,** The use of transposon Tn5 mutagenesis in the rapid generation of correlated physical and genetic maps of DNA segments cloned into multicopy plasmids — a review, *Gene,* 27, 131, 1983.

42. **Bagdasarian, M. M., Amann, E., Lurz, R., Rückert, B., and Bagdasarian, M.,** Activity of the hybrid *trp-lac (tac)* promoter of *Escherichia coli* in *Pseudomonas putida.* Construction of broad-host-range, controlled-expression vectors, *Gene,* 26, 273, 1983.

43. **Ankenbauer, R., Hanne, L. F., and Cox, C. D.,** Mapping of mutations in *Pseudomonas aeruginosa* defective in pyoverdin production, *J. Bacteriol.,* 167, 7, 1986.

44. **Hohnadel, D., Haas, D., and Meyer, J. M.,** Mapping of mutations affecting pyoverdine production in *Pseudomonas aeruginosa, FEMS Microbiol. Lett.,* 36, 195, 1986.

45. **De Weger, L. A., van Boxtel, R., van der Burg, B., Gruters, R., Geels, F. P., Schippers, B., and Lugtenberg, B.,** Siderophores and outer membrane proteins of antagonistic plant-growth-stimulating, root-colonizing *Pseudomonas* spp., *J. Bacteriol.,* 165, 585, 1986.

46. **Marugg, J. D., de Weger, L. A., Nielander, H. B., Oorthuizen, M., Recourt, K., Lugtenberg, B., van der Hofstad, G. A. J. M., and Weisbeek, P. J.,** Cloning and characterization of a gene encoding an outer membrane protein required for siderophore-mediated uptake of Fe^{3+} in *Pseudomonas putida* WCS358, *J. Bacteriol.,* 171, 2819, 1989.

47. **Spaink, H. P., Okker, R. J. H., Wijffelman, C. A., Pees, E., and Lugtenberg, J. J.,** Promoters in the nodulation region of the *Rhizobium leguminosarum* Sym plasmid pRL1JI, *Plant Mol. Biol.,* 9, 27, 1987.

48. **Inouye, S., Nakazawa, A., and Nakazawa, T.,** Nucleotide sequence of the promoter region of the *xylDEGF* operon on TOL plasmid of *Pseudomonas putida, Gene,* 29, 323, 1984.

49. **Miller, J. H., Ed.,** *Experiments in Molecular Genetics,* Cold Spring Harbor Laboratory, Cold Spring Harbor, NY, 1972.

50. **Nozaki, M.,** Metapyrocatechase *(Pseudomonas), Methods Enzymol.,* 17A, 522, 1970.

51. **Magazin, M. D., Moores, J. C., and Leong, J.,** Cloning of the gene coding for the outer membrane receptor protein for ferric pseudobactin, a siderophore from a plant growth-promoting *Pseudomonas* strain, *J. Biol. Chem.,* 261, 795, 1986.

52. **Meyer, J. M., Mock, M., and Abdallah, M. A.,** Effect of iron on the protein composition of the outer membrane of fluorescent pseudomonads, *FEMS Microbiol. Lett.,* 5, 395, 1982.

53. **Neilands, J. B.,** Microbial envelope proteins related to iron, *Annu. Rev. Microbiol.,* 36, 285, 1982.

54. **Buyer, J. S. and Leong, J.,** Iron transport-mediated antagonism between plant growth-promoting and plant-deleterious *Pseudomonas* strains, *J. Biol. Chem.,* 261, 791, 1986.

55. **De Weger, L. A., van Arendonk, J. J. C. M., Recourt, K., van der Hofstad, G. A. J. M., Weisbeek, P. J., and Lugtenberg, B.,** Siderophore-mediated uptake of Fe^{3+} by the plant growth-stimulating *Pseudomonas putida* strain WCS358 and by other rhizosphere microorganisms, *J. Bacteriol.,* 170, 4693, 1988.

56. **Heller, K. and Kadner, R. J.,** Nucleotide sequence of the gene for the vitamin B12 receptor protein in the outer membrane of *Escherichia coli, J. Bacteriol.,* 161, 904, 1985.

57. **Pressler, U., Staudenmaier, H., Zimmerman, L., and Braun, V.,** Genetics of the iron dicitrate transport system of *Escherichia coli, J. Bacteriol.,* 170, 2716, 1988.

58. **Lundrigan, M. D. and Kadner, R. J.**, Nucleotide sequence of the gene for the ferrienterochelin receptor FepA in *Eschericia coli*, *J. Biol. Chem.*, 261, 10797, 1986.

59. **Coulton, J. W., Mason, P., Cameron, D. R., Carmel, G., Jean, R., and Rode, H. N.**, Protein fusions of β-galactosidase to the ferrichrome-iron receptor of *Escherichia coli* K-12, *J. Bacteriol.*, 165, 181, 1986.

60. **Krone, W. J. A., Stegehuis, F., Koningstein, G., van Doorn, C., Rosendaal, B., de Graaf, F., and Oudega, B.**, Characterization of the pColV-K30 encoded cloacin DF13/aerobactin outer membrane receptor protein of *Escherichia coli*; isolation and purification of the protein and analysis of its nucleotide sequence and primary structure, *FEMS Microbiol. Lett.*, 26, 153, 1985.

61. **Earhart, C. F.**, Ferrienterobactin transport in *Escherichia coli*, in *Iron Transport in Microbes, Plants and Animals*, Winkelman, G., van der Helm, D., and Neilands, J., Eds., VCH Publishers, Weinheim, West Germany, 1987, 67.

62. **Braun, V., Hantke, K., Eick-Helmerich, K., Köster, W., Pressler, U., Sauer, M., Schäffer, S., Schöffler, H., Staudenmaier, H., and Zimmerman, L.**, Iron transport systems in *Escherichia coli*, in *Iron Transport in Microbes, Plants and Animals*, Winkelmann, G., van der Helm, D., and Neilands, J., Eds., VCH Publishers, Weinheim, West Germany, 1987, 35.

63. **Crosa, J. H.**, Genetics and molecular biology of siderophore-mediated iron transport in bacteria, *Microbiol. Rev.*, 53, 517, 1989.

64. **Marugg, J. D.**, manuscript in preparation.

65. **Leong, J.**, unpublished results.

66. **van der Hofstad, G.**, unpublished results.

67. **Bitter, W., Marugg, J. D., de Weger, L. A., Tommassen, J., and Weisbeck, P. J.**, The ferric-pseudobactin receptor Pup A *Pseudomonas putida* WCS358: homology to TonB-dependent *Escherichia coli* receptors and specificity of the protein, *Molec. Microbiol.*, in press, (1991).

AQUEOUS SOLUTION EQUILIBRIUM AND KINETIC STUDIES OF IRON SIDEROPHORE AND MODEL SIDEROPHORE COMPLEXES

Alvin L. Crumbliss

INTRODUCTION

The emphasis in this review is on chemical studies which are relevant to siderophore-mediated iron uptake by microorganisms. Since almost all chelators which have been established to function as siderophores contain either the hydroxamate (**I**) or catecholate (**II**) Fe^{3+} binding group, we will focus our attention on solution equilibria, ligand exchange kinetics, and oxidation-reduction reactions of Fe_{aq}^{3+} with siderophores and synthetic chelators which contain the hydroxamate and/or catecholate binding group.

The processes involved in microbial iron assimilation are solubilization, transport to the cell, and deposition at an appropriate site within the cell. The important chemical characteristics of the natural iron chelators which are related to these processes are affinity and selectivity for Fe^{3+}, and lability at a specific site (e.g., cell wall or interior). This lability occurs either through ligand exchange, which often involves ternary complex intermediates, and/or a reduced oxidation state for iron (i.e., Fe^{2+}).

In this review we first treat solution equilibria and the associated Fe^{3+} formation or stability constants for hydroxamate and catecholate siderophores and their synthetic analogues. Then we will consider the kinetics and mechanism of Fe_{aq}^{3+} chelation and dechelation processes and ligand exchange reactions for these complexes. Finally, we will briefly review the relevant literature dealing with Fe^{3+}-siderophore oxidation-reduction reactions. The reader is also referred to other recent related reviews which treat the synthesis, solution chemistry, and biochemistry of this interesting class of compounds.[1-8]

IRON-SIDEROPHORE COMPLEX STABILITY IN AQUEOUS SOLUTION

Formation constants can serve to define the stability of an Fe-siderophore complex in solution. In addition, these constants are a measure of the siderophore ligand's selectivity for Fe_{aq}^{3+} in the presence of competing metal ions and its ability to solubilize Fe and prevent precipitation by hydrolysis. The Fe_{aq}^{3+} stability constant may also be the deciding factor in determining the mechanism for iron release from the siderophore, by influencing the Fe dissociation rate constant and Fe(III/II) reduction potential.

METHODS OF EXPRESSING IRON-SIDEROPHORE COMPLEX STABILITY

Since both the hydroxamate and catecholate groups are bidentate ligands, multiple complexation equilibria must be considered for mono- or bis-hydroxamates or catecholates in order to saturate the Fe^{3+} coordination number of six. These equilibrium steps are usually expressed as follows, where L^{n-} represents a bidentate ligand and coordinated H_2O has been omitted for clarity. The constants K_n and β_n are defined as stepwise and overall formation or stability constants, respectively. It is usually true that $K_1 > K_2 > K_3$.

$$Fe_{aq}^{3+} + L^{n-} \rightleftharpoons FeL^{(3-n)+} \tag{3}$$

$$K_1 = \frac{[FeL^{(3-n)+}]}{[Fe_{aq}^{3+}][L^{n-}]} = \beta_1 \tag{4}$$

$$FeL^{(3-n)+} + L^{n-} \rightleftharpoons FeL_2^{(3-2n)+} \tag{5}$$

$$K_2 = \frac{[FeL_2^{(3-2n)+}]}{[FeL^{(3-n)+}][L^{n-}]} \tag{6}$$

$$\beta_2 = K_1K_2 = \frac{[FeL_2^{(3-2n)+}]}{[Fe_{aq}^{3+}][L^{n-}]^2} \tag{7}$$

$$FeL_2^{(3-2n)+} + L^{n-} \rightleftharpoons FeL_3^{(3-3n)+} \tag{8}$$

$$K_3 = \frac{[FeL_3^{(3-3n)+}]}{[FeL_2^{(3-2n+}][L^{n-}]} \tag{9}$$

$$\beta_3 = K_1K_2K_3 = \frac{[FeL_3^{(3-3n)+}]}{[Fe_{aq}^{3+}][L^{n-}]^3} \tag{10}$$

In general, the greater the basicity of a siderophore ligand, the greater is that ligand's affinity for Fe_{aq}^{3+}. Since both Fe_{aq}^{3+} and H^+ are hard acids there is a competitive equilibrium (Equation 11) for the basic form of the siderophore ligand. This competitive equilibrium can be quantified by using equilibrium constants (acidity constants, K_a) for competing ligand protonation reactions to define an effective metal-ligand complex stability constant, K_{eff}.

$$LH + Fe_{aq}^{3+} \rightleftharpoons FeL^{2+} + H^+ \tag{11}$$

In the absence of competing metal ions and Fe_{aq}^{3+} hydrolysis, K_{eff} may be defined as follows:[9,10]

$$\text{Log } K_{eff} = \log \beta_{FeL} - n \log \alpha_L^{-1} \tag{12}$$

where β_{FeL} is the conventional metal-ligand binding constant

$$Fe_{aq}^{3+} + L^{n-} \rightleftharpoons FeL^{(3-n)+} \tag{13}$$

$$\beta_{FeL} = \frac{[FeL^{(3-n)+}]}{[Fe_{aq}^{3+}][L^{n-}]} \tag{14}$$

and α_L^{-1} is a correction for ligand proton basicity

$$\alpha_L^{-1} = [1 + [H^+]\beta_1^H + [H^+]^2\beta_2^H + \ldots + [H^+]^n\beta_n^H] \tag{15}$$

$$\beta_n^H = \frac{[H_nL]}{[H^+]^n[L]} \tag{16}$$

The β or K values are the normal mode for tabulating metal-ligand stability constants; however, due to the different acidities of various siderophores these values are not necessarily a good measure of differing affinities of the siderophores for Fe_{aq}^{3+} at a given (physiological) pH. The K_{eff} calculations described above are an attempt to deal with this problem. A more pragmatic or empirical approach which does not require knowledge of the ligand K_a values (which is important in the case of the catechols) is the pM value for a particular ligand.

$$K_n^* = \frac{[FeL][H^+]^n}{[H_nL][Fe_{aq}^{3+}]} \tag{17}$$

pM for the case of Fe^{3+} is defined[6] as the negative log of the free or uncomplexed Fe_{aq}^{3+} concentration, $pM = -\log[Fe_{aq}^{3+}]$, calculated from the observed H^+-dependent formation constant K_n^* for a fixed set of experimental conditions: pH 7.4; $[H_nL]_{tot} = 10~\mu M$ and $[Fe^{3+}]_{tot} = 1~\mu M$. The use of pM values also avoids the inherent problems of comparing K_n^* K_{eff}, or β values for ligands of differing denticity. The larger the pM value for a particular ligand, the more stable is the metal complex at the conditions described above. Precipitation as ferric hydroxide is indicated by a pM value below the limit set by the K_{sp} of $Fe(OH)_3$.

METHODS OF MEASURING STABILITY CONSTANTS

Experimental and computational methods for stability constant measurements prior to the advent of computer methods have been described in detail in a classic work by Rossotti and Rossotti.[11] Since then several reviews have appeared which include computer programs for data handling. Some reviews provide a critical comparison of computer programs and pitfalls in their use.[12,13] There are also several compilations of stability constants available; the most useful for the subject matter at hand is the critical stability constant series edited by Martell and Smith.[14] A recent book by Martell and Motekaitis[15] discusses various methods for determining stability constants along with an analysis of their applicability to different cases, and also includes a disk of computer programs for data manipulation. The most common methods for determining Fe-siderophore stability constants include potentiometric titration, electrochemistry, and spectrophotometry.[15,16]

Spectrophotometry is a useful technique for determining the equilibrium position of a reaction which involves strongly absorbing species with characteristic spectra. This is certainly a characteristic of hydroxamic acid and catechol complexes with Fe^{3+}. The spectra are both qualitatively and quantitatively diagnostic of the L/Fe ratio. For the tris-hydroxamate complexes (L/Fe = 3) $\lambda_{max} = 420$ to 430 nm, $\epsilon = circa~2500~M^{-1}~cm^{-1}$; for bis-hydroxamate complexes (L/Fe = 2) $\lambda_{max} = 460$ to 480 nm, $\epsilon = circa~1800~M^{-1}~cm^{-1}$; for monohydroxamate complexes (L/Fe = 1) $\lambda_{max} = 510$ to 520 nm, $\epsilon = circa~1000~M^{-1}$ cm^{-1}.[17] For the tris-catecholate complexes (L/Fe = 3) $\lambda_{max} = 480$ nm; for bis-catecholate complexes (L/Fe = 2) $\lambda_{max} = 560$ nm; for monocatecholate complexes (L/Fe = 1) $\lambda_{max} = 680$ nm.[18]

Spectral characteristics of the Fe^{3+}/hydroxamates have been utilized to determine the number of hydroxamate groups bound to Fe^{3+} for synthetic (**III**)[19] and natural (**IV**, rhodotorulic acid)[20] dihydroxamic acids, thereby establishing the pH-dependent equilibrium shown in Equation 18 where Fe_2L_3 is a dimer with a tris-hydroxamate Fe^{3+} coordination shell and FeL a monomer with a bis-hydroxamate Fe^{3+} coordination shell.

$$Fe_2L_3 + 2~H^+ \rightleftharpoons 2~FeL^+ + LH_2 \tag{18}$$

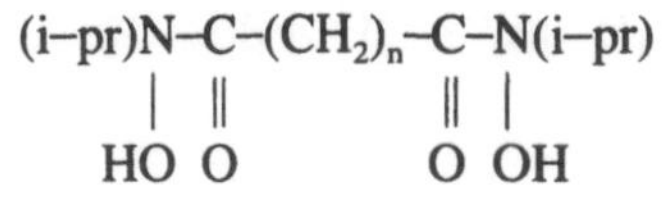

$$(\text{i-pr})N-C-(CH_2)_n-C-N(\text{i-pr})$$

HO O   O OH

$$(n = 3 \longrightarrow 6)$$

III

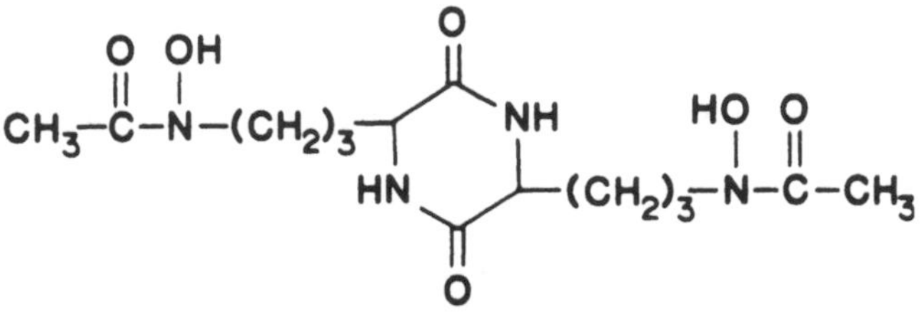

$$CH_3-C-N-(CH_2)_3$$

O OH

NH

HN

O

$$-(CH_2)_3-N-C-CH_3$$

HO O

O

Rhodotorulic Acid

IV

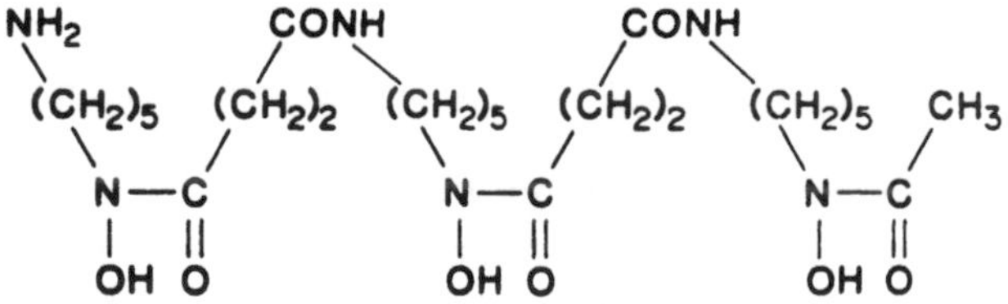

NH$_2$ CONH   CONH

(CH$_2$)$_5$ (CH$_2$)$_2$ (CH$_2$)$_5$ (CH$_2$)$_2$ (CH$_2$)$_5$ CH$_3$

N—C   N—C   N—C

OH O   OH O   OH O

Deferriferrioxamine B (H$_3$DFB)

V

These characteristic spectral shifts have also made it possible to spectroscopically identify intermediates whose existence is implied by kinetic studies in the stepwise dissociation of Fe^{3+} from deferriferrioxamine B,[21-23] a linear trihydroxamic acid siderophore (**V**), and the dissociation of the first, second, and last hydroxamate ligand from tris-synthetic hydroxamate complexes.[17]

In order to use spectrophotometry to calculate a precise stability constant there needs to be a measurable amount of both reactants and products present. This is difficult for the siderophores of interest since their affinity for Fe^{3+} is so high that the concentration of reactant, free Fe^{3+}_{aq}, is very low. This problem may be overcome by the use of competition equilibrium experiments whereby in effect the equilibrium reaction being measured can be manipulated so that comparable quantities of reactants and products are present. It is then possible to calculate the stability constant for the Fe^{3+}/siderophore complex, provided the equilibrium constant for the competing ligand with Fe^{3+}_{aq} is known.

In practice the usual competing ligand utilized is EDTA, whose pK_a values and Fe^{3+} binding constant are well characterized. Furthermore, the β^{EDTA}_{110} value $(10^{25})^{24}$ (in β_{xyz} nomenclature the subscripts designate the metal [x], ligand [y], and acidic proton [z] stoichiometry) is sufficiently close to that of the Fe^{3+}/hexadentate siderophore complexes that the competition equilibrium position can be manipulated in the midpoint range in order that convenient spectral changes may be observed. In addition, the FeEDTA absorption spectrum does not interfere with that of the Fe^{3+}/siderophore.

A relevant example may be obtained from Raymond's laboratory[19] where competition equilibrium experiments were carried out within a pH range where the complex stoichiometry had been established. The equilibrium of interest is Equation 19, where L represents a series of synthetic di-hydroxyamic acids **III**.

$$2 \, Fe^{3+} + 3 \, L^{2-} \xrightleftharpoons{\beta_{230}} Fe_2L_3 \tag{19}$$

The competition equilibrium constant, K_c, actually measured corresponds to the following reaction at equilibrium.

$$Fe_2L_3 + 2 \, EDTA^{4-} \xrightleftharpoons{K_c} 2 \, FeEDTA^- + 3 \, L^{2-} \tag{20}$$

$$K_c = \frac{[L^{2-}]^3[FeEDTA]^2}{[Fe_2L_3][EDTA]^2} \tag{21}$$

K_c may be written as

$$K_c = \frac{[FeEDTA^-]^2}{[Fe_{aq}^{3+}]^2[EDTA]^2} \cdot \frac{[Fe_{aq}^{3+}]^2[L^{2-}]^3}{[Fe_2L_3]} = \frac{(\beta_{110}^{EDTA})^2}{\beta_{230}} \tag{22}$$

By knowing the total concentrations of EDTA, L, and Fe^{3+}, the pKa values for EDTA and LH_2, and the concentration of Fe_2L_3 (from the absorption spectrum), a value for K_c can be calculated. Since β_{110}^{EDTA} is known, β_{230} can be calculated from Equation 23.

$$\beta_{230} = \frac{(\beta_{110}^{EDTA})^2}{K_c} \tag{23}$$

The log β_{230} value determined in this way for **III** (n = 3) is 62.1.[19] For comparison purposes the log of the corresponding constant for rhodotorulic acid (**IV**) is 62.2.[20]

PROTONATION EQUILIBRIA

Ferric siderophores can undergo solution protonation equilibria which may or may not involve a coordination number change. Protonation constants may be obtained from equilibrium potentiometric or spectrophotometric data as well as kinetic data. Ligand-to-ferric ion charge transfer bands are often sensitive to complex protonation. If there are only two such absorbing species in solution, the absorbance data may be treated according to the method of Schwarzenbach[6,25] in order to determine the number of protons involved in the equilibrium (n) and the thermodynamic protonation constant (K_{MH_nL}).

$$FeL^{m-} + nH^+ \xrightleftharpoons{K_{MH_nL}} FeH_nL^{(m-n)-} \tag{24}$$

$$A_{obs} = (A_o - A_{obs})/K_{MH_nL} \, [H^+]^n + \epsilon_{MH_nL} \, C_T \tag{25}$$

In Equation 25 A_{obs} is the absorbance at the λ_{max} of the unprotonated species FeL^{m-} at each $[H^+]$, C_T is the total concentration of siderophore ($[FeH_nL^{(m-n)-}] + [FeL^{m-}]$), ϵ_{MH_nL} is the extinction coefficient of the protonated species, and A_o is the initial absorbance at λ_{max}. When the chemically relevant value for n is used a linear plot of A_{obs} vs. $(A_o - A_{obs})/[H^+]^n$ is obtained and K_{MH_nL} may be obtained from the slope.

The protonation constants of several trihydroxamate siderophore complexes are listed

TABLE 1
Hydroxamate Siderophore
Protonation Constants

Siderophore	log K_{MHL}	Ref.
Coprogen	0.50	26
Ferrioxamine B	10.40,[a] 0.94[b]	25, 27
Ferricrocin	0.53	26
Ferrichrome	1.49	25

[a]   Log K_{MHL} as shown in Equation 26.
[b]   Log K_{MH_2L} as shown in Equation 27.

in Table 1. In each case, protonation involves dechelation of a hydroxamate group. The small values for log K_{MHL} illustrate the fact that the ferric siderophore complexes are hexacoordinate at pH 7 and are only protonated in strongly acidic medium. Protonation constants for two processes (Equations 26 and 27) are listed in Table 1 for ferrioxamine B. The first protonation occurs at the free amine group and does not involve a dechelation process. This protonation constant is large and ferrioxamine B exists as the cationic species Fe(HDFB)$^+$ (VII) at pH 7. The second protonation equilibrium constant is consistent with the other siderophores listed in Table 1.

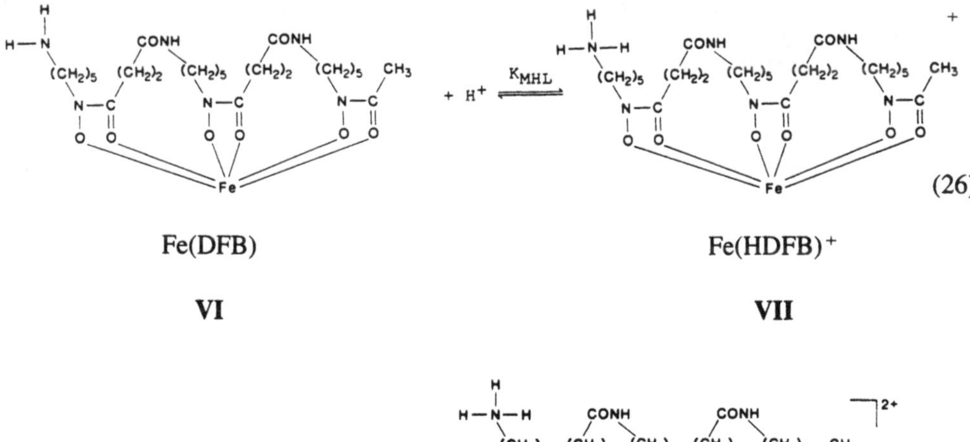

$$\text{Fe(DFB)} \qquad\qquad\qquad \text{Fe(HDFB)}^+ \tag{26}$$

VI   **VII**

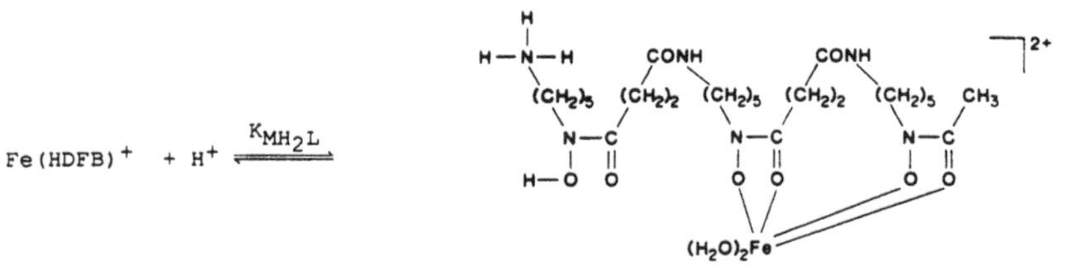

$$\text{Fe(H}_2\text{DFB)}^{2+} \tag{27}$$

VIII

CLASSIFICATION OF SIDEROPHORE STRUCTURES

There are three components to consider in the design or evolution of a Fe^{3+}-specific (or any other metal) high affinity chelating agent. These are (1) the type of Fe^{3+} binding group(s); (2) the number of binding groups in a particular molecule; and (3) the stereochemical arrangement of the binding groups. Over 100 siderophores have been isolated to date[1] and almost all are hexadentate ligands which contain either or both of the bidentate chelating

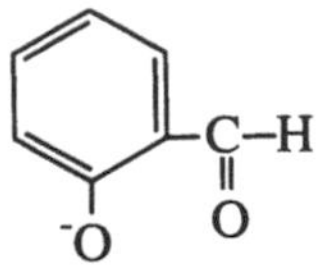

C–H
O
⁻O
IX

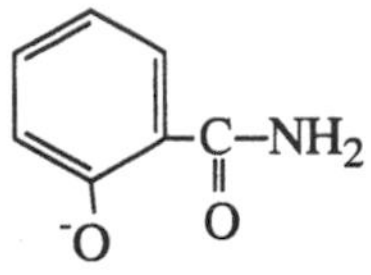

C–NH₂
O
⁻O
X

groups: hydroxamic acid (**I**) and catechol (**II**). For the most part, these ligands provide six coordinate chelate complexes with three stable five-membered ring chelate groups connected by a flexible backbone. We will now individually consider each of the above aspects of siderophore ligand design and structure.

THE HYDROXAMATE AND CATECHOLATE CHELATING GROUPS

A hydroxamic acid is the combination of a sterically efficient neutral carbonyl donor $\ce{>C=O}$ and an anionic O^- donor $(\ce{-N-O^-})$. The N atom of the hydroxamate moiety adjacent to the carbonyl functional group increases the coordinating ability of the carbonyl O atom to Fe^{3+}. This can be seen by comparing the Fe^{3+} binding constants for the salicylate (**IX**; $\log K_1 = 8.75$) and salicylamide (**X**; $\log K_1 = 10.02$) anions.[3,28] Iron(III) binding increases by 1.27 log units. This represents an increase in carbonyl O atom donor ability since proton binding at the phenolic oxygen increases only 0.8 log units.[28]

The presence of a negatively charged O atom donor in both the hydroxamate and catecholate Fe^{3+} chelating group undoubtedly plays a major role in producing thermodynamically stable complexes with Fe^{3+}. Iron(III) has a high affinity for the OH^- ion, the archetypal anionic oxygen donor ligand. This can be seen from the relatively large hydrolysis constant $(\log K_h = -2.56)$ for the $Fe(H_2O)_6^{3+}$ ion.[29] A linear correlation between pK_h for a number of different metal ions and the product of metal ion electronegativity times charge $(\chi_M \cdot Z)$ exists,[10] which suggests that the high affinity that Fe^{3+} exhibits for OH^- is a result of the high charge and electronegativity of the ion.

$$\ce{Fe(H2O)6^{3+} <=>[K_h] Fe(H2O)5OH^{2+} + H+} \tag{28}$$

Figure 1 is a plot of $\log K_1^L$ vs. $\log K_1^{OH}$ for some mono- and polyhydroxamate and catecholate ligands.[27] A good linear relationship exists for these data. This plot demonstrates that the high affinity of Fe^{3+} for the catecholate and hydroxamate chelate groups is a result of the Fe^{3+} ion being a strong hard Lewis acid. Furthermore, due to the fact that $\log K_1^{OH}$ correlates with $\chi_M Z$, the selectivity of the catechol and hydroxamic acid siderophores for Fe^{3+} (as illustrated by the slope of the lines in Figure 1) is apparently a result of the high electronegativity and charge of the Fe^{3+} ion, particularly in comparison to other biological or environmental metal ions which tend to have a +2 charge and lower electronegativity.[10] Figure 1 also demonstrates that Fe^{3+} selectivity increases with increasing number of oxygen donors in the ligand.

$$\ce{Fe(H2O)6^{3+} + L^{n-} <=>[K_1^L] Fe(L)(H2O)_x^{(3-n)+}} \tag{29}$$

$$\ce{Fe(H2O)6^{3+} + OH^- <=>[K_1^{OH}] Fe(H2O)5OH^{2+}} \tag{30}$$

The selectivity of the catecholate and hydroxamate chelating moieties for metals with high charge and high electronegativity has been exploited for various purposes. Ferrioxamine

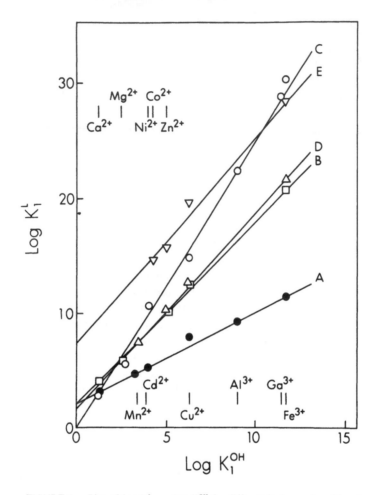

FIGURE 1. Plot of log K_1^L vs. log K_1^{OH} for different hydroxamic acid and catechol ligands. (A) Acetohydroxamic acid ($CH_3C(O)N(OH)H$); (B) 1,8-octanedihydroxamic acid (Structure **XXXVIII**, n = 8 in text); (C) deferri-ferrioxamine B (Structure **V** in text); (D) catechol (Structure **II** in text, R = H); (E) LICAMS (Structure **LVIII** in text). Data are from References 3, 6, 14, and 27. Linear correlations are defined for more metal ion data points than are shown.[3,27]

B has been used for chelation therapy for iron and also aluminum.[30-33] This is because ferrioxamine B is fairly specific for the +3 metal ions, as opposed to the relatively toxic chelating agents EDTA and DTPA[34] which have high affinities for Ca^{2+} as well as Zn^{2+} and Cu^{2+}.[35] The highly charged Ga^{3+} and In^{3+} ions in the form of the radionuclides ^{67}Ga and ^{111}In have found extensive use as tumor-imaging agents.[36,37] One of the problems associated with this imaging method is the distribution of these metals via iron metabolic routes in the body which results in a high background radiation level.[6] An important goal then is the elimination of excess metal by removal from transferrin. The synthetic tricatechols MECAMS and 3,4-LICAMS are found to have higher binding constants for Ga^{3+} than transferrin.[6,38] Synthetic and naturally occurring hydroxamic acids and catechols are also being investigated as possible high oxidation state metal (e.g., Pu^{4+}) decorporation pharmaceuticals.[39]

STEREOCHEMISTRY

Siderophore structures can vary considerably with respect to the placement of the bidentate hydroxamate or catecholate binding groups. Linear and cyclic siderophore structures

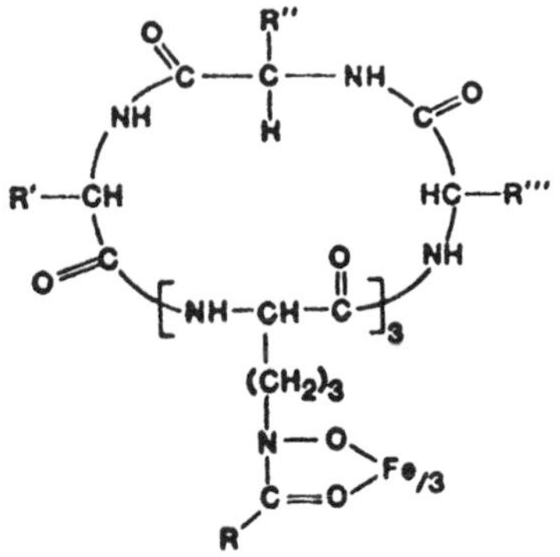

Ferrichrome (R = CH$_3$, R' = R" = R''' = H)

Ferricrocin (R = CH$_3$, R' = R''' = H, R" = CH$_2$OH)

Ferrichrome A (R = (trans)CH=C(CH$_3$)—CH$_2$CO$_2$H

R' = R" = CH$_2$OH, R''' = H)

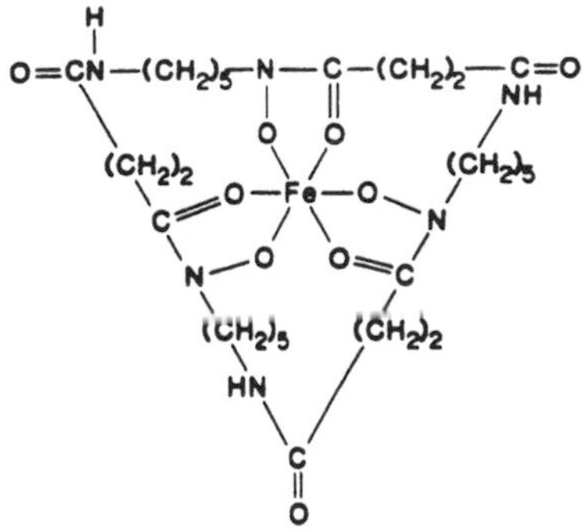

Ferrioxamine E

XII

XIII

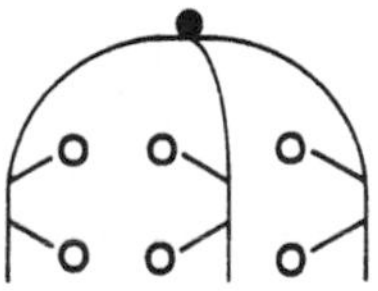

XIV

XV

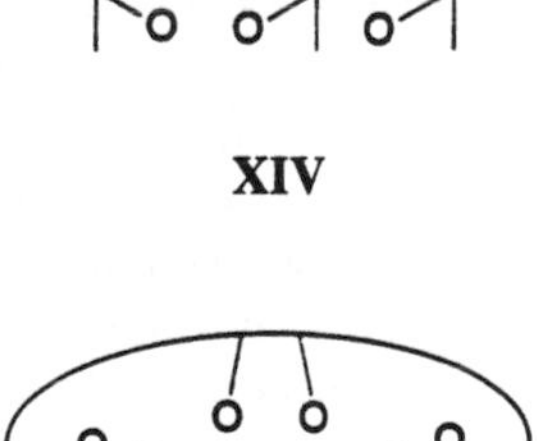

XVI

may be placed in three broad classifications: (1) acyclic (e.g., ferrioxamine B [V]); (2) exocyclic (e.g., ferrichrome [XI]); and (3) endocyclic (e.g., ferrioxamine E [XII]). The hydroxamate siderophores provide examples of each type of structure, while no endocyclic structure has been identified for catecholate siderophores. Synthetic strategies have been developed for siderophore analogues which place the catecholate and hydroxamate Fe^{3+} binding groups in acyclic (XIII), tripodal (XIV), exocyclic (XV), macrocyclic (endocyclic) (XVI), and macrobicyclic (XVII) structures. We shall look at the Fe^{3+} affinity, and ligand exchange and redox activity of representative examples of these structures for the natural and synthetic siderophores.

Mixed hexadendate ligands which incorporate either the hydroxamate or phenolate donor groups with other hard O and/or N donor atoms have also been identified as siderophores.

Aerobactin

XVII   **XVIII**

These include the hexadentate chelators aerobactin (**XVIII**), schizokinen, and arthrobactin, each of which contain two hydroxamate and one α-hydroxycarboxylate groups, and mycobactin which contains two hydroxamate groups in conjunction with a phenolate O and oxazoline N atom donor. Agrobactin and parabactin incorporate five phenolate O and one oxazoline N atoms to form a hexadentate chelator.[40] Pseudobactin and pseudobactin A form hexadentate chelators by incorporating one hydroxamate, two phenolate, and one α-hydroxycarboxylate group.[41,42] In all of these mixed chelate examples there has either been a hydroxamate or phenolate Fe^{3+} binding group.

There are also examples of siderophores which are less than hexadentate. Ferrioxamine H (two hydroxamates and one carboxylate) is a pentadentate ligand where the sixth coordination site on Fe^{3+} is apparently occupied by H_2O.[43] Rhodotorulic acid (**IV**) is a tetradentate siderophore with two hydroxamate binding groups that forms both 1:1 and 3:2 complexes with Fe^{3+}.[20] Bidentate natural siderophores, which form 3:1 complexes, are the monohydroxamates and thiohydroxamates thioformin (*N*-methyl-thioformohydroxamate),[44] *N*-methyl-phenylacetohydroxamic acid,[45] and L-α-aspartyl-L-α-*N*-hydroxy-aspartyl-D-cycloserine.[46]

There are some siderophores which contain neither the hydroxamate nor catecholate Fe^{3+} binding groups. Rhizobactin contains α-hydroxycarboxylate and ethylenediamine Fe^{3+} binding moieties. Many proposed phytosiderophores fall in this class. Such a compound is mugeneic acid, a hexadentate chelator with carboxylamine, hydroxyl, and amine Fe^{3+} binding groups.[47]

ADVANTAGES OF POLYDENTATE LIGAND STRUCTURES
Chelate Effect

The formation constants for Fe^{3+}/hydroxamate complexes seem to depend only on the number of hydroxamate groups bound to Fe^{3+}, and not on whether they are part of a multichelate molecule or independent ligands. This apparent lack of a chelate effect for the hydroxamate siderophores, which was first pointed out by Schwarzenbach and Schwarzenbach[48] and later emphasized by Carrano et al.,[20] may be illustrated by comparing log β_2 and log β_3 values for mono-hydroxamates with log β values for di- and trihydroxamates. Relevant equilibria for the simple monohydroxamic acid acetohydroxamic acid are shown in Equations 3, 5, and 8 where $L^{n-} = CH_3C(O)N(O)H^-$ and the corresponding log K_n and log β_n values are listed in Table 2. The expected trend in K_n values ($K_1 > K_2 > K_3$) is observed. The log β_3 for acetohydroxamic acid is nearly equivalent to the log β value for the linear trihydroxamic acid deferriferrioxamine B (log $\beta = 30.60$).[48]

TABLE 2
Stepwise and Overall Stability Constants for Mono-, Bis-, and Tris-Acetohydroxamatoiron(III) Complex Formation in Aqueous Solution[a]

$\log K_1$	11.41
$\log K_2$	9.68
$\log K_3$	7.20
$\log \beta_2$	21.09
$\log \beta_3$	28.29

[a] Reference 48, 20°C; $I = 0.1 \ M$ (NaClO$_4$).

TABLE 3
Log β and pM Values for the Linear Dihydroxamic Acids (*III*) and Rhodotorulic Acid (*IV*)

Ligand	$\log \beta_{230}$[a]	$\log \beta_{110}$[b]	pM[c]	Ref.
III n = 3	62.1	22.84	20.69	19
III n = 4	62.2	22.83	20.39	19
III n = 5	62.4	22.76	20.24	19
III n = 6	62.3	22.63	20.23	20
IV	62.2	21.55	21.76	20

[a] Log stability constant for Equation 19. To obtain $\log \beta_{230}$ per Fe^{3+}, divide by 2.

[b] Log stability constant for Equation 32.

[c] $pM = -\log [Fe^{3+}]$; $pH = 7.4$; $[Fe^{3+}]_t = 10^{-6} \ M$, $[L]_t = 10^{-5} \ M$.

$$Fe^{3+}_{aq} + HDFB^{2-} \overset{\beta}{\rightleftharpoons} FeHDFB^+ \tag{31}$$

A similar comparison can be made for the linear dihydroxamic acid siderophore rhodotorulic acid (**IV**) and a series of linear synthetic dihydroxamic acids (**III**). Both rhodotorulic acid and the synthetic dihydroxamic acids are involved in equilibrium reactions to form the 1:1 tetracoordinate complex FeL$^+$ (Equation 32) and 3:2 hexacoordinate complex Fe$_2$L$_3$ (Equation 19).[19,20]

$$Fe^{3+} + L^{2-} \overset{\beta_{110}}{\rightleftharpoons} FeL^+ \tag{32}$$

The relevant $\log \beta_{110}$ and $\log \beta_{230}$ values are listed in Table 3. The $\log \beta_{110}$ values are relatively constant in the table and within 1.7 log units of $\log \beta_2$ for acetohydroxamic acid (Table 2). For the case of rhodotorulic acid the stability increase over acetohydroxamic acid is less than 0.5 log unit. This small increase in $\log \beta_{110}$ can easily be accounted for by the expected increase in stability arising from the alkyl group on the N atom (see below).[49] A comparison of $^1/_2 \log \beta_{230}$ (stability constant per Fe atom in Fe$_2$L$_3$) values in Table 3 with $\log \beta_3$ for acetohydroxamic acid (Table 2) again shows an enhancement (3 log units) which is consistent with N-alkylation[49] and shows no evidence of a chelate effect.

The nearly equivalent affinity of the mono- and polyhydroxamate chelators for Fe^{3+}_{aq}

suggests that the hydroxamate groups function independently as a result of the long chains separating them. The connecting chains result in the formation of rather large chelate rings (e.g., 14 atoms in the case of ferrioxamine B [**VII**]). In the case of **III**, however, the log β values appear to be independent of the connecting chain length (Table 3) and when n = 3 the length of the chain is only eight atoms. However, it is possible that in this case the steric bulk of the isopropyl groups may be playing a role.

If there is little or no stability enhancement for polyhydroxamates relative to the monohydroxamates, then why are most siderophores polyhydroxamates rather than monohydroxamates (or catecholates)? First, there is a concentration effect which favors ligands of higher denticity.[50] Consider the rearranged forms of the concentration quotient expressions which define β_3 and β for a monohydroxamate (A^-) and trihydroxamate (L^{3-}) (or mono- and tricatecholate). One can see from these expressions that for a given ratio of hexacoordinated iron to free

$$[A^-] = \left(\frac{[FeA_3]}{[Fe^{3+}]\beta_3} \right)^{1/3} \tag{33}$$

$$[L^{3-}] = \frac{[FeL]}{[Fe^{3+}]\beta} \tag{34}$$

Fe_{aq}^{3+}, and comparable (large) values of β_3 and β, the concentration of free monohydroxamate to achieve that given ratio will be several orders of magnitude greater than the concentration of free trihydroxamate. For example, 10^{20} times more acetohydroxamate than deferriferrioxamine B (on a moles/liter basis) is required to produce a hexacoordinate to free Fe^{3+} ratio of 10:1. The influence of the concentration effect may also be seen in the pM values for acetohydroxamic acid (18.3) and ferrioxamine B (27.5) calculated at pH 7, $[Fe^{3+}]_T = 10^{-6}$ M and $[ligand]_T = 10^{-3}$ M. The concentration effect means that fewer moles of polyhydroxamate (or polycatecholate) siderophore need be excreted by an organism than a monohydroxamate (or monocatecholate). This statement also applies on a gram basis, since as illustrated above, when β and β_3 are large the effect is greater than molecular weight differences between siderophore structures of various denticity.

In some cases a polydentate chelate structure may also afford a higher Fe^{3+} affinity due to a preorganization effect as described in the next section. It is also likely that the polyhydroxamates (or catecholates) are more kinetically inert with respect to ternary complex formation and ligand exchange than the monohydroxamates or catecholates, as discussed below in the section on kinetics.

Preorganization

As noted above, long connecting chains between hydroxamate groups provide unfavorable entropy effects for metal complexation which may be responsible for the absence of a chelate effect in the di- and trihydroxamic acid siderophores. However, certain types of connecting chains may serve to favorably preorient the Fe^{3+} binding groups and minimize the unfavorable entropy effect.

The endocyclic hydroxamate siderophores may exhibit some degree of preorganization. Ferrioxamine E (**XII**) has an Fe_{aq}^{3+} binding constant which is over 4 log units greater than β_3 for acetohydroxamic acid (see Tables 2 and 4). Furthermore, inspection of Table 4 reveals that endocyclic ferrioxamine E has a higher affinity for Fe_{aq}^{3+} than the other hydroxamate siderophores with linear or exocyclic structures.

It may be that the triester group of enterobactin (**XIX**) is preorganized in a configuration which is optimal for hexacoordination to Fe^{3+}.[6] For example, the exocyclic model triscatechol chelators CYCAM and CYCAMS (**XX**) have log K values of 40 and 38, respec-

TABLE 4
Summary of Fe^{3+}-Siderophore Stability Constants in Aqueous Solution

Siderophore	log β_{110}	pM[a]	Ref.
Aerobactin	22.5	23.3	57
Coprogen	30.2	27.3	38
Ferrichrysin	29.96	25.8	25
Ferricrocin	30.4	26.5	58
Ferrichrome	29.07	25.2	25, 58
Ferrichrome A	32.0		59
Ferrioxamine B	30.60[b]	26.6	48
	30.99[c]		27
Ferrioxamine D_1	30.76		25
Ferrioxamine E	32.49	27.7	60
TAFC[d]	31.8		61
Rhodotorulic Acid	62.2[e]	21.8	20
	21.55[f]		
Enterobactin	52	35.5	62

[a] Data taken from Reference 1; $pM = -\log [Fe^{3+}_{aq}]$ when
    $[Fe^{3+}]_T = 10^{-6} M$, [siderophore] $= 10^{-5} M$; pH = 7.4.
[b] Log β for $Fe^{3+}_{aq} + HDFB^{2-} \rightleftharpoons Fe(HDFB)^+$.
[c] Log β for $Fe^{3+}_{aq} + DFB^{3-} \rightleftharpoons Fe(DFB)$.
[d] N,N',N''-triacetylfusarinine C; data at 30°C.
[e] Log β_{230} for Reaction 19.
[f] Log β_{110} for Reaction 32.

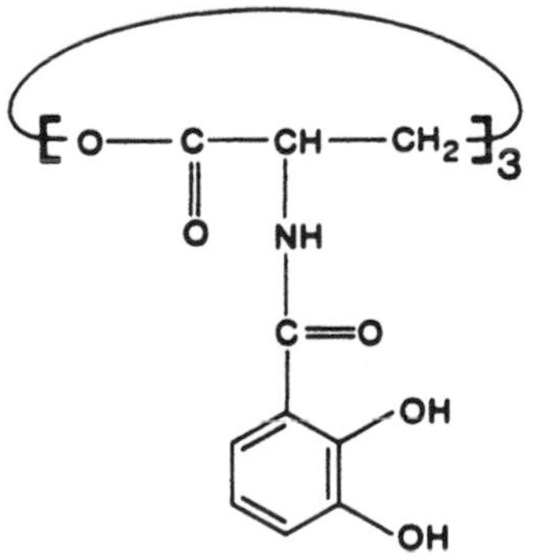

Enterobactin

XIX

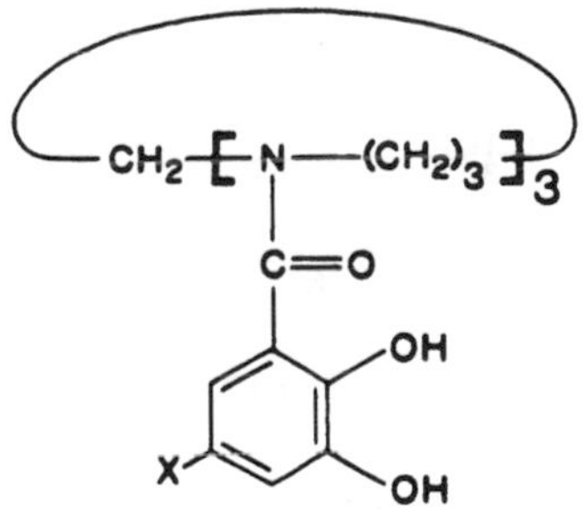

X = H, CYCAM

X = SO_3^-, CYCAMS

XX

tively,[51] which are significantly lower than the corresponding value (52)[62] for enterobactin. These lower values illustrate the ligand strain imposed by complete encapsulation of the Fe^{3+}.

Kiggen and Vögtle[52,53] report that the macrobicyclic tris-catechol **XXI**, in which the O donor atoms are rigidly preorganized, has a significantly higher affinity for Fe^{3+} (log K = 59) than its more flexible open chain analogue **XXII**.

Macrobicyclic structures, however, do not always produce higher binding affinities than their open chain or tripodal analogues. For example, bicapped TRENCAM (**XXIII**) has a log K value of 43,[39,54] while the more flexible open tripodal structural analogue TRENCAM

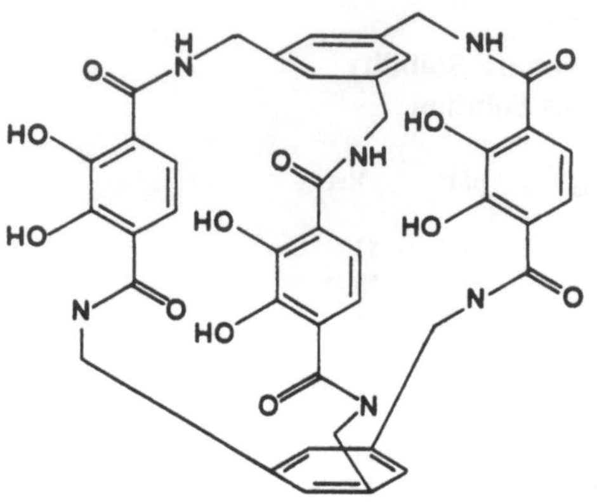

XXI

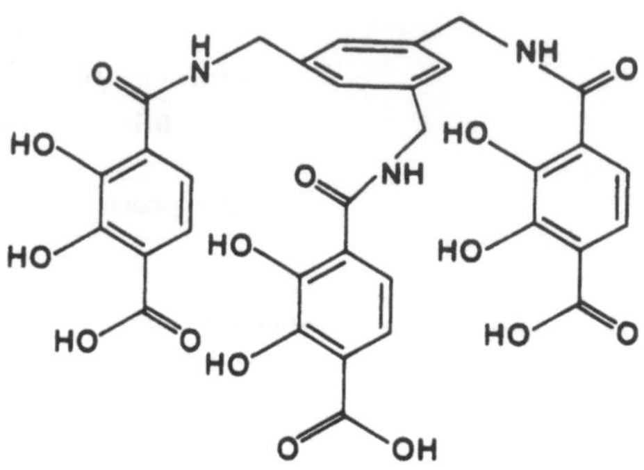

XXII

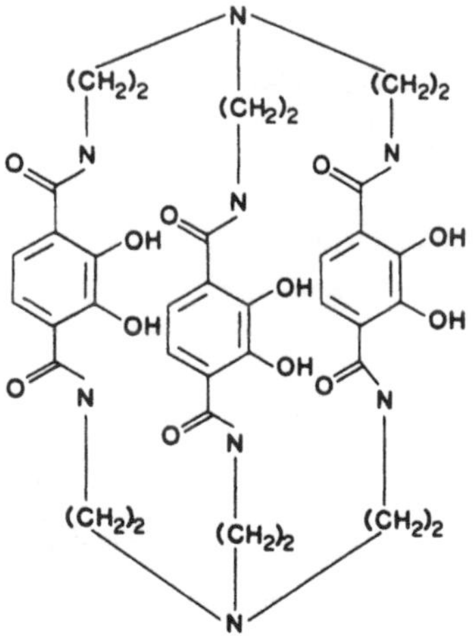

Bicapped TRENCAM

XXIII

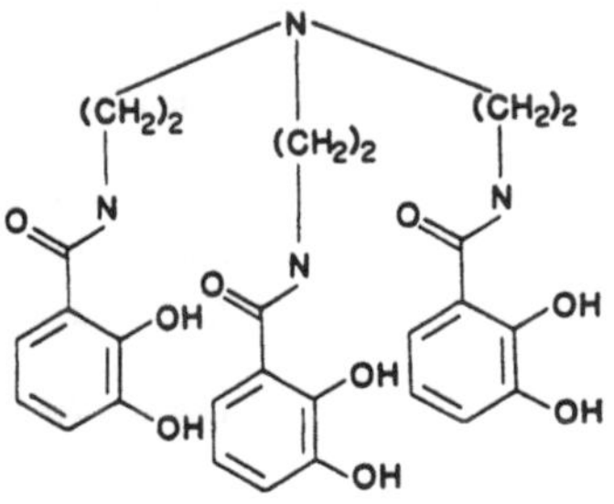

TRENCAM

XXIV

(**XXIV**) has a log binding constant of 43.6.[55] The X-ray crystal structure determination of Fe(bicapped TRENCAM) shows an unusual trigonal prismatic coordination for Fe^{3+} [56] Comparison of the expected conformation of metal-free bicapped TRENCAM with the crystal structure for Fe(bicapped TRENCAM) illustrates that preorganizational effects are not operative. The pM values for bicapped TRENCAM (30.7)[39] and TRENCAM (27.8),[55] however, make the macrobicyclic compound a more effective Fe^{3+} chelator due to its lower pK_a values.

SUMMARY OF IRON-SIDEROPHORE STABILITY CONSTANTS

The siderophores exhibit a range of stabilities with respect to Fe^{3+}_{aq}. Inspection of Table 4 reveals a 30 order of magnitude variation in β values for the hexadentate siderophores, from enterobactin (**XIX**), a tris-catecholate (log β = 52), to aerobactin (**XVIII**), a bis-hydroxamate-mono α-hydroxy carboxylate (log β = 22.5). The pM values for these two siderophores cover a range of 10 orders of magnitude. The hexadentate trihydroxamate siderophores have log β values in the high but narrow range of 29.1 to 32.5, and pM values in the narrow range 25.2 to 27.7. Enterobactin has the highest binding constant for Fe_{aq}^{3+} yet reported in the literature,[62] with the exception of the synthetic tris-catechol **XXI**.[52,53]

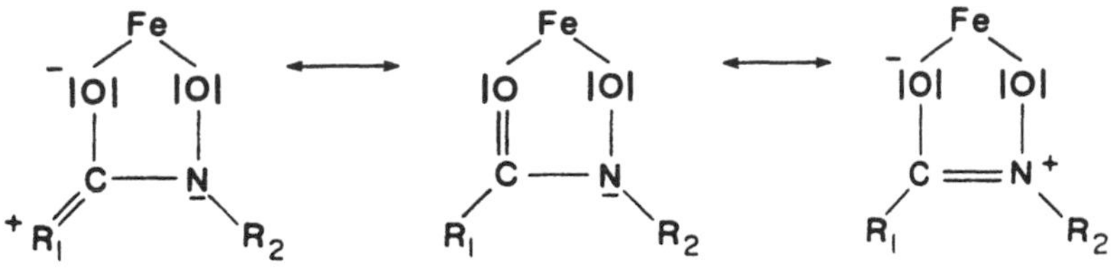

SOLUTION EQUILIBRIA WITH SYNTHETIC SIDEROPHORE ANALOGUES

Synthetic analogues of the hydroxamic acid and catechol classes of siderophores have been prepared and their Fe_{aq}^{3+} affinities established. These investigations were carried out in an attempt to better understand the reasons behind the high and specific affinity of different siderophore structures for Fe_{aq}^{3+} and for the development of new high specificity chelators for Fe^{3+} and other high valent metal ions in applications such as enhanced metal ion removal from human subjects.

HYDROXAMIC ACIDS
Monohydroxamic Acids
A series of 18 synthetic monohydroxamic acids were synthesized to study the influence of the substituents at either end of the hydroxamate group (carbonyl, R_1; nitrogen, R_2; R_1 and R_2 are either H, phenyl, substituted phenyl, or alkyl) on Fe^{3+} complex stability.[49,63,64]

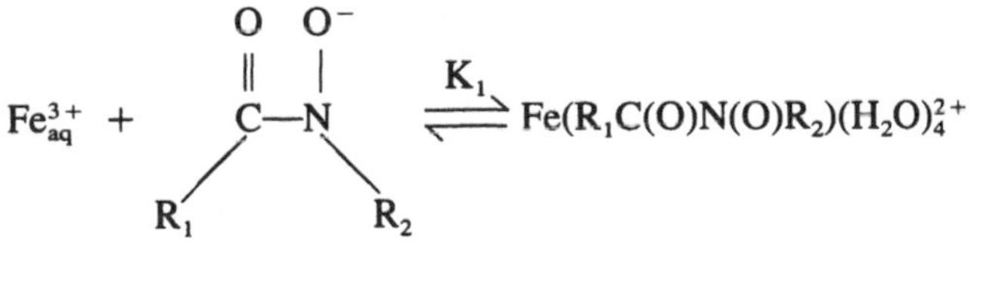

$$Fe_{aq}^{3+} + \underset{R_1 \quad\quad R_2}{\overset{\overset{O \quad O^-}{\underset{\|}{}\,\underset{|}{}}}{C-N}} \overset{K_1}{\rightleftharpoons} Fe(R_1C(O)N(O)R_2)(H_2O)_4^{2+} \tag{35}$$

Both the R_1 and R_2 groups were found to influence Fe^{3+} complex stability. Linear correlations between log K_1 and Hammett sigma parameters were found for substituted phenyl groups in the R_1 or R_2 positions. Three relevant canonical forms are shown below which illustrate that both inductive and resonance donors and acceptors influence stability in the R_1 position, while in the R_2 position inductive donors and acceptors, but only resonance acceptors, influence stability. Although both R_1 and R_2 groups influence complex stability, the R_2 group plays the dominant role through inductive stabilization of N atom lone electron pair delocalization into the C–N bond, which increases the formal negative charge on the carbonyl oxygen atom (**XXVII**). The most stable monohydroxamate complexes are those with an alkyl group in the R_2 position, which is capable of inductively stabilizing a formal positive charge on the adjacent N atom.

Hydroxamate derivatives of amino acids form stable complexes with Fe_{aq}^{3+}. Brown and co-workers[65] have investigated the solution equilibria and kinetics associated with the $Fe^{3+}/$ glycine hydroxamic acid system. Potentiometric titration data are consistent with several solution species being present over the pH range from 2 to 11. These include $FeHA^{3+}$, FeA^{2+}, FeA_2H^{2+}, FeA_2^+, FeA_3H^+, FeA_3, $FeA_3H_{-1}^-$, and $FeA_3H_{-2}^{2-}$, with FeA^{2+}, FeA_2^+, and FeA_3 representing the major species (A = glycine hydroxamate anion). Indirect evidence

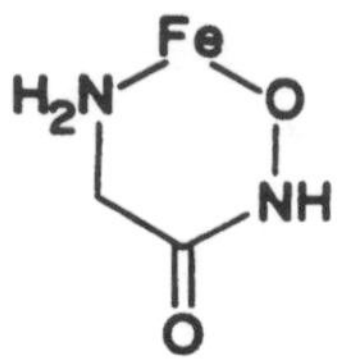

XXVIII 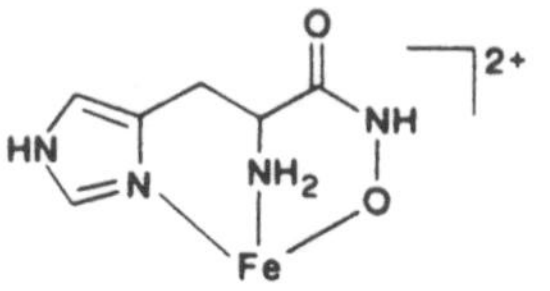　　　　　　**XXIX** 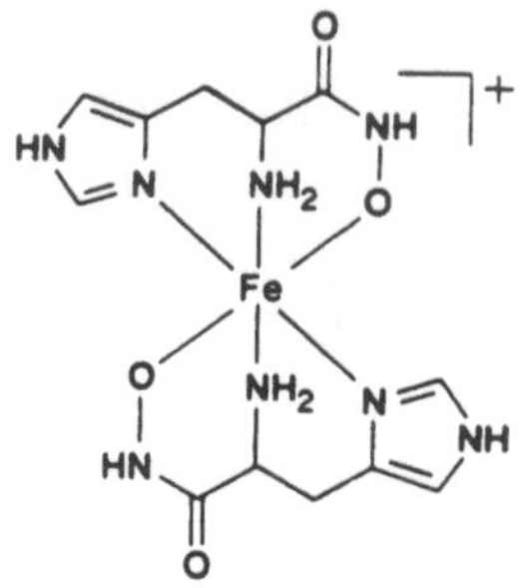　　　　　　**XXX**

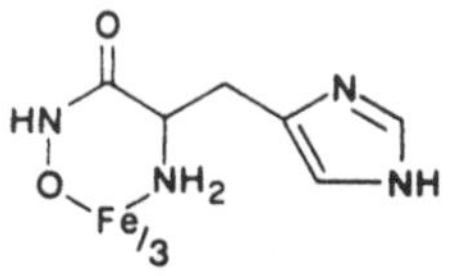

$H_2NCH_2-(C(O)NHCH_2)_n-C-NH$

with $\| \quad |$ and $O \quad OH$ below

n = 1,2

XXXI　　　　　　　　**XXXII**

based on a potentiometric determination of the number of H^+ ions liberated suggests to these authors that coordination of glycine hydroxamic acid to Fe^{3+} occurs through the hydroxylamine O and amine N atoms (**XXVIII**). Biruš and co-workers[66] have investigated tris-Fe^{3+} complex equilibria for aceto-, glycine-, and betaine-hydroxamic acid. They conclude that amine group coordination does not occur and that differences in complex stability can be explained on the basis of the electronic effect of the substituent on the carbonyl group. The coordination mode issue is further complicated by the fact that crystal structures of Cu^{2+}, Ni^{2+}, and Co^{3+} complexes of glycine hydroxamic acid show bidentate chelation through both the amine and hydroxamate N atoms in the solid state.[67-69]

Brown and Sekhon[70] have also investigated the solution equilibria and kinetics associated with the Fe^{3+}/histidine hydroxamic acid system. Potentiometric titration data are consistent with several solution species over the pH range 3 to 11. These include FeH_2A^{4+}, $FeHA^{3+}$, FeA^{2+}, $FeH_2A_2^{3+}$, $FeHA_2^{2+}$, FeA_2^+, FeA_3, and FeA_3OH^- (A^- = histidine hydroxamate anion). Based on potentiometry, NMR, IR, and UV-visible spectroscopy, the major species FeA^{2+}, FeA_2^+, and FeA_3 are proposed with Structures **XXIX**, **XXX**, and **XXXI**. Log β_3 for the tris complex, FeA_3, is 21.61 (25°C, 0.15 M NaCl).[70]

In contrast with the work of Brown and Sekhon[70], Pribanić, and co-workers[71] conclude on the basis of UV-visible spectroscopy and comparison with nonamino acid hydroxamate complexes that complexation of histidine hydroxamic acid does not involve amine coordination at acidic pH values. At pH values above 5 they propose that imidazole ring coordination occurs along with C(O) and N(O) hydroxamate coordination.

Monohydroxamic acids of the peptides glycylglycine (**XXXII**, n = 1) and triglycine (**XXXII**, n = 2) form stable Fe^{3+} complexes.[72] Several species were identified by potentiometric titration for diglycinehydroxamic acid (Fe_2A^{5+}, $FeA_2H_2^{3+}$, $FeA_3H_2^{2+}$, FeA_2^+, FeA_3, $FeA_2(OH)_2^-$) and triglycinehydroxamic acid (Fe_2A^{5+}, $FeA_2H_2^{3+}$, FeA^{2+}, $FeA_3H_3^{3+}$, FeA_2H^{2+}, $FeA_3H_2^{2+}$, FeA_2^+, FeA_3, $FeA_2(OH)_2^-$) over the pH range 2 to 11.[72] The tris complex (FeA_3) is the predominant form over the pH range 8 to 10 for **XXXII** (n = 1) and 7.5 to 9 for **XXXII** (n = 2). Log β_3 values (0.2 M KCl, 25°C) for glycine-, diglycine-, and triglycine-hydroxamic acid diminish with increasing chain length: 26.50, 22.22, and 20.50. There was no evidence presented for amine coordination.

$$\text{XXXIII}$$

$$\text{XXXIV}$$

$$-\overset{\parallel}{\underset{O}{C}}-\overset{|}{\underset{OH}{N}}\overset{\sim\!\!\sim}{}\overset{\parallel}{\underset{O}{C}}-\overset{|}{\underset{OH}{N}}-$$

$$-\overset{|}{\underset{OH}{N}}-\overset{\parallel}{\underset{O}{C}}\overset{\sim\!\!\sim}{}\overset{\parallel}{\underset{O}{C}}-\overset{|}{\underset{OH}{N}}-$$

$$-\overset{\parallel}{\underset{O}{C}}-\overset{|}{\underset{OH}{N}}\cdots\overset{|}{\underset{OH}{N}}\ \ \overset{\parallel}{\underset{O}{C}}$$

$$\text{II}\ \overset{|}{\underset{HO}{N}}-\overset{\parallel}{\underset{O}{C}}-(CH_2)_n-\overset{\parallel}{\underset{O}{C}}-\overset{|}{\underset{OH}{N}}-H$$

$$n = 3,4,6,7,8$$

| **XXXV** | **XXXVI** | **XXXVII** | **XXXVIII** |

Chang and co-workers have investigated the kinetics and equilibrium thermodynamics of Fe^{3+} complexation by the peptide hydroxamic acid Z-gly-NHOH, **XXXIII**.[73] The log K_1, K_2, and K_3 values are 11.4, 10.1, and 8.7, respectively, at 25°C. Since the stepwise formation constants are in the expected order $K_1 > K_2 > K_3$, and comparable to the acetohydroxamate system (see Table 2), there is no evidence for any cooperativity effects between peptide chains. Although precipitation occurs at pH > 6 for the Fe^{3+}/Z-gly-NHOH system, the tris(Z-ala-gly-NHOH)iron(III) complex (**XXXIV**) is stable with respect to precipitation up to pH 10. This is presumably due to a well-shielded structure which resists hydrolysis.

Dihydroxamic Acids

There are several features to consider in the di- and trihydroxamic acids which are not present in the monohydroxamic acid systems. These include the influence of the length of the spacer link between hydroxamic acid groups, the rigidity or orientation of the spacer link (preorganization), and the relative orientation of the linked hydroxamic acid groups as shown in **XXXV** to **XXXVII**. The latter point also involves the effect of N-alkylation on Fe^{3+} binding strength (see above). Solution equilibrium studies of the dihydroxamic acids are complicated by two major types of equilibrium reactions to produce tetracoordinated (FeL^+) (Equation 32) and hexacoordinated (Fe_2L_3) (Equation 19) iron.

For glutarodihydroxamic acid (**XXXVIII**, n = 3), potentiometric titrations and UV-visible spectra show that the ligand-bridged complex Fe_2L_3 is the dominant species over the pH range 5.8 to 8.8.[74] The log β_{230} value (Equation 19) for Fe_2L_3 formation is 48.6 (25°C; 0.15 $NaNO_3$). This should be compared with the corresponding value for rhodotorulic acid (**IV**) $(62.2)^{[20]}$ and N-isopropyldihydroxamic acid (**III**, n = 3) $(62.1^{[19]}$; Table 3), which illustrates the importance of N-substitution for thermodynamic stability. Several other species were also identified by analytical potentiometry over the pH range 2 to 10: FeL^+, $FeLH^{2+}$, $Fe_3L_4^{2-}$, Fe_2L_3, $Fe_2L_3H_{-1}^-$, $Fe_2L_3H_{-2}^{2-}$.[74]

Martell and co-workers investigated the longer chain dihydroxamates (**XXXVIII**, n = 4, 6, 7, 8) at conditions where only the 1:1 complex FeL^+ was formed.[27] Data for log β_{110} (Equation 32) are given in Table 5 and show an increase in stability with increasing chain length, with the largest incremental increase evident on increasing the chain length from 6 to 7. This is somewhat shorter than the nine atom spacing between hydroxamate groups in deferriferrioxamine B, a linear trihydroxamic acid (**V**). The necessity for a longer chain length in deferriferrioxamine B may be a result of the more sterically demanding presence of three rather than two hydroxamate groups. These data are in contrast to results reported by Raymond and co-workers[19,20] for the Fe^{3+} complexes of the N-isopropyl-substituted dihydroxamic acids of varying chain length (**III**). In that case almost no variation in Fe^{3+}

TABLE 5
Log β Values for the
Linear Dihydroxamic
Acids *XXXVIII*[a]

Ligand	log β_{110}[b]
XXXVIII n = 4	17.60
XXXVIII n = 6	18.01
XXXVIII n = 7	20.08
XXXVIII n = 8	20.30

[a] Reference 27.
[b] Log stability constant for Equation 32 at 25°C, 1 M KNO$_3$.

$$\text{H}_3\text{C}-\overset{\underset{|}{\text{HO}}}{\text{N}}-\overset{\underset{\|}{\text{O}}}{\text{C}}-(\text{CH}_2)_m-\text{NHC(O)}-(\text{CH}_2)_n-(\text{O})\text{CNH}-(\text{CH}_2)_m-\overset{\underset{\|}{\text{O}}}{\text{C}}-\overset{\underset{|}{\text{OH}}}{\text{N}}-\text{CH}_3$$

n = 2; m = 1, 2, 3

n = 4; m = 1, 2, 3

XXXIX

binding constants were observed for varying chain lengths (see Table 3). This may come about as a result of a balance of opposing effects, namely, increased stability arising from longer chain length and steric crowding brought about by the presence of the *N*-isopropyl groups.

The inductive effect of *N*-alkyl substitution (**XXVII**) may also be seen in comparing log β_{110} values for the N–H dihydroxamic acids **XXXVIII** and *N*-isopropyl hydroxamic acids **III** in Tables 3 and 5. The substitution of an alkyl group on the N atom appears to increase β_{110} by four to five orders of magnitude.

Desaraju and Winston have also synthesized a series of dihydroxamic acids (**XXXIX**) in an effort to determine if changes in the spacing between hydroxamate groups will influence Fe_{aq}^{3+} affinity.[75] Mole ratio plots show that above pH 4 three hydroxamate groups are bound to Fe^{3+}, presumably in a complex of stoichiometry Fe_2L_3. At pH 2 the complexes are completely converted to FeL^+, with two hydroxamate groups bound per Fe^{3+} as shown in Equation 18.[75] This is in contrast to behavior observed by Winston and co-workers for polymeric hydroxamic acids[76,77] where 3:1 hydroxamate/Fe binding is observed at pH 2. Competition experiments with EDTA show no difference in Fe^{3+} stability constants for different hydroxamate group spacings in the dihydroxamic acids (**XXXIX**).[75]

Dihydroxamic acid derivatives of the polyamine carboxylate ligands EDTA and DTPA (**XL, XLI,** and **XLII**) have been synthesized.[78,79] Stable, water-soluble complexes with Fe^{3+} are formed which may be viewed as synthetic analogues of the siderophore aerobactin (**XVIII**).[78] As can be seen in Table 6, **XLI** and **XLII** form more stable ferric complexes than the natural siderophores rhodotorulic acid (**IV**) and aerobactin (**XVIII**) (Table 4). In fact, the ferric binding constants are in the range expected for *tri*hydroxamate chelators. Derivatization of EDTA to form **XLI** produces a ligand with an Fe^{3+} binding constant five orders of magnitude higher than that of EDTA, whereas **XLII** exhibits only two orders of magnitude higher affinity for Fe_{aq}^{3+} than DTPA. The N-alkylated hydroxamic acid **XLI** forms much more stable complexes with Fe_{aq}^{3+} than does the corresponding unsubstituted compound

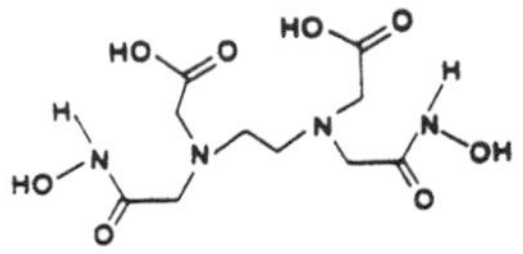

EDTA-DX   di-isopropyl EDTA-DX   di-isopropyl DTPA-DX

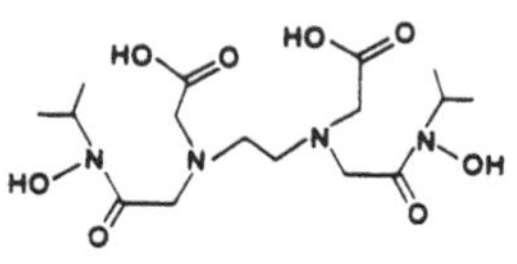

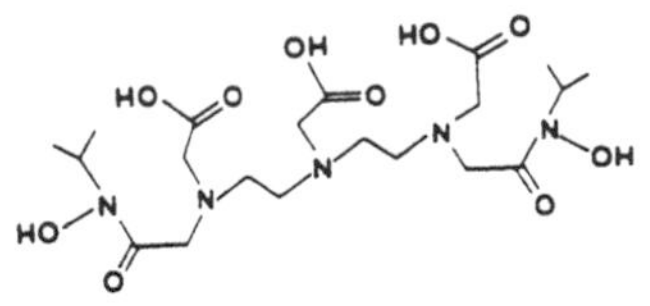

XL   **XLI**   **XLII**

TABLE 6
Log β and pM Values for EDTA, DTPA, and Their Hydroxamate Derivatives

Ligand	log β	pM[a]	Ref.
XL EDTA-DX	22.15		79
XLI (i-Pr)$_2$EDTA-DX	30.2	25.0	78
XLII (i-Pr)$_2$DTPA-DX	29.7	23.6	78
EDTA	25.0		35
DTPA	27.8		35

[a] Conditions: $[Fe^{3+}] = 1 \ \mu M$; $[L] = 10 \ \mu M$; pH 7.40; 25°C.

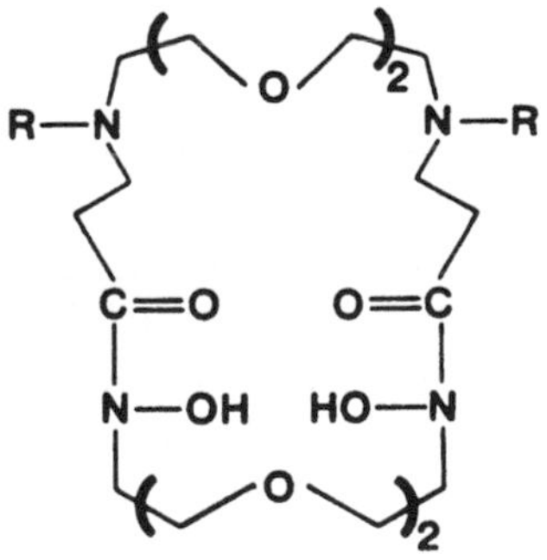

A   R = H

B   R = CH$_2$COOH

XLIII

(**XL**), as noted above for other mono- and dihydroxamic acids. However, the eight orders-of-magnitude increase in stability is quite large and may be partially a result of N–H hydrogen bonding with the tertiary amine in the unsubstituted ligand **XL** which must be broken for Fe^{3+} complexation.[78]

Martell and co-workers have prepared two endocyclic dihydroxamic acids with head-to-head hydroxamate groups separated by an eight-atom chain, **XLIIIA** and **XLIIIB**.[80] In the case of both of these macrocyclic ligands bis complexes are formed with Fe^{3+}_{aq} as determined by spectroscopy and log β values are listed in Table 7. For compound **XLIIIA**

TABLE 7
Log β and pM Values for the Endocyclic
Dihydroxamic Acids *XLIIIA* and *XLIIIB*[a]

Ligand	Equilibrium Rxn	log β	pM[b]
XLIIIA	$Fe_{aq}^{3+} + HL^- \rightleftharpoons$ $FeHL^{2+}$	15.78	13.0
XLIIIB	$Fe_{aq}^{3+} + HL^{3-} \rightleftharpoons$ $FeHL$	17.28	15.2
	$Fe_{aq}^{3+} + L^{4-} \rightleftharpoons$ FeL^-	24.45	

[a]   Reference 80; 25°C; 0.100 M KNO$_3$; determined potentiometrically.
[b]   pH = 7.2; $[Fe^{3+}]_t = 10^{-3}\ M$; 10% excess ligand.

$$CH_3C-N-(CH_2)_2C(O)NH(CH_2)_5-C-N-(CH_2)_2C(O)NH(CH_2)_5-C-N-(CH_2)_2C(O)NHCH_2C(O)NHPh$$
$$\ \ \| \ |\qquad\qquad\qquad\qquad\qquad \| \ |\qquad\qquad\qquad\qquad\qquad \| \ |$$
$$\ \ O\ OH\qquad\qquad\qquad\qquad\qquad O\ OH\qquad\qquad\qquad\qquad\qquad O\ OH$$

XLIV

$$CH_3C-N-(CH_2)_5NHC(O)(CH_2)_2-C-N-(CH_2)_5NHC(O)(CH_2)_2-C-N-(CH_2)_5NHC(O)Ph$$
$$\ \ \| \ |\qquad\qquad\qquad\qquad\qquad \| \ |\qquad\qquad\qquad\qquad\qquad \| \ |$$
$$\ \ O\ OH\qquad\qquad\qquad\qquad\qquad O\ OH\qquad\qquad\qquad\qquad\qquad O\ OH$$

Benzoyl-DFB

XLV

stability is less than might be expected for a dihydroxamate in comparison with log β_2 (21.09, Table 2) for bis(acetohydroxamato)iron(III). This may in part be due to coulombic effects of the protonated amines and the head-to-head arrangement of the hydroxamate groups in **XLIIIA**. The addition of carboxylic acid groups in **XLIIIB** has only a small influence on stability, probably due to coulombic effects. This macrocyclic carboxylate derivative **XLIIIB** has the same functional groups as **XLI**; however, the Fe^{3+} binding constant for **XLIIIB** is five orders of magnitude lower than for **XLI** (Tables 6 and 7). Steric bulk in the macrocycle may be responsible for this difference.

Trihydroxamic Acids

Akiyama and co-workers have synthesized a trihydroxamic acid anilide **XLIV** that bears a good structural resemblance to *N*-benzoyl-deferriferrioxamine B (benzoyl-DFB; **XLV**).[81] The common feature of these two chelators is that they are both linear trihydroxamic acids with nine-atom spacing between hydroxamate groups; the difference is in the position of the connecting amide groups. Benzoyl-DFB (**XLV**) has previously been shown to form a stable complex with Fe^{3+} [82] and the formation constant of a similar Fe^{3+} complex with acetyl-DFB is on the order of 10^{30}.[25] For both **XLIV** and **XLV** stable 1:1 tris-hydroxamate complexes (λ_{max} = 435 nm) are formed in 50% aqueous DMF. This illustrates that the nine-atom spacing between hydroxamate groups appears to be adequate to produce hexacoordination in a trihydroxamic acid ligand, regardless of the position of the amide connecting groups.

A series of tripodal trihydroxamic acid chelators (**XLVI**) incorporating chiral amino acids in the pendant arms has been prepared.[83,84] These chelators have been proposed as

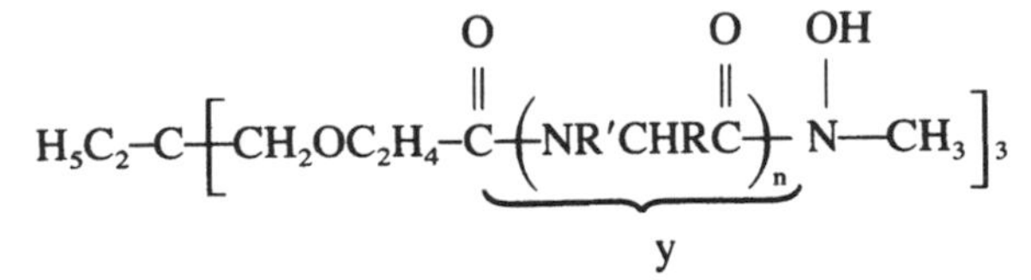

A   n = 0

B   n = 1   y = L–leu   (R = i–Bu; R' = H)

C   n = 1   y = L–pro   (R–R' = –(CH$_2$)$_{\overline{3}}$)

D   n = 1   y = L–ala   (R = CH$_3$; R' = H)

XLVI

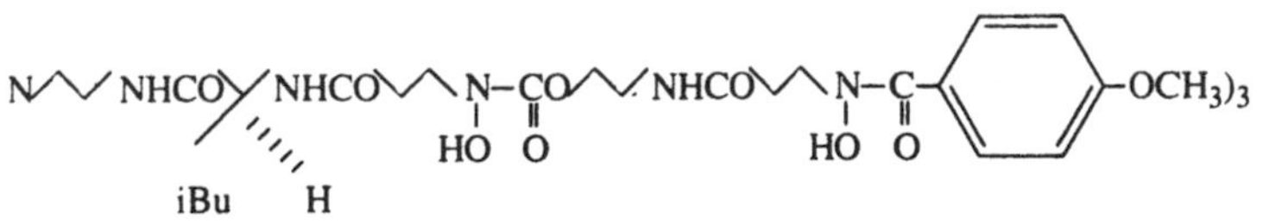

XLVII

synthetic ferrichrome (**XI**) analogues and have been shown to exhibit growth promotion activity for *Arthrobacter flavescens*. Relative Fe^{3+} binding strengths were determined in methanol/0.1 M aqueous sodium acetate (8:2) containing chloride ion and found to be of the same order of magnitude (relative stabilities in parentheses): **XLVIA** (1.0); **XLVIB** (0.8; **XLVIC** (1.1), **XLVID**) (0.65).[83] Fe^{3+} exchange rates (Equation 36) in the same solvent were found to be of the same order of magnitude in the sequence **XLVIA** > **XLVID** > **XLVIB** > **XLVIC**, where the second-order rate constant for **XLVIC** is 1.8×10^{-2} M^{-1} s^{-1}.

$$FeL + EDTA \longrightarrow FeEDTA^- + L \qquad (36)$$

Growth promotion shows a wide activity range in the order (relative activity in parenthesis): **XLVID** (100) > **XLVIC** (80) > **XLVIB** (1) ~**XLVIA** (1), with **XLVID** being equivalent to ferrichrome. Changing the lipophilicity of the side chains does not appear to influence the thermodynamics or kinetics of iron binding. Lipophilicity may influence growth activity, but other factors are also certainly operative, such as recognition. Apparently the hexapeptide anchor for the pendant hydroxamate binding groups in ferrichrome (see **XI**) does not play a role in the receptor-chelate recognition process in *A. flavescens,* since **XLVID** exhibits the same growth promotion activity as does ferrichrome.

The tripodal chiral double Fe^{3+} chelator **XLVII** was synthesized by Libman and Shanzer.[85,86] This is a triple-stranded helical structure that incorporates two Fe^{3+} ions in a stepwise fashion, each with a delta-*cis* tris-hydroxamate coordination shell. Release of Fe^{3+} from the complex also occurs in a stepwise fashion upon reaction with EDTA. No conditions are given, but two distinct rate constants for Fe^{3+} exchange are reported (6.9 M^{-1} s^{-1} and 2.0 M^{-1} s^{-1}).[85] Thus, this tripodal chelator possesses two nonidentical Fe^{3+} binding sites which are loaded and unloaded sequentially.

Raymond and co-workers[87] have synthesized the tripodal trihydroxamic acid TREN-DROX (**XLVII**), based on the flexible TREN backbone. The Fe^{3+} stability constant for this ligand is compared with constants for the more rigid ligands based on a phenyl platform (**XLIX**) in Table 8. The affinity of **XLVIII** for Fe^{3+} is significantly higher. Comparison of

TABLE 8
Log β and pM Values for the
Tripodal Trihydroxamic Acids
XLVIII **and** *XLIX*

Ligand	log β	pM[a]	Ref.
XLVIII	32.9	27.8[a]	87
XLIXA	26.32	21.5[b]	88, 89
XLIXB		25[c]	89
XLIXC		25[c]	89

[a] Calculated for 10 μM ligand, 1 μM Fe^{3+}, pH 7.4.
[b] Assuming FeL is soluble at 1 μM concentration.
[c] Estimated values from Reference 89.

$$N\text{+}CH_2CH_2NHC(O)CH_2CH_2C\text{--}N\text{---}C_6H_4CH_3)_3$$
$$\qquad\qquad\qquad\qquad\quad \|\ \ |$$
$$\qquad\qquad\qquad\qquad\quad O\ \ OH$$

TRENDROX

XLVIII

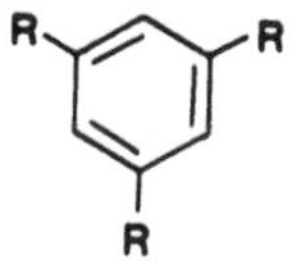

A   R = C(O)NHCH$_2$CH$_2$C–NH
$\qquad\qquad\qquad\qquad\ \|\ \ |$
$\qquad\qquad\qquad\qquad\ O\ \ OH$

B   R = CH$_2$C(O)NHCH$_2$C–NH
$\qquad\qquad\qquad\qquad\quad\ \|\ \ |$
$\qquad\qquad\qquad\qquad\quad\ O\ \ OH$

C   R = C(O)NHCH$_2$CH$_2$CH$_2$C–NCH$_3$
$\qquad\qquad\qquad\qquad\qquad\ \|\ \ |$
$\qquad\qquad\qquad\qquad\qquad\ O\ \ OH$

XLIX

pM values for **XLVIII, XLIXA,** and ferrioxamine B (**V,** Table 4) shows **XLVIII** to have the highest Fe^{3+} affinity at physiological pH. However, Fe(**XLVIII**) precipitates in aqueous solution above pH 3.[87]

Polyhydroxamic Acids

Winston and co-workers have investigated the Fe^{3+} chelation properties of a series of methacrylate polymers with hydroxamic acid side chains of differing lengths.[76,77] Representative structures are shown in **L** and **LI**. Mole ratio plots were used to establish tris-

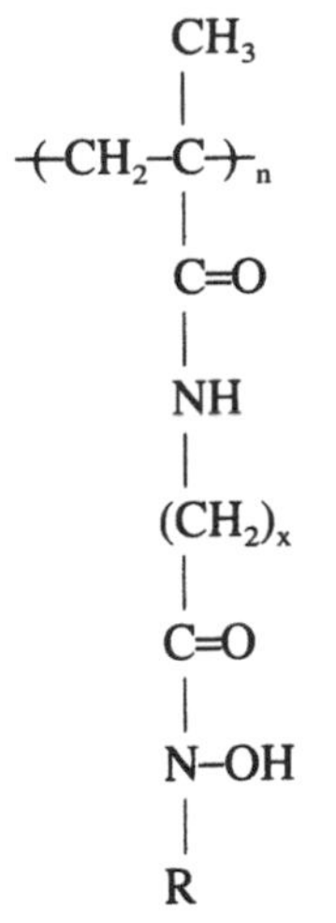

$$
\begin{array}{c}
CH_3 \\
| \\
\text{—}CH_2\text{-}C\text{—}_n \\
| \\
C{=}O \\
| \\
NH \\
| \\
(CH_2)_x \\
| \\
C{=}O \\
| \\
N\text{–}OH \\
| \\
R
\end{array}
\qquad\qquad
\begin{array}{c}
CH_3 \\
| \\
\text{—}CH_2\text{-}C\text{—}_n \\
| \\
C{=}O \\
| \\
NH \\
| \\
CH_2 \\
| \\
C{=}O \\
| \\
NH \\
| \\
CH_2 \\
| \\
C{=}O \\
| \\
N\text{–}OH \\
| \\
CH_3
\end{array}
$$

x = 1, 2, 3

R = CH$_3$, H

L   **LI**

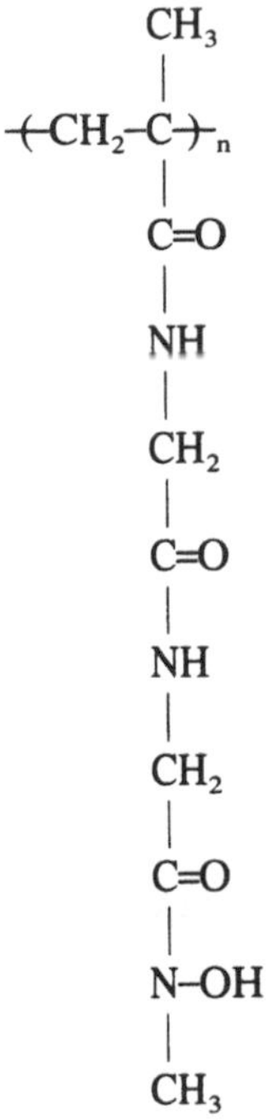

hydroxamate complexation to Fe^{3+}. Stability constants were determined in aqueous solution spectrophotometrically by measuring the position of the competition equilibrium (Equation 37):

$$FeZ + L \rightleftharpoons FeL + Z \qquad (37)$$

where Z = EDTA or DTPA and L the polyhydroxamic acid. These data are compiled in Table 9 and illustrate maximum stability at an 11-atom separation between hydroxamate groups, as opposed to 9 in ferrioxamine B. Steric strain in the polymer presumably lowers the stability when the separation is only nine atoms. *N*-alkylation also slightly enhances stability.

Polyhydroxamic acids based on six-membered carbon ring backbone structures have been synthesized by Winston and co-workers.[90] Structures based on β-cyclodextrin (**LII**) and cyclohexanone (**LIII**) are shown below. Mole ratio plots at an unreported pH were used to establish that three hydroxamate groups are bound per Fe^{3+}. Chelation competition equilibrium experiments with EDTA were carried out at pH 2 and qualitative estimates of the log K for **LII** and **LIII** are reported as 29 to 30.

CATECHOLS

Raymond and co-workers[91] have investigated the Fe^{3+} chelating ability of the bidentate substituted monocatechol ligands *N,N'*-dimethyl catechoylamide (**LIV**) and 2,3-dihydroxyterephthalamide (**LV**). These ligands are of interest as **LIV** is the Fe^{3+} binding subunit of the exocyclic siderophore enterobactin (**XIX**) and the synthetic triscatechol TRENCAM (**XXIV**), and **LV** is the Fe^{3+} binding subunit of the macrobicyclic bicapped TRENCAM (**XXIII**) and the endocyclic ethane trimer **LVI**.[92]

Fe^{3+} stability constants for **LIV** and a series of **LV** ligands are given in Table 10.[91] The catechoylamide **LIV** is a weaker acid than the 2,3-di-hydroxyterephthalamide **LV**, which

TABLE 9
Log K Values for the Polymeric
Hydroxamic Acids *L* and *LI*

Ligand	Spacing[a]	log K	Ref.
L x = 1; R = CH$_3$	9	28.6	76
L x = 2; R = CH$_3$	11	29.7	76
L x = 2; R = H	11	29.2	77
L x = 3; R = CH$_3$	13	29.4	76
LI	15	29.0	77

[a]   Number of atoms between hydroxamate groups.

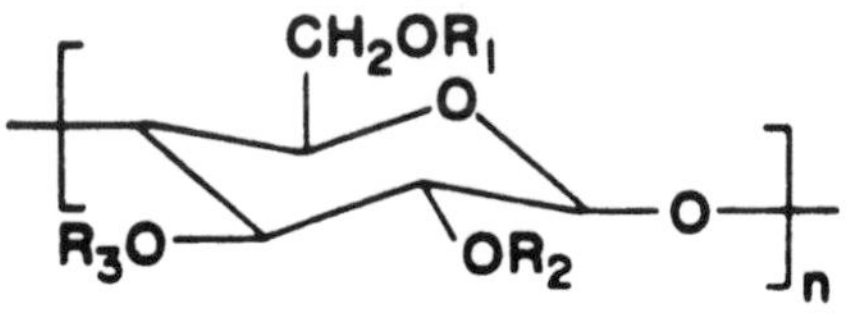

$R_1 = -C(O)(CH_2)_2\ C-N-CH_3$

          $\overset{\|}{O}\ \overset{|}{OH}$

and if $R_2 = R_1$, then $R_3 = H$

or if $R_3 = R_1$, then $R_2 = H$

LII

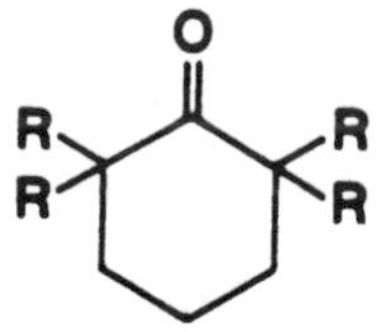

$R = -(CH_2)_2C-N-CH_3$

          $\overset{\|}{O}\ \overset{|}{OH}$

LIII

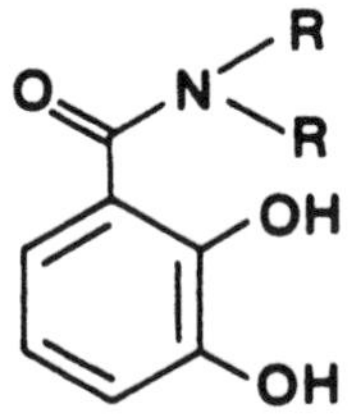

R = CH$_3$

LIV

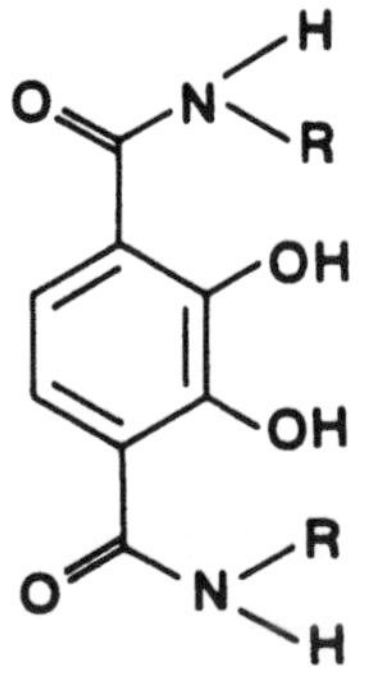

R = CH$_3$, C$_2$H$_5$, C$_3$H$_7$

LV

is why the monocomplex formation constant (log K_{110}) is greater for **LIV** than **LV**. However, there is less of a decrease in stepwise formation constants for the subsequent chelation steps (K_{120}, K_{130}) for **LV**, and as a result, log β_{130} is larger for **LV**. The unusual combination of increased ligand acidity coupled with increased stepwise formation constants for formation

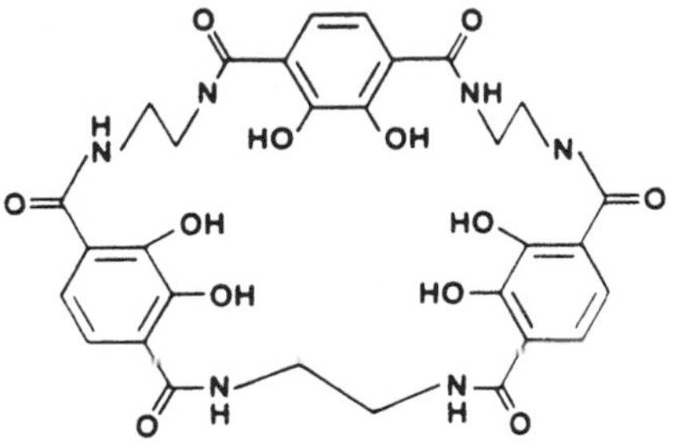

LVI

TABLE 10
Protonation and Formation Constants for the
Monocatechol Ligands *LIV* and *LV*[a]

Stability constant	Ligand			
			LV	
	LIV	R = CH$_3$	R = C$_2$H$_5$	R = C$_3$H$_7$
log K$_{110}$	17.8	16.4	16.3	16.0[b]
log K$_{120}$	13.9	14.5	14.4	15.2
log K$_{130}$	8.5	10.9	11.5	11.9
log β$_{130}$	40.2	41.8	42.2	43.1
pM[c]	15.0	21.1	21.6	22.7
log K$_{011}$	12.1[b]	11.1	11.1	11.0
log K$_{012}$	8.42	6.1	6.0	6.0

[a] Reference 91.
[b] Estimate.
[c] pM = $-$log [Fe] at pH 7.4; [L]$_T$ = 10^{-5} *M*; [Fe]$_T$ = 10^{-6} *M*.

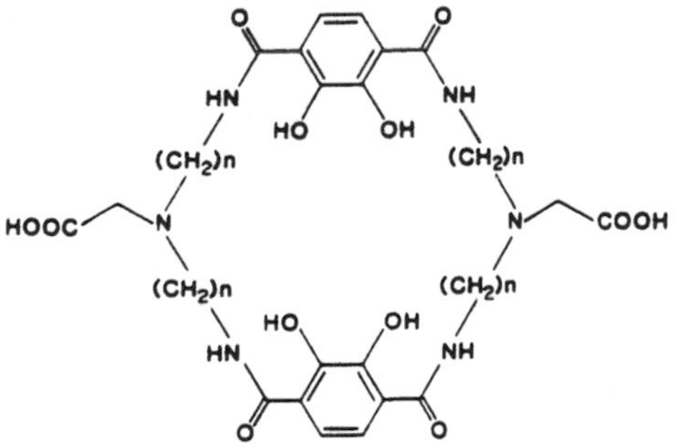

n = 2,3

LVII

of the bis and tris complexes results in higher pM values for the terephthalamides, which are the highest reported for a bidentate chelator.[91] The efficient Fe^{3+} binding capability of the terephthalamides is most likely a result of the charge delocalization capabilities of the substituents on the aromatic ring.

Martell and co-workers[93] recently synthesized two endocyclic dicatechol chelators (**LVII**) of differing ring size. Iron(III) chelation results may be compared with the endocyclic tricatecholate **LVI** prepared by Raymond and co-workers.[92] Stability constants with Fe^{3+}

TABLE 11
Log β and pM Values for
the Cyclic Catechol Ligands
***LVII* and *LVIII*[a]**

Ligand	log β[b]	pM[c]
LVI	38.7	27.4
LVII n = 2	37.6	28.0
LVII n = 3	36.0	29.3

[a]   Data from References 92 and 93, 25°C; I = 0.1 M KNO$_3$.
[b]   β = [FeL^{3-}]/[Fe^{3+}][L^{6-}].
[c]   pH 7.4; [H$_n$L]$_T$/[Fe]$_T$ = 10.

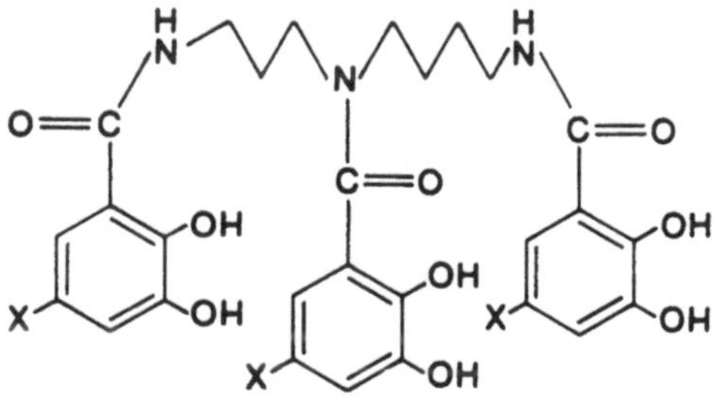

3,4 LICAM   X = H

3,4 LICAMS   X = SO$_3^-$

LVIII

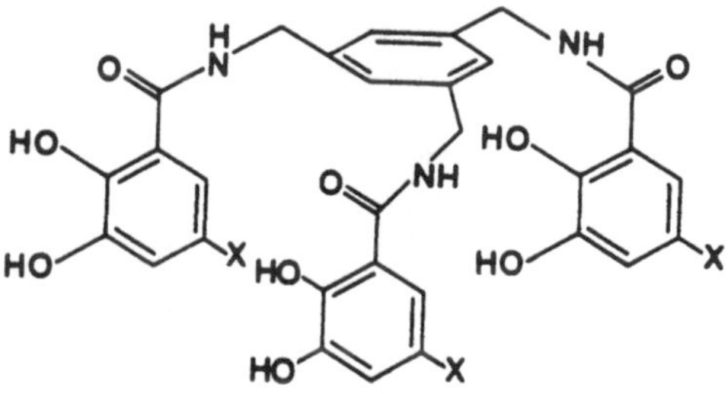

MECAM   X = H

MECAMS   X = SO$_3^-$

LIX

were determined by competition with EDTA and the results are collected in Table 11. Ligand **LVI** has a higher Fe^{3+} stability constant, but due to its higher basicity it has a lower pM value (see table). All three ligands, **LVI** and **LVII** (n = 2, 3), have lower iron binding constants than enterobactin. This is presumably due to a less than optimum ring size, steric hindrance, and strain associated with acting as a hexadentate ligand.

Raymond and co-workers have synthesized a number of tricatechol ligands as enterobactin analogues in an attempt to mimic their large binding constant. This work has been reviewed recently,[1,39,94] and is only summarized here. Tripodal tris-catecholate ligands have been built on an amine linking atom (LICAM[S] **LVIII** and TRENCAM **XXIV**), a benzene platform (MECAM[S], **LIX** and TRIMCAM[S], **LX**), and an exocyclic structure (CY-CAM[S],**XX**). Macrobicyclic versions of these structures have also been prepared. These tris-catechol structural types and their various derivatives (including sulfonates) have log β$_{110}$ values in the range 40 to 46 and pM values 23 to 29. Thus, they have a very high affinity for Fe^{3+}, but less than that of enterobactin.

Ferric ion affinity does seem to be sensitive to changes in the structure of these chelators. Consider, for example, MECAMS (**LIX**) and TRIMCAMS (**LX**) which differ only by an isomeric structural variation. Log β for both ligands is 41, but the pM for MECAMS is 29.1 and 25.1 for TRIMCAMS.[6] The structural change does not influence the Fe^{3+} binding

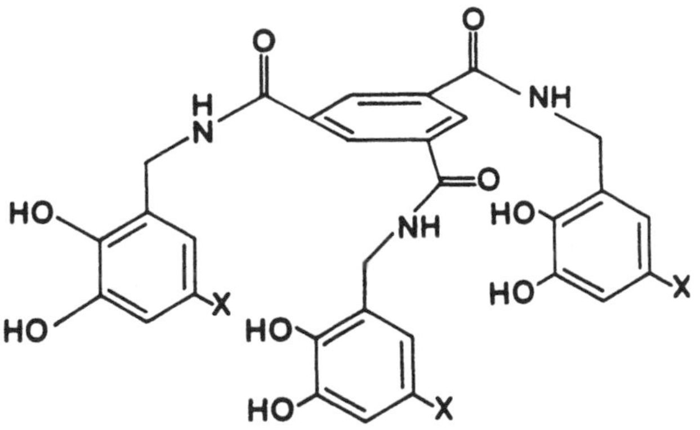

TRIMCAM   X = H

TRIMCAMS   X = SO$_3^-$

LX

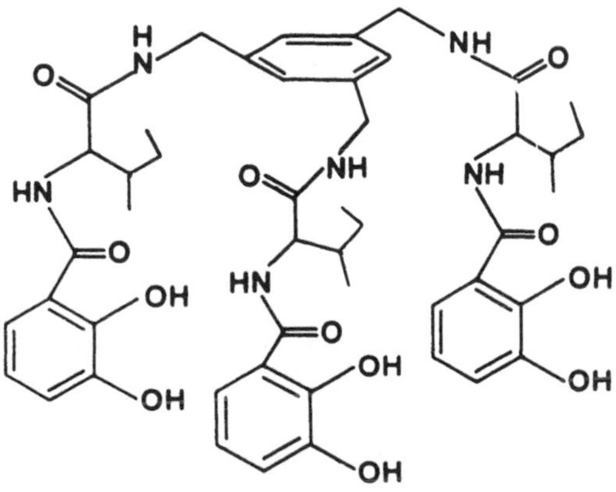

LXI

of the two synthetic ligands (log β values), but differences in protonation constants between the two make MECAMS a better Fe^{3+} ligand at pH 7.4 than TRIMCAMS.

Raymond and co-workers[95] have utilized the structural difference in MECAM (**LIX**) and TRIMCAM (**LX**) to probe molecular recognition by the ferric enterobactin receptor in *Escherichia coli*. The key difference between MECAM and TRIMCAM is the position of the $>$C=O group, being adjacent to the catechol ring or adjacent to the platform. Work with these synthetic analogues has shown that the Fe^{3+} catechol coordination shell and the position of the $>$C=O group are essential for recognition and uptake of iron.[95]

Shanzer and co-workers[96] have prepared a chiral tricatecholate analogue of enterobactin, **LXI**, which forms a delta-*cis* 1:1 complex with Fe^{3+}, like enterobactin. The Fe^{3+} binding strength is somewhat greater for **LXI** than for MECAM (**LIX**), but less than for enterobactin (**XIX**). The small advantage of **LXI** over **LIX** implies less steric strain for the former, or that less conformational entropy is lost in binding.

KINETICS AND MECHANISMS OF IRON SIDEROPHORE AND MODEL SIDEROPHORE REACTIONS

Kinetic studies involving Fe^{3+}-siderophores can be approached in three different ways:

1. Ferric chelate formation/aquation studies involving siderophores and model sidero-phores in which uncomplexed Fe_{aq}^{3+} is consumed or produced. Due to the proclivity of Fe_{aq}^{3+} to hydrolyze forming insoluble polymers,[10,29] this type of study must be carried out in strongly acidic medium. These data are of importance in determining inner coordination shell reactivity patterns and ligand dechelation/chelation steps without additional complications arising from a competing ligand. However, the direct appli-cation of such studies to biological systems is limited, since free Fe_{aq}^{3+} does not occur in siderophore-mediated iron uptake due to hydrolysis of the aquo ion at physiological pH.

2. Kinetic studies of the exchange of iron between two ferric complexes monitored using $^{55}Fe^{3+}$ labeling techniques as shown below, where L and L′ are different deferri-siderophores or model synthetic chelators.

$$^{55}FeL + FeL' \rightleftharpoons FeL + {}^{55}FeL' \tag{38}$$

These studies may be carried out at physiological pH since free Fe_{aq}^{3+} is not pro-duced.

3. Kinetic studies of the exchange of Fe^{3+} from a siderophore (or model siderophore) complex to a free ligand.

$$FeL + L' \rightleftharpoons FeL' + L \tag{39}$$

This process may be monitored using spectrophotometric or $^{55}Fe^{3+}$ labeling techniques and may also be studied at physiological pH.

Since iron removal from its siderophore complex *in vivo* must be a relatively facile process at the appropriate environmental conditions, we are also interested in processes which will enhance the Fe^{3+} dissociation rate, such as H^+ catalysis, the labilizing effect of anions or low molecular weight chelators, and/or oxidation-reduction.

COMPLEX FORMATION/AQUATION STUDIES

Since Fe_{aq}^{3+} has a strong tendency to hydrolyze, forming monomeric, dimeric, and polymeric OH^--containing species in aqueous solution,[10,29] studies involving Fe_{aq}^{3+} as a reactant or product are confined to strongly acidic medium, usually pH 2 or below. Although this is not an environment likely to be found *in vivo*, important information may be obtained concerning the mechanisms and reactivity patterns at an iron(III) center with hydroxamate or catecholate chelating groups.

Model Monohydroxamic Acids

Crumbliss and co-workers have investigated the kinetics and mechanism of Fe_{aq}^{3+} che-lation and dechelation for a series of synthetic monohydroxamic acids with different sub-stituents R_1 and R_2.[49,63,64,97]

$$Fe(H_2O)_6^{3+} + R_1C(O)N(OH)R_2 \rightleftharpoons Fe(R_1C(O)N(O)R_2)(H_2O)_4^{2+} + H^+ \tag{40}$$

These data will serve to illustrate reactivity patterns of Fe^{3+} and the influence of hydroxamate chelating group substituent effects on these patterns. Reaction kinetics were studied as a function of $[H^+]$ below pH 2 in both the forward and reverse direction. A parallel path reaction (Scheme 1) was found to be operative involving both $Fe(H_2O)_6^{3+}$ and $Fe(H_2O)_5OH^{2+}$.

$$Fe(H_2O)_6^{3+} + R_1C(O)N(OH)R_2 \underset{k_{-1}}{\overset{k_1}{\rightleftharpoons}} Fe(R_1C(O)N(O)R_2)(H_2O)_4^{2+} + H^+ + 2H_2O \quad (41)$$

Path 1

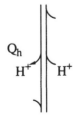

Q_h

$H^+ \quad H^+$

Path 2

$$Fe(H_2O)_5OH^{2+} + R_1C(O)N(OH)R_2 \underset{k_{-2}}{\overset{k_2}{\rightleftharpoons}} Fe(R_1C(O)N(O)R_2)(H_2O)_4^{2+} + 2H_2O \quad (42)$$

SCHEME 1

For a series of 18 synthetic monohydroxamic acids, small variations in k_1 and k_2 were observed with k_1 on the order of $1\ M^{-1}\ s^{-1}$ and k_2 $10^3\ M^{-1}\ s^{-1}$ (25°C, I = 2.0 M NaClO$_4$/ HClO$_4$). The average ratio for k_2/k_1 is 720, with a range of 300 to 2300. The ratio of H$_2$O exchange rate constants for $Fe(H_2O)_5OH^{2+}$ and $Fe(H_2O)_6^{3+}$ is 750.[98-100] This comparison suggests that complexation by the hydroxamate group is dominated by inner coordination shell water exchange dynamics in an interchange (I) process. Activation parameters were obtained and linear isokinetic plots of $\Delta H^\ddagger$ vs. $\Delta S^\ddagger$ were observed for both the forward and reverse processes in Paths 1 and 2.[49,63,64] These isokinetic relationships suggest that all of the hydroxamic acids studied react by the same mechanism. Comparison of $\Delta H_1^\ddagger$ for complexation via Path 1 with $\Delta H^\ddagger$ for H$_2$O exchange at $Fe(H_2O)_6^{3+}$, and the significantly negative $\Delta S_1^\ddagger$ values, suggest an associative interchange, I_a, mechanism for complexation via Path 1. A similar analysis of the parameters for complexation via Path 2 suggests an associative character for that path, but this may also be influenced by H-bonding in the encounter complex between coordinated OH$^-$ and the acidic proton of the hydroxamic acid. In both paths, initial bond formation at the carbonyl O atom is rate limiting with rapid ring closure.

Dissociation kinetics were found to be sensitive to the R_1 and R_2 substituents, with the R_2 group playing the dominant role.[49,63] This is consistent with the influence of the R_2 group on electron density at the carbonyl O atom (O_1) as shown in resonance form **XXVII** and a mechanism whereby initial bond cleavage occurs at Fe–O$_2$ followed by rate-limiting Fe–O$_1$ bond cleavage. The overall mechanism for complexation and dissociation is shown in Scheme 2. The important features of the mechanism are initial bond formation at Fe–O$_1$, initial bond cleavage at Fe–O$_2$, the role of the cis H$_2$O ligand in proton transfer to the dissociating hydroxamate ligand in the acid-independent dissociation path (path 2), and the influence of the R_2 substituent on the dissociation kinetics.

Das and co-workers have investigated the complexation and dissociation kinetics of a series of substituted benzohydroxamic acids ($XC_6H_4C(O)N(OH)H$; X = ortho-H, OH, Cl, CH$_3$, NH$_2$) and found the parallel path mechanism described above in Scheme 1 to be operative.[101] Isokinetic plots are consistent with earlier reports[49,63,64] and comparison of

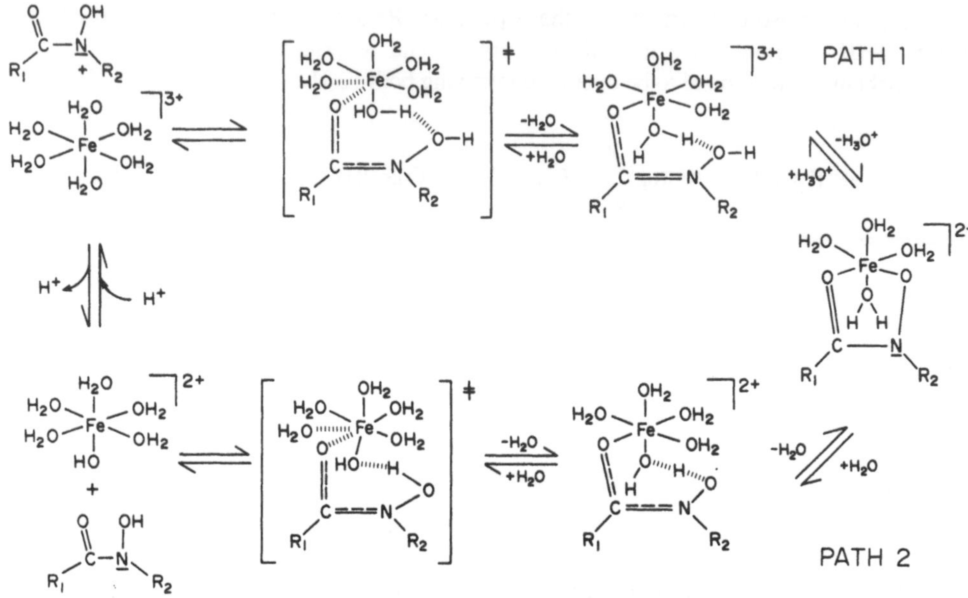

SCHEME 2

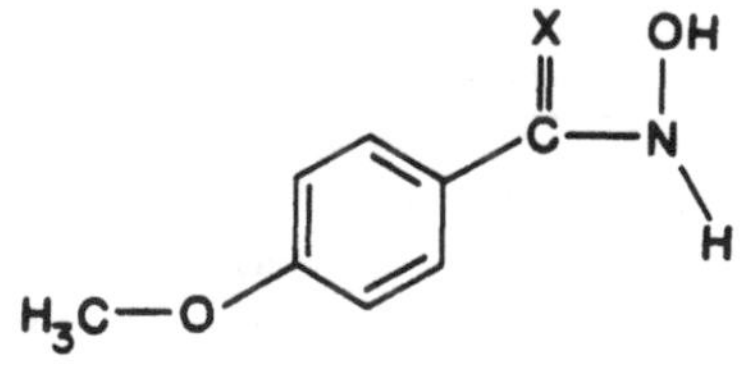

A	X = S
B	X = O

LXII

activation parameters with those for water exchange suggests an I_a mechanism for path 1 and I_d path 2.

A synthetic mono*thio*hydroxamic acid (**LXII.A**) and its *O*-hydroxamate counterpart (**LXII.B**) were used in a kinetic study of Fe_{aq}^{3+} chelation and dissociation at low pH.[64] The thiohydroxamic acid was found to react in the same way as the *O*-hydroxamate counterpart, via the parallel path reaction scheme shown above. The intimate mechanism is apparently also the same for the *S*- and *O*-hydroxamates. Complexation rate constants are similar and $\Delta H^\ddagger$, $\Delta S^\ddagger$ values follow the same isokinetic relationship for the forward and reverse reaction as the set of 18 *O*-hydroxamic acids.[49,63,64] Ratios of dissociation rate constants for the *O*-hydroxamate to *S*-hydroxamate are 50:1 for both the acid-dependent and acid-independent paths. This additional kinetic stability for the *S*-hydroxamate is consistent with an increased contribution from resonance form **XXVII** when an S atom is substituted for the carbonyl O atom. This is expected since sulfur-carbon double bonds are less likely to form than oxygen-carbon double bonds, and thio-amides have a greater degree of carbon-nitrogen double bond character than their oxygen counterparts.[102-106]

The similarity of the Fe^{3+} chelation/dissociation mechanism for the *O*- and *S*-hydroxamates, the enhanced kinetic stability of the *S*-hydroxamate chelate of Fe^{3+}, and the more positive redox potential for ferric S-hydroxamates relative to ferric O-hydroxamates (see below)[107-109] all suggest that the thiohydroxamic acids are an effective Fe^{3+} binding group and imply a siderophore role for such compounds isolated from microbial culture broths.

Kinetic studies of the depolymerization of ferric polymers by monohydroxamic acids have biological relevance, since one of the functions of a siderophore is the solubilization of environmental iron. The kinetics of ferric citrate depolymerization by aceto-, glycine-, and histidine hydroxamic acid were investigated at 25°C and pH 8.5. Acetohydroxamic acid is the most efficient of the hydroxamates and is approximately eight times as fast in solubilizing the ferric citrate polymer as EDTA under these conditions.[70,110,111]

Model Monocatechols

Relatively few kinetic studies with simple catechol systems have been investigated. This is due at least partially to the fact that internal redox processes (to yield a semiquinone and Fe^{2+}) may be operative, which contribute to iron complex instability. A number of years ago, Mentasti and Pelizzetti[112] reported the kinetics of Fe^{3+} complexation by catechol in acid medium considering the usual parallel path scheme (Scheme 3; LH_2 = catechol). These authors only observed complex formation via path 2 and reported $k_2 = 3.1 \times 10^3\ M^{-1}\ s^{-1}$ at 25°C, I = 1.0 M $NaClO_4$. Complexation of $Fe(H_2O)_5OH^{2+}$ by phenol proceeds with a rate constant $k_2 = 1.5 \times 10^3\ M^{-1}\ s^{-1}$,[113] which is comparable to the catechol reaction and suggests that ring closure is rapid in the later case. The overall reaction involves formation of Fe^{2+} and orthoquinone,[114] and this subsequent redox step may be responsible for Mentasti and co-workers[112] not being able to observe complexation via $Fe(H_2O)_6^{3+}$. Hider and co-workers[112] have proposed that an internal redox occurs in the formation of the monocatechol complex.[115,116]

$$Fe(H_2O)_6^{3+} + LH_2 \underset{k_{-1}}{\overset{k_1}{\rightleftharpoons}} Fe(H_2O)_4L^+ + 2H^+ + 2H_2O \qquad (43)$$

Path 1

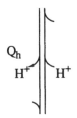

Q_h

$H^+ \quad H^+$

Path 2

$$Fe(H_2O)_5OH^{2+} + LH_2 \underset{k_{-2}}{\overset{k_2}{\rightleftharpoons}} Fe(H_2O)_4L^+ + H^+ + 2H_2O \qquad (44)$$

SCHEME 3

McArdle and Coffey report the kinetics of Fe_{aq}^{3+} complexation by benzazepine (**LXIII**) in acid medium.[117] Complexation by this substituted catechol was found to follow the parallel path mechanism shown above with $k_1 = 20\ M^{-1}\ s^{-1}$ and $k_2 = 830\ M^{-1}\ s^{-1}$ at 25°C, I = 1.0 M ($NaClO_4$). Chelation by Path 2 is comparable to that observed in the case of unsubstituted catechol[112] and phenol[118] and establishes ring closure to be rapid. The ratio $k_2/k_1 = 40$ is low when compared to the corresponding ratio of rate constants for water exchange at $Fe(H_2O)_5OH^{2+}$ and $Fe(H_2O)_6^{3+}$.[98-100] This suggests that water exchange dynamics may

LXIII

not be rate controlling. These data are consistent with some I_a character for catechol substitution at $Fe(H_2O)_6^{3+}$ and I_d at $Fe(H_2O)_5OH^{2+}$.

Jordan and Xu report the Fe_{aq}^{3+} chelation and dissociation kinetics for 2,3-dihydroxybenzoic acid (DHB) and 1,2-dihydroxy-3,5-benzenedisulfonate (Tiron).[118] A parallel path mechanism as described above is proposed. For Tiron (H_2L^{2-}) $k_1 < 5\ M^{-1}\ s^{-1}$ and $k_2 = 3.3 \times 10^3\ M^{-1}\ s^{-1}$ (25°C; 1 M $NaClO_4/HClO_4$). For DHB a salicylate mode of chelation is proposed and $k_1 < 5\ M^{-1}\ s^{-1}$ and $k_2 = 4.8 \times 10^3\ M^{-1}\ s^{-1}$. At lower acidities complexation via the dimer $Fe_2(OH)_2^{4+}$ was observed with a second-order rate constant $8.8 \times 10^3\ M^{-1}\ s^{-1}$ at 25°C.

FERRIOXAMINE B CHELATION AND AQUATION STUDIES

Deferriferrioxamine B is a linear trihydroxamic acid siderophore whose structure is shown in **V**. The kinetics and mechanism of ferrioxamine B formation and aquation reactions are probably the most studied of all the iron siderophore complexes.[21-23,119-123] This is likely due to the stability and ready availability of the ligand as a result of its use as a therapeutic agent in the treatment of individuals with acute[30] or chronic[31] iron poisoning and iron overload associated with β-thalassemia (Cooley's anemia).[32,33] Relevant ligand exchange reactions with ferrioxamine B have also been investigated.[124,125] A diferrioxamine B complex in which one Fe^{3+} is coordinated to two adjacent hydroxamate groups and a second Fe^{3+} is coordinated to the third hydroxamate group has been observed[126] and studied kinetically.[23,121,123,126,127]

The dechelation of ferrioxamine B in strong acid is not of direct biological relevance. However, an elucidation of the dechelation mechanism under experimental conditions where complete dechelation can occur (strong acid) provides a basis for understanding the dechelation process in neutral or weakly acid media. Furthermore, these data will also enhance our understanding of biologically relevant ligand interchange processes and catalyzed iron release from many different siderophore structures. The ferrioxamine B aquation reaction serves to illustrate the stepwise process of Fe^{3+}-siderophore dechelation and the variation in microscopic rate constants that are associated with each step. The most complete investigations of the stepwise dissociation of Fe^{3+} from ferrioxamine B in acidic medium in the absence of coordinating anions were performed by Monzyk and Crumbliss,[21,22] and Pribanić and co-workers.[23] These two studies are largely in agreement, although as outlined below some significant differences are also evident. Both studies were carried out at acidic conditions to prevent complications from hydroxy complex formation and utilized a sudden decrease in pH to drive the dechelation process.

There is general agreement between both laboratories[21-23] that there are four kinetically detectable ferrioxamine B dechelation steps operable with rate constants which span five orders of magnitude (from 10^2 to $10^{-3}\ M^{-1}\ s^{-1}$) and with intermediate protonated structures which are summarized in Scheme 4. An important feature of this process is that dechelation is initiated at the protonated amine end of the molecule, based on analogy with the synthetic monohydroxamic acid model complex kinetics described above.[22,49,63] The first step in the scheme is the dissociation and protonation of the first hydroxamate group to form the

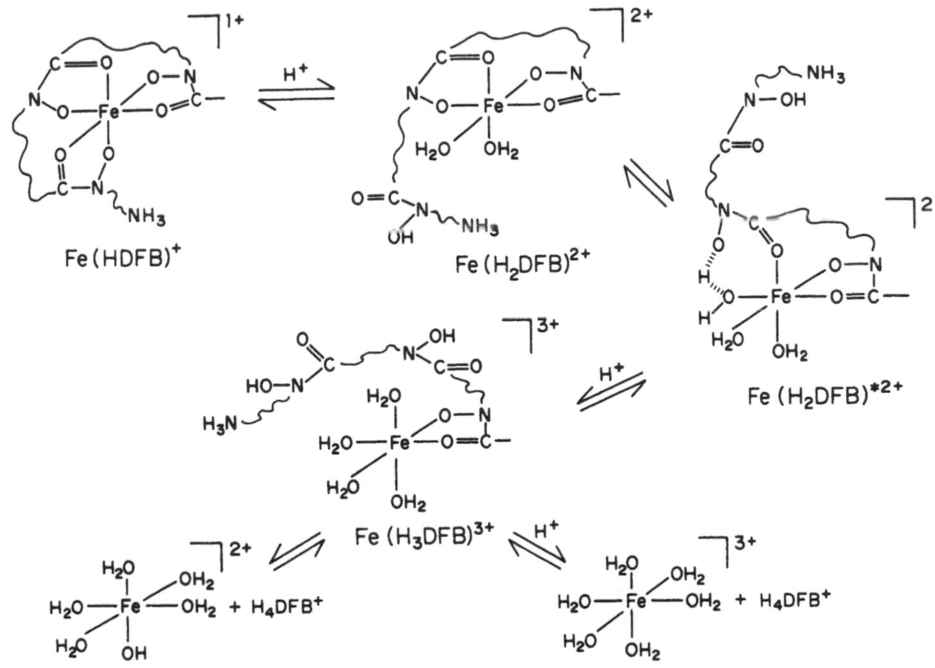

SCHEME 4

tetradentate complex. A discrepancy between the two reports[22,23] concerns whether the protonation occurs before or after the rate-limiting process in the production of the tetra-coordinate intermediate, $Fe(H_2DFB)^{2+}$. Although the structures of the tetradentate $(Fe(H_2DFB)^{2+})$ and bidentate $(Fe(H_3DFB)^{3+})$ intermediates[22,23] may be inferred from proton stoichiometries and spectral comparisons with synthetic monohydroxamate complexes of Fe^{3+},[17,63] the structure of the intermediate formed in the proton-independent step between the tetra- and bidentate forms of the dechelating complex is speculative. The final dechelation step to yield free Fe_{aq}^{3+} from the bidentate form of the complex is the slowest and proceeds via parallel acid-dependent and acid-independent paths to produce $Fe(H_2O)_6^{3+}$ and $Fe(H_2O)_5OH^{2+}$. It is noteworthy that even at high acid concentration the dechelation reaction does not go to completion.

Comparison of these results with suitable model complex reactions is now in order, and subsequently we will consider the biomechanistic implications of this dechelation process which provides, in a stepwise fashion, vacant coordination sites suitable for labile ternary complex formation.

COMPARISON OF FERRIOXAMINE B AND MODEL SYSTEMS

The kinetics of hydrolysis of the tris(acetohydroxamato)Iron(III) complex, $Fe(CH_3C(O)N(O)H)_3$, have been investigated. The reaction occurs in three kinetically detectable stages, which correspond to the following steps (HA = $CH_3C(O)N(OH)H$; coordinated H_2O omitted for clarity):

$$FeA_3 + H^+ \underset{k_{-1}}{\overset{k_1}{\rightleftharpoons}} FeA_2^+ + HA \tag{45}$$

$$FeA_2^+ + H^+ \underset{k_{-2}}{\overset{k_2}{\rightleftharpoons}} FeA^{2+} + HA \tag{46}$$

$$FeA^{2+} + H^+ \underset{k_{-3}}{\overset{k_3}{\rightleftharpoons}} Fe^{3+} + HA \tag{47}$$

$$FeA^{2+} \underset{k_{-4}}{\overset{k_4}{\rightleftharpoons}} FeOH^{2+} + HA \tag{48}$$

It will be instructive to compare results obtained for this scheme with the ferrioxamine B system. Two laboratories have studied the reaction kinetics beginning with the tris complex (FeA_3),[17,128] and several laboratories the monocomplex $(FeA(H_2O)_4^{2+})$.[49,63,129-132] A complete study of the equilibria, kinetics, and mechanism for tris(acetohydroxamato)iron(III) (FeA_3) aquation in aqueous acidic solution at conditions identical to the ferrioxamine B aquation studies was carried out by Pribanić and co-workers.[17] The acetohydroxamate ligand serves as a model for each hydroxamate group of ferrioxamine B. As noted in the section on stability constants $\log \beta_3$ for tris(acetohydroxamato)iron(III) (Table 2) is comparable to $\log \beta$ for ferrioxamine B (Table 4), which lends support to the use of the acetohydroxamic acid system as a kinetic model for deferriferrioxamine B.

We now compare the kinetics of each step of the ferrioxamine B dechelation reaction with the stepwise hydrolysis of $Fe(CH_3C(O)N(O)H)_3$ at 25°C (H_2O omitted for clarity). Pribanić et al. report a value for $k_1' = 380\ M^{-1}\ s^{-1}$ [23] and $k_1 = 1.0 \times 10^5\ M^{-1}\ s^{-1}$.[17]

$$FeA_3 + H^+ \underset{k_{-1}}{\overset{k_1}{\rightleftharpoons}} FeA_2^+ + HA \tag{45}$$

$$Fe(HDFB)^+ + H^+ \underset{k_{-1}'}{\overset{k_1'}{\rightleftharpoons}} Fe(H_2DFB)^{2+} \tag{45a}$$

Monzyk and Crumbliss[22] report an acid-independent rate constant ($290\ s^{-1}$) for the dechelation of $Fe(HDFB)^+$ to form the tetradentate species $Fe(H_2DFB)^{2+}$. Apparently dissociation of a hydroxamate group from $Fe(CH_3C(O)N(O)H)_3$ is a kinetically more facile process than from $Fe(HDFB)^+$. This may reflect the influence of the hydrophobic connecting chain(s) in $Fe(HDFB)^+$ which hinder attack by the hydrated proton.[23]

If we now consider further dechelation to form the bidentate chelated ferrioxamine B from the half chelated intermediate and the corresponding acetohydroxamate/Fe^{3+} dissociation reaction, Pribanić et al. report a value for $k_2' = 2.3 \times 10^{-2}\ M^{-1}\ s^{-1}$ [23] and $k_2 = 1.4 \times 10^3\ M^{-1}\ s^{-1}$.[17] Monzyk and Crumbliss report a value for the corresponding process as $k_2' < 7.2 \times 10^{-2}\ M^{-1}\ s^{-1}$.[22] The proton-assisted dechelation of the second hydroxamate residue from ferrioxamine B is five orders of magnitude slower than dissociation of the second hydroxamate group from $Fe(CH_3C(O)N(O)H)_3$. Again the hydrophobicity and/or steric constraint of the siderophore backbone may be playing a role.

$$FeA_2^+ + H^+ \underset{k_{-2}}{\overset{k_2}{\rightleftharpoons}} FeA^{2+} + HA \tag{46}$$

$$Fe(H_2DFB)^{*2+} + H^+ \underset{k_{-2}'}{\overset{k_2'}{\rightleftharpoons}} Fe(H_3DFB)^{3+} \tag{46a}$$

The slowest step in the overall ferrioxamine B dechelation process is the dissociation of the last hydroxamate group. Comparison with the corresponding acetohydroxamate reaction is again instructive. This final step proceeds by parallel paths to produce $Fe(H_2O)_6^{3+}$ and $Fe(H_2O)_5OH^{2+}$. Pribanić and co-workers report values of 5×10^{-4} M^{-1} s^{-1} and 9.3×10^{-4} s^{-1} for k_3' and k_4', respectively, which were obtained from equilibrium data and formation rate studies.[23] Monzyk and Crumbliss report corresponding values of 1.9×10^{-3} M^{-1} s^{-1} and 2.1×10^{-3} s^{-1} for k_3' and k_4' by studying the dissociation reaction directly,[22] which is in reasonable agreement with Pribanić and co-workers results. The corresponding values for acetohydroxamatic acid dissociation are 1.1×10^{-1} M^{-1} s^{-1} for k_3 and 7.9×10^{-2} s^{-1} for k_4.[63] These data again show that dissociation of an acetohydroxamate group from Fe^{3+} is faster than dechelation of a hydroxamate moiety in ferrioxamine B, this time by two orders of magnitude.

$$FeA^{2+} + H^+ \underset{k_{-3}}{\overset{k_3}{\rightleftharpoons}} Fe^{3+} + HA \tag{47}$$

$$FeA^{2+} \underset{k_{-4}}{\overset{k_4}{\rightleftharpoons}} FeOH^{2+} + HA \tag{48}$$

$$Fe(H_3DFB)^{3+} + H^+ \underset{k_{-3}'}{\overset{k_3'}{\rightleftharpoons}} Fe^{3+} + H_4DFB^+ \tag{47a}$$

$$Fe(H_3DFB)^{3+} \underset{k_{-4}'}{\overset{k_4'}{\rightleftharpoons}} FeOH^{2+} + H_4DFB^+ \tag{48a}$$

The existence of parallel dissociation paths for the final step in ferrioxamine B aquation is a result of the strong tendency for the aquo ferric ion, $Fe(H_2O)_6^{3+}$, to hydrolyze.[10,29] The mechanism of dissociation of the hydroxamate group by initial cleavage of the Fe—O(N) bond, with synchronous protonation from solution (Path 1, Scheme 2) or a cis coordinated water molecule (Path 2, Scheme 2), is based on a kinetic investigation of Fe_{aq}^{3+} chelation and dissociation by a series of synthetic monohydroxamic acids $R_1C(O)N(OH)R_2$ as described above.[49,63,64] Scheme 2 is directly applicable to the final stage of ferrioxamine B aquation (Equations 47a and 48a), where $R_1 = CH_3$ and $R_2 = -[(CH_2)_5NHCO(CH_2)_2CONOH]_2(CH_2)_5NH_3$. Hydrolysis of partially chelated ferric ion is relatively unimportant in acidic medium, so the parallel acid-independent dechelation step involving an intramolecular proton transfer from coordinated H_2O is only observed in the final step.

The greater kinetic efficiency of the final hydroxamate group dissociation in the acetohydroxamic acid complex relative to ferrioxamine B may again be, in part, due to the hydrophobic effect of the connecting backbone in the siderophore. However, the acid independent intramolecular H^+ transfer path is also two orders of magnitude greater for the acetohydroxamate system than for ferrioxamine B (Equations 48 and 48a). This suggests that something other than connecting backbone hydrophobicity is likely to be operative.

The structure of the linear trihydroxamate deferriferrioxamine B (**V**) is such that the substituent on the N atom of each hydroxamate moiety is an electron-releasing alkyl group. Comparison of the kinetic parameters for the final hydroxamate dissociation step between ferrioxamine B and a synthetic monohydroxamate ($R_1C(O)N(OH)R_2$) may be more appropriate for the case where R_2 is an alkyl group rather than H, as is the case with acetohydroxamic acid. For example, rate constants for the acid-dependent and acid-independent dissociation of the mono N-methylacetohydroxamate complex of Fe^{3+} are 2.8×10^{-3} M^{-1}

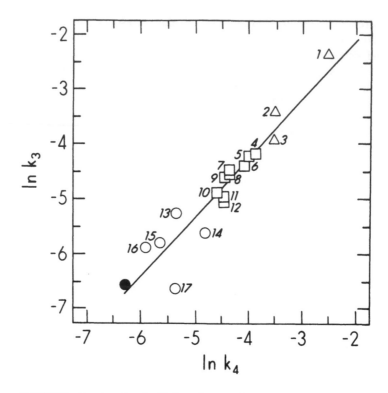

FIGURE 2.   Log-log plot of acid-dependent aquation rate constant (k_3, Reaction 47) as a function of the acid-independent aquation rate constant (k_4, Reaction 48) for a series of monohydroxamatoiron(III) complexes, $Fe(H_2O)_4(R_1C[O]N[O]R_2)^{2+}$, and the last stage of ferrioxamine aquation (k_3', Reaction 47a; k_4', Reaction 48a). $\triangle$: R_2 = H, R_1 = CH_3 (1), C_6H_5 (2), 4-$CH_3OC_6H_4$ (3). $\square$: R_2 = C_6H_4X (X = 3-CN (4), 4-C[O]CH$_3$ (5), 4-CN (6), 4-I (7), 4-Cl (8), 3-I (9), 4-CH$_3$ (10), H (11); R_1 = CH_3. R_2 = C_6H_5, R_1 = C_6H_5 (12). $\circ$: R_2 = CH_3; R_1 = CH_3 (13), C_6H_4X (X = 4-NO$_2$ (14), 4-CH$_3$ (15), H (16), 4-CH$_3$O (17)). $\bullet$: Ferrioxamine B, $Fe(H_3DFB)^{3+}$. Data from References 21, 22, 49, 63, and 64.

s^{-1} and 2.7×10^{-3} s^{-1}, respectively.[63] This is comparable to the k_3' and k_4' values reported for the final step of ferrioxamine B dechelation.[22,23]

Figure 2 shows the logarithm of the acid-dependent rate constant (ln k_3) plotted as a function of the acid-independent rate constant (ln k_4) (Reactions 47 and 48), for the aquation of a series of 18 mono(hydroxamato)tetraaquoiron(III) complexes, $Fe(R_1C(O)N(O)R_2)$ $(H_2O)_4^{2+}$, where R_1 and R_2 are different alkyl, aryl, and H substituents.[49,63,64] This linear correlation with a slope of unity suggests that all of the aquation reactions proceed via the same mechanism and that the transition states for the two parallel paths differ by a H^+.[63] Closer inspection of the plot shows that both the R_1 and R_2 substituents influence aquation rates via both paths, but that the dominant substituent is the one at the R_2 position on the N atom. When R_2 = H (independent of the R_1 substituent) the data are collected in the upper right portion of the plot where aquation rate constants are the largest. When R_2 = alkyl (independent of the R_1 substituent) the data are collected in the lower left portion of the plot where the aquation rate constants are small. The R_2 = aryl hydroxamic acids are grouped together in the middle of the plot. Note that the data point for the final stage of the ferrioxamine B aquation (k_3' and k_4' in Equations 47a and 48a) falls on this line in the region for the synthetic hydroxamates where R_2 = alkyl, but at the extreme low end.[21] Conformity to this linear relationship suggests that the final dechelation step of ferrioxamine B proceeds by the same mechanism as the dissociation of the synthetic monohydroxamate

TABLE 12
Comparison of ΔH° and ΔS° for Overall Chelation Equilibria between Fe_{aq}^{3+} and Acetohydroxamic Acid and Deferriferrioxamine B

Equilibrium reaction	$\Delta H^{\circ a}$	$\Delta S^{\circ a}$	Ref.
49	7.3	31	63, 133
50	−3.1	3	133
51	−5.9	112	63, 133, 134
52	−19.8	71	133, 135

[a] Units are kcal/mol and cal/mol/degree.

group from $Fe(R_1C(O)N(O)R_2)(H_2O)_4^{2+}$. Additional evidence for this mechanistic similarity is that the activation parameters ($\Delta H^\ddagger$, $\Delta S^\ddagger$) for the forward and reverse parallel path process in the final ferrioxamine B dechelation step[133] follow the same linear isokinetic $\Delta H^\ddagger$-$\Delta S^\ddagger$ relationship defined by 18 synthetic hydroxamic acids.[49,63,64]

The localization of the hydroxamic acids with R_2 = alkyl in the lower left region of Figure 2 has been discussed (see above) in terms of the inductive stabilization of the formal positive charge on N resulting from the delocalization of the N atom lone electron pair into the C–N bond. This places additional electron density on the carbonyl O atom as shown in resonance form **XXVII**.

Our analysis suggests that the small rate constant for the final dissociation step of ferrioxamine B via the acid-dependent and acid-independent paths may be due to the electronic influence of the N-alkyl substituent. This also suggests that electronic factors may at least be partially responsible (in addition to possible hydrophobic and steric effects) for the smaller rate constants in the earlier stages of ferrioxamine B dechelation relative to the acetohydroxamic acid complex.

Finally, we return to a comparison of the solution equilibria between Fe_{aq}^{3+} and acetohydroxamic acid and deferriferrioxamine B. The corresponding ΔH° and ΔS° values are listed in Table 12. The more positive entropy change for the tris acetohydroxamate complex (Equation 51) is probably due to the larger entropy of desolvation for three A^- anions relative to one $HDFB^{2-}$ anion. The more exothermic enthalpy change for Equation 52 may be a result of the larger inductive electron donor strength of the alkyl R_2 substituent in H_4DFB^+ than in acetohydroxamic acid (Equation 51) where R_2 = H.

$$Fe^{3+} + 3HA \rightleftharpoons FeA_3 + 3H^+ \qquad (49)$$

$$Fe^{3+} + H_4DFB^+ \rightleftharpoons Fe(HDFB)^+ + 3H^+ \qquad (50)$$

$$Fe^{3+} + 3A^- \rightleftharpoons FeA_3 \qquad (51)$$

$$Fe^{3+} + HDFB^{2-} \rightleftharpoons Fe(HDFB)^+ \qquad (52)$$

TERNARY COMPLEXES IN SIDEROPHORE DECHELATION AND EXCHANGE REACTIONS

Although ferrioxamine B dechelation in strongly acidic medium as described above is not directly comparable to siderophore-mediated iron transport at biological conditions, an understanding of the process provides some insight into possible catalytic pathways for a whole range of siderophores which may be operative *in vivo*. As dechelation of ferrioxamine

B occurs, vacant coordination sites on Fe^{3+} become available for competing ligands, and metal-free binding sites on the deferriferrioxamine B become available for chelation to other metal ions. The latter case has been illustrated by Pribanić and Wilkins and co-workers[23,121,123,126,127,133] in the reaction of Fe^{3+}_{aq} and deferriferrioxamine B (H_4DFB^+), or the hydrolysis of ferrioxamine B ($Fe(HDFB)^+$) in the presence of excess Fe^{3+}_{aq} to produce diferrioxamine B, $Fe_2(HDFB)^{4+}$. Such a species is not likely to be important in terms of siderophore-mediated iron bioavailability, but it does suggest the possibility for excess metal ions of some sort taking the place of a H^+ ion in preventing ring reclosure during the dechelation process in ferrioxamine B, or another hydroxamate siderophore. A similar example may be seen in the citrate-promoted reductive removal of iron from rhodotorulic acid, ferrichrome, ferrioxamine B, triacetyl fusarinine C, and ferrichrome A, which occurs only in the presence of Ga^{3+}.[136] Apparently Ga^{3+} plays the role of H^+ in preventing ring closure of the dissociating siderophore and enhances the accessibility of the Fe^{3+} to the citrate reducing agent.

Chloride Ion

An example of the case described above where a competing ligand binds to a vacant (aquated) Fe^{3+} coordination site is the ferrioxamine B chelation/dechelation studies carried out in chloride ion-containing media. The first step of ferrioxamine B chelation or the last step of the reverse dechelation process is represented in Equations 47a and 48a. The rate acceleration for each of the microscopic rate constants on changing the reaction medium from $I = 2.0\,M$ $NaClO_4$ to $I = 1.0\,M$ NaCl at 25°C is as follows (the number in parentheses represents the ratio of k'_n in Cl^- medium to k'_n in ClO_4^- medium):[23,122] k'_3 (1.3×10^3); k'_3 (1.4×10^2); k'_4 (1.7×10^1); k'_4 (1.1). The observed rate constant increase is too large to be ascribed to a medium effect and clearly the biggest effect is the non-OH^- ligand path. Presumably Cl^- enters the coordination shell of Fe^{3+} and labilizes the dissociating donor atoms, which are either the hydroxamate group in the dechelation step or coordinated H_2O in the chelation step. This is similar to the labilizing effect of OH^- and probably explains why Cl^- has a lesser effect on the k'_4/k'_{-4} path (Equation 48a).

Pribanić and co-workers[132] have also investigated the influence of Cl^- on mono-(acetohydroxamato) and (betaine hydroxamato)iron(III) complex formation and dissociation rates. Using the parallel path reaction Scheme 1, rate accelerations in $1.0\,M$ Cl^- for acetohydroxamic acid and betaine hydroxamic acid ($^\oplus(CH_3)_3NCH_2C(O)N(OH)H$) are (the numbers in parentheses represent the ratio of k_n in Cl^- medium to k_n in ClO_4^- medium; acetohydroxamic acid rate constant ratios are listed first): k_1 (1.4×10^2, 8.6×10^2); k_{-1} (3.8×10^2, 1.4×10^3); k_2Q_h (2.0, 3.0); k_{-2} (5.2, 4.6). The kinetic data support the formation of $FeCl^{2+}$ and $FeCl_2^+$ as reactive species in the complex formation reaction.[132] As observed in the ferrioxamine B system, Cl^- accelerates complex formation and dissociation via Path 1 in Scheme 1, but has little effect on Path 2 where the labilizing influence of OH^- in the inner coordination shell apparently plays a dominant role.

El-Ezaby and co-workers have obtained kinetic and stability constants for the complexation of Fe^{3+}_{aq} by a series of monohydroxamic acids in the presence of Cl^- ($0.15\,M$, $1.0\,M$): glycine-;[137] methionine-;[138] aceto-;[139] histidine-;[139] serine-;[140] and salicyl-hydroxamic acid.[141] The kinetic results yield microscopic rate constants for the chelation reactions that are larger than expected based on comparison with other hydroxamic acid complexation reactions in ClO_4^- or NO_3^- containing media with no Cl^- present. This suggests the presence of the labilizing Cl^- ligand in the inner coordination sphere of iron. Kazmi and McArdle[120] observed faster hydroxamic acid complexation rates in the presence of formate ion, presumably due to entry of formate into the coordination shell of Fe^{3+}. These studies also serve to emphasize the importance of utilizing innocent noncoordinating ions when controlling the pH and/or ionic strength of the reaction medium in Fe^{3+}_{aq} kinetic studies.

Low M.W. Ligands

Kinetic results obtained in the presence of chloride or formate ions illustrate the labilizing influence that small molecules or ions (ligands) may have in siderophore ligand exchange reactions. The kinetics and mechanism of the ferrioxamine B/EDTA exchange reaction (Equation 53)[124,125] and the influence of small ligands on the exchange process[125] has been investigated. The addition of any one of several monohydroxamic acids ($CH_3C(O)N(OH)H$, $C_6H_5C(O)N(OH)H$, or $CH_3C(O)N(OH)CH_3$) was found to catalyze the exchange reaction.[125] A detailed kinetic study which included spectrophotometric characterization of reactive intermediates was carried out at pH 5.4. The mechanism for the monohydroxamate-catalyzed process involves four parallel paths, three of which include ternary complex formation to produce $Fe(H_2DFB)A^+$, $Fe(H_3DFB)A_2^+$, and FeA_3 (A = hydroxamate anion) which lead to $FeEDTA^-$ product formation. Different stabilities of the various ternary complexes are reported which are consistent with the R_1 and R_2 substituents in the monohydroxamic acid catalyst. Catalysis presumably occurs as a result of the labilizing effect of the monohydroxamate in the inner coordination shell, which also prevents ferrioxamine B ring closure. Furthermore, the entering monohydroxamic acid may provide a proton to the dissociating ferrioxamine B via a H^+ transfer within the inner coordination shell which prevents ring closure. Although a redox process may be involved in *in vivo* iron release from ferrioxamine B, the monohydroxamic acid catalysis of Fe^{3+} release from ferrioxamine B illustrates a potentially biologically relevant route to intracellular Fe^{3+} release from its siderophore carrier.

$$Fe(HDFB)^+ + H_xEDTA^{(4-x)-} \rightleftharpoons FeEDTA^- + H_4DFB^+ \qquad (53)$$

Ternary complex formation was also found to accelerate the dissociation of Fe^{3+} from its benzohydroxamic acid complex in aqueous solution.[101] The presence of the tetradentate anion nitrilotriacetate, Nta^{3-}, accelerates the rate of benzohydroxamic acid dissociation by a factor of 5×10^3.

Competing Ligands

Contrary to early assumptions that iron siderophore ligand exchange will be rapid due to the presence of the normally labile high spin d^5 Fe^{3+} ion,[142,143] ligand exchange reactions are often slow and involve stepwise ligand dissociation and protonation, and ternary complex formation.

Popov and co-workers[144] have investigated Fe^{3+} ligand exchange between a monohydroxamato complex and EDTA at 25°C (I = 0.1 $NaClO_4/HClO_4$). The second-order rate constant for Reaction 54 is relatively insensitive to R for R = $C_1 \rightarrow C_7$ hydrocarbons, $-CH_2Cl$, benzyl, $-C(O)N(OH)H$, and $-CH_2-C(O)N(OH)H$, but sensitive to $[H^+]$. This suggests that more than one protonated form of EDTA can form a ternary complex and displace the hydroxamate ligand from the inner coordination shell.

$$Fe(RC(O)N(O)H)(H_2O)_4^{2+} + H_xEDTA^{(4-x)-} \rightleftharpoons FeEDTA^-$$
$$+ RC(O)N(OH)H + (x - 1)H^+ \qquad (54)$$

^{55}Fe exchange between ferrioxamine B (**V**, $FeHDFB^+$) and ferrichrome A (**XI**, Fe-DFC^{3-}) is observed to be slow (Reaction 55).[145] At conditions where there is a 5% excess of H_4DFB^+ over $FeHDFB^+$, the $t_{1/2}$ for exchange for equimolar concentrations (4.0 m*M*) of the two complexes at 25°C and pH 7.4 is *circa* 220 h. A two-step chain reaction mechanism utilizing the excess ligand is proposed (Reactions 56 and 57).[145] The kinetics reveal an apparent first-order dependence of the rate on each of the siderophore complexes. It is proposed that both reaction steps 56 and 57 involve ternary complex formation. The exchange

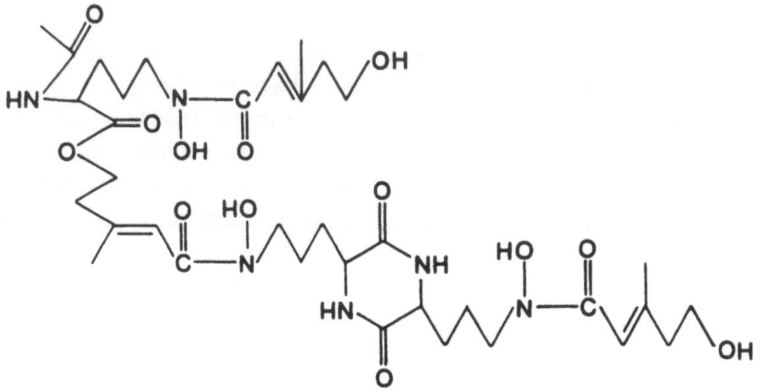

Coprogen

LXIV

rate is accelerated with increasing H^+ concentration, which is consistent with a protonation equilibrium step which lowers the denticity of the dissociating siderophore ligand, thereby enhancing the rate of subsequent multidentate ligand exchange.

$$^{55}\text{Fe(HDFB)}^+ + \text{Fe(DFC)}^{3-} \rightleftharpoons \text{Fe(HDFB)}^+ + {}^{55}\text{Fe(DFC)}^{3-} \tag{55}$$

$$\text{FeDFC}^{3-} + \text{H}_4\text{DFB}^+ \underset{k_{-1}}{\overset{k_1}{\rightleftharpoons}} \text{FeHDFB}^+ + \text{H}_3\text{DFC}^{3-} \tag{56}$$

$$^{55}\text{FeHDFB}^+ + \text{H}_3\text{DFC}^{3-} \underset{k_1}{\overset{k_{-1}}{\rightleftharpoons}} {}^{55}\text{FeDFC}^{3-} + \text{H}_4\text{DFB}^+ \tag{57}$$

In contrast with the slow rate observed for Reaction 55 above, the exchange reaction involving $^{55}\text{Fe(3,4-LICAMS)}$ (**LVIII**) is relatively rapid. A $t_{1/2}$ of *circa* 10 h was observed for 1.0 mM concentration of each complex at pH 7.4 and 25°C.[6,146]

$$^{55}\text{FeHDFB}^+ + \text{Fe(3,4-LICAMS)} \rightleftharpoons \text{FeHDFB}^+ + {}^{55}\text{Fe(3,4-LICAMS)} \tag{58}$$

The importance of both leaving and entering ligand in siderophore ligand exchange reactions may be seen in the following trihydroxamic acid ligand exchange reaction investigated at pH 7.4 and 25°C where L = ferrichrome (**XI**),

$$\text{FeHDFB}^+ + \text{L} \underset{k_r}{\overset{k_f}{\rightleftharpoons}} \text{FeL} + \text{H}_4\text{DFB} \tag{59}$$

ferricrocin (**XI**), ferrichrome A (**XI**), or coprogen (**LXIV**). Data for the second-order rate constants k_f and k_r are given in Table 13.[6] The 60-fold variation in k_r with leaving group roughly correlates with the affinity of L for Fe^{3+} as measured by log β_{110} (Table 4). However, the variation in k_f with L for a constant H_4DFB^+ leaving ligand suggests the importance of ternary complex formation. The reason(s) for the variations in kinetic efficiency of L in removing iron from FeHDFB^+ remain obscure; k_f values do not correlate with log β_{110} values for L. A true understanding of reactivity patterns in these systems is difficult since

TABLE 13
Rate Constants for Siderophore Ligand
Exchange Reaction 59[a]

FeL	k_f $(M^{-1} s^{-1} \times 10^2)$	k_r $(M^{-1} s^{-1} \times 10^2)$
Ferrichrome	0.67	6.3
Ferricrocin	0.41	1.2
Coprogen	4.2	0.46
Ferrichrome A	2.7	0.089

[a] Reference 6; 25°C, pH 7.4.

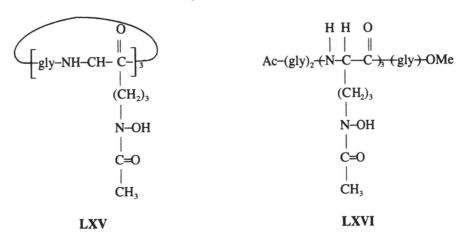

LXV　　　　　　　　　　**LXVI**

rates of reactions involving pathways with ternary complexes may be influenced by the ease of ternary complex formation, its relative stability (and therefore concentration), and lability.

When L = 3,4-MECAMS (**LIX**) in Reaction 59 a complex pH dependence is observed.[6,146] Acid catalysis at low pH supports the importance of H^+ in influencing iron-ligand exchange reaction kinetics. At pH 7.4 the reaction was found to be second order and generally faster than the case where L is a trihydroxamic acid. Data such as these have led Raymond and co-workers to conclude that catechols can more effectively compete for hydroxamate-bound Fe^{3+} than can the hydroxamates.[1,6]

THE INFLUENCE OF POLYHYDROXAMIC ACID STRUCTURE ON LIGAND EXCHANGE KINETICS

Akiyama and co-workers have investigated the influence that structural changes in the basic ferrichrome structure have on the kinetic and thermodynamic stability of the Fe^{3+} complex.[147] Their approach is to investigate both cyclic (**LXV**) and linear (**LXVI** to **LXVIII**) sequence-modified analogues of deferriferrichrome (**XI**). In all cases the Fe^{3+} coordination shell contains three hydroxamate groups. The relative lability of Fe^{3+} was determined by measuring initial rates for the exchange Reaction 60 (L = ferrichrome (**XI**) or synthetic analogues (**LXV** to **LXVIII**)) at pseudo first-order conditions ([EDTA] ≫ [FeL]).

$$FeL + EDTA \xrightarrow{k_f} FeEDTA + L \qquad (60)$$

The calculated pseudo first order rate constants are given in Table 14. The cyclic chelates (**XI, LXV**) are about 20 times less reactive than their linear analogues (**LXVI, LXVII**) with respect to Fe^{3+} dissociation ($k_f^{LXVI}/k_f^{XI} \sim k_f^{LXVII}/k_f^{LXV} \sim 20$). This may reflect less accessibility

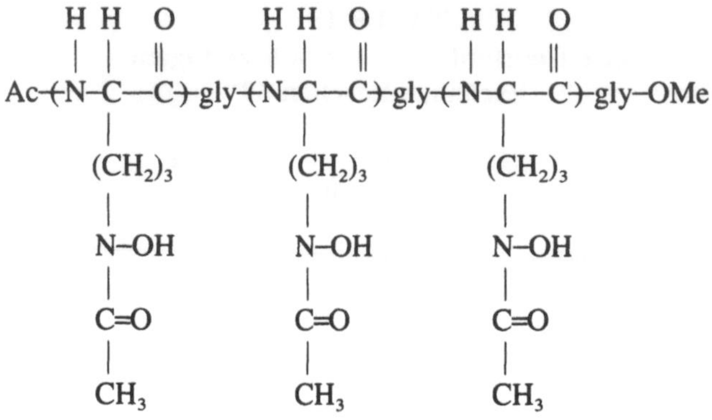

LXVII

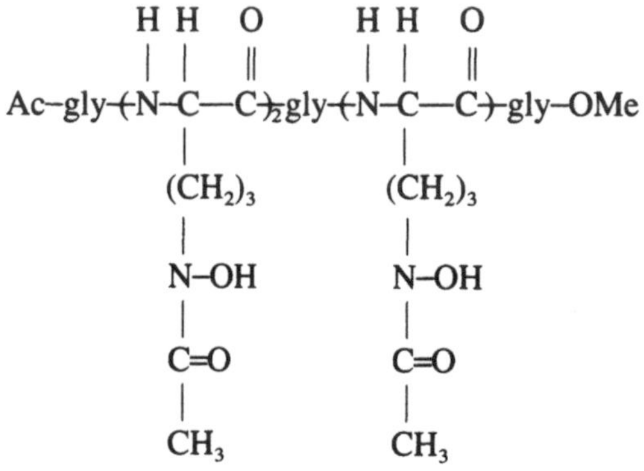

LXVIII

TABLE 14
**Pseudo First-Order Rate Constants for Reaction 60 with
Ferrichrome and Related Synthetic Trihydroxamic Acids[a]**

Trihydroxamic acid	K_r^b s^{-1}	Relative rate
XI ferrichrome (cyclic, asymmetric)	6.1×10^{-4}	1.0
LXV (cyclic, symmetric)	2.0×10^{-4}	0.3
LXVI (linear analogue of ferrichrome (**XI**))	1.2×10^{-2}	20
LXVII (linear analogue of (**LXV**))	4.6×10^{-3}	7.5
LXVIII (intermediate between (**LXVI**) and (**LXVII**))	1.9×10^{-2}	30

[a] Reference 147.
[b] Pseudo first-order rate constant. pH 5.4 (AcOH/NaOAc) 25°C, $I = 0.1$; $[FeL]_0 = 3.8 \times 10^{-4} M$; $[EDTA]_0 = 7.6 \times 10^{-3} M$.

of Fe^{3+} to attack by EDTA to form a ternary complex in the cyclic chelate structure. The structurally "one-sided" ferrichrome complex (**XI**) has a slightly larger ligand exchange rate constant than the more symmetric sequence-modified analogue **LXV**. This is consistent with the lower thermodynamic stability of ferrichrome (**XI**) (log K = 29.1, Table 4) relative to the symmetric cyclic analogue **LXV** (log K = 31.8).[147] A qualitative order of stability with respect to attack by H^+ (**XI** > **LXVII** > **LXV** > **LXVI** > **LXVIII**) or OH^- (**XI** > **LXVII** > **LXVI** > **LXV** > **LXVIII**) shows ferrichrome to be the most stable over the

$$H_2N(CH_2)_5-A-(CH_2)_2C(O)N(H)(CH_2)_5-A-(CH_2)_2C(O)N(H)-(CH_2)_5-A-(CH_2)_2CO_2H$$

LXIX   Deferriferrioxamine G (DFG)   A = $-N-C-$
                                                       | ‖
                                                       OH O

LXX   A = $-C-N-$
                                                       ‖ |
                                                       O OH

$$\overline{HN(CH_2)_5-A-(CH_2)_2C(O)N(H)(CH_2)_5-A-(CH_2)_2C(O)N(H)-(CH_2)_5-A-(CH_2)_2C(O)}$$

XII   Deferriferrioxamine E (DFE)   A = $-N-C-$
                                                      | ‖
                                                      OH O

LXXI   A = $-C-N-$
                                                        ‖ |
                                                        O OH

widest range of pH values. These limited results demonstrate that within a narrow reactivity range, backbone sequence changes in siderophore structures can influence Fe^{3+} lability, but the largest effect is observed on going from a cyclic to an open chain structure.

Shimizu and Akiyama have synthesized analogues of the linear trihydroxamate siderophore deferriferrioxamine G (**LXIX**) and cyclic trihydroxamate deferriferrioxamine E (**XII**) in which the direction of the hydroxamate moiety has been reversed (**LXX** and **LXXI**, respectively).[148] UV-visible absorption spectra show that an Fe^{3+} complex is formed for both **LXX** and **LXXI** with three hydroxamate groups in the inner coordination shell. An attempt was made to measure Fe^{3+}_{aq} affinity by measuring the exchange kinetics for Reaction 61 in both directions at pH 5.3 (HOAc/NaOAc) and 25°C.

$$FeL + EDTA \xrightleftharpoons[k_r]{k_f} FeEDTA + L \qquad (61)$$

Unfortunately, no quantitative kinetic comparisons can be made since the exchange kinetics were investigated at the conditions where $k_{exp} = k_f + k_r$. However, the cyclic ligand **LXXI** was reported to have the highest affinity for Fe^{3+} relative to compound **LXX** and deferriferrioxamine B, since at 25°C, pH 5.3, [Fe-**LXXI**] = $3.2 \times 10^{-4}\ M$ and [EDTA] = $8.3 \times 10^{-3}\ M$, 15% of Fe^{3+} remains complexed to **LXXI**, while at the same conditions exchange of Fe^{3+} to EDTA was 100% complete for Fe-**LXX** and ferrioxamine B.[148] Deferriferrioxamine G (**LXIX**) and E (**XII**) were not investigated.

N,N',N''-triacetylfusarinine C, TAFC (**LXXII**), is a cyclic siderophore which consists of three monomeric hydroxamic acids, N-acetylfusarinine, linked together by ester bonds rather than peptide bonds as found in other linear and cyclic trihydroxamate siderophores. TAFC is produced by *Mycelia sterilia* EP-76 and has been found to function as a siderophore for this organism. The proposed mechanism for iron assimilation by *M. sterilia* involves three distinctive pathways:[61] (1) a Fe-TAFC siderophore system involving intracellular hydrolysis of the siderophore ester bonds for iron release; (2) an exchange mechanism between complexed iron and extracellular TAFC, followed by uptake of Fe-TAFC; and (3) a ferrichrome transport system. The hydrolysis mechanism is somewhat surprising, since the Fe-

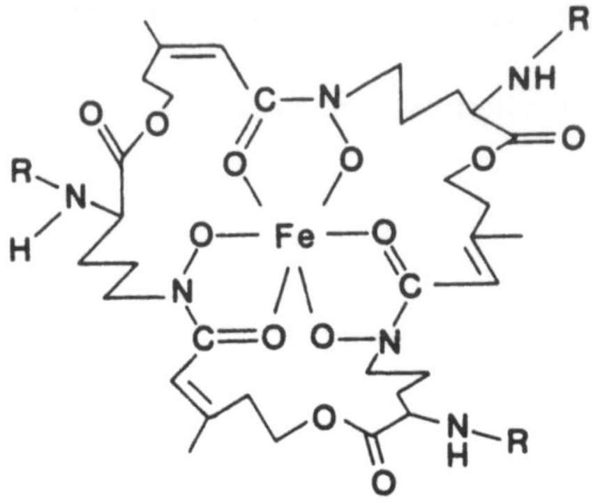

R = H, Fusarinine C

R = –COCH$_3$, TAFC

LXXII

TABLE 15
Iron Exchange Kinetics (Reaction 62) between TAFC and Other Siderophore Ligands

		% Fe exchange[b]			
L[a]	L'[a]	0.5 h[c]	12 h[c]	0.5 h[d]	12 h[d]
Deferriferrichrome	TAFC	42	90	4	6
Deferriferrioxamine B	TAFC	8	74	1	19
Deferriferrichrome A	TAFC	4	64	10	38
Deferri-rhodotorulic acid	TAFC	82	85[e]		
Citric acid	TAFC	63	90[e]		
N-Acetylfusarinine	TAFC	92	99		
N-Acetylfusarinine	Deferriferrichrome	8	86		

[a] As defined in Reaction 62.
[b] Data from Reference 61.
[c] Reaction 62 proceeding in forward direction.
[d] Reaction 62 proceeding in reverse direction.
[e] Determined at 16 h.

TAFC redox potential is high enough for a reductive delivery mechanism without chemical modification of the siderophore[61] (see next section).

TAFC appears to differ from many other trihydroxamate siderophores in its ability to participate in the rapid exchange of Fe^{3+} at neutral pH. Adjimani and Emery[61] investigated Fe^{3+} exchange between TAFC and other siderophores *in vitro* using equimolar amounts of an isotopically labeled iron complex and a competing ligand at pH 6.8 and 30°C. These data are summarized in Table 15 and demonstrate that TAFC is very efficient at removing Fe^{3+} from the TAFC monomer, rhodotorulic acid, and citrate, but not ferrioxamine B. These data support mechanism (2) described above for iron assimilation by *M. sterilia*. The chemical reason for the efficiency of TAFC in removing Fe^{3+} from other siderophores is not apparent. It has been suggested that the flexibility of the ring may play a role,[61] which may be true in comparison with other cyclic hydroxamate siderophore structures, but not open chain structures.

$$^{59}FeL + L' \rightleftharpoons {}^{59}FeL' + L \qquad (62)$$

$$R = -C(O)N(H)CH_2(CH_2OCH_2)_nN-C-CH_3$$

n = 1, 2, 3   HO O

LXXIII   LXXIV

TABLE 16
Fe^{3+}-Hydroxamate (*LXXIII* and *LXXIV*)
Exchange Kinetics with EDTA[a]

Fe^{3+}-hydroxamate	k_f/s^{-1} [b]	Relative rate
LXXIII n = 1	1.3×10^{-3}	4.6
LXXIII n = 2	2.8×10^{-4}	1.0
LXXIII n = 3	9.0×10^{-4}	3.2
LXXIV n = 1	1.0×10^{-2}	36
LXXIV n = 2	1.5×10^{-2}	54
LXXIV n = 3	1.6×10^{-2}	57

[a] Reference 149.
[b] Pseudo first-order rate constants for Reaction 63 cal-
     culated from initial rates. $[FeL]_0 = 9 \times 10^{-5}\ M$;
     $[EDTA]_0 = 1.7 \times 10^{-3}\ M$; pH = 6.75 (tris-HCl);
     25°C.

Akiyama and co-workers have investigated the kinetic reactivity of a series of Fe^{3+} complexes with synthetic di- and trihydroxamic acids which are based on a combination of a rigid lipophilic benzene platform connecting flexible hydrophilic oligo(ethyleneoxy) chains with terminal hydroxamate groups.[149] Mole ratio plots show that at pH 4 and 7 (50% aqueous DMF) **LXXIII** forms a tris hydroxamate complex with Fe^{3+}, whereas **LXXIV** forms a bis hydroxamate complex at pH 4 and a tris hydroxamate complex at pH 7. No data are reported for the metal/ligand stoichiometry for the dihydroxamic acid (**LXXIV**) complex, but presumably a monomer-dimer equilibrium similar to that shown in Equation 18 for rhodotorulic acid[20] is involved.

Relative kinetic stabilities of the Fe^{3+} complexes of **LXXIII** and **LXXIV** were determined by measuring initial rates for the exchange Reaction 63 at pseudo first-order conditions ([EDTA] ≫ [FeL]).[149] Pseudo first-order rate constants and relative rates are listed in Table 16. Since these exchange rate constants were obtained from initial rate data they may be taken as pseudo first-order rate constants (k_f) for Reaction 63 as written.

$$FeL + EDTA \xrightarrow{k_f} L + FeEDTA \qquad (63)$$

The exchange kinetics were found to be strongly pH dependent with faster rates obtained at lower pH. The rate constants for the trihydroxamic acids are roughly an order of magnitude less than those observed for the dihydroxamic acids. Small variations are seen for both di- and trihydroxamate ligands with changes in chain length. It is difficult to say if this difference is due to the difference in ligand denticity or the difference in reactivity between a mono-nuclear and dinuclear complex. Although the authors claim these data demonstrate a more labile metal hydroxamate complex than ferrioxamine B,[148,149] comparison with other studies must be made with extreme caution due to the presence of Cl^- ion in the buffer system. Chloride ion is known to enhance Fe^{3+} ligand exchange reaction rates (see above), although Pribanić and co-workers claim not to have seen such an effect in an exchange reaction with EDTA.[132]

SIDEROPHORE-MEDIATED IRON REMOVAL FROM TRANSPORT PROTEINS

Siderophore-mediated removal of iron from transferrin serves as an illustration of the kinetics of Fe^{3+} complexation by a siderophore. In addition, such studies may elucidate the mechanism whereby microbes acquire their necessary iron in fluids containing this iron binding protein. This iron acquisition process is involved in bacterial infection.[150] Bacteria can proliferate with increased availability of iron, and transferrin may prevent bacterial infection by complexing iron.[151-155] However, the energetics of these Fe^{3+}-siderophore for-mation reactions may well be dominated by the dissociating transferrin ligand and therefore provide only minimal information about Fe^{3+}-siderophore complex formation.

$$Fe(transferrin) + deferrisiderophore \longrightarrow Fe(siderophore) + apotransferrin \qquad (64)$$

Neilands and co-workers have demonstrated that enterobactin (**XIX**) and aerobactin (**XVIII**) can remove iron from serum,[156,157] ovo-, and lacto-transferrin *in vitro*.[158] Results are reported which show that both siderophores can remove iron from transferrin in buffer and can transfer the iron to *E. coli*. Removal from lacto- and ovo-transferrin was found to be slower than from serum transferrin. In serum, enterobactin was found to be less active than aerobactin, possibly due to binding to serum proteins.[156,157] However, enterobactin was more efficient than aerobactin for Fe^{3+} removal from ovo- and lacto-transferrin *in vitro*.[158] This exchange process was found to be catalyzed by pyrophosphate. Neilands and co-workers have spectrophotometric evidence for ternary complex formation in the aerobactin removal of iron from serum transferrin[156] and ovo-transferrin.[158] The authors suggest that the rate of iron exchange between Fe^{3+}transferrin and aerobactin is controlled by the rate of transferrin conformational change and the rate of dissociation of the ternary complex.

Raymond and co-workers have shown through *in vitro* experiments that enterobactin (**XIX**) and the synthetic catechol siderophore analogues (MECAM [**LIX**], 3,4-LICAMS [**LVIII**], Me$_3$MECAMS, and 3,4-LICAMC) can remove Fe^{3+} from human transferrin at an equivalent rate at neutral pH and 25°C.[159-163] A mechanism involving ternary complex formation is proposed. The kinetic efficiency of the catechols is significantly greater than the trihydroxamate siderophore deferriferrioxamine B(**V**). For example, after only 30 min 5% of the Fe^{3+} was removed from transferrin by deferriferrioxamine B at a 100:1 siderophore/transferrin ratio, while 50% removal was effected by 3,4-LICAMS at a 40:1 ligand/transferrin ratio.

OXIDATION-REDUCTION

A possible mechanistic step in the release of iron from a siderophore complex in microbial iron assimilation is reduction of Fe^{3+} to Fe^{2+}. Siderophore stability constants are much lower for Fe^{2+}, which has a lower charge-to-radius ratio, and ligand exchange reactions for high spin Fe^{2+} are faster than for Fe^{3+}. Consequently, studies of oxidation-reduction re-actions of iron siderophore and model siderophore complexes by electrochemical and chem-ical methods are of importance.

Electrochemical techniques are useful for determining redox potentials, which are im-

TABLE 17
Siderophore Redox Potentials

Siderophore	$E_{1/2}$[a] (V vs. NHE)	Ref.
Coprogen	−0.447	58
Ferrioxamine B	−0.468	82, 164
Ferricrocin	−0.412	58
Ferrichrome A	−0.440	165
Ferrichrome	−0.400	165
Aerobactin	−0.336	1, 57
Rhodotorulic acid	−0.359	1, 20
Fe-TAFC[b]	−0.468	61
Fe-TAFC monomer	> −0.333	61
Enterobactin	−0.750	166, 167
2,3-Dihydroxybenzoylserine	−0.350	168
Parabactin	−0.673	169
Parabactin A	−0.400	169

[a] pH 7.
[b] N,N',N''-Triacetylfusarinine C.

portant in assessing whether Fe(III/II) reduction is thermodynamically feasible and therefore a viable process which may be involved in the mechanism for iron assimilation. Siderophore redox potentials must fall in the range of biological reductants for Fe(III/II) reduction to be a part of the iron transport mechanism. Electrochemical techniques may also be used to determine ligand substituent effects and biologically relevant environmental effects (pH, hydrophobic/hydrophilic medium) on the redox potential. Stability constants may be determined by electrochemical methods[16] and the pH dependence of the redox potential can be used to determine the proton stoichiometry of the electrode reaction and equilibrium protonation constants for the system.[1]

SIDEROPHORE REDOX MECHANISMS IN MICROBIAL SYSTEMS
Hydroxamic Acids

Table 17 is a summary of available data for siderophore redox potentials. The hydroxamate siderophores have $E_{1/2}$ values in a range distinctly more positive than the catechol siderophores. Several reports have been made of reductants (e.g., NADH, NADPH, or flavins) in cell-free extracts which are capable of reducing hydroxamate siderophore-bound Fe^{3+}.[170-173] This suggests that under physiological conditions reductive release of iron from its hydroxamate siderophore complex is a viable process, particularly in the presence of an Fe^{2+} trap such as porphyrin. Several years ago, Emery set up a model for ferrichrome transport of iron across an organic phase (membrane) followed by reduction and complexation of Fe^{2+}.[174] These experiments support the chemical viability of a reductive mechanism for release of iron in ferrichrome-mediated iron transport.

In principle, the Fe(III/II) reduction process can occur inside or outside the cell. Emery and co-workers propose a reductive mechanism for both ferrichrome- and ferrichrome A-mediated iron uptake in *Ustilago sphaerogena*.[175-178] This is illustrated in Scheme 5,[175] which may also be generally applicable to other siderophore/microbe systems.[179] For *U. sphaerogena* cultured under Fe-*deficient* conditions both ferrichrome A (**XI**) and ferrichrome (**XI**) are excreted, with ferrichrome in tenfold excess. In Fe-*sufficient* media only ferrichrome is excreted.[178] Emery argues that this demonstrates that ferrichrome is the true iron siderophore and ferrichrome A is only necessary for extracellular iron solubilization (assisted by three acid groups) under extreme conditions of iron deficiency. EPR and isotopic labeling studies show that Fe^{3+} in ferrichrome A is reduced at the cell surface with deferriferrichrome A

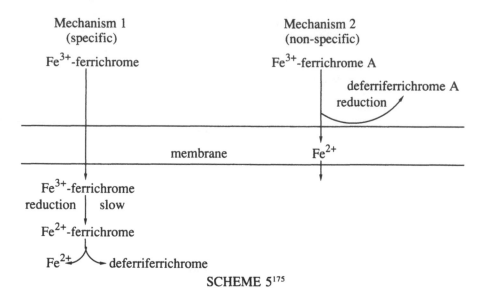

SCHEME 5[175]

remaining external to the cell. Ferrichrome used as a control experiment entered the cell intact and reduction of Fe^{3+} lagged far behind iron uptake.[180]

In the above scheme proposed for *U. sphaerogena*, but which may be treated as a more general iron assimilation scheme, two paths are presented. One path involves intracellular reductive release of iron (Mechanism 1) and another path involves the reductive transport of iron across the cell membrane (Mechanism 2). In Mechanism 1 the carrier is a specific Fe^{3+} chelating substance synthesized by the organism. Synthesis of the specific carrier may be partly or completely repressed depending on the availability of iron in the environment. This is the highly specific siderophore-mediated system, as illustrated by ferrichrome in *U. sphaerogena* or enterobactin in *E. coli*. It has been shown in EPR studies for *U. sphaerogena* that cytoplasmic reduction of the Fe^{3+} is necessary for removal of iron from the siderophore, but not for the uptake of the intact Fe-siderophore complex.[180] In Mechanism 2 the Fe(III/II) reduction is a prerequisite for transfer of iron across cellular membranes to gain access to the cytoplasm, or to cellular organelles such as mitochondria. Mechanism 2 is not specific for a certain carrier. This is illustrated by ferrichrome A in *U. sphaerogena*. The parallel path hypothesis states that microbes purposely evolved a specific and nonspecific mechanism for iron uptake.[175] Further details of the last step or ultimate disposition of the iron in either path remain unspecified. The existence of parallel specific and nonspecific uptake systems can be used to explain many seemingly perplexing or contradictory results in the iron transport literature.[175]

Another illustration of Mechanism 2 in the above scheme comes from *Rhodotorula pilimanae*. Rhodotorulic acid (**IV**) readily donates iron to *R. pilimanae,* but the ligand remains extracellular.[181] Indirect support for a reductive mechanism for *R. piliminae* comes from the observation that the organism cannot utilize iron from trihydroxamate siderophores, such as ferrioxamine B.[175] This may be due to the less favorable reduction potential for ferrioxamine B (see Table 17).

Although the Fe-TAFC (**LXXII**) redox potential is high enough for a reductive delivery mechanism, apparently this mechanism is not operative *in vivo*. Hydrolysis of the siderophore to give the Fe-TAFC monomer complex shifts the redox potential to a more positive value at pH 8 (see Table 17).[61]

Catechols

The tricatecholate siderophores have negative redox potentials as shown in Table 17. Raymond and co-workers have also measured the redox potentials of Fe^{3+} complexes of synthetic tripodal, exocyclic, macrocyclic, and macrobicyclic tricatechols and found them to be in the range -0.8 to -1.1 V (NHE).[1,6,39,55,94] Direct determination of tricatecholate redox potentials at pH 7 by cyclic voltammetry is complicated by proton transfer reactions which lead to irreversible waves. Estimated values may be obtained by extrapolation of data obtained at higher pH values where reversible cyclic voltammetric waves are observed.[6] The redox potentials for enterobactin and several of the model catecholate systems are very pH sensitive. For example, the formal electrochemical potentials for iron(III/II) enterobactin are (vs. NHE): -0.99 (pH >10.4); -0.79 (pH 7.4); -0.57 (pH 6.0).[166]

The pH 7 redox potential values for the catechol siderophores are beyond the range of normal physiological reducing agents. This suggests that cellular iron release may occur by a reductive mechanism only after one or both of the following processes occur: (A) chemical transformation of the Fe^{3+}-siderophore to a form with an appropriate redox potential; and/or (B) a significant drop in pH to bring about an increase in Fe^{3+}-siderophore redox potential. Enterobactin serves as an illustration for both processes A and B.

It is generally agreed that iron acquisition by *E. coli* involves reductive release of iron within the cell. There are some problems with this idea, however, since the reduction potential of ferric enterobactin is extremely low,[166] beyond the range of physiological reducing agents (see Table 17). However, since a soluble enterobactin esterase has been found in this organism,[6,168,182] hydrolysis of the ligand bound to iron may result in a complex that is more easily reduced, as evidenced by the shift to a more positive redox potential for the tris iron complex of 2,3-dihydroxy-benzoylserine (see Table 17). This suggests the possibility of process A. However, structural analogues of enterobactin with no hydrolyzable ester bonds have also been found to deliver iron to *E. coli*[183-185] and *B. subtilis*.[186] The redox potentials for these synthetic tricatechol complexes of Fe^{3+} are too negative for cellular reduction to occur.[1,6,39,55,94] This brings into question the applicability of process A for enterobactin *in vivo*. Process B, however, may be applicable since the enterobactin redox potential undergoes a strong anodic shift with decreasing pH and may be as high as $+170$ mV (NHE) when the pH is dropped to 4.[167] Protonation equilibria, intramolecular redox, the structure of the protonated enterobactins, and their relationship to iron uptake in *E. coli* have all been a source of controversy in the literature.[1,6,115,116,187,188] This topic has been reviewed elsewhere,[1,187] along with the mechanism for iron uptake in *E. coli*.[1]

Another illustration of process A described above may be seen in the case of parabactin and parabactin A.[169] Parabactin has too negative a redox potential for iron removal via a reductive mechanism (see Table 17). However, upon hydrolysis of the oxazoline group of parabactin to produce parabactin A the redox potential is shifted to a more positive value that is within the physiological range for Fe^{3+} reduction. Therefore, *in vivo* oxazoline ring cleavage followed by Fe^{3+} reduction may be involved in metabolic iron removal in *Paracoccus denitrificans*.[115,169]

Phytosiderophores

The shift to a more favorable redox potential with decreasing pH as described in process B above may also be illustrated in the siderophore-mediated uptake of iron by plants. Although definitive evidence that plants produce siderophores may be lacking,[175] iron siderophore complexes have been shown to be effective sources of iron for plants.[189,190] Plant roots are known to excrete acidic compounds (e.g., humic[191] and mugineic acids[192-194]) and phenolic compounds which can act as Fe^{3+} reducing agents.[195] This group of compounds has been called phytosiderophores[192] and suggests that acidification, chelation, and reduction are involved in iron uptake by plants.[196,197]

It has been proposed that nitrate reductase may play a role in iron transport in plants by reducing siderophore-bound Fe^{3+}.[198] The optimum pH for nitrate reductase reduction of siderophores (ferrioxamine B, ferrichrome, ferrichrome A, rhodotorulic acid, and schiko-

TABLE 18
Electrochemical Parameters for Fe^{3+} Complexes of the
Trihydroxamic Acids *LXIV*, *XLV*, and *LXXIII*

Fe(III/II)-hydroxamate	$E_{1/2}$[a] (V vs. NHE)	ΔEp[b] (mv)	i_c/i_a[c]	Ref.
XLIV	−0.462	35	1.2	81
XLV	−0.469	35	1.3	81
LXXIII n = 1	−0.374	30	1.9	149
LXXIII n = 2	−0.399	45	1.3	149
LXXIII n = 3	−0.389	40	1.9	149

[a]   pH = 8.0, carbon electrode.
[b]   Cyclic voltammogram peak-peak separation.
[c]   Ratio cathodic/anodic peak currents.

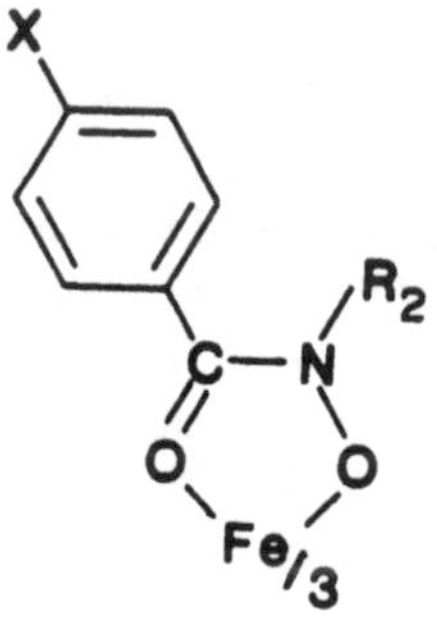

R_2 = H, CH_3, C_6H_5

X = NO_2, Cl, H, CH_3, OCH_3, OH

LXXV

zinen) is 4 to 5. This suggests that the role of acidification by plant roots may not be merely to aid in iron solubilization, but to facilitate a reductive uptake mechanism which utilizes nitrate reductase.

Mugineic acid is a plant siderophore excreted from the roots of gramineous plants which aids in Fe^{3+} solubilization from the soil and reduction of Fe^{3+} to Fe^{2+}.[194] The Fe^{3+} complex reduction potential (−0.102 V [NHE]) is higher than those of the bacterial siderophores suggesting a reductive iron release mechanism.

IRON-HYDROXAMIC ACID COMPLEX ELECTROCHEMISTRY

The electrochemical properties of the trihydroxamate-iron(III) complexes of linear (**XLIV** and **XLV**) and exocyclic (**LXXIII**) structures were determined by cyclic voltammetry.[81,149] Unsymmetrical nonreversible redox waves were obtained and the parameters are summarized in Table 18. These results suggest adsorption on the carbon electrode surface. The nonreversibility of the Fe(III/II) couple makes interpretation of the $E_{1/2}$ values difficult, but it does appear that Fe^{3+} in the exocyclic trihydroxamate structure (**LXXIII**) is more easily reduced than in the linear compounds (**XLIV** and **XLV**).

Ghosh and Chakravorty[199] have investigated substituent effects on the Fe(III/II) redox potential in tris(substituted-benzohydroxamic acid) complexes of Fe^{3+} (**LXXV**). Reversible cyclic voltammograms were obtained in acetonitrile for R_2 = CH_3 and C_6H_5. Redox potentials fall in the range −0.524 to −0.786 V (NHE). A linear correlation is observed between

$E_{1/2}$ and the Hammett substituent constant for X (3σ), with the more electron donating groups shifting the potential to more negative values. For any given substituent X the potential is more negative for $R_2 = CH_3$ than for $R_2 = C_6H_5$, consistent with the greater electron donating effects of an N-alkyl group (see above). When $R_2 = H$ only the cathodic peak is observed, which varies linearly with the substituent X in the same way as observed when $R_2 = CH_3$ or C_6H_5.[199] These results are consistent with an earlier report by Murray and co-workers,[107,108] who also report that Fe^{3+} complexes of primary hydroxamic acids exhibit irreversible electrochemistry.

The redox potential of the acetohydroxamato complex of Fe^{3+} in DMF was found to shift to more positive values with decreasing ligand/metal ratio.[200] This illustrates the possibility for increasing the ease of reduction of Fe^{3+} as a hydroxamate siderophore dissociates to produce a bis- or monohydroxamate coordinated iron. Raymond and co-workers[109] have investigated the pH dependence of the Fe(III/II) redox potential of tris(acetohydroxamato)iron(III) in aqueous media and found a shift to more positive values with decreasing pH. These data suggest that regions of low pH may influence Fe(III/II) reduction in the hydroxamate siderophores as well as the catecholate systems.

Murray and co-workers[107,108] and Raymond and co-workers[109] have investigated the electrochemistry of a series of related O- and S-hydroxamic acid complexes of Fe^{3+}. Changing the donor atoms from O,O to S,O leads to approximately a 0.3 V anodic shift in the reduction potential; i.e., the S-hydroxamate complex is more easily reduced to the Fe^{2+} state than the corresponding O-hydroxamate complex. This further supports a siderophore role for microbially synthesized compounds containing the thiohydroxamic acid functional group.

The redox potentials of the dimeric complexes Fe_2L_3 (L = **III**) have been investigated in a methanol/water mixture by cyclic voltammetry in an attempt to obtain some evidence for interaction between the two iron sites.[19] When the hydroxamate groups are separated by six to ten methylene groups (**III**, n = 6 to 10) a quasi-reversible one-electron reduction occurs at *circa* -0.60 V (NHE). This indicates that the Fe^{3+} atoms are independently reduced at identical or very similar potentials. When n = 4 in **III** a quasi-reversible two-electron process is observed at a slightly more positive potential. An interaction between the two iron sites at smaller n values in Fe_2L_3 is suggested from Mössbauer and EPR spectroscopic results. Cyclic voltammetric results obtained for the dihydroxamic acid **III** when n = 3 may be interpreted in a similar manner to the n = 6 to 10 cases or as a very irreversible two-electron process. When n = 5 the system is complicated by the formation of a methoxide bridged species.

Chemical Oxidation-Reduction

In vitro chemical oxidation-reduction reactions with siderophores and model siderophore complexes are of interest because the results can provide insight and information concerning Fe^{3+} reduction rates, access of the siderophore-bound Fe^{3+} to various reducing agents, and environmental effects (e.g., pH) on redox rates.

Although their inner coordination shell geometries and redox potentials are quite similar, the three hydroxamate siderophores ferrioxamine B (**V**), ferrichrome (**XI**), and ferrichrome A (**XI**) display a variety of electron transfer reactivities.[201] Ferrioxamine B and ferrichrome, linear and cyclic trihydroxamates, respectively, are Fe^{3+} ionophores which are transported intact into the cell where reduction occurs. Ferrichrome A, an exocyclic trihydroxamate with pendant hydrophilic carboxylate groups, remains external to the cell where reduction presumably occurs at the membrane surface (see above). The oxidation-reduction kinetics of these and other siderophore systems have been investigated by McArdle and co-workers: (1) reduction of ferrioxamine B by Cr^{2+}, V^{2+}, and $S_2O_4^{2-}$;[202] (2) reduction of ferrichrome and ferrichrome A by Cr^{2+}, V^{2+}, Eu^{2+}, and $S_2O_4^{2-}$;[201] and (3) reduction of ferrioxamine E and $Fe(CH_3C(O)N(O)H)_3$ by Cr^{2+}, Eu^{2+}, V^{2+}, and $S_2O_4^{2-}$.[203] All of these reactions are outer sphere with the exception of Cr^{2+} reductions of ferrioxamine B,[202] ferrioxamine E,[203] and ferrichrome,[201] and the reduction of ferrioxamine E by Eu^{2+} [203] and V^{2+}.[203]

Apparently ferrioxamine B (**V**) and E (**XII**) provide a trihydroxamate coordination environment that is favorable to rapid electron transfer. The Fe(III/II) self-exchange rate for ferrioxamine B and ferrioxamine E ($3.0 \times 10^5 M^{-1} s^{-1}$) is an order of magnitude higher than for tris(acetohydroxamato)iron(III) ($1.3 \times 10^4 M^{-1} s^{-1}$).[203] These results were obtained by application of the Marcus equation to dithionite reductions of ferrioxamine B, ferrioxamine E, and tris(acetohydroxamato)iron(III).

The ferrichromes have the least accessible redox site of the siderophores investigated by McArdle and co-workers.[201] Protonation renders ferrioxamine B and ferrichrome susceptible to attack by inner-sphere reducing agents. Inner-sphere reduction of ferrichrome by Cr^{2+} proceeds at a much slower rate than ferrioxamine B, and ferrichrome A is sufficiently sterically hindered that the inner-sphere pathway is unavailable. On the other hand, the small outer-sphere reductant SO_2^- shows similar reactivity with ferrichrome, ferrichrome A, and ferrioxamine B ($k \sim 10^4 M^{-1} s^{-1}$).[201] Apparently, SO_2^- has equal access to each of the three redox active sites. This suggests the possibility of a common, small, highly reactive *in vivo* reductant.[201]

The diminished accessibility of Fe^{3+} in ferrichrome A to inner-sphere attack by external reagents may be partly responsible for the small range in reactivity observed for ascorbate reduction of several siderophore complexes at pH 5.4 (relative rate in parenthesis): rhodotorulic acid (18), ferrichrome (4.4), ferrioxamine B (4.4), triacetylfusarinine C (3.5), and ferrichrome A (1.0).[136] The overall reaction is shown in Equation 65:

$$Fe^{III}L + \text{ascorbate} + Ga^{3+} + Z \longrightarrow Fe^{II}Z + GaL \qquad (65)$$

where L = deferrisiderophore and Z is the Fe^{2+}-specific chelator ferrozine. No significant reduction of the Fe^{3+} siderophore by ascorbate occurs in the absence of Ga^{3+}. The Ga^{3+} most likely plays the role of H^+ in preventing ring closure of the dissociating siderophore and enhances the accessibility of the Fe^{3+} to the reducing agent (see above).

Ulstrup and co-workers propose that the organic backbone of a trihydroxamate siderophore produces a diffuse double layer around the siderophore surface and that this double layer may influence access of potential reactant molecules to the iron core.[204] This is similar to the earlier discussion of possible reasons for the decrease in H^+-dependent dechelation rate constants for ferrioxamine B relative to tris(acetohydroxamato)iron(III) (see above). As a model system for this hypothesis the chemical and electrochemical reduction of $Fe(BHA)_3$ (BHA = $C_6H_5C(O)N(O)H^-$) in a SDS micelle has been investigated.[204] The negatively charged groups on the micelle surface provide a double layer in the vicinity of the hydroxamate-Fe^{3+} coordination shell.

There is some evidence that the SDS micelle may influence the stability constant for binding the third benzohydroxamate group to Fe^{3+}.[204] Although the anionic head group of the SDS surfactant may play a role, it may also be the influence of the hydrophobic environment provided by the SDS tail, since K_3 in Reaction 66 is similar for a SDS micelle solution and a 2:3 (v/v) acetonitrile/water mixture. (The K_3 values for H_2O, 2:3 [v/v] acetonitrile/water solvent mixture and SDS micelle, are $2.5 \times 10^7 M^{-1}$, $1.2 \times 10^6 M^{-1}$, and $1.8 \times 10^6 M^{-1}$, respectively.[204] The last two values were calculated assuming the same pK_a for benzohydroxamic acid as that found in H_2O.)

$$Fe(BHA)_2^+ + BHA^- \overset{K_3}{\rightleftharpoons} Fe(BHA)_3 \qquad (66)$$

Ulstrup and co-workers have observed an inverse relationship between (HBHA) concentration and rate of reduction of $Fe(BHA)_3$ by ascorbic acid in 3:1 (v/v) dioxane/water and SDS micelle media. They conclude that BHA dissociation is a requisite first step to expose Fe^{3+} to direct attack by ascorbic acid and extrapolate their results to suggest that the micelle double layer influences this predissociation step.

ACKNOWLEDGMENT

The author acknowledges the donors of the Petroleum Research Fund, administered by the American Chemical Society, for their support of our research in this area. The author also wishes to thank Alain Gaudemer and the members of the Laboratoire de Chimie Bioorganique et Bioinorganique, Université Paris-SUD for their hospitality during a period when part of this review was written.

REFERENCES

1. **Matzanke, B. F., Müller-Matzanke, G., and Raymond, K. N.**, *Physical Bioinorganic Chemistry Series: Iron Carriers and Iron Proteins*, Loehr, T. M., Ed., VCH Publishers, New York, 1989, chap. 1.
2. **Miller, M. J.**, *Chem. Rev.*, 89, 1563, 1989.
3. **Hancock, R. D. and Martell, A. E.**, *Chem. Rev.*, 89, 1875, 1989.
4. **Winkelmann, G., van der Helm, D., and Neilands, J. B., Eds.**, *Iron Transport in Microbes, Plants and Animals*, VCH Publishers, Weinheim, West Germany, 1987.
5. **Neilands, J. B.**, *Annu. Rev. Plant Physiol.*, 37, 187, 1986.
6. **Raymond, K. N., Müller, G., and Matzanke, B. F.**, *Topics in Current Chemistry*, Vol. 123, Boscheke, F. L., Ed., Springer-Verlag, New York, 1984, 49.
7. **Bergeron, R. J.**, *Chem. Rev.*, 84, 587, 1984.
8. **Chimiak, A., Ed.**, *Siderophores from Microorganisms and Plants, Structure and Bonding*, Vol. 58, Springer-Verlag, New York, 1984.
9. **Martell, A. E.**, *Development of Iron Chelators for Clinical Use*, Martell, A. E., Anderson, W. F., and Badman, D. G., Eds., Elsevier, New York, 1981, 67.
10. **Crumbliss, A. L. and Garrison, J. M.**, *Comments Inorg. Chem.*, 8, 1, 1988.
11. **Rossotti, F. C. and Rossotti, H.**, *The Determination of Stability Constants*, McGraw-Hill, New York, 1961.
12. **Hofman, T. and Krzyzanowska, M.**, *Talanta*, 33, 851, 1986.
13. **Izquierdo, A. and Beltran, J. L.**, *Anal. Chim. Acta*, 181, 87, 1986.
14. **Martell, A. E. and Smith, R. M., Eds.**, *Critical Stability Constants*, Vol. 1 to 6, Plenum Press, New York, 1974—1977, 1982, and 1989.
15. **Martell, A. E. and Motekaitis, R. J.**, *The Determination and Use of Stability Constants*, VCH Publishers, New York, 1988.
16. **Meloun, M., Havel, J., and Högfeldt, E.**, *Computation of Solution Equilibria*, Ellis Horwood, Halstead Press, New York, 1988.
17. **Biruš, M., Bradić, Z., Kujundžić, N., Pribanić, M., Wilkins, P. C., and Wilkins, R. G.**, *Inorg. Chem.*, 24, 3980, 1985.
18. **Marakami, Y. and Nakamura, K.**, *Bull. Chem. Soc. Jpn.*, 36, 1408, 1963.
19. **Barclay, S. J., Huynh, B. H., and Raymond, K. N.**, *Inorg. Chem.*, 23, 2011, 1984.
20. **Carrano, C. J., Cooper, S. R., and Raymond, K. N.**, *J. Am. Chem. Soc.*, 101, 599, 1979.
21. **Monzyk, B. and Crumbliss, A. L.**, *Inorg. Chim. Acta*, 55, L5, 1981.
22. **Monzyk, B. and Crumbliss, A. L.**, *J. Am. Chem. Soc.*, 104, 4921, 1982.
23. **Biruš, M., Bradić, Z., Krzanarić, G., Kujundžić, N., Pribanić, M., Wilkins, P. C., and Wilkins, R. G.**, *Inorg. Chem.*, 26, 1000, 1987.
24. **Martell, A. E. and Smith, R. M., Eds.**, *Critical Stability Constants*, Vol. 3, Plenum Press, New York, 1974, 301-305.
25. **Anderegg, G., L'Eplattenier, F., and Schwarzenbach, G.**, *Helv. Chim. Acta*, 46, 1409, 1963.
26. **Wong, G. B., Kappel, M. J., Raymond, K. N., Matzanke, B., and Winkelmann, G.**, *J. Am. Chem. Soc.*, 105, 810, 1983.
27. **Evers, A., Hancock, R. D., Martell, A. E., and Motekaitis, R. J.**, *Inorg. Chem.*, 28, 2189, 1989.
28. **Martell, A. E. and Smith, R. M., Eds.**, *Critical Stability Constants*, Vol. 3, Plenum Press, New York, 1977, 194 and 199.

29. **Kragten, J.,** *Atlas of Metal-Ligand Equilibria in Aqueous Solution*, Halstead Press, New York, 1978.
30. **Ackrill, P., Ralston, A. J., Day, J. P., and Hooge, K. C.,** *Lancet*, 2, 692, 1980.
31. **Barry, M., Flynn, D. M., Letsky, E. A., and Risdon, R. A.,** *Br. Med. J.*, 2, 16, 1974.
32. **Weatherall, D. J., Pippard, M. J., and Callender, S. T.,** *N. Engl. J. Med.*, 308, 456, 1983.
33. **McLaren, G. D., Muir, W. A., and Kellermeyer, R. W.,** *CRC Crit. Rev. Chem. Lab. Sci.*, 19, 205, 1983.
34. **Taylor, G. N., Williams, J. L., Roberts, L., Albertan, D. R., and Shabestari, L.,** *Health Phys.*, 27, 285, 1974.
35. **Martell, A. E. and Smith, R. M., Eds.,** *Critical Stability Constants*, Vol. 1, Plenum Press, New York, 1974.
36. **Edwards, G. L. and Hayes, R. L.,** *J. Nucl. Med.*, 10, 103, 1969.
37. **Harper, P. V.,** *Int. J. Appl. Radiat. Isot.*, 28, 5, 1977.
38. **Pecoraro, V. L., Wong, G. B., and Raymond, K. N.,** *Inorg. Chem.*, 21, 2209, 1982.
39. **Raymond, K. N. and Garrett, T. M.,** *Pure Appl. Chem.*, 60, 1807, 1988.
40. **Neilands, J. B., Peterson, T., and Leong, S. A.,** *ACS Symp. Ser.*, 140, 263, 1980.
41. **Teintze, M., Houssain, M. B., Barnes, C. I., Leong, J., and van der Helm, D.,** *Biochemistry*, 20, 6446, 1981.
42. **Teintze, M. and Leong, J.,** *Biochemistry*, 20, 6457, 1981.
43. **Adapa, S., Huber, P., and Keller-Schierlein, W.,** *Helv. Chim. Acta*, 65, 1818, 1982.
44. **Bell, S. J., Friedman, S. A., and Leong, J.,** *Antimicrob. Agents Chemother.*, 15, 384, 1979.
45. **Hulcher, F. H.,** *Biochemistry*, 21, 4491, 1982.
46. **McCullough, W. G. and Merkal, R. S.,** *Bacteriology*, 137, 243, 1979.
47. **Sugiura, Y. and Tanaka, H.,** *J. Am. Chem. Soc.*, 103, 6979, 1981.
48. **Schwarzenbach, G. and Schwarzenbach, K.,** *Helv. Chim. Acta*, 46, 1390, 1963.
49. **Brink, C. P. and Crumbliss, A. L.,** *Inorg. Chem.*, 23, 4708, 1984.
50. **Crumbliss, A. L., Palmer, R. A., Sprinkle, K. A., and Whitcomb, D. R.,** *Development of Iron Chelators for Clinical Use*, Anderson, W. F. and Hiller, M. C., Eds., Proc. Symp. September 22, 1975, DHEW Publ. No. (NIH) 76-994, National Institutes of Health, Public Health Service, U.S. Department of Health, Education and Welfare, 1976, 175.
51. **Harris, W. R., Raymond, K. N., and Weitl, F. L.,** *J. Am. Chem. Soc.*, 103, 2667, 1981.
52. **Kiggen, W. and Vögtle, F.,** *Angew. Chem. Int. Ed. Engl.*, 23, 714, 1984.
53. **Stutte, P., Kiggen, W., and Vögtle, F.,** *Tetrahedron*, 43, 2065, 1987.
54. **McMurry, T. J., Rodgers, S. J., and Raymond, K. N.,** *J. Am. Chem. Soc.*, 109, 3451, 1987.
55. **Rodgers, S. J., Lee, C.-W., Ng, C. Y., and Raymond, K. N.,** *Inorg. Chem.*, 26, 1622, 1987.
56. **McMurry, T. J., Hosseini, M. W., Garrett, T. M., Hahn, F. E., Reyes, Z. E., and Raymond, K. N.,** *J. Am. Chem. Soc.*, 109, 7196, 1987.
57. **Harris, W. R., Carrano, C. J., and Raymond, K. N.,** *J. Am. Chem. Soc.*, 101, 2722, 1979.
58. **Wong, G. B., Kappel, M. J., Raymond, K. N., Mutzanke, B., and Winkelmann, G.,** *J. Am. Chem. Soc.*, 105, 810, 1983.
59. **Anderson, B. F., Buckingham, D. A., Robertson, G. B., Webb, J., Murray, K. S., and Clark, D. E.,** *Nature*, 262, 772, 1976.
60. **Anderegg, G., L'Eplattenier, F., and Schwarzenbach, G.,** *Helv. Chim. Acta*, 46, 1400, 1963.
61. **Adjimani, J. P. and Emery, T.,** *J. Bacteriol.*, 169, 3664, 1987.
62. **Harris, W. R., Carrano, C. J., and Raymond, K. N.,** *J. Am. Chem. Soc.*, 101, 2213, 1979.
63. **Monzyk, B. and Crumbliss, A. L.,** *J. Am. Chem. Soc.*, 101, 6203, 1979.
64. **Fish, L. L. and Crumbliss, A. L.,** *Inorg. Chem.*, 24, 2198, 1985.
65. **Brown, D. A., Chidambaram, M. V., and Glennon, J. D.,** *Inorg. Chem.*, 19, 3260, 1980.
66. **Biruš, M., Kujundžić, N., Pribanić, M., and Tabor, Z.,** *Croat. Chem. Acta*, 57, 313, 1984.
67. **Brown, D. A., Roche, A. L., Pakkanen, T. A., Pakkanen, T. T., and Smolander, K.,** *J. Chem. Soc. Chem. Commun.*, p. 676, 1982.
68. **de Miranda-Pinto, C. O. B., Paniago, E. B., Cavalho, S., Tabak, M., and Mascarenhas, Y. P.,** *Inorg. Chim. Acta*, 137, 145, 1987.
69. **Pakkanen, T. T., Pakkanen, T. A., Smolander, K., Brown, D. A., Glass, W. K., and Roche, A. L.,** *J. Mol. Struct.*, 162, 313, 1987.
70. **Brown, D. A. and Sekhon, B. S.,** *Inorg. Chim. Acta*, 91, 103, 1984.
71. **Kujundžić, N., Bačić, V., and Pribanić, M.,** *Acta Pharm. Jugosl.*, 35, 221, 1985.
72. **Brown, D. A. and Mageswaran, R.,** *Inorg. Chim. Acta*, 161, 267, 1989.
73. **Chang, C. A., Sekhar, V. C., Garg, B. S., Guziec, F. S., Jr., and Russo, T. C.,** *Inorg. Chim. Acta*, 135, 11, 1987.
74. **Brown, D. A., Geraty, R., Glennon, J. D., and Choileain, N. N.,** *Inorg. Chem.*, 25, 3792, 1986.
75. **Desaraju, P. and Winston, A.,** *J. Coord. Chem.*, 14, 241, 1986.
76. **Winston, A. and Kirchner, D.,** *Macromolecules*, 11, 597, 1978.

77. Varaprasad, D. V. P. R., Rosthauser, J., and Winston, A., *J. Polym. Sci. Polym. Chem. Ed.*, 22, 2131, 1984.
78. Turowski, P. N., Rodgers, S. J., Scorrow, R. C., and Raymond, K. N., *Inorg. Chem.*, 27, 474, 1988.
79. Motekaitis, R. J., Murase, I., and Martell, A. E., *J. Coord. Chem.*, 1, 77, 1971.
80. Sun, Y., Martell, A. E., and Motekaitis, R. J., *Inorg. Chem.*, 24, 4343, 1985.
81. Shimizu, K., Nakayama, K., and Akiyama, M., *Bull. Chem. Soc. Jpn.*, 59, 2421, 1986.
82. Bickel, H., Hall, G. E., Keller-Schierlein, W., Prelog, V., Vischer, E., and Wettstein, A., *Helv. Chim. Acta*, 43, 2129, 1960.
83. Shanzer, A., Libman, J., Lazar, R., Tor, Y., and Emery, T., *Biochem. Biophys. Res. Commun.*, 157, 389, 1988.
84. Shanzer, A., Libman, J., Lazar, R., and Tor, Y., *Pure Appl. Chem.*, 61, 1529, 1989.
85. Libman, J., Tor, Y., and Shanzer, A., *J. Coord. Chem.*, 18, 241, 1988.
86. Libman, J., Tor, Y., and Shanzer, A., *J. Am. Chem. Soc.*, 109, 5880, 1987.
87. Ng, C. Y., Rodgers, S. J., and Raymond, K. N., *Inorg. Chem.*, 28, 2062, 1989.
88. Yoshida, I., Murase, I., Motekaitis, R. J., and Martell, A. E., *Can. J. Chem.*, 61, 7740, 1983.
89. Martell, A. E., Motekaitis, R. J., Murase, I., Sala, L. F., Stoldt, R., Ng, C. Y., Rosenkrantz, J., and Metterville, J. J., *Inorg. Chim. Acta*, 138, 215, 1987.
90. Varaprasad, D. V. P. R., Desaraju, P., and Winston, A., *Bioorg. Chem.*, 14, 8, 1986.
91. Garrett, T. M., Miller, P. W., and Raymond, K. N., *Inorg. Chem.*, 28, 128, 1989.
92. Rodgers, S. J., Ng, C. Y., and Raymond, K. N., *J. Am. Chem. Soc.*, 107, 4094, 1985.
93. Sun, Y., Martell, A. E., and Motekaitis, R. J., *Inorg. Chem.*, 25, 4780, 1986.
94. Raymond, K. N., McMurry, T. J., and Garrett, T. M., *Pure Appl. Chem.*, 60, 545, 1988.
95. Ecker, D. J., Loomis, L. D., Cass, M. E., and Raymond, K. N., *J. Am. Chem. Soc.*, 110, 2457, 1988.
96. Tor, Y., Libman, J., Shanzer, A., and Lifson, S., *J. Am. Chem. Soc.*, 109, 6517, 1987.
97. Crumbliss, A. L., Brink, C. P., and Fish, L. L., *Inorg. Chim. Acta*, 79, 218, 1983.
98. Swaddle, R. W. and Merbach, A. E., *Inorg. Chem.*, 20, 4212, 1981.
99. Grant, M. and Jordan, R. B., *Inorg. Chem.*, 20, 55, 1981.
100. Dodgen, H. W., Liu, G., and Hunt, J. P., *Inorg. Chem.*, 20, 1002, 1981.
101. Das, P. K., Bhattacharya, S. G., Banerjee, R., and Banerjea, D., *J. Coord. Chem.*, 19, 311, 1989.
102. Sandström, J., *Tetrahedron Lett.*, p. 639, 1979.
103. Walter, W., Schaumann, E., and Voss, J., *Org. Magn. Reson.*, 3, 733, 1971.
104. Bernardi, F., Lunazzi, L., and Zanivato, P., *Tetrahedron*, 33, 1337, 1977.
105. Piccini-Leopardi, C., Fobre, O., Zimmermann, D., and Reisse, J., *Can. J. Chem.*, 55, 2649, 1977.
106. Walter, W. and Schaumann, E., *Justus Liebigs Ann. Chem.*, 743, 154, 1971.
107. Brockway, D. J., Murray, K. S., and Newman, P. J., *J. Chem. Soc. Dalton Trans.*, p. 1112, 1980.
108. Murray, K. S., Newman, P. J., Gatehouse, B. M., and Taylor, D., *Aust. J. Chem.*, 31, 983, 1978.
109. Abu-Dari, K., Cooper, S. R., and Raymond, K. N., *Inorg. Chem.*, 17, 3394, 1978.
110. Spiro, T. G., Pape, L., and Saltman, P., *J. Am. Chem. Soc.*, 89, 5555, 1967.
111. Brown, D. A., Chidambaram, M. V., Clarke, J. J., and McAleese, D. M., *Bioinorg. Chem.*, 9, 255, 1978.
112. Mentasti, E. and Pelizzetti, E., *J. Chem. Soc.*, p. 2605, 1973.
113. Cavaseno, F. P. and DiDio, E., *J. Chem. Soc. A*, p. 1151, 1970.
114. Mentasti, E., Pelizzetti, E., and Saini, G., *J. Chem. Soc. Dalton Trans.*, p. 2609, 1973.
115. Hider, R. C., Mohd-Nor, A. R., Silver, J., Morrison, I. E. G., and Rees, L. V. G., *J. Chem. Soc. Dalton Trans.*, p. 609, 1981.
116. Hider, R. C., Bickar, D., Morrison, I. E. G., and Silver, J., *J. Am. Chem. Soc.*, 106, 6983, 1984.
117. McArdle, J. V. and Coffey, J. C., *Recl. Trav. Chim. Pays Bas*, 106, 248, 1987.
118. Jordan, R. B. and Xu, J.-H., *Pure Appl. Chem.*, 60, 1203, 1988.
119. Lentz, D. J., Henderson, G. H., and Eyring, E. M., *Mol. Pharmacol.*, 6, 514, 1973.
120. Kazmi, S. A. and McArdle, J. V., *Inorg. Biochem.*, 15, 153, 1981.
121. Biruš, M., Bradić, Z., Kujundžić, N., and Pribanić, M., *Inorg. Chim. Acta*, 56, L43, 1981.
122. Biruš, M., Bradić, Z., Kujundžić, N., and Pribanić, M., *Croat. Chem. Acta*, 56, 61, 1983.
123. Biruš, M., Bradić, Z., Kujundžić, N., and Pribanić, M., *Inorg. Chem.*, 23, 2170, 1984.
124. Tufano, T. P. and Raymond, K. N., *J. Am. Chem. Soc.*, 103, 6617, 1981.
125. Monzyk, B. and Crumbliss, A. L., *J. Inorg. Biochem.*, 19, 19, 1983.
126. Biruš, M., Bradić, Z., Kujunžić, N., and Pribanić, M., *Inorg. Chim. Acta*, 78, 87, 1983.
127. Biruš, M., Bradić, Z., Kujundžić, N., and Pribanić, M., *Acta Pharm. Jugosl.*, 32, 163, 1983.
128. Kazmi, S. A. and McArdle, J. V., *J. Inorg. Nucl. Chem.*, 43, 3031, 1981.
129. Kujundžić, N. and Pribanić, M., *J. Inorg. Nucl. Chem.*, 40, 729, 1978.
130. Biruš, M., Kujundžić, N., and Pribanić, M., *Inorg. Chim. Acta*, 55, 65, 1980.
131. Funahashi, S., Ishihara, K., and Tanaka, M., *Inorg. Chem.*, 22, 2070, 1983.
132. Biruš, M., Krznarić, G., Pribanić, M., and Uršić, S., *J. Chem. Res.(S)*, p. 4, 1985.

133. **Biruš, M., Krznarić, G., Kujundžić, N., and Pribanić, M.,** *Croat. Chem. Acta,* 61, 33, 1988.
134. **Monzyk, B. and Crumbliss, A. L.,** *J. Org. Chem.,* 45, 4670, 1980.
135. **Gould, B. L. and Langermann, N.,** *Arch. Biochem. Biophys.,* 215, 148, 1982.
136. **Emery, T.,** *Biochemistry,* 25, 4629, 1986.
137. **El-Ezaby, M. S. and Hassan, M. M.,** *Polyhedron,* 4, 429, 1985.
138. **El-Ezaby, M. S., Marafie, H. M., Hassan, M. M., and Abu Soud, H. M.,** *Inorg. Chim. Acta,* 123, 53, 1986.
139. **Shuaib, N. M., Marafie, H. M., Hassan, M. M., and El-Ezaby, M. S.,** *J. Inorg. Biochem.,* 31, 171, 1989.
140. **Hayat, L. J., Marafie, H. M., Hassan, M. M., and El-Ezaby, M. S.,** *Polyhedron,* 7, 747, 1988.
141. **Shuaib, N. M., El-Ezaby, M. S., and Al-Hussaini, O.,** *Polyhedron,* 8, 1477, 1989.
142. **Lowenberg, W., Buchanan, B. B., and Rabinowitz, J. C.,** *J. Biol. Chem.,* 236, 3899, 1963.
143. **Emery, T. and Hoffer, P. B.,** *J. Nucl. Med.,* 21, 935, 1980.
144. **Popov, V. A., Norikov, V. T., and Myshko, E.,** *Izv. Vyssh. Uchebn. Zaved. Khim. Tekhnol.,* 31(11), 18, 1988; *Chem. Abstr.,* 110, 219815t, 1989.
145. **Tufano, T. P. and Raymond, K. N.,** *J. Am. Chem. Soc.,* 103, 6617, 1981.
146. **Raymond, K. N. and Tufano, T. P.,** *The Biological Chemistry of Iron,* Dunford, H. B., Dolphin, D., Raymond, K. N., and Seiker, L., Eds., D. Reidel Publishing, Boston, 1982, 85.
147. **Akiyama, M., Katoh, A., and Mutoh, T.,** *J. Org. Chem.,* 53, 6089, 1988.
148. **Shimizu, K. and Akiyama, M.,** *J. Chem. Soc. Chem. Commun.,* p. 183, 1985.
149. **Akiyama, M., Katoh, A., and Ogawa, T.,** *J. Chem. Soc. Perkin Trans. 2,* p. 1213, 1989.
150. **Sawatzki, G.,** *Iron Transport in Microbes, Plants, and Animals,* Winkelmann, G., van der Helm, D., and Neilands, J. B., Eds., VCH Publishers, Weinheim, West Germany, 1987, chap. 25.
151. **Weinberg, E. D.,** *Physiol. Rev.,* 64, 65, 1984.
152. **Tranter, H. S. and Board, R. G.,** *J. Appl. Bacteriol.,* 55, 67, 1982b.
153. **Garibaldi, J. A.,** *Appl. Microbiol.,* 20, 558, 1970.
154. **Valenti, P., Antonini, G., Rossi Fanelli, M. R., Orsi, N., and Antonini, E.,** *Antimicrob. Agents Chemother.,* 21, 840, 1982.
155. **Valenti, P., Antonini, G., Von Humolstein, C., Visca, P., Orsi, N., and Antonini, E.,** *Int. J. Tissue React.,* 5, 97, 1983.
156. **Konopka, K., Bindereif, A., and Neilands, J. B.,** *Biochemistry,* 21, 6503, 1982.
157. **Konopka, K. and Neilands, J. B.,** *Biochemistry,* 23, 2122, 1984.
158. **Neilands, J. B., Konopka, K., Schwyn, B., Coy, M., Francis, R. T., Paw, B. H., and Bagg, A.,** *Iron Transport in Microbes, Plants, and Animals,* Winkelmann, G., van der Helm, D., and Neilands, J. B., Eds., VCH Publishers, Weinheim, West Germany, 1987, chap. 1.
159. **Carrano, C. J. and Raymond, K. N.,** *J. Am. Chem. Soc.,* 101, 5401, 1979.
160. **Pecoraro, V. L., Weitl, F. L., and Raymond, K. N.,** *J. Am. Chem. Soc.,* 103, 5133, 1981.
161. **Weitl, F. L., Raymond, K. N., and Durbin, P. W.,** *J. Med. Chem.,* 24, 203, 1981.
162. **Raymond, K. N. and Chung, T. D. Y.,** *Inorg. Chim. Acta,* 79, 62, 1983.
163. **Kretchmar, S. A. and Raymond, K. N.,** *Inorg. Chem.,* 27, 1436, 1988.
164. **Cooper, S. R., McArdle, J. V., and Raymond, K. N.,** *Proc. Natl. Acad. Sci. U.S.A.,* 75, 3551, 1978.
165. **Wawrousek, E. F. and McArdle, J. V.,** *J. Inorg. Biochem.,* 17, 169, 1982.
166. **Lee, C. W., Ecker, D. J., and Raymond, K. N.,** *J. Am. Chem. Soc.,* 107, 6920, 1985.
167. **Pecoraro, V. L., Harris, W. R., Wong, G. B., Carrano, C. J., and Raymond, K. N.,** *J. Am. Chem. Soc.,* 105, 4623, 1983.
168. **O'Brien, I. G., Cox, G. B., and Gibson, F.,** *Biochem. Biophys. Acta,* 237, 537, 1971.
169. **Robinson, J. P. and McArdle, J. V.,** *J. Inorg. Nucl. Chem.,* 43, 1951, 1981.
170. **Straka, J. B. and Emery, T.,** *Biochim. Biophys. Acta,* 569, 277, 1979.
171. **Ernst, J. F. and Winkelmann, G. F.,** *Biochim. Biophys. Acta,* 500, 27, 1977.
172. **Arceneaux, J. E. L. and Byers, B. R.,** *J. Bacteriol.,* 141, 715, 1980.
173. **McCreary, K. A. and Ratledge, C.,** *J. Gen. Microbiol.,* 113, 67, 1979.
174. **Emery, T.,** *Biochim. Biophys. Acta,* 363, 219, 1974.
175. **Emery, T.,** *Iron Transport in Microbes, Plants and Animals,* Winkelmann, G., van der Helm, D., and Neilands, J. B., Eds., VCH Publishers, Weinheim, West Germany, 1987, chap. 14.
176. **Arceneaux, J. E. L., Davis, W. B., Downer, D. N., Hayden, A. H., and Byers, B. R.,** *J. Bacteriol.,* 115, 919, 1973.
177. **Emery, T.,** *Biochemistry,* 10, 1483, 1971.
178. **Ecker, D. J., Passavant, C. W., and Emery, T.,** *Biochim. Biophys. Acta,* 720, 242, 1982.
179. **Winkelmann, G.,** *FEBS Lett.,* 97, 43, 1979.
180. **Ecker, D. J., Lancaster, J. R., Jr., and Emery, T.,** *J. Biol. Chem.,* 257, 8623, 1982.
181. **Carrano, C. J. and Raymond, K. N.,** *J. Bacteriol.,* 136, 69, 1978.
182. **Langman, L., Young, I. G., Trost, G., Rosenberg, H., and Gibson, F.,** *J. Bacteriol.,* 112, 1142, 1972.

183. Heidinger, S., Braun, V., Pecoraro, V. L., and Raymond, K. N., *J. Bacteriol.*, 153, 109, 1983.
184. Collins, D. J., Lewis, C., and Swan, J. M., *Aust. J. Chem.*, 27, 2593, 1974.
185. Hollifield, W. C., Jr. and Neilands, J. B., *Biochemistry*, 17, 1922, 1978.
186. Lodge, G. S., Gaines, C. G., Arcenaux, J. E. L., and Byers, B. R., *Biochem. Biophys. Res. Commun.*, 97, 1291, 1980.
187. Cass, M. E., Garrett, T. M., and Raymond, K. N., *J. Am. Chem. Soc.*, 111, 1677, 1989.
188. Hider, R. C., Silver, J., Neilands, J. B., Morrison, I. E. G., and Rees, L. V. G., *FEBS Lett.*, 102, 325, 1979.
189. Orlando, J. A. and Neilands, J. B., *Chemistry and Biology of Hydroxamic Acids*, Kehl, H., Ed., S. Karger, New York, 1982, 123.
190. Miller, G. W., Pushnik, J. C., Brown, J. C., Emery, T. F., Jolley, V. D., and Warnick, K. Y., *J. Plant Nutr.*, 8, 249, 1985.
191. Wallace, A., *Current Topics in Plant Nutrition*, Edwards Brothers, Ann Arbor, MI, 1986.
192. Suqiura, Y. and Nomoto, K., *Structure and Bonding*, Vol. 58, Chimrak, A., Ed., Springer-Verlag, New York, 1984, 107.
193. Nomoto, K., Suqiura, Y., and Takagi, S., *Iron Transport in Microbes, Plants, and Animals*, Winkelmann, G., van der Helm, D., and Neilands, J. B., Eds., VCH Publishers, Weinheim, West Germany, 1987, chap. 22.
194. Mino, Y., Ishida, T., Ota, N., Inoue, M., Nomoto, K., Takemoto, T., Tanaka, H., and Sugiura, Y., *J. Am. Chem. Soc.*, 105, 4671, 1983.
195. Olsen, R. A. and Brown, J. C., *J. Plant Nutr.*, 2, 629, 1980.
196. Römheld, V., *Iron Transport in Microbes, Plants, and Animals*, Winkelmann, G., van der Helm, D., and Neilands, J. B., Eds., VCH Publishers, Weinheim, West Germany, 1987, chap. 19.
197. Bienfait, F., *Iron Transport in Microbes, Plants, and Animals*, Winkelmann, G., van der Helm, D., and Neilands, J. B., Eds., VCH Publishers, Weinheim, West Germany, 1987, chap. 18.
198. Castignetti, D. and Smarrelli, J., Jr., *Biochem. Biophys. Res. Commun.*, 125, 52, 1984.
199. Ghosh, P. and Chakravorty, A., *Inorg. Chim. Acta*, 56, L77, 1981.
200. Escot, M.-T., Martre, A.-M., Pouillen, P., and Martinet, P., *Bull. Soc. Chim. Fr.*, No. 3, 316, 1989.
201. Kazmi, S. A., Shorter, A. L., and McArdle, J. V., *Inorg. Chem.*, 23, 4332, 1984.
202. Kazmi, S. A., Shorter, A. L., and McArdle, J. V., *J. Inorg. Biochem.*, 17, 269, 1982.
203. Kazmi, S. A., Shorter, A. L., and McArdle, J. V., *New Trends in Natural Products Chemistry 1986: Studies in Organic Chemistry*, Vol. 26, Rahman, A. and LeQuesne, P. W., Eds., Elsevier, Amsterdam, 1986, 185.
204. Kristijánsson, I., Nielson, F. H., and Ulstrup, J., *Inorg. Chem.*, 23, 3674, 1984.

ISOLATION AND SPECTROSCOPIC IDENTIFICATION OF FUNGAL SIDEROPHORES

Mahbubul A. F. Jalal and Dick van der Helm

INTRODUCTION

Siderophores are a group of natural compounds, produced by aerobic microorganisms to acquire and transport ferric iron under iron-stress conditions. The iron(III) chelating groups in siderophores are commonly of two types, hydroxamate and catecholate. Both of these groups are highly specific for ferric iron. Occasionally other less specific chelating groups are also involved.

A large number of siderophores belonging to various structural families have been reported from fungi in the last two decades.[1-3] A common feature of these eukaryotic siderophores is that they are hydroxamate-type compounds, based on the amino acid, ornithine. The hydroxamate functionality originates in these siderophores through subsequent hydroxylation and acylation of the δ-amino group of ornithine.[3,4] The N^δ-acyl groups are derived from various carboxylic acids. Out of nine different acyl groups found so far, only three (R = a, b, and f, Scheme 1) are most commonly observed. The acylated hydroxyornithine units are joined with each other in various fashions to form the dihydroxamate (dimeric) and the trihydroxamate (trimeric) siderophores. Siderophores are generally made of L-ornithine, although D-ornithine has been found in one case.[5]

The families of fungal siderophores differ in the mode of conjugation of the monomeric hydroxyacyl-ornithine units. Almost all of the fungal siderophores fall in one of the following three families; (1) the ferrichromes, (2) the coprogens, and (3) the fusarinines. The ferrichromes (1 to 21)[6-20] represent a group of siderophores in which the three hydroxyacylornithine units form a tripeptide. This tripeptide makes a larger hexapeptide ring in combination with three other neutral amino acids, one of which is always glycine. This glycine occupies a specific position adjacent to the amino end of the triornithyl region. The other two amino acids in the hexapeptide ring may be a combination of glycine, serine, or alanine. An exceptional member of this family tetraglycylferrichrome (22), contains a cyclic heptapeptide ring with an additional glycine residue.[21] In another siderophore, desdiserylglycylferrirhodin (DDF), (23), the hexapeptide ring and the three nonornithine amino acids are absent and the ornithine residues form a linear tripeptide.[22] Seven (R = a-g, Scheme 1) out of the nine N^δ-acyl residues have been found in the ferrichrome siderophores.

In the dihydroxamate members of the coprogen family, two of the acylated hydroxyornithine units join to form a cyclic dipeptide with a diketopiperazine ring in the middle (24, 25).[23,24] In the trihydroxamate members (26 to 36), another unit of the monomer is linked to the dimer via an ester bond. Only three different acyl groups (R = a, b, and h, Scheme 1) have been found so far in this family. In most trihydroxamate coprogens, the N^α-amino group is acetylated.[25-30]

In the case of fusarinines (37 to 45), the hydroxyacylornithine units are joined head-to-tail via ester bonds to form a linear dimer and linear and cyclic trimers. Only two different kinds of N^δ-acyl groups R = f and i, Scheme 1) have been reported from this family. In some members of this family, the N^α-amino group is acetylated.[5,31-35]

Ornithine-based hydroxamate siderophores have not been isolated from any prokaryotic microorganisms so far. Bacteria and actinomycetes also produce hydroxamate siderophores, such as ferrioxamines and schizokinens, but these compounds are usually made of hydroxylated and acylated alkylamines.[36-38] A number of the prokaryotic siderophores are of the catecholate type. Many of these compounds, on the other hand, contain a combination of

	R_1	R_2	R_3	R_4	R_5
(1) Ferrichrome	H	H	CH_3	CH_3	CH_3
(2) Ferrichrome A	CH_2OH	CH_2OH	d	d	d
(3) Ferrichrome C	H	CH_3	CH_3	CH_3	CH_3
(4) Sake Colorant A	CH_2OH	CH_3	CH_3	CH_3	CH_3
(5) Ferricrocin	H	CH_2OH	CH_3	CH_3	CH_3
(6) Ferrichrysin	CH_2OH	CH_2OH	CH_3	CH_3	CH_3
(7) Malonichrome	H*	CH_3*	e	e	e
(8) Ferrirubin	CH_2OH	CH_2OH	b	b	b
(9) Ferrirhodin	CH_2OH	CH_2OH	f	f	f
(10) Asperchrome A	CH_2OH	CH_3	b	b	b
(11) Asperchrome B1	CH_2OH	CH_2OH	CH_3	b	b
(12) Asperchrome B2	CH_2OH	CH_2OH	b	CH_3*	b*
(13) Asperchrome B3	CH_2OH	CH_2OH	b	b*	CH_3*
(14) Asperchrome C	CH_2OH	CH_2OH	c	b	b
(15) Asperchrome D1	CH_2OH	CH_2OH	b	CH_3	CH_3
(16) Asperchrome D2	CH_2OH	CH_2OH	CH_3	b*	CH_3*
(17) Asperchrome D3	CH_2OH	CH_2OH	CH_3	CH_3*	b*
(18) Asperchrome E	CH_2OH	CH_2OH	f*	b*	b*
(19) Asperchrome F1	CH_2OH	CH_2OH	g*	b*	b*
(20) Asperchrome F2	CH_2OH	CH_2OH	b*	g*	b*
(21) Asperchrome F3	CH_2OH	CH_2OH	b*	b*	g*
(22) Tetraglycylferrichrome **	H	H	CH_3	CH_3	CH_3

(23) Des-diserylglycylferrirhodin (DDF)

(24) Rhodotorulic acid R = CH_3
(25) Dimerum acid R = b

*   Position not confirmed.
**   An additional glycyl residue present in the peptide ring.

	R_1	R_2	R_3	R_4
(26) Coprogen	H	$-COCH_3$	b	b
(27) Coprogen B	H	H	b	b
(28) Neocoprogen I	H	$-COCH_3$	CH_3	b
(29) Isoneocoprogen I	H	$-COCH_3$	b	CH_3
(30) Neocoprogen II	H	$-COCH_3$	CH_3	CH_3
(31) N^α-Dimethylcoprogen	CH_3	CH_3	b	b
(32) N^α-Dimethylneocoprogen I	CH_3	CH_3	CH_3	b
(33) N^α-Dimethylisoneocoprogen I	CH_3	CH_3	b	CH_3
(34) Hydroxycoprogen	H	$-COCH_3$	b	h
(35) Hydroxyneocoprogen I	H	$-COCH_3$	CH_3	h
(36) Hydroxyisoneocoprogen I	H	$-COCH_3$	h	CH_3

	R	n
(37) cis-Fusarinine	H	1
(38) N^α-Acetyl-cis-fusarinine	$-COCH_3$	1
(39) Fusarinine A	H	2
(40) N^α-Diacetylfusarinine A	$COCH_3$	2
(41) Fusarinine B	H	3
(42) N^α-Triacetylfusarine B	$COCH_3$	3

(43) Fusarinine C (fusigen)   R = H
(44) N^α-Triacetylfusarinine C   R = $-COCH_3$
(45) Neurosporin***   R = $-COCH_3$

*** N^δ-Acyl group = i, (Scheme 1) instead of f and D-ornithine instead of L-ornithine.

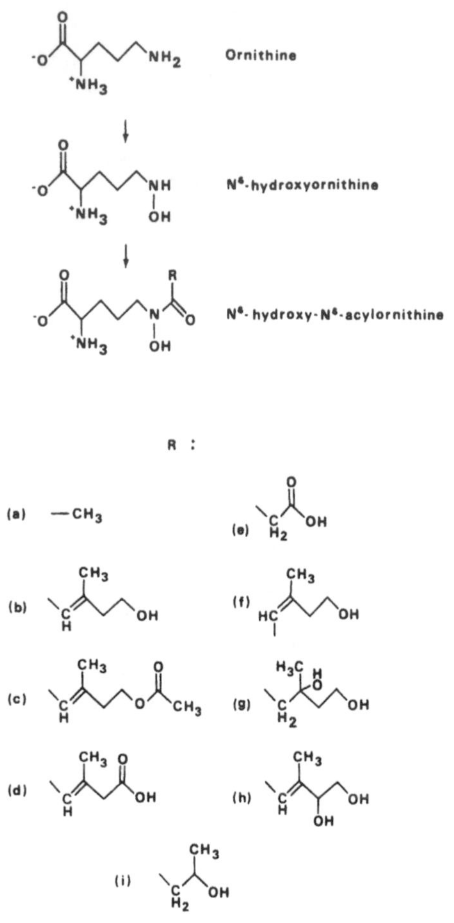

SCHEME 1.   Origin of the basic unit of fungal siderophores, N^δ-hydroxy-N^δ-acylornithine. (a to i) The N^δ-acyl groups discovered so far in various fungal siderophores.

hydroxamate, catecholate, *o*-hydroxyphenyloxazolone, *o*-hydroxyphenylthiazoline, or α-hydroxycarboxylic acid functionalities as their three Fe(III)-chelating groups.[2,39,40]

Representative members of each of the three families of fungal siderophores have been crystallized and their complete structures and conformations have been determined by X-ray diffraction (Figures 1 to 3).[2] However, a large number of these siderophores have not yielded crystals. Alternative structure elucidation methods, such as UV/visible, mass, and NMR spectroscopy and chemical degradation, have been very useful in determining their structures. A comprehensive review of the X-ray structures of siderophores has been published recently.[2] In the present article, a discussion of the isolation of fungal siderophores and their structure elucidation by spectroscopic means is presented.

EXTRACTION FROM CULTURE BROTH

Siderophores are produced inside the fungal cells under iron-deficient conditions, and then excreted out into the medium. Sometimes the concentration of the accummulated siderophores in the medium exceeds 1 g/l.[18] In a few cases, fungi produce a single type of siderophore, e.g., *Rhodotorula rubra* produces only rhodotorulic acid. The majority of fungi, however, produce a large number of siderophores, some of which may belong to different

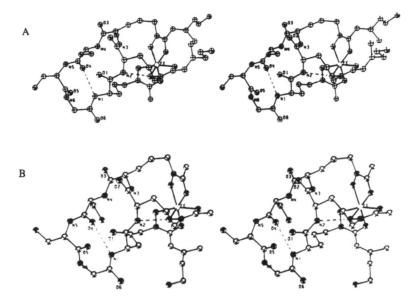

FIGURE 1. X-ray structures (stereo view) of asperchrome B1 (A) and asperchrome D1 (B), two ferrichrome-type siderophores. (Reproduced from Jalal, M. A. F., Hossain, M. B., van der Helm, D., and Barnes, C. L., *Biol. Metals*, 1, 77, 1988. With permission of Springer Verlag.)

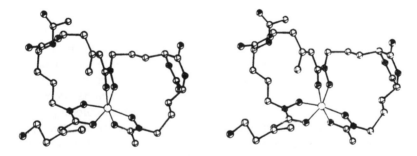

FIGURE 2. X-ray structure (stereo view) of neocoprogen I. (Reproduced from van der Helm, D., Jalal, M. A. F., and Hossain, M. B., *Iron Transport in Microbes, Plants and Animals*, Winkelmann, G., van der Helm, D., and Neilands, J. B., Eds., VCH Verlagsgessellschaft, Weinheim, West Germany, 1987, 135. With permission of VCH Verlagsgesellschaft.)

families.[15,30,41-44] Efficient extraction and separation methods are needed to obtain the compounds in pure form, which is an essential prerequisite for the structure elucidation work.

The siderophores vary widely in their physical and chemical properties. This is why a single method of extraction may not be suitable for all of the siderophores produced by fungi, or even by a single fungus. The efficiency of the extraction procedure should be checked in microscale before the method is employed for large-scale use.

The ferric complexes of the siderophores are usually more stable and easy to follow (for their red color), compared to their corresponding deferri compounds. After removal of mycelia by centrifugation or filtration, the first step in the extraction process is usually the formation of the ferric complexes of the siderophores present in the culture filtrate, by addition of 2% aqueous $FeCl_3$ solution. Occasionally, precipitate is formed during this step, which may be removed by centrifugation. The clear red supernatant is then subjected to one of the following extraction procedures.

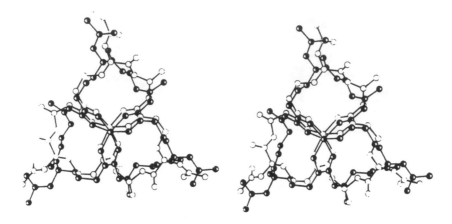

FIGURE 3. X-ray structures (stereo view, superimposed) of two members of the fusarinine family, triacetylfusarinine C and neurosporin (open circle), separated from each other by a translation of 0.5 Å. (Reproduced from van der Helm, D., Jalal, M. A. F., and Hossain, M. B., *Iron Transport in Microbes, Plants and Animals*, Winkelmann, G., van der Helm, D., and Neilands, J. B., Eds., VCH Verlagsgessellschaft, Weinheim, West Germany, 1987, 135. With permission of VCH Verlagsgessellschaft.)

THE CHLOROFORM-PHENOL-ETHER-WATER EXTRACTION

The aqueous supernatant is extracted with chloroform-phenol (1:1) solution.[18,45] (This solution is usually made by dissolving 500 g of pure solid phenol with 500 ml of chloroform.) The two phases are allowed to settle and then the lower organic phase with extracted siderophores is taken out from the separating funnel. Repetitions of this step are usually needed to achieve complete extraction of siderophores. Phase separation may be a problem if the density of the culture filtrate is high, but can be achieved by centrifugation in a bench centrifuge. The chloroform-phenol extract is then diluted with four to six volumes of diethyl ether and extracted three or four times with a small amount of water until the siderophores transfer completely to the aqueous phase. The aqueous extract is then partitioned four times with diethyl ether to remove all traces of phenol. Sometimes the chloroform-phenol mixture is replaced by benzyl alcohol in the initial extraction step. However, the density of benzyl alcohol is closer to that of the aqueous medium and the phase separation problem is more severe. To overcome this problem, ammonium sulfate is added to the medium (500 g/l) prior to extraction with benzyl alcohol.[46] On the other hand, a chloroform-phenol solution is a better solvent of siderophores compared to benzyl alcohol and the transfer of siderophore to the organic phase is easier with this solvent mixture.

THE XAD-2 EXTRACTION

This extraction procedure is much simpler and easier than the chloroform-phenol-ether-water extraction method. However, the siderophore should have enough lipophilic character to be able to bind with the resin XAD-2 and should be soluble in an organic solvent such as methanol or ethanol. Most fungal siderophores have these properties and are amenable to this extraction procedure.[29,30,47] A column is prepared with XAD-2, which is prewashed with water, methanol, and water again. The aqueous culture supernatant containing the ferric siderophores is slowly passed through the column. The ferric siderophores bind to the top of the column and a pale-colored supernatant elutes out. The column is then washed with five to ten bed volumes of water to remove all unbound components of the medium. Solvent is then changed to methanol, which elutes out the ferric siderophores as a single band. A small fraction of siderophores may adsorb strongly to the resin and take longer time to elute out. When the column is left to equilibrate with methanol, all siderophores dissolve in the solvent and a complete recovery of siderophores is possible.

The two extraction methods described above may not be suitable for certain siderophores. A case, in hand, is the major siderophores of *Gliocladium virens*. This fungus produces two monohydroxamates, cis and trans fusarinine, and a dihydroxamate, dimerum acid, along with the minor amounts of some trihydroxamates.[41] The monohydroxamates and the dihydroxamate compounds form a large number of homogeneous and mixed ferric complexes after $FeCl_3$ is added to the culture filtrate. When the chloroform-phenol extraction is tried with this culture supernatant, a large amount of the ferric siderophores remains in the aqueous phase even after repeated trials. The ferric complexes of these siderophores are very polar and those of the monohydroxamates are even insoluble in methanol. For this reason, the XAD-2 method of extraction cannot be applied successfully with these ferric siderophores.

An alternative method of extraction may be used for these polar compounds. First, the medium containing the siderophores in their deferri form is evaporated *in vacuo* into an oily gum, which is dissolved in methanol. The precipitate formed is filtered out and the methanol extract is evaporated again. The residue is redissolved in methanol and some more precipitate is formed, which is filtered out.[31] The procedure is repeated once or twice more, until no more precipitate forms in the methanolic extract. This extract is finally evaporated to dryness and redissolved in water. Prewashed XAD-2 resin is added to this aqueous extract and slowly shaken overnight. The suspension is then transferred into a column and water is drained out. Washing with water removes the monohydroxamate compounds from the resin, while methanol elutes out the dihydroxamate and the trihydroxamate siderophores.[41]

CHROMATOGRAPHIC ISOLATION

INITIAL GEL FILTRATION IN BIOGEL P2

The aqueous extract obtained from the chloroform-phenol extraction usually contains the ferric siderophores mixed with a large amount of colorless impurities. This crude extract does not even produce clean spots in silica gel thin layer chromatography as the streak produced by the impurity often masks or distorts the spots. Quality of XAD-2 extract is much better and usually good thin layer separation is obtained with these extracts. A very efficient procedure used to remove the nonsiderophore impurities from these crude extracts, especially from the one obtained through the chloroform-phenol extractions, is by gel filtration with Biogel P2. The swollen gel is washed, degassed, and packed in a column with water. The concentrated ferric siderophore solution in minimum water is loaded onto the column and eluted with water. Siderophores usually elute out as a single band, although in a few cases they produce more than one band.

Gel filtration through Biogel P2 is not suitable for the ferric complexes of some monohydroxamates, as they move extremely slowly and get deferriated on the column.

THIN LAYER CHROMATOGRAPHY

After the gel filtration in Biogel P2, the extract is suitable for analysis by silica gel thin layer chromatography, which gives some valuable information on the kinds and possible identities of siderophores present in the extract, their approximate ratios, their polarity, and their mobility in organic solvent systems that may be employed for preparative column chromatography for large-scale separation. Thin layer chromatography is also necessary to follow the purification steps and to analyze the composition of the separated fractions obtained from the preparative column chromatography.

A very useful thin layer chromatographic system for siderophores is chloroform-methanol-water, 35:12:2, using silica gel thin layers. A large number of compounds separate in this system and produce discrete spots.[9,18,28-30,48,49] The solvent mixture is also suitable for preparative silica gel column chromatography. Butanol-acetic acid-water (4:1:5, upper phase) is also a generally used system for silica gel layers. Some other less frequently used solvent systems are chloroform-methanol-water-acetic acid, 65:24:4:3; *n*-butanol-acetic acid-water, 4:1:1; chloroform-benzyl alcohol-methanol, 2:1:1; 1-propanol-acetic acid-water, 4:1:1; water-saturated *n*-butanol and *n*-butanol-ethanol-water, 4:1:5.[49-51] Various ratios of chloroform and methanol with or without hexane are also useful to investigate the homogeneity of a siderophore sample. Freshly prepared solvents should be used, as the R_f values in these solvent systems depend on when the solvents have been prepared.

No detection reagent is necessary for ferric siderophores, because of their obvious red color on chromatograms. However, when thin layer chromatography is carried out on deferri siderophores or their breakdown products, detection of the spot by some means is necessary. Silica gel plates made with fluorescent dyes (e.g., Whatman KGF or Merck F254 plates) allow for easy nondestructive detection of the spots, which are almost always absorbent purple against a fluorescent green background under UV light. Deferri siderophores may be readily revealed by $FeCl_3$ spray (1% $FeCl_3$ in 0.001 *M* HCl). Shades of color produced by this spray may even distinguish between the monohydroxamates, the dihydroxamates, and the trihydroxamate compounds.

Preparative thin layer chromatography on UV fluorescent silica gel plates is sometimes useful in separating siderophore mixtures. Chloroform-methanol-water, 35:12:2, is again the best solvent used for this purpose. After development the separated bands (detected under UV) are scraped off the plates. The compounds may be eluted with methanol. The eluted siderophore solution may contain impurities from the silica gel support material and should be purified further by gel filtration (see Chromatographic Isolation — Initial Gel Filtration in Biogel P2) before spectroscopy.

ELECTROPHORESIS

Paper electrophoresis is an important tool employed for the analysis of a siderophore extract or to determine the nature of charge present or absent in a ferric siderophore. Routine electrophoresis is carried out on Whatman 3M paper with 1000 V field strength for 1 h using pH 5.0 (pyridine-acetic acid-water, 14:10:930) and pH 2.0 (4% formic acid).[31,35] The neutral ferric siderophores do not show significant mobility in these pH values. Compounds having a free or alkylated amino group (e.g., coprogen B and N^α-dimethylcoprogens) bear a positive charge at these pH values and move toward the negative electrode.[29] The compound which contains free carboxyl groups (e.g., ferrichrome A) is negatively charged at pH 5.0 and moves toward the positive electrode. At pH 2.0, this compound loses its charge and behaves as a neutral compound. A ferric siderophore such as DDF contains a free amino group as well as a free carboxyl group, and this compound can be distinguished from the other types of compounds on the basis of its mobility. At pH 5.0, its net charge is zero and the compound behaves as a neutral molecular species. At pH 2.0, on the other hand, the net charge is 1+, and the compound moves toward the negative electrode.[22]

Electrophoresis at pH 2.0 provides additional information on the Fe(III) binding nature of the siderophore. The ferric monohydroxamate and the dihydroxamate complexes usually form deep-purple and pink-purple colors, respectively, while the ferric trihydroxamates remain red.

The actual rate of mobility of a ferric siderophore at a certain pH depends on a number of factors including the net charge present in the molecule and the molecular weight of the compound. The monomer, dimer, linear trimer, and the cyclic trimer of various fusarinines produced by the red *Fusarium* sp. differ in their net charge to molecular weight ratio at pH 5.0. The compounds separate from one another during electrophoresis at this pH.[32] The information provided by electrophoresis helps in planning the purification procedure to be used for that extract. The charged ferric siderophores may be separated from one another and from neutral species by ion-exchange chromatography employing Sephadex or cellulose ion exchangers.

ION-EXCHANGE CHROMATOGRAPHY

Compounds, which are found to be ionic in the initial paper electrophoretic analysis, may be separated from one another (or from other neutral components of the mixture) by ion-exchange chromatography. Choice of support material depends primarily on the type of charge present in the compounds. The Sephadex ion exchangers provide good separation of the range of charged siderophores. CM-Sephadex (C-25) cation exchangers may be used for the positively charged siderophores, such as coprogen B, N^α-dimethyl-coprogens, and fusarinines, while DEAE-Sephadex (A-25) is useful for anionic siderophores such as ferri-chrome A (2) and malonichrome (7). CM-cellulose and DEAE-cellulose may also be used instead of the Sephadex ion exchangers.[9,15]

The support material is swelled in water, washed thoroughly, and degassed before loading into a column of a suitable size. Deionized water is a good mobile phase, although one may choose to use a buffer system to avoid the fluctuation of pH during operation. The choice of the buffer system and its precise pH depends on the type of compounds to be separated. Phosphate buffer pH 6.8 (0.01 to 0.1 M) may be used as a general purpose buffer.

The siderophore mixture is applied to the column as a solution in the mobile phase solvent. After the solution enters into the column, it may be washed with four to six bed volumes of water (or buffer) to elute the unbound impurities as well as the neutral and the oppositely charged siderophores (cationic siderophores in an anion exchange column and vice versa), if any are present.

Elution of bound siderophore from the ion-exchange column may be carried out with a salt gradient (e.g., 0 to 1 M NH$_4$Cl or NaCl). The gradient may be fine tuned to accomplish maximum resolution of the separated bands, by using a narrow range of salt concentration, by adding an isocratic step in the gradient, or by changing the mixing rate in the gradient.

Purity of the eluted bands may be checked by paper electrophoresis or thin layer chromatography after removal of salts used in the gradient and buffer. Salt may be removed by either gel filtration (see Chromatographic Isolation — Initial Gel Filtration in Biogel P2) or solvent extraction (see Extraction From Culture Broth).

PREPARATIVE SILICA GEL COLUMN CHROMATOGRAPHY

The siderophore extract obtained after initial purification from the Biogel P2 column or ion-exchange column may be separated in a preparative silica gel column. The solvent of choice is usually chloroform-methanol-water, 35:12:2.[9,48] Best separation is obtained when thin layer chromatography-grade silica gel (E. Merck) is used as the support material instead of the usual column chromatography-grade silica gel (230-400 mesh). Individual bands coming out of the silica gel column do not always contain a single compound. Often they contain mixtures of a small number of siderophores. All three isomeric asperchromes (e.g., asperchrome B1-B3 or asperchrome D1-D3 produced by *Aspergillus ochraceous*) move as single bands in this system. When the siderophore extract of *Curvularia lunata* is subjected to this chromatographic system, neocoprogen II moves as a band ahead of another band which is composed of neocoprogen I and coprogen.[28] When a more polar, slower moving compound is to be separated from the other faster moving siderophores, it is better to use the coarser (column chromatography)-grade silica gel. Using this adsorbent, desdiserylgly-cylferrirhodin, DDF, is easily separated from all other (more than a dozen) siderophores of *A. ochraceous*.

Overlapping of bands is not unusual in this type of chromatography. However, the overlapped fractions may be rechromatographed in a similar column to obtain complete separation of the bands. In all types of chromatography, the quality of separation depends on a number of factors, such as the particle size of the support material, the flow rate of the solvent, the length and the diameter of the column, etc.

Repetition of silica gel column chromatography using a different solvent system (e.g., chloroform-methanol, 3:1, or chloroform-hexane-methanol, 1:1:1) sometimes removes the minor impurities from a band and yields the ferric siderophore in pure form. In other cases, reversed-phase chromatography has been found very useful for separation of individual compounds, which move together in a single band in the silica gel column.

REVERSED-PHASE CHROMATOGRAPHY

Low pressure preparative reversed-phase chromatography using C_{18}-silica gel (e.g., Whatman LRP1) and a water-methanol gradient (0 to 40% methanol) separates individual compounds from various ferric siderophore mixtures.[48] Best results are obtained when this system is used for the siderophore bands (separated in silica gel columns) containing a few siderophores. The most hydrophilic ferric siderophore is usually eluted first and the most lipophilic one comes out last from the column. The separation of some structural isomers (e.g., asperchrome F1, F2, and F3) (*vide infra*) is also achieved in this system. The order of elution of ferric siderophores from a C_{18} reversed-phase column (with the water-methanol gradient) has been found to be (approximately) inversely related to their mobility on silica gel thin layers developed in butanol-acetic acid-water (4:1:5, upper phase).

Most of the neutral siderophores form discrete bands in this reversed-phase system. However, some ionic siderophores such as coprogen B and N^α-dimethylcoprogens isolated from *Alternaria longipes* and *F. dimerum* produce elongated slow-moving bands. The ferric complexes of dihydroxamates (e.g., dimerum acid) and monohydroxamates (e.g., cis and trans fusarinines) should not be loaded in this type of column, because deferration occurs in the column, making the column discolored and the recovery of the compounds difficult. The purified siderophore obtained from the reversed-phase column may be finally passed through a Biogel P2 column before lyophilization.

An efficient and fast analytical reversed-phase HPLC separation system for the fungal siderophores has been described recently.[49] The system has separated 16 fungal siderophores including ferrichromes, coprogens, and triacetyl fusarinine C using C_{18} (Nucleosil) and C_8 (Spherisorb) reversed-phase material with a water-acetonitrile gradient (Figure 4). The method is useful in quick analysis of siderophore production in fungi.

CHEMICAL DEGRADATION

Partial chemical degradation of purified siderophores offers a range of valuable information essential for the structure elucidation work. Out of an array of various degradation methods, three are discussed below for their general usefulness and applicability. These chemical degradations should preferably be done on deferrisiderophores, as the Fe(III) atom in the molecule may interfere with the reaction process.

REDUCTIVE HYDROLYSIS WITH HI

This method has been developed for siderophores by Emery and Neilands[6] to produce and quantitate ornithine. The method has been used successfully for ferrichromes,[6,22] fusarinines,[31,32] and coprogens.[28] The siderophore sample (3 to 5 mg) is dissolved in 0.5 ml of 50% HI. The tube containing the solution is sealed and heated at 110°C for 24 h. After cooling, the tube is opened and the contents are dried *in vacuo* over NaOH pellets. The residue is dissolved in 0.5 ml water, followed by the addition of 0.2 ml of 0.02 N sodium thiosulfate. The solution is then transferred to a volumetric flask and diluted to 25 ml with water. Identity of ornithine may be established by chromatography with an authentic sample and its amount may be determined by the spectrophotometric method of Chinard.[52]

ACID HYDROLYSIS WITH 6 N HCL

Hydrolysis of the hydroxamic acid group is carried out with 6 N HCl to release $^\delta N$-hydroxyornithine and the acylating acids.[6,30,53] To 1 ml of an aqueous solution of the

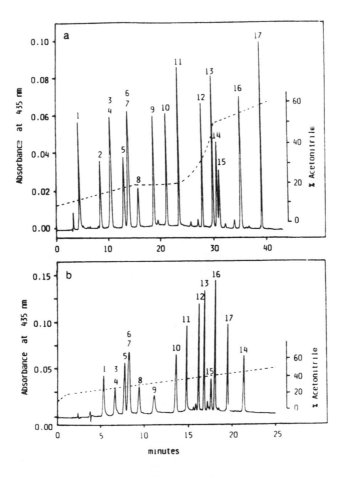

FIGURE 4. Separation of standard ferric siderophore mixtures by
HPLC. (a) Nucleosil C_{18}, (b) Spherisorb C_8 using a gradient of 10 mM
phosphate buffer, pH 3, and acetonitrile. Flow rate: 1 ml/min. Sample
volume, 30 μl. (1) Coprogen B; (2) ferrioxamine B; (3) (Gly)4-fer-
richrome; (4) neocoprogen II; (5) ferricrocin; (6) ferrichrysin; (7) ferrichrome; (8) ferrichrome C; (9) neocoprogen I; (10) asperchrome D_1;
(11) coprogen; (12) asperchrome B_1; (13) ferrirubin; (14) DDF; (15)
ferrichrome A; (16) ferrirhodin; (17) triacetylfusarinine C. (Reproduced
from Konetschny-Rapp, S., Huschka, H.-G., Winkelmann, E., and
Jung, G., *Biol. Metals*, 1, 9, 1988. With permission of Springer Verlag.)

siderophore, 1 ml of 12 N HCl is added. The tube is sealed and heated at 110°C for 24 h.
When cooled, the tube is opened and the contents dried over NaOH *in vacuo*. The residue
is dissolved in water or methanol and analyzed for the hydrolysis products.

Detection of δN-Hydroxyornithine

The hydrolyzed residue is dissolved in 100 μl of water and 10-μl aliquots are analyzed
by electrophoresis on Whatman No. 1 paper with pyridine-acetic acid-citric acid-water,
40:30:12:930, pH 4.9, as the buffer system. The dried paper shows δN-hydroxyornithine as
a ninhydrin-positive spot with a cathodic migration of about 11 cm at 35 V/cm for 1 h.[6] It
also gives a red color with tetrazolium reagent.[54] Another ninhydrin-positive spot near the
origin corresponding to the amino acids, glycine and serine, is observed, when a ferrichrome-
type siderophore is hydrolyzed.

Detection of ⁸N-Acylating Acids

6 *N* HCl hydrolysis also releases the ⁸N-acylating acid from the siderophore. When hydrolysis is carried out in aqueous solution, it releases the free acid. However, when a methanolic solution of the siderophore is added to HCl during the procedure described above, the ⁸N-acylating acids are released in their methyl ester forms. The acylating acids or their methyl esters may be extracted from the aqueous solution of the hydrolysate by ether or ethyl acetate. Further purification may be achieved with preparative thin layer chromatography on silica gel with chloroform-methanol-water, 35:12:2. The methyl ester forms of the acylating acids may be eluted from the scraped-off silica gel with chloroform and used for NMR and mass spectroscopy without further purification. Using this procedure, the methyl esters of trans-4-hydroxy-3-methyl-2-pentenoic acid (trans-anhydromevalonic acid) and trans-4,5-dihyroxy-3-methyl-2-pentenoic acid, produced by the acid hydrolysis of a methanolic solution of hydroxycoprogen, have been separated and purified.[30] Methanol may be needed to elute the free forms of the ⁸N-acylating acids from the silica gel support material. The eluted solution is likely to contain impurity from silica gel and may be purified further by some purification methods such as gel filtration. Lipophilic Sephadex, LH20, and ethanol offer a good gel filtration system for such compounds.

DEGRADATION WITH AMMONIACAL METHANOL

This method is especially useful for locating the ⁸N-acyl moieties in isomeric coprogens. The procedure is mild and selective. It breaks the ester bond between the middle ⁸N-acyl group and the terminal ornithine moiety in trihydroxamate coprogens. As a result of the reaction, the dimer containing the diketopiperazine ring is released free. The ester-linked ⁸N-acylornithine moiety is released in the methyl ester form. This last fragment may also be present in its free carboxylic acid form. Purification and subsequent spectroscopy of the two parts of the compound unequivocally establish the constitution of their acyl groups.[26,27,29,30]

The degradation procedure is carried out by adding 10 ml of concentrated NH₄OH to 10 ml of the methanolic solution of the siderophore (5 to 20 mg). Refluxing the mixture drives the reaction faster. A small aliquot of the solution may be withdrawn at regular intervals and chromatographed on silica gel thin layers with chloroform-methanol-water, 35:12:2, to monitor the progress of the reaction.

The breakdown products may be revealed either by UV light (using fluorescent silica gel plates) or by spraying with 1% FeCl₃ in 0.001 *N* HCl. At the completion of the reaction, the solution is dried *in vacuo* to remove NH₄OH completely. The residue is dissolved in a small volume of methanol and the individual products are separated by preparative TLC in the above system. The individual bands are eluted with methanol and purified by gel filtration through a Sephadex LH20 column in ethanol. The purified products are then characterized by NMR and other spectroscopic methods.

SPECTROSCOPIC IDENTIFICATION

If the purified ferric siderophore is a known compound, it may be provisionally identified by cochromatography with an authentic sample. However, cochromatography alone is sometimes misleading, because two or more compounds may have very similar mobility in some solvent systems. More reliable proofs of their identity come from various spectroscopic measurements, such as the NMR and the mass spectral data. Structural information provided by various spectroscopic methods is presented in the following sections.

VISIBLE AND CD SPECTROSCOPY

The ferric hydroxamates absorb in the visible region producing a broad absorption band centered at the 420- to 440-nm region. The actual position of the maximum depends on the type of iron coordination center present in the ferric siderophore, and also on the pH of the aqueous solution in which the siderophore is dissolved. The ferric complexes of trihydroxamate siderophores, such as ferrichrome, ferricrocin, or ferrichrysin in which all three N^δ-acyl groups are derived from acetic acid, produce absorption bands with maxima in the region of 420 to 428 nm at neutral pH. The ferric siderophores which contain three N^δ-anhydromevalonic acid residues (e.g., coprogen and ferrirubin) usually produce absorption bands with maxima between 430 and 438 nm. This bathochromic shift of the absorption band in the latter compounds is caused by the double bonds in the three N^δ-anhydromevalonoyl groups, which are in conjugation with the hydroxamate functions that bind the iron atom. When the number of these conjugated double bonds are reduced to two and one (as in neocoprogen I and neocoprogen II, respectively), the absorption maximum also shifts upfield proportionately.[28]

The pH dependence of the absorption maximum has been used successfully to distinguish between the ferric complexes of the monohydroxamates, the dihydroxamates, and the trihydroxamates.[32,41,55] The trihydroxamate siderophores (e.g., ferrichrome and coprogen) produce 1:1 (iron/ligand) (FeL) complexes, which are stable up to a pH of 2.0. Their absorption maximum shows no significant shift when pH of the solution is varied in the pH range of 2.0 to 7.0. The dihydroxamate siderophores (e.g., dimerum acid, rhodotorulic acid, etc.) usually produce 2:3 complexes (Fe_2L_3 species) at neutral pH, which start to dissociate into FeL^+ species in the pH range 4.0 to 5.0. The monohydroxamate siderophores (e.g., cis and trans fusarinine), on the other hand, produce 1:3 tris complexes (FeL_3 species) which dissociate into FeL_2^+ and FeL^{2+} species at lower pH values. The visible absorption maxima of these ferric mono- and dihydroxamate siderophores accordingly show a large downfield shift (to 500 to 520 nm) when the pH of the solutions drops to 2.0.

The circular dichroism (CD) spectroscopy is useful in determining the chirality of the metal center in ferric siderophores.[56] Each of the three hydroxamate groups forms a five-membered ring with the Fe(III) atom in the center. The planes of these five-membered rings may form a right-handed propellor (Δ) or a left-handed propellor (Λ). The ferric trihydroxamate siderophores may prefer to form only one type of configuration in their metal center; the ferric dihydroxamates and the monohydroxamates form both types but usually one type predominates over the other in solution.

The Λ and Δ isomers of ferric monothiohydroxamate, which are stable in solution, have been separated and their CD spectra determined.[57] The Λ isomer characteristically shows a positive CD band in the lower wavelength (maximum in the region of 455 nm) and a negative CD band in the higher wavelength (minimum in the region of 350 nm). A mirror image of this spectra is produced by the Δ isomer.

The Λ configuration is generally observed in the members of the ferrichrome family, while in the coprogen family the Δ configuration is observed in the solid state,[28] but a mixture of Λ and Δ may occur in solution.[58,59] Ferric triacetylfusarinine C, a cyclic trihydroxamate member of the fusarinine family, on the other hand, produces either the Λ or the Δ configuration depending on the solvent and crystallization conditions.[35]

FAST ATOM BOMBARDMENT MASS SPECTROMETRY

The ferric siderophores contain multiple functional groups in the molecule in addition to the Fe(III) atom, which render them unsuitable for EI mass spectrometry. The ferric as well as the deferri compounds decompose during thermal ionization and the resulting EI mass spectra provide little information on the structure of the compound. The problem of decomposition is overcome when Fast Atom Bombardment (FAB) technique is used.[60] The method produces sufficient ionization of both ferric and deferri siderophores.

The ferric siderophore produces two conspicuous ion peaks, corresponding to the MH^+ and the MNa^+ ions. Considerable variation in the ratio of the two peaks has been observed. However, the MH^+ peak is usually the base peak of the spectrum, while the MNa^+ peak

TABLE 1

m/z and Relative Abundance of the Molecular Ions in the FAB Mass Spectra of Some Fungal Siderophores

Compound	MH$^+$	MNa$^+$
Coprogen	822 (100)	844 (16)
Neocoprogen I	752 (100)	774 (19)
Neocoprogen II	682 (100)	704 (17)
N$^\alpha$-Dimethylcoprogen	808 (100)	830 (43)
N$^\alpha$-Dimethylneocoprogen I	738 (100)	760 (4)
N$^\alpha$-Dimethylisoneocoprogen I	738 (100)	760 (7)
Hydroxycoprogen	838 (100)	860 (17)
Hydroxyneocoprogen I	768 (100)	790 (17)
Hydroxyisoneocoprogen I	768 (100	790 (18)
Ferrichrysin	801 (100)	823 (28)
Ferricrocin	771 (64)	793 (60)
Ferrirubin	1011 (100)	1033 (15)
Asperchrome F1	1029 (29)	1051 (7)
Asperchrome F2	1029 (52)	1051 (7)
Asperchrome F3	1029 (100)	1051 (20)
Deferriferrichrysin	748 (100)	770 (55)
Deferriferrirubin	958 (100)	980 (42)
Deferriasperchrome F1	976 (35)	998 (100)
Deferriasperchrome F2	976 (22)	998 (100)
Deferriasperchrome F3	976 (45)	998 (100)

is of much lower intensity. The deferri siderophores, on the other hand, produce a much stronger MNa$^+$ ion (Table 1). The strong MH$^+$ peak is almost always accompanied by some minor peaks one or two mass units above and below it (Figures 5 and 6).

The high resolution FAB mass spectrometry can provide a measure of the accurate mass of the MH$^+$ or the MNa$^+$ ion peak which may be used to determine the molecular formula of the siderophore.[29,30] In addition to providing information on the molecular weight and the molecular formula of the ferric and deferri siderophore, FAB mass spectroscopy provides information on the purity of a compound and the composition of a siderophore mixture. Hydroxycoprogens elute out as a single band from the silica gel column chromatography with the chloroform-methanol-water system. FAB mass spectra of this band show two pairs of ion peaks: one pair (m/z 838 and 860) corresponding to the MH$^+$ and the MNa$^+$ ions of one compound with molecular formula $C_{35}H_{53}N_6O_{14}Fe$, and the other pair (m/z 768 and 790) corresponding to the MH$^+$ and MNa$^+$ ions of another compound having a molecular formula $C_{31}H_{47}N_6O_{13}Fe$. Reversed-phase chromatography with C_{18} silica gel separated three compounds from this band, which are hydroxycoprogen (MW 837), hydroxyneocoprogen I, and its isomer (both having MW 767). The structural isomers of siderophores are difficult to distinguish on the basis of the FAB mass spectra.

NMR SPECTROSCOPY

Ferric iron present in the ferric siderophore exists in a paramagnetic high spin d^5 state which causes severe line broadening of the NMR signals. For this reason, iron is removed from the ferric siderophore prior to NMR spectroscopy. To obtain the spectra of the siderophore in the metal chelated form, usually the Ga(III) or the Al(III) siderophore analogues are employed.

Deferration of ferric siderophores may be carried out by extraction of iron with 8-hydroxyquinoline.[22,61,62] The ferric siderophore is dissolved in a small volume of water, to which excess recrystallized 8-hydroxyquinoline is added. The suspension is slowly shaken

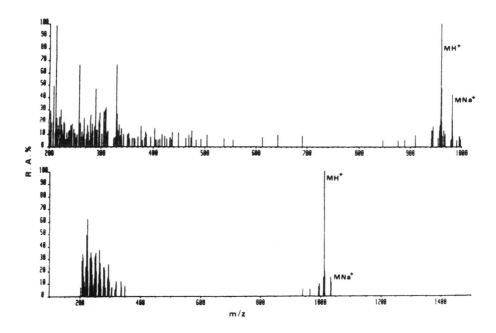

FIGURE 5. FAB mass spectra of ferrirubin (lower) and deferriferrirubin (upper).

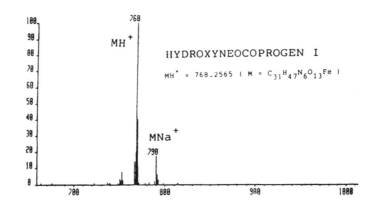

FIGURE 6. FAB mass spectra of hydroxyneocoprogen I.

overnight at room temperature. The greenish-black iron complex of 8-hydroxyquinoline as well as the unused reagent is removed by chloroform extraction (four to five times). Some siderophores take only a couple of hours to get completely deferriated. Others require much longer time or a second incubation period with 8-hydroxyquinoline. Warming the aqueous suspension (at 45°C) helps deferration in some cases. Addition of methanol or other organic solvents in the incubation mixture often causes insufficient deferration.

Deferri siderophores tend to retain moisture in the samples, which interfere with ^{1}H NMR spectroscopy. Samples should be lyophilized and thoroughly dried prior to dissolving in a suitable deuteriated solvent. D_2O is not regarded as a suitable solvent for deferri siderophores. In this solvent, the hydroxamate group exists in a mixture of cis and trans conformations and the NMR signals produced by the two conformers of the siderophore are more complex (Figure 7).[41] CD_3OD is a good solvent used in the NMR spectroscopy of siderophores. This solvent can be easily removed to recover the sample. The spectra obtained are simpler, because the NH, OH, and other exchangeable protons do not show up in the

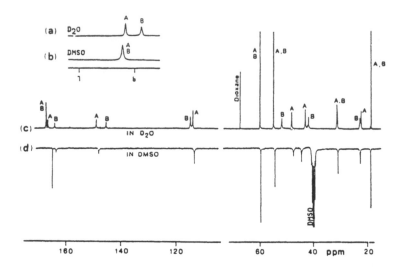

FIGURE 7. Effect of solvent on the spectra of dimerum acid. (a and b) ^{1}H-NMR signal of the =CH proton in D_2O (two conformers, A and B) and in DMSO (one conformer only). (c and d) ^{13}C NMR spectra in D_2O (two conformers, A and B) and in DMSO (one conformer only). (^{13}C assignment is given in Table 4.) (Reproduced from Jalal, M. A. F., Love, S. K., and van der Helm, D., *J. Inorg. Biochem.*, 28, 417, 1986. With permission of Elsevier Science Publishing Co., Inc.)

spectra. Full and detailed spectra are obtained with $(CD_3)_2SO$ (d_6-DMSO) and the information provided by the spectra in this solvent is very useful in the complete identification of the siderophore. Subsequent discussion of the NMR results of siderophores in the following sections is based on their spectra taken in $(CD_3)_2SO$.

Basic Units
Ornithyl Residues

As we discussed earlier, fungal siderophores are multimers of a basic structural unit, which is N^δ-hydroxy-N^δ-acylornithine. The first step in the interpretation of the NMR spectra involves the identification of the signals produced by the individual protons of this unit (Figure 8). The ornithyl part of this basic unit is represented by a single spin system composed of three CH_2 groups, the α-CH group, the COOH, and the NH_2 protons. In most siderophores, the COOH and the NH_2 groups are substituted, and the production of NMR signals by these groups depends on the type of substitution present. The β- and the γ-CH_2 protons produce a complex multiplet centered at δ 1.60. The multiplet produced by the δ-CH_2 group appears near δ 3.5. Position of the α-CH proton signal is more dependent on the substitution of the nearby amino and carboxyl functions. In a majority of the siderophores it appears in the chemical shift of δ 3.8 to 4.5.

Each of the β-, γ-, and δ-CH_2 groups in various siderophores usually produce ^{13}C signals in a narrow chemical shift range. The ^{13}C signal of the γ-CH_2 is most upfield (δ 22.0 to 24.0) the β-CH_2 resonates immediately downfield (δ 26.0 to 30.0), and the δ-CH_2 produces a signal in the region of δ 46.0 to 48.0. The position of the α-CH carbon varies greatly with the type of substitution present on the adjacent carboxyl and the amino group. In the majority of siderophores it appears in the vicinity of δ 52.0 to 55.0.

N^δ-Acyl Groups

A number of N^δ-acyl groups have been found in various fungal siderophores. The most common N^δ-acyl groups are those derived from acetic acid (a, Scheme 1), cis-anhydromevalonic acid (f), and trans-anhydromevalonic acid (b). The CH_3 group of the N^δ-acetyl

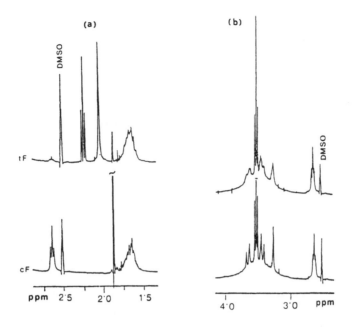

FIGURE 8. Part of the ^{1}H-NMR spectra showing (a) the difference in the signals of cis-fusarinine (lower spectra) and *trans*-fusarinine (upper spectra, a trace of cis-fusarinine present) (b) the low-field region in cis-fusarinine (upper spectra), and the same region after ornithyl β and γ methylene signals are decoupled to reveal a singlet for the α-CH and an AB pattern for the δ-CH$_2$ protons (lower spectra) (^{1}H assignment is given in Table 2). (Reproduced from Jalal, M. A. F., Love, S. K., and van der Helm, D., *J. Inorg. Biochem.*, 28, 417, 1986. With permission of Elsevier Science Publishing.)

moiety produces a proton signal near δ 1.97 and a ^{13}C signal around δ 20.3. The two isomers of anhydromevalonoyl moiety may be distinguished easily on the basis of their proton and ^{13}C NMR signals. The protons in the =CH– and the –CH$_2$O– groups in the two isomers do not show significant difference in their signals, which appear near δ 6.22 and δ 3.53, respectively, in both isomers. The CH$_3$ and the =C–CH$_2$– signals in the two isomers, however, differ greatly in chemical shift.[41] The CH$_3$ signal in the cis form appears further upfield (δ 1.87) and the –CH$_2$– protons appear further downfield (δ 2.64) compared to the respective signals in the trans form (δ 2.03 and 2.24, respectively) (Figure 8). The ^{13}C signals of these two groups also differ considerably.[22] In the trans form, the methylene carbon in =C–CH$_2$– and the methyl carbon produce signals at δ 43.8 and 18.0, respectively, while these carbons in the cis form appear near δ 36.3 and 25.2, respectively. Effect of cis-trans isomerism on the NMR signals of the other ^{13}C nuclei in the anhydromevalononoyl group is minimal.

In some siderophores the CH$_2$OH group of the N^5-anhydromevalonoyl moiety is esterified with a carboxylic acid moiety. In asperchrome C (**14**),[20] the ester-linked acid is acetic acid, whereas in coprogens and fusarinines, it is another acylated ornithine group. Esterification of the CH$_2$OH group causes deshielding of the –CH$_2$–O– and the =C–CH$_2$– protons, which move to δ 4.18 (Δδ = 0.68 ppm) and δ 2.39 (Δδ = 0.15 ppm), respectively, in the trans isomer. In the cis isomer, the –CH$_2$–O– protons move to δ 4.18, while the =C–CH$_2$– protons do not show any significant shift. The ^{13}C signal of the –CH$_2$–O– in both isomers moves downfield (3 to 4 ppm), while that of the methylene carbon in =C–CH$_2$– moves upfield (3 to 4 ppm).

TABLE 2
NMR Chemical Shifts of Some Members of the
Fusarinine Family

Structural group	37	^{1}H tF[a]	44	^{13}C 44[b]
α CH	3.20 (m)	3.20 (m)	4.18 (m)	52.5
β CH$_2$	1.62 (m)	1.62 (m)	1.62 (m)	29.3
γ CH$_2$	1.62 (m)	1.62 (m)	1.62 (m)	23.3
δ CH$_2$	3.42 (m)	3.42 (m)	3.48 (m)	48.1
–COO–	—	—	—	170.9[c]
N$^\delta$-Acyl –CH=	6.27 (s)	6.27 (s)	6.22 (s)	118.4
N$^\delta$-Acyl –CH$_2$–	2.64 (t)	2.23 (t)	2.64 (t)	32.4
N$^\delta$-Acyl –CH$_2$O	3.53 (m)	3.53 (m)	4.18 (m)	62.9
N$^\delta$-Acyl –CH$_3$	1.86 (s)	2.03 (s)	1.87 (s)	24.4
N$^\delta$-Acyl =C<	—	—	—	149.1
N$^\alpha$-Acetyl CH$_3$	—	—	1.84 (s)	22.9
N$^\alpha$-Acetyl >C=O	—	—	—	172.0[c]
Hydroxamic >C=O	—	—	—	172.0

[a]   tF = trans-fusarinine (all ^{1}H spectra in d$_6$-DMSO, 23°C).
[b]   After Reference 50 (in CDCl$_3$, 46°C).
[c]   May be interchanged.

Families of Siderophores

The families of fungal siderophores differ in the arrangements of the individual monomer units and represent various types of substitution and bond formation between the units. Based on these substitutions, patterns of NMR signals have emerged in various families. These patterns are discussed in the following sections.

Fusarinines

The two common monomeric siderophores, trans- and cis-fusarinine (N$^\delta$-acyl group = b and f), belong to this family. Their spectral features have been discussed in the previous section. In the dimeric and trimeric members, the trans-fusarinine unit has not been observed so far. The monomer units are joined by an ester linkage formed by the carboxyl group of one unit and the terminal hydroxyl group of another unit. The formation of the ester bond causes the following changes in the NMR signals: (1) the α-CH proton and the acyl-CH$_2$O signals shift downfield to δ 4.18 (Table 2); (2) the ^{13}C signal for the –CH$_2$O– group appears at δ 62.9. The free –CH$_2$OH produces methylene NMR signals at δ 3.53 (^{1}H) and δ 59.6 (^{13}C), respectively.

Coprogens

The salient structural feature of the coprogen family is the formation of a diketopiperazine ring by the two monomer units. The signals arising from the diketopiperazine ring are common to all members of the coprogen family (Table 3).[28-30] The CH signals of this ring occur at δ 3.82 (^{1}H) and δ 53.7 (^{13}C), the NH protons produce a sharp singlet around δ 8.15, and the ^{13}C signals for the carbonyls appear as a singlet at δ 167.8 (Table 4) (Figures 9 and 10).

The third monomer unit in the trihydroxamates is linked to the dimer by an ester bond as in the fusarinine family. The N$^\delta$-acyl moiety which contributes the OH group of the ester bond is always a trans-anhydromevalonoyl group. The substitution enables one to distinguish the signals of this middle N$^\delta$-acyl group from the similar N$^\delta$-acyl groups that may be present in the two ends (Table 3 and 4). Scheme 2 shows the ^{1}H and ^{13}C NMR signals of the various types of acyl groups found in the coprogen family.

TABLE 3

Proton NMR Chemical Shifts of the Deferri Forms of Some Members of the Coprogen Family (d$_6$-DMSO, 23°C)

Structural group	24	25	26	28	30	31	34	35
Diketopiperazine Ring								
>CH–	3.82 (m)	3.82 (m)	3.82 (m)	3.82 (m)	3.82 (m)	3.82 (m)	3.82 (m)	3.82 (m)
–NH	8.15 (s)	8.15 (s)	8.14 (s)	8.14 (s)	8.14 (s)	8.15 (s)	8.14 (s)	8.14 (s)
Ornithyl Residues								
>CH–	—	—	4.18 (m)	4.18 (m)	4.18 (m)	3.13 (t)	4.18 (m)	4.18 (m)
–NH–	—	—	8.27 (d)	8.27 (d)	8.27 (d)	—	8.27 (d)	8.27 (d)
–CH$_2$–CH$_2$–	1.61 (m)	1.62 (m)	1.60 (m)	1.60 (m)	1.60 (m)	1.60 (m)	1.60 (m)	1.60 (m)
–CH$_2$N<	3.45 (m)	3.52 (m)	3.49 (m)	3.49 (m)	3.49 (m)	3.50 (m)	3.50 (m)	3.50 (m)
–NOH–	9.70 (s)	9.65 (s)	9.70 (s)	9.70 (s)	9.70 (s)	9.73 (s)	9.68 (s)	9.70 (s)
N$^\alpha$-Acetyl Residue								
–CH$_3$	—	—	1.84 (s)	1.84 (s)	1.84 (s)	—	1.84 (s)	1.84 (s)
N$^\alpha$-Dimethyl Residues								
–N(CH$_3$)$_2$	—	—	—	—	—	2.21 (s)	—	—
Middle N$^\delta$-Anhydromevalonoyl								
–CH=	—	—	6.22 (s)	6.22 (s)	6.22 (s)	6.22 (s)	6.22 (s)	6.22 (s)
–CH$_2$–	—	—	2.39 (t)	2.39 (t)	2.39 (t)	2.42 (t)	2.39 (t)	2.39 (t)
–CH$_2$O–	—	—	4.18 (m)	4.18 (m)	4.18 (m)	4.20 (m)	4.18 (m)	4.18 (m)
–CH$_3$	—	—	2.03 (s)	2.03 (s)	2.03 (s)	2.02 (s)	2.02 (s)	2.03 (s)

TABLE 3 (continued)
Proton NMR Chemical Shifts of the Deferri Forms of Some Members of the Coprogen Family (d_6-DMSO, 23°C)

Structural group	24	25	26	28	30	31	34	35
				Terminal N^{δ}-Acyl Residues				
$-CH=$	—	6.21 (s)	6.22 (s)	6.22 (s)	—	6.22 (s)	6.22 (s)	—
$-CHOH-$	—	—	—	—	—	—	6.48 (s)	6.48 (s)
	—	—	—	—	—	—	3.92 (t)	3.92 (t)
$-CH_2-$	—	2.24 (t)	2.25 (t)	2.25 (t)	—	2.24 (t)	2.24 (t)	—
$-CH_2O-$	—	3.52 (m)	3.53 (m)	3.53 (m)	—	3.53 (m)	3.50 (m)	3.50 (t)
$-CH_3$	—	2.02 (s)	2.03 (s)	2.03 (s)	—	2.02 (s)	2.02 (s)	—
	—	—	—	—	—	—	1.95 (s)	1.95 (s)
$-CHOH-$	—	—	—	—	—	—	5.08 (s)	5.08 (s)
$-CH_2OH$	—	4.58 (s)	4.59 (s)	4.56 (s)	—	4.58 (s)	4.56 (s)	—
	—	—	—	—	—	—	4.64 (s)	4.64 (s)
$CO-CH_3$	1.97 (s)	—	—	1.97 (s)	1.97 (s)	—	—	1.97 (s)

TABLE 4

^{13}C NMR Chemical Shifts of the Deferri Forms of Some Members of Coprogen Family (d$_6$-DMSO, 23°C)

Structural group	25	26	28	30	31	34	35
Diketopiperazine Ring							
>C=O	167.8	167.8	167.8	167.7	167.8	167.7	167.8
>CH-	53.7	53.7	53.7	53.7	53.7	53.8	53.7
Ornithyl Residues							
>C=O	—	169.5	169.5	169.5	171.4	169.5	169.6
>CH- (α)	—	51.9	51.9	51.9	65.9	52.0	51.9
-CH$_2$- (β)	30.3	30.3 (2)	30.3 (2)	30.2 (2)	30.3 (2)	30.5 (2)	30.3 (2)
		28.0 (1)	28.0 (1)	27.9 (1)	26.2 (1)	28.1 (1)	28.0 (1)
-CH$_2$- (γ)	22.1	22.1 (2)	22.1 (2)	21.9 (2)	22.1 (2)	22.3 (2)	2.2 (2)
	—	23.1 (1)	23.0 (1)	23.0 (1)	23.2 (1)	23.2 (1)	23.1 (1)
-CH$_2$- (δ)	46.7	46.7 (2)	46.7 (2)	46.7 (2)	46.7 (3)	46.9 (2)	46.7 (2)
	—	46.4 (1)	46.4 (1)	46.3 (1)	—	46.5 (1)	46.4 (1)
N$^\alpha$-Acetyl Residue							
>C=O	—	172.1	172.0	172.0	—	172.0	172.1
-CH$_3$	—	22.0	22.1	22.1	—	22.2	22.0
N$^\alpha$-Dimethyl Groups							
N(CH$_3$)$_2$	—	—	—	—	41.1	—	—
Middle N$^\delta$-Anhydromevalonoyl Residue							
>C=O	—	166.1	166.1	166.1	166.4	166.2	166.2
-CH=	—	117.4	117.1	117.1	117.1	117.1	117.1

TABLE 4 (continued)
^{13}C NMR Chemical Shifts of the Deferri Forms of Some Members of Coprogen Family (d_6-DMSO, 23°C)

Structural group	25	26	28	30	31	34	35
$=C<$	—	148.8	148.8	148.8	148.6	148.6	148.7
$-CH_2-$	—	40.0ᵃ	40.0	40.0	40.0	40.0	40.0
$-CH_2O-$	—	62.3	62.2	62.2	61.5	62.4	62.3
$-CH_3$	—	18.0	18.0	17.9	18.0	18.1	18.0
Terminal N^δ-Acyl Residues							
$>C=O$	166.4	166.4	166.4	—	166.4	166.5	—
						166.8	166.8
$=CH-$	116.2	116.2	116.1	—	116.2	116.2	114.9
						114.9	
$=C<$	151.1	151.0	151.0	—	150.9	151.0	—
						153.5	153.2
$-CHOH-$						76.6	76.6
$-CH_2-$	43.8	43.8	43.7	—	43.8	43.8	—
$-CH_2OH$	59.1	59.1	59.1	—	59.1	59.2	—
						64.6	64.6
$-CH_3$	18.2	18.2	18.2	—	18.2	18.4	14.8
						15.0	
$-CO-CH_3$			20.2	20.2			20.3
$-CO-\underline{C}H_3$			170.2	170.2			170.2

ᵃ   Under solvent peak

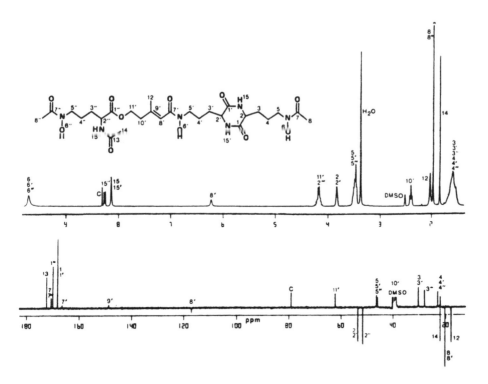

FIGURE 9. ¹H and ¹³C NMR spectra of deferrineocoprogen II. ¹³C NMR spectra are obtained using an APT pulse sequence program to show the CH_3 and CH carbon signals down and the CH_2, C, and —C=O signals up. C, residual chloroform used during deferration.

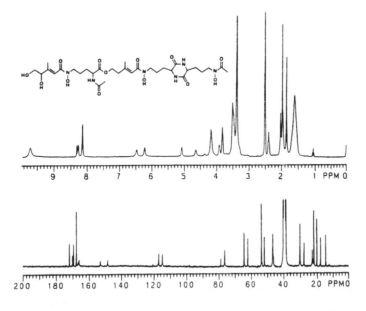

FIGURE 10. ¹H and ¹³C NMR spectra of deferrihydroxyneocoprogen I (assignments are given in Tables 3 and 4).

In many members of the coprogen family, the α-NH₂ group is acetylated. The signals arising from the Nα-acetyl group are distinct from the terminal Nδ-acetyl signals present in the siderophores such as neocoprogens (Figure 9). The α-NH signal of this ester-linked ornithine unit almost always produces a doublet (δ 8.27) near the singlet produced by the NH groups of the diketopiperazine ring.

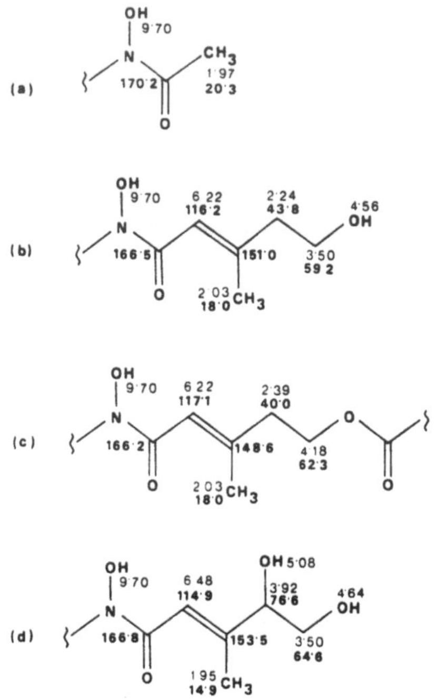

(a)

OH
9·70
N
170·2
CH₃
1·97
20·3
O

(b)

OH
9·70
N
166·5
6·22
116·2
2·24
43·8
4·56
OH
151·0
3·50
59·2
O
2·03 CH₃
18·0

(c)

OH
9·70
N
166·2
6·22
117·1
2·39
40·0
148·6
4·18
62·3
O
2·03 CH₃
18·0
O

(d)

OH
9·70
N
166·8
6·48
114·9
OH 5·08
3·92
76·6
4·64
OH
153·5
3·50
64·6
O
1·95 CH₃
14·9

SCHEME 2.   ¹H and ¹³C chemical shifts of various types of N⁸-acyl groups found in the coprogen family.

Ferrichromes

In this family, the N⁸-hydroxy-N⁸-acyl units are joined as a tripeptide unit and the N⁸-acyl ends are usually free. DDF, which is closely related to ferrichromes, contains nothing more than this tripeptide structure; the NMR signals for the α-CH groups are produced at three chemical shift values depending on their position in the tripeptide chain (Table 5).[19] In the members of the ferrichrome family, however, three more amino acids are present to form the well-known hexapeptide ring. For this reason, the NMR signals of the ferrichromes are more complex than the coprogens or fusarinines. A comprehensive study on the NMR signals of the ferrichromes, especially those containing three N⁸-acetyl groups, has been carried out by Llinas and co-workers.[61,63-68] We are limiting our discussion only to the use of NMR spectroscopy as a tool for identification of siderophores and not going into the details of conformational analysis.

The six NH proton signals of the hexapeptide ring appear in the chemical shift range of δ 7.5 to 8.7 (Figure 11). The NH proton signal of the glycyl residue, which is present in all ferrichromes, appears in a comparatively downfield position. One of the CH₂ protons of this glycyl residue produces a clear quartet centered at δ 3.76, while the other CH₂ proton is buried under the complex multiplets at δ 3.5. The five α-CH protons produce a complex pattern in the chemical shift range of δ 3.9 to 4.25. The exact position of the NH and the CH proton signals are dependent on the types of the two amino acid residues other than the three ornithyl and the glycyl residues. Some minor effect is also exerted by the type of the N⁸-acyl residues. The seryl CH₂-proton signals are not easily identifiable because part of its multiplet is buried underneath the complex system present at δ 3.5. However, the seryl OH signals appear separately around δ 4.96 and δ 5.10 and may be used to determine the number of seryl residues present in the molecule. The alanyl CH₃ protons produce a doublet near δ 1.25 and may be easily identified.

The ¹³C signals of the amino acid residues (as well as the N⁸-acyl moieties) are well separated and may be easily identified (Figures 11 and 12). The ornithyl and seryl α-CH

TABLE 5
Proton Chemical Shifts of the Deferri Forms of Some Members of Ferrichrome Family (d$_6$-DMSO, 23°C)

Structural group	1	8	9	10	11	14	15	18	19	23
Glycyl										
CH$_2$ (α)	3.60– 3.90	3.60– 3.90	3.60– 3.90	3.60– 3.90	3.60– 3.90	3.60– 3.90	3.60– 3.90	3.60– 3.90	3.60– 3.90	— —
NH	8.12 8.33 8.50	— 8.53	— 8.51	— 8.64	— 8.50	— 8.55	— 8.49	— 8.48	— 8.49	— —
Seryl										
CH (α)	—	3.90– 4.15	3.90– 4.15	4.08 —	3.90– 4.15	3.90– 4.15	3.90– 4.15	3.90– 4.15	3.90– 4.15	—
CH$_2$ (β)	—	3.50– 3.70	3.50– 3.70	3.50– 3.70	3.50– 3.70	3.50– 3.70	3.50– 3.70	3.50– 3.70	3.50– 3.70	— —
OH	—	4.96 5.12	4.95 5.08	— 5.04	4.96 5.12	4.97 5.12	4.96 5.12	4.95 5.10	4.96 5.12	—
NH	—	8.02 8.08	7.92 8.0	7.75 —	7.96 8.02	8.02 —	7.97 —	7.95 —	8.0 —	— —
Alanyl										
CH (α)	—	—	—	4.12	—	—	—	—	—	—
CH$_3$ (β)	—	—	—	1.25	—	—	—	—	—	—
NH	—	—	—	8.32	—	—	—	—	—	—
Ornithyl										
CH (α)	4.00 4.12 4.25	3.90– 4.20 —	3.90– 4.20 —	3.93 4.12 4.25	3.90– 4.20 —	3.90– 4.20 —	3.90– 4.20 —	3.90– 4.20 —	3.90– 4.20 —	3.73 3.93 4.23
CH$_2$ (β)	1.60– 1.90	1.60– 1.90	1.60– 1.90	1.60– 1.90	1.60– 1.90	1.60– 1.90	1.60– 1.90	1.60– 1.90	1.60– 1.90	1.60– 1.90
CH$_2$ (γ)	1.45– 1.75	1.45– 1.75	1.45– 1.75	1.45– 1.75	1.45– 1.75	1.45– 1.75	1.45– 1.75	1.45– 1.75	1.45– 1.75	1.45– 1.75
CH$_2$ (δ)	3.45– 3.65	3.45– 3.65	3.45– 3.65	3.45– 3.65	3.45– 3.65	3.45– 3.65	3.45– 3.65	3.45– 3.65	3.45– 3.65	3.40– 3.62
NH	7.97 8.12	7.50 7.98	7.50 7.92	7.54 7.90	7.50 7.98	7.54 8.00	7.50 7.98	7.50 8.00	7.50 8.00	7.65 8.67

TABLE 5 (continued)
Proton Chemical Shifts of the Deferri Forms of Some Members of Ferrichrome Family (d$_6$-DMSO, 23°C)

Structural group	1	8	9	10	11	14	15	18	19	23
NOH	8.33, 9.87	8.34, 9.75	8.28, 9.72	8.07, 9.66	8.32, 9.75	8.34, 9.73	8.28, 9.75	8.30, 9.67	8.30, 9.70	—, 10.00
N$^\delta$-acyl										
-CH=	—	6.22	6.26	6.23	6.22	6.22	6.22	6.22, 6.26	6.22	6.27
-CH-	—	—	—	—	—	—	—	—	2.56	—
=C-CH$_3$	—	2.03	1.87	2.03	2.03	2.03	2.03	2.03, 1.87	2.03	1.87
-C-CH$_3$	—	—	—	—	—	—	—	—	1.14	—
-C-OH	—	—	—	—	—	—	—	—	4.38	—
=C-CH$_2$	—	2.25	2.64	2.25	2.24	2.25, 2.40	2.24	2.24, 2.64	2.20	2.64
-C-CH$_2$-	—	—	—	—	—	—	—	—	1.70	—
-CH$_2$O-	—	3.53	3.52	3.55	3.53	3.53	3.53	3.53	3.53	3.54
-CH$_2$OH	—	4.62	4.73	4.57	4.56	4.18, 4.61	4.56	4.56, 4.72	4.60	4.76
-O-CO-CH$_3$	—	—	—	—	—	1.99	—	—	—	—
-N-CO-CH$_3$	1.97	—	—	—	1.97	—	1.97	—	—	—

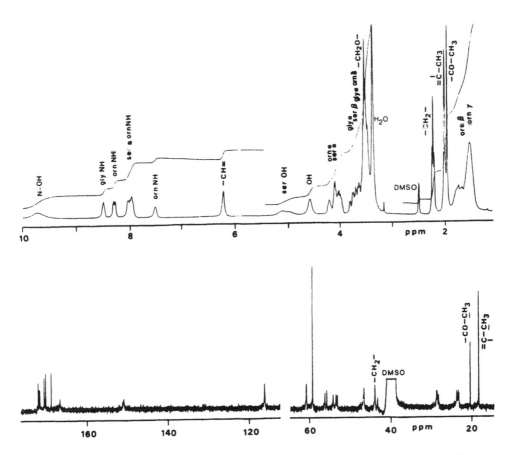

FIGURE 11. ¹H and ¹³C NMR spectra of deferriasperchrome B1 (complete assignment given in Tables 5 and 6).

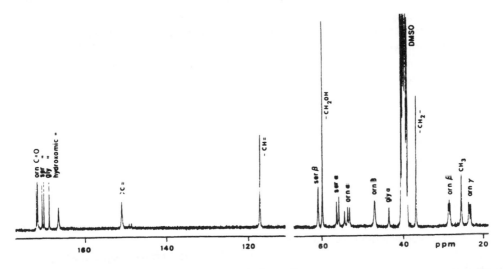

FIGURE 12. ¹³C NMR spectra of deferriferrirhodin. (Note difference in the group signals of ferrirhodin and asperchrome B₁ in Figure 11.)

carbon atoms produce separate signals in the chemical shift ranges of δ 52.5 to 54.5 and δ 55.0 to 57.0, respectively. The alanyl CH signal appears at δ 49.0. The glycyl CH_2 signal is also well separated and may be identified without much difficulty (Table 6).

The sequence of the hexapeptide ring may be resolved by the position and multiplicity of the NH protons of the A1(III)-complexed forms of ferrichromes. Llinas and Neilands[66] used this technique to determine the sequence of two alanine-containing siderophores, ferrichrome C and Sake colorant A (Figure 13).

ISOLATION AND CHARACTERIZATION OF FOUR NOVEL FERRICHROMES

A large number of ferrichrome-type siderophores have been isolated from the fungus *Aspergillus ochraceous* ATCC 58722 and the majority of these have been reported as novel compounds.[18-20,22,48] Characterization of four more novel ferrichromes isolated from this fungus, based on some of the methods of extraction, purification, and spectroscopic analysis discussed earlier, is presented below.

Two percent $FeCl_3$ solution is added to the cell-free culture medium of *A. ochraceous* to form the ferric complexes of the siderophores. The ferric siderophores are extracted according to the chloroform-phenol-ether-water extraction method (see Extraction From Culture Broth — The Chloroform-Phenol-Ether-Water Extraction). The crude extract is subjected to gel filtration once through a Biogel P2 column (see Chromatographic Isolation — Initial Gel Filtration in Biogel P2). Paper electrophoresis of this extract shows that most of the siderophores are neutral compounds at pH 5.0 (only one positively charged species at pH 2.0, which has been identified as DDF[22]). Thin layer chromatography of the extract on 0.25 mm silica gel with chloroform-methanol-water, 35:12:2, shows nine spots, out of which one is in much larger quantity (identified as ferrirubin[16]) (Figure 14). Similar to the TLC result, the compounds resolve into nine bands in the preparative silica gel column chromatography in the same solvent system (see Chromatographic Isolation — Preparative Silica Gel Column Chromatography). Components of each band separate into pure compounds after reversed-phase chromatography (see Chromatographic Isolation — Reversed-Phase Chromatography) in 0 to 40% methanol-water gradient. One of the bands (no. 5, Figure 14) obtained from the preparative silica gel column yields asperchrome D1, D2, and D3[19] and an unknown siderophore, termed asperchrome E, after the reversed-phase column. Band 8, which runs in-between ferrirubin (band 7) and DDF (band 9), separated into three components in the reversed-phase column. These compounds have been termed asperchrome F1, F2, and F3.

Asperchrome E (**18**) behaved as a neutral compound on paper electrophoresis at pH 5.0 and 2.0. Its visible absorption spectrum is stable in the pH range of 2.0 to 7.0. The NMR data obtained on its deferri form (Tables 5 and 6) clearly show that it is a ferrichrome-type compound having the following structural features: (1) it contains a hexapeptide ring made of three ornithines, two serines, and a glycine, as found in the major siderophores of the fungus, ferrirubin and ferrichrysin. The chemical shifts and the coupling constants of the hexapeptide ring protons are identical to those of ferrirubin, indicating that the sequence of the amino acids is similar to ferrirubin; (2) two of the N^δ-acyl groups are derived from trans-anhydromevalonic acid and the other one from its cis-isomer.

Asperchromes F1 (**19**), F2 (**20**), and F3 (**21**) also behaved neutral on paper electrophoresis at pH 5.0 and 2.0 (see Chromatographic Isolation — Electrophoresis). They have r_f values of 0.22, 0.23, and 0.23 in chloroform-methanol-water, 35:12:2, and 0.11, 0.12, and 0.13 in butanol-acetic acid-water, 4:1:5, upper phase, respectively. Under similar conditions DDF (**23**) and ferrirubin (**8**) have r_f of 0.13 and 0.36 in the former solvent system and 0.08 and 0.18 in the latter solvent system, respectively. The visible absorption spectra of asperchrome F1, F2, and F3 are stable in the pH range of 2.0 to 7.0, indicating that they are trihydroxamate siderophores.

TABLE 6

^{13}C NMR Chemical Shifts of Deferri Forms of Some Members of Ferrichrome Family (d$_6$-DMSO, 23°C) (Number of Carbons in Parentheses, When More Than One)

Structural group	8	9	10	11	14	15	18	19	23
Glycyl									
C=O	169.2	168.6	169.2	169.2	168.6	168.6	169.1	168.9	—
CH$_2$ (α)	43.0	43.1	43.3	43.2	43.1	43.1	43.1	43.0	—
Seryl									
C=O	170.9 170.4	170.6 170.0	— 170.3	170.6 170.0	170.9 170.4	170.3 169.9	170.8 170.4	170.6 170.2	—
CH (α)	56.2 55.6	56.1 55.5	55.8	56.6 56.3	56.2 55.6	56.1 55.5	56.2 55.6	56.3 55.6	—
CH$_2$ (β)	60.8 (2)	60.7 60.6	60.8	60.8 (2)	60.8 (2)	60.6 (2)	60.7 (2)	60.7 (2)	—
Alanyl									
C=O	—	—	172.8	—	—	—	—	—	—
CH (α)	—	—	49.0	—	—	—	—	—	—
CH$_3$ (β)	—	—	17.6	—	—	—	—	—	—
Ornithyl									
C=O	172.4 172.0 (2) —	172.4 172.1 172.0	172.3 172.2 172.0	172.4 172.2 (2) —	172.4 172.0 (2) —	171.8 171.4 (2) —	172.4 172.0 (2) —	172.3 172.0 (2) —	173.4 170.1 169.6
CH (α)	54.0 53.4 52.8	54.1 53.3 52.8	54.3 53.0 52.7	54.4 53.4 52.8	54.0 53.4 52.8	54.0 53.3 52.6	54.0 53.3 52.8	54.0 53.4 52.8	53.0 (2) 52.5 —
CH$_2$ (β)	28.4 28.3 28.0	28.4 28.2 28.0	28.5 28.2 (2) —	28.4 28.3 28.0	28.4 28.2 28.0	28.3 28.2 27.9	28.4 28.3 28.0	28.4 (2) 28.0 —	29.4 (2) 29.2 —
CH$_2$ (γ)	23.6 23.4 23.0	23.5 23.2 22.9	23.7 23.5 23.2	23.4 23.3 23.0	23.6 23.4 23.0	23.5 23.3 22.8	23.6 23.4 23.0	23.6 23.5 23.1	22.9 22.8 22.4
CH$_2$ (δ)	46.6 (3) — —	46.6 (3) — —	46.5 (3) — —	46.6 (3) — —	46.6 (3) — —	46.5 (3) — —	46.6 (3) — —	46.5 (3) — —	47.4 46.9 46.6

TABLE 6 (continued)
^{13}C NMR Chemical Shifts of Deferri Forms of Some Members of Ferrichrome Family (d$_6$-DMSO, 23°C) (Number of Carbons in Parentheses, When More Than One)

Structural group	8	9	10	11	14	15	18	19	23
N$^\delta$-acyl									
C=O	167.2 (3)	166.4 (3)	167.2 (3)	167.3 (2); 170.3	167.2 (3)	166.5; 170.2 (2)	167.2 (3)	166.8 (3)	166.4 (3)
=CH-	116.2 (3)	117.0 (3)	116.2 (3)	116.6 (3)	116.2 (2); 117.0	116.1	116.2 (2); 117.0	116.3 (2)	117.5; 117.2; 117.0
-CH-	—	—	—	—	—	—	—	41.7	—
=C<	151.2 (3)	151.2 (3)	151.2 (3)	151.8 (2)	151.2 (2)	151.1	151.2 (3)	151.2 (2)	151.0; 150.6; 150.2
>C(OH)-	—	—	—	—	—	—	—	70.5	—
-CH$_2$-	43.8 (3)	36.4 (3)	43.8 (3)	43.8 (2)	43.8 (2)	43.7	43.8 (2); 36.4	43.8 (2); 40.0	36.3 (3)
=C-CH$_3$	18.2 (3)	25.3 (3)	18.2 (3)	18.2 (2)	18.2 (2); 17.9	18.2	18.2 (2); 25.3	18.2 (2)	25.2 (3)
-C(OH)-CH$_3$	—	—	—	—	—	—	—	27.4	—
-CH$_2$OH	59.1 (3)	59.6 (3)	59.1 (3)	59.3 (2)	59.1 (2); 61.6; 40.0	59.1	59.1 (2); 59.6	59.2 (2); 57.2	59.6 (3)
-CH$_2$-O-	—	—	—	—	—	—	—	—	—

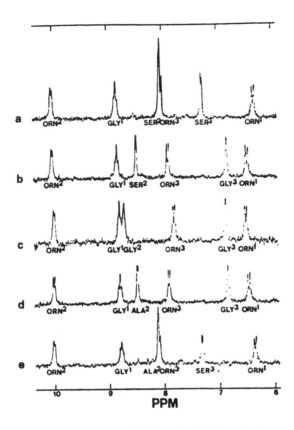

FIGURE 13. Sequence analysis of the hexapeptide ring using the amide NH proton resonances (solvent d₆-DMSO, 50°C). (a) Alumichrysin (cyclic orn-orn-orn-ser-ser-gly); (b) alumicrocin (cyclic orn-orn-orn-gly-ser-gly); (c) alumichrome (cyclic orn-orn-orn-gly-gly-gly); (d) alumichrome C (cyclic orn-orn-orn-gly-ala-gly); and (e) alumisake (cyclic orn-orn-orn-ser-ala-gly). (Reproduced from Llinas, M. and Neilands, J. B., *Biophys. Struct. Mechanism*, 2, 105, 1976. With permission of Springer Verlag.)

The FAB mass spectra of their ferric and deferri forms (Table 1) show that all three compounds are isomers. They show MH⁺ ions at 1029 (ferric complex) and 976 (deferri form) corresponding to the molecular formulas $C_{41}H_{66}N_9O_{18}Fe$ and $C_{41}H_{69}N_9O_{18}$, respectively. The ¹H and ¹³C NMR data of their deferri forms (Tables 5 and 6) are similar to each other and show that they are also members of the ferrichrome family with the following structural features: (1) similar to ferrirubin, their hexapeptide ring is made of three ornithines, two serines, and one glycine in the same sequence; (2) two of the N⁵-acyl groups in each compound are made of trans-anhydromevalonic acid residues; (3) the third N⁵-acyl group in each of the compounds is derived from mevalonic acid (g, Scheme 1). The CO—CH₂— protons in the mevalonic acid residue appear at δ 2.56 (δ 41.7 for ¹³C), while the CH₃ protons appear at δ 1.14 (δ 27.4 for ¹³C). The middle methylene protons in —C(OH)(CH₃)—CH₂—CH₂OH produce signals at δ 1.70 (δ 40.0 for ¹³C), while the terminal methylene proton signals appear at δ 3.53 (δ 57.2 for ¹³C). The ¹³C signal for the quaternary carbon is observed at δ 70.5. A signal corresponding to the —C—OH proton appears at δ 4.38.

Mevalonic acid, which acts as a precursor in the biosynthesis of a large number of metabolites including anhydromevalonic acid, has not been reported from any other siderophore before. Asperchrome F1, F2, and F3 differ from each other on the basis of the position of the mevalonic acid residue in the three ornithine moieties.

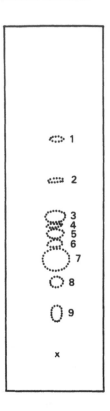

FIGURE 14. A schematic diagram of the siderophore spots observed after TLC separation of a crude siderophore extract of *Aspergillus ochraceous*; solvent, chloroform-methanol-water, 35:12:2. (Spot no. 1 and 2, unidentified; 3, mixture of ferrichrysin and asperchrome C; 4, asperchrome A; 5, mixture of asperchrome D1-D3 and E; 6, mixture of asperchrome B1-B3; 7, ferrirubin; 8, mixture of asperchrome F1-F3; 9, DDF.)

NMR studies have limitations when it comes to the identification of the exact position of acyl groups in the siderophore molecule. The proton and the ^{13}C spectra of deferri siderophores do not indicate which ornithine contains the odd N^{δ}-acyl group in isomeric asperchromes (such as F1, F2, and F3) and coprogens (such as neocoprogen I and isoneocoprogen I). In the case of coprogens, partial hydrolysis coupled with the NMR spectroscopy of the hydrolyzed products has been used successfully to solve this problem. Hydrolysis of the ester bond in coprogens is carried out with ammoniacal methanol (see Chemical Degradation — Degradation With Ammoniacal Methanol). A similar method is not available to distinguish isomeric asperchromes and single crystal X-ray diffraction is required to establish the structure of each isomer.[19]

CONCLUSIONS

This article is not a comprehensive review of the existing literature; it is, rather, a general guideline based on laboratory experience. Persons engaged or getting ready to engage in the analysis of siderophores may find this article useful for practical purposes. Every method has its limitations and exceptions and may not be useful for certain compounds not covered in this article. However, the extraction, separation, and spectroscopic identification methods described here have been used successfully for a large number of known as well as novel compounds. Spectroscopic methods sometimes do not yield unequivocal determination of structures, such as the case of isomeric asperchromes. For cases like this and also for a complete three-dimensional structure of the molecule, single crystal X-ray diffraction remains the method of choice.

REFERENCES

1. **Neilands, J. B.**, Microbial iron compounds, *Annu. Rev. Biochem.*, 50, 715, 1981.
2. **van der Helm, D., Jalal, M. A. F., and Hossain, M. B.**, The crystal structures, conformations and configurations of siderophores, in *Iron Transport in Microbes, Plants and Animals*, Winkelmann, G., van der Helm, D., and Neilands, J. B., Eds., VCH Verlagsgessellschaft, Weinheim, West Germany, 1987, 135.
3. **Neilands, J. B.**, Siderophores, *Adv. Inorg. Biochem.*, 5, 137, 1983.
4. **Emery, T.**, Initial steps in the biosynthesis of ferrichrome: incorporation of δ-N-hydroxyornithine and δ-N-acetyl-δ-N-hydroxyornithine, *Biochemistry*, 5, 3649, 1966.
5. **Eng-Wilmot, D. L., Rahman, A., Mendenhall, J. V., Grayson, S. L., and van der Helm, D.**, Molecular structure of ferric neurosporin, a minor siderophore-like compound containing N^8-hydroxy-D-ornithine, *J. Am. Chem. Soc.*, 106, 1285, 1984.
6. **Emery, T. F. and Neilands, J. B.**, Structure of ferrichrome compounds, *J. Am. Chem. Soc.*, 83, 1626, 1961.
7. **Rogers, S. and Neilands, J. B.**, Synthetic experiments in the ferrichrome series, *Biochemistry*, 2, 1850, 1964.
8. **Zalkin, A., Forrester, J. D., and Templeton, D. H.**, Ferrichrome-A tetrahydrate. Determination of crystal and molecular structure, *J. Am. Chem. Soc.*, 88, 1810, 1966.
9. **van der Helm, D., Baker, J. R., Eng-Wilmot, D. L., Hossain, M. B., and Loghry, R. A.**, Crystal structure of ferrichrome and a comparison with the structure of ferrichrome A, *J. Am. Chem. Soc.*, 102, 4224, 1980.
10. **Llinas, M. and Neilands, J. B.**, Structure of ferricrocin, *Bioionorg. Chem.*, 2, 159, 1972.
11. **Barnes, C. L., Eng-Wilmot, D. L., and van der Helm, D.**, Ferricrocin ($C_{29}H_{44}FeN_9O_{13}\cdot7H_2O$), an iron(III)-binding peptide from *Aspergillus versicolor*, *Acta Crystallogr.*, C40, 922, 1984.
12. **Norrestam, R., Stensland, B., and Branden, C. I.**, On the conformation of cyclic iron-containing hexapeptides: the crystal and molecular structure of ferrichrysin, *J. Mol. Biol.*, 99, 501, 1975.
13. **Keller-Schierlein, W.**, Stoffwechselprodukte von Mikroorganismen. XLV. Mitteilung, uber die Konstitution von Ferrirubin, Ferrirhodin und Ferrichrome A, *Helv. Chim. Acta*, 46, 1920, 1963.
14. **Keller-Schierlein, W. and Deer, A.**, Stoffwechselprodukte von Mikroorganismen. XLIV. Mitteilung zur Konstitution von Ferrichrysin und Ferricrocin, *Helv. Chim. Acta*, 46, 1907, 1963.
15. **Emery, T.**, Malonichrome, a new iron chelate from *Fusarium rosium*, *Biochim. Biophys. Acta*, 629, 382, 1980.
16. **Barnes, C. L., Hossain, M. B., Jalal, M. A. F., Eng-Wilmot, D. L., Grayson, S. I., Benson, B. A., Agarwal, S. K., Mocharla, R., and van der Helm, D.**, Ferrirubin two crystal forms,, $C_{41}H_{64}O_{17}N_9Fe\cdot10H_2O$ (1) and $C_{41}H_{64}O_{17}N_9FeCH_3CNH_2O$ (2), *Acta Crystallogr.*, C41, 341, 1985.
17. **Fidelis, K., Hossain, M. B., Jalal, M. A. F., and van der Helm, D.**, Structure and molecular mechanics of ferrirhodin, Acta Crystallogr., C46, 1612, 1990.
18. **Jalal, M. A. F., Mocharla, R., Barnes, C. L., Hossain, M. B., Powell, D. R., Eng-Wilmot, D. L., Grayson, S. I., Benson, B. A., and van der Helm, D.**, Extracellular siderophores of *Aspergillus ochraceous*, *J. Bacteriol.*, 158, 683, 1984.
19. **Jalal, M. A. F., Hossain, M. B., van der Helm, D., and Barnes, C. L.**, Structure of ferrichrome-type siderophores with dissimilar N^8-acyl groups: asperchrome B_1, B_2, B_3, D_1, D_2 and D_3, *Biol. Metals*, 1, 77, 1988.
20. **Jalal, M. A. F., Mocharla, R., Barnes, C. L., Hossain, M. B., Powell, D. R., Benson, B. A., and van der Helm, D.**, Iron-binding cyclic hexapeptides of *Aspergillus ochraceous*, in *Peptides, Structure and Function*, Hruby, V. J. and Rich, D. J., Eds., Pierce Chemical Company, Rockford, 1983, 503.
21. **Deml, G., Voges, K., Jung, G., and Winkelmann, G.**, Tetraglycylferrichrome — the first heptapeptide ferrichrome, *FEBS Lett.*, 173, 53, 1984.
22. **Jalal, M. A. F., Galles, J. F., and van der Helm, D.**, Structure of des-(diserylglycyl)ferrirhodin, DDF, a novel siderophore from *Aspergillus ochraceous*, *J. Org. Chem.*, 50, 5642, 1985.
23. **Diekmann, H.**, Stoffwechselprodukte von Mikroorganismen. LXXXI. Mitteilung. Vorkommen und Strukturen von Coprogen B und Dimerumsaure, *Arch. Mikrobiol.*, 73, 65, 1970.
24. **Atkin, C. L. and Neilands, J. B.**, Rhodotorulic acid, a diketopiperazine dihydroxamic acid with growth factor activity. I. Isolation and characterization, *Biochemistry*, 7, 3734, 1968.
25. **Keller-Schierlein, W. and Diekmann, H.**, Stoffwechselprodukte von Mikroorganismen. Zur Konstitution des Coprogens, *Helv. Chim. Acta*, 53, 2035, 1970.
26. **Frederick, C. B., Bentley, M. D., and Shive, W.**, Structure of triornicin, a new siderophore, *Biochemistry*, 20, 2436, 1981.
27. **Frederick, C. B., Bentley, M. D., and Shive, W.**, The structure of the fungal siderophore, isotriornicin, *Biochem. Biophys. Res. Commun.*, 105, 133, 1982.

28. **Hossain, M. B., Jalal, M. A. F., Benson, B. A., Barnes, C. L., and van der Helm, D.,** Structure and conformation of two coprogen type siderophores: (i) Neocoprogen and (ii) Neocoprogen II, *J. Am. Chem. Soc.,* 109, 4948, 1987.

29. **Jalal, M. A. F., Love, S. K., and van der Helm, D.,** N^α-dimethylcoprogens. Three novel trihydroxamate siderophores from pathogenic fungi, *Biol. Metals,* 1, 4, 1988.

30. **Jalal, M. A. F. and van der Helm, D.,** Siderophores of highly phytopathogenic *Alternaria longipes*: structure of hydroxycoprogens, *Biol. Metals,* 2, 11, 1989.

31. **Emery, T.,** Isolation, characterization and properties of fusarinine, a δ-hydroxamic acid derivative of ornithine, *Biochemistry,* 4, 1410, 1965.

32. **Sayer, J. M. and Emery, T. F.,** Structre of naturally occurring hydroxamic acids, fusarinines A and B, *Biochemistry,* 7, 184, 1968.

33. **Diekmann, H. and Zahner, H.,** Konstitution von fusigen und dessen abbau zu Δ^2-anhydromevalon-saurelacton, *Eur. J. Biochem.,* 3, 213, 1967.

34. **Moore, R. E. and Emery, T.,** N^α-acetylfusarinines: isolation, characterization and properties, *Biochemistry,* 15, 2719, 1976.

35. **Hossain, M. B., Eng-Wilmot, D. L., Loghry, R. A., and van der Helm, D.,** Circular dichroism, crystal structure and absolute configuration of the siderophore ferric, N,N'N''-triacetylfusarinine, $FeC_{39}H_{57}N_6O_{15}$, *J. Am. Chem. Soc.,* 102, 5766, 1980.

36. **Keller-Schierlein, W., Mertens, P., Prelog, V., and Walser, A.,** Stoffwechselprodukte von Mikroorganismen. Die Ferrioxamine A_1, A_2 und D_2, *Helv. Chim. Acta,* 48, 710, 1965.

37. **van der Helm, D. and Poling, M.,** The crystal structure of ferrioxamine E, *J. Am. Chem. Soc.,* 98, 82, 1976.

38. **Hossain, M. B., Jalal, M. A. F., and van der Helm, D.,** The structure of ferrioxamine D_1-ethanol-water (1/2/1), *Acta Crystallogr.,* C42, 1305, 1986.

39. **Teintze, M., Hossain, M. B., Barnes, C. L., Leong, J., and van der Helm, D.,** Structure of ferric pseudobactin, a siderophore from plant growth promoting *Pseudomonas, Biochemistry,* 20, 6446, 1981.

40. **Jalal, M. A. F., Hossain, M. B., van der Helm, D., Sanders-Loehr, J., Actis, L. A., and Crosa, J. H.,** Structure of anguibactin, a unique plasmid-related bacterial siderophore from the fish pathogen *Vibrio anguillarum, J. Am. Chem. Soc.,* 111, 292, 1989.

41. **Jalal, M. A. F., Love, S. K., and van der Helm, D.,** Siderophore mediated iron(III) uptake in *Gliocladium virens*. I. Properties of *cis*-fusarinine, *trans*-fusarinine, dimerum acid and their ferric complexes, *J. Inorg. Biochem.,* 28, 417, 1986.

42. **Jalal, M. A. F., Love, S. K., and van der Helm, D.,** Siderophore mediated iron(III) uptake in *Gliocladium virens*. II. Role of ferric mono and dihydroxamates as iron transport agents, *J. Inorg. Biochem.,* 29, 259, 1987.

43. **Horowitz, N. H., Charlang, G., Horn, G., and Williams, N. P.,** Isolation and identification of the conidial germination factor of *Neurospora crassa, J. Bacteriol.,* 127, 135, 1976.

44. **Huschka, H., Naegeli, H. U., Leuenberger-Ryf, H., Keller-Schierlein, W., and Winkelmann, G.,** Evidence for a common siderophore transport system but different siderophore receptors in *Neurospora crassa, J. Bacteriol.,* 162, 715, 1985.

45. **Zahner, H., Keller-Schierlein, W., Hutter, R., Hess-Leisinger, K., and Deer, A.,** Stoffwechselprodukte von Mikroorganismen. XL. Mitt. Sideramine aus *Aspergillaceen, Arch. Mikrobiol.,* 45, 119, 1963.

46. **Pidacks, C., Whitehill, A. R., Pruess, L. M., Hesseltine, C. W., Hutchings, B. L., Bohonos, N., and Williams, J. H.,** Coprogen, the isolation of a new growth factor required by *pilobolus* species, *J. Am. Chem. Soc.,* 75, 6064, 1953.

47. **Kappner, M., Hasenbohler, A., and Zahner, H.,** Stoffwechselprodukte von Mikroorganismen. CLXVI. Mitt. Optimierung der Desferr-Ferricrocinbildung bei *Aspergillus viridi-nutans* Ducker & Thrower, *Arch. Mikrobiol.,* 115, 323, 1977.

48. **Jalal, M. A. F., Mocharla, R., and van der Helm, D.,** Separation of ferrichromes and other hydroxamate siderophores of fungal origin by reversed phase chromatography, *J. Chromatogr.,* 301, 247, 1984.

49. **Konetschny-Rapp, S., Huschka, H.-G., Winkelmann, G., and Jung, G.,** High performance liquid chromatography of siderophores from fungi, *Biol. Metals,* 1, 9, 1988.

50. **Charlang, G., Ng, B., Horowitz, N. H., and Horowitz, R. M.,** Cellular and extracellular siderophores of *Aspergillus nidulans* and *Penicillium chrysogenum, Mol. Cell. Biol.,* 1, 94, 1981.

51. **Burt, W. R.,** Identification of coprogen B and its breakdown products from *Histoplasma capsulatum, Infect. Immun.,* 35, 990, 1982.

52. **Chinard, F. P.,** Photometric estimation of proline and ornithine, *J. Biol. Chem.,* 199, 91, 1952.

53. **Emery, T. and Neilands, J. B.,** Contribution to the structure of the ferrichrome compounds: characterization of the acyl moieties of the hydroxamate functions, *J. Am. Chem. Soc.,* 82, 3658, 1960.

54. **Snow, G. A.,** Mycobactin. A growth factor for *Mycobacterium johnei*. II. Degradation, and identification of fragments, *J. Chem. Soc.,* p. 2588, 1954.

55. **Schwarzenbach, G. and Schwarzenbach, K.**, Hydroxamatkomplexe. I. Die Stabilitat der Eisen(III)-Komplexe einfacher Hydroxamsauren und des Ferrioxamins B, *Helv. Chim. Acta*, 46, 1390, 1963.
56. **Matzanke, B. F., Muller-Matzanke, G., and Raymond, K. N.**, Siderophore-mediated iron transport, in *Iron Carriers and Iron Proteins*, Loehr, T. M., Ed., VCH Verlagsgesellschaft, Weinheim, West Germany, 1989, 1.
57. **Abu-Dari, K. and Raymond, K. N.**, Coordination isomers of biological iron transport compounds. The resolution of tris(hydroxamato) and tris(thiohydroxamato) complexes of high-spin iron(III), *J. Am. Chem. Soc.*, 99, 2003, 1977.
58. **Wong, G. B., Kappel, M. J., Raymond, K. N., Matzanke, B., and Winkelmann, G.**, Coordination chemistry of microbial iron transport compounds. XXIV. Characterization of coprogen and ferricrocin, two ferric hydroxamate siderophores, *J. Am. Chem. Soc.*, 105, 810, 1983.
59. **Fidelis, K.** Partial atomic charges in ferric hydroxamates and the determination of relative potential energies in a series of optical and geometrical isomers of neocoprogen I, Ph.D. thesis, University of Oklahoma, 1989.
60. **Dell, A., Hider, R. C., Barber M., Bordoli, R. S., Sedgwick, R. D., Tyler, A. N., and Neilands, J. B.**, Field desorption and fast atom bombardment mass spectroscopy of hydroxamate containing siderophores, *Biomed. Mass Spectrom.*, 9, 158, 1982.
61. **Llinas, M., Klein, M. P., and Neilands, J. B.**, Solution conformation of ferrichrome, a microbial iron transport cyclohexapeptide, as deduced by high resolution proton magnetic resonance, *J. Mol. Biol.*, 52, 399, 1970.
62. **Wiebe, C. and Winkelmann, G.**, Kinetic studies on the specificity of chelate iron uptake in *Aspergillus*, *J. Bacteriol.*, 123, 837, 1975.
63. **Llinas, M., Klein, M. P., and Neilands, J. B.**, Solution conformation of the ferrichromes. II. Proton magnetic resonances of metal-free ferricrocin and ferrichrysin, conformational implications, *Int. J. Pept. Protein Res.*, 4, 157, 1972.
64. **Llinas, M., Klein, M. P., and Neilands, J. B.**, Solution conformation of ferrichromes. III. A comparative proton magnetic resonance study of glycine and serine containing ferrichromes, *J. Mol. Biol.*, 68, 265, 1972.
65. **Llinas, M., Wilson, D. M., and Neilands, J. B.**, Peptide strain. Conformation dependence of the carbon-13 nuclear magnetic resonance chemical shifts in the ferrichromes, *J. Am. Chem. Soc.*, 99, 3631, 1977.
66. **Llinas, M. and Neilands, J. B.**, The structure of two alanine containing ferrichromes: sequence determination by proton magnetic resonance, *Biophys. Struct. Mechanism*, 2, 105, 1976.
67. **Demarco, A., Llinas, M., and Wuthrich, K.**, Analysis of the ^{1}H-NMR spectra of ferrichrome peptides. I. The non-amide protons, *Biopolymers*, 17, 617, 1978.
68. **Demarco, A., Llinas, M., and Wuthrich, K.**, Analysis of the ^{1}H-NMR spectra of ferrichrome peptides. II. The amide resonances, *Biopolymers*, 17, 637, 1978.

SYNTHESIS OF CATECHOLAMIDE AND HYDROXAMATE SIDEROPHORES

Raymond J. Bergeron and James S. McManis

INTRODUCTION

Although iron has a variety of oxidation states available to it ranging from -2 to $+6$, the $+2$ and $+3$ valences are of the greatest consequence in biological systems. These two oxidation states, characterized by their d^6 and d^5 ground-state configurations, respectively, are exquisitely sensitive to both pH and the nature of the ligating functionality.[1] This sensitivity has been exploited at a cellular level inasmuch as the metal serves as both an electron acceptor and as an electron donor. In fact, iron is an essential cofactor in a variety of biological redox systems, e.g., cytochromes, oxidases, peroxidases, ribonucleotide reductase.[2,3] However even though it is the second most abundant metal on earth, living systems have had to evolve rather clever methods for acquiring iron.[4-10]

The most primitive forms of life developed about 3.5 billion years ago, life forms which existed in an essentially anaerobic atmosphere. In this atmosphere, in the absence of molecular oxygen, iron was largely in the $+2$ oxidation state, a form of the metal which is far more water soluble and accessible to biological systems than Fe(III).[11] Along with the evolution of the blue-green algae, there arose problems associated with the availability of this essential micronutrient. The oxygen produced from these algae by photosynthesis resulted in the conversion of Fe(II) to Fe(III), a species which is highly insoluble in an aqueous environment. In fact, the solubility product of Fe(III) under physiological conditions, K_{sp} $= 2 \times 10^{-38}\ M$, indicates a solution concentration of the free cation of approximately $1 \times 10^{-18}\ M$.[12] Iron(III) forms insoluble ferric hydroxide polymers under most conditions which exist in the biosphere. In spite of the abundance of the metal, primitive life forms now had to develop methods for rendering it utilizable.

Bacteria ultimately adapted to this problem by producing a group of relatively low molecular weight, virtually iron-specific ligands for the purpose of accessing and utilizing the transition metal.[13-15] The microorganisms biosynthesize and release large quantities of these molecules into their environment. Under conditions of low iron availability, some microorganisms actually secrete several times their own dry weight of ligand each day.[16] These chelators form soluble complexes with the metal, and the bacteria are then bathed in iron chelates which the organism can utilize. One of the key questions regarding these "utilizable iron complexes" is related to how the microorganism processes them. These ligands, siderophores (from the Greek *sidero* and *phore*, literally iron carrier), frequently bind iron very tightly, with formation constants as high as $10^{52}\ M^{-1}$ (enterobactin-Fe[III] complex).[17] The question then becomes: how does the microorganism manage to extract iron from a molecule which binds it so tightly? The answer to such questions demands studies of the nature of siderophore receptor binding and postreceptor binding events. Such assessments are, of course, facilitated by the availability of analogues and homologues of the parent siderophore and the corresponding radiolabeled materials. Access to these compounds calls for the design and development of synthetic methods for entry into such systems.

The synthetic methods will also make the siderophores and their analogues available for additional biological evaluations. The importance of such synthetic methods becomes even more amplified in view of the search for iron chelators to treat Haemochromatosis,[18] recent successes in utilizing siderophores to control the growth of tumor cell,[19-21] as well as the application of siderophores as vectors for delivering antibiotics to pathogenic organisms.[22-25]

In this review we will discuss the total synthesis and synthetic methods developed for

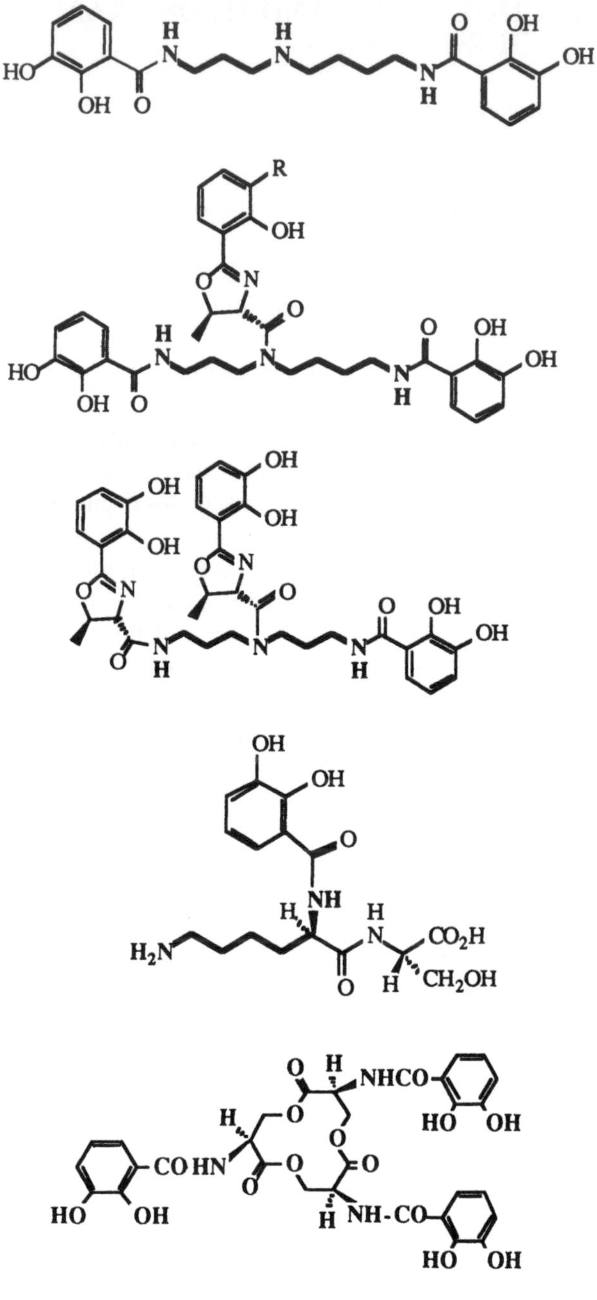

Compound II

L-Parabactin R=H

L-Agrobactin R=OH

Vibriobactin

Chrysobactin

Enterobactin

FIGURE 1. Catecholamide siderophores. Polyamine backbones are highlighted by darkened bonds.

the production of a number of siderophores. While the review will make no attempt to describe all of the syntheses of these natural products, it will focus on schemes which it is hoped will be of general value in accessing these catecholamide and hydroxamate chelators.

Although there is a rather large number of siderophores that have been isolated, they can, for the most part, be separated into two basic structural groups, the hydroxamates and the catecholamides. In the catecholamide family (Figure 1), we will describe the synthesis of the natural products: bis-N^1,N^8-(2,3-dihydroxybenzoyl)spermidine (initially referred to as Compound II), parabactin, agrobactin, vibriobactin, enterobactin, and chrysobactin. In the hydroxamate group (Figure 2), we will describe the synthesis of desferrioxamine B, des-

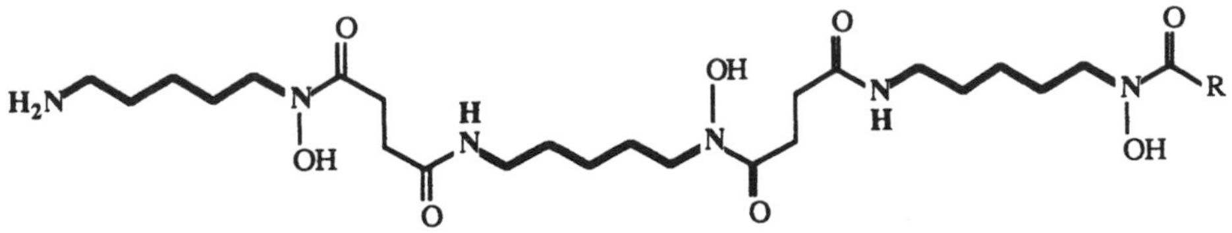

Desferrioxamine B   R=CH₃

Desferrioxamine G   R=(CH₂)₂CO₂H

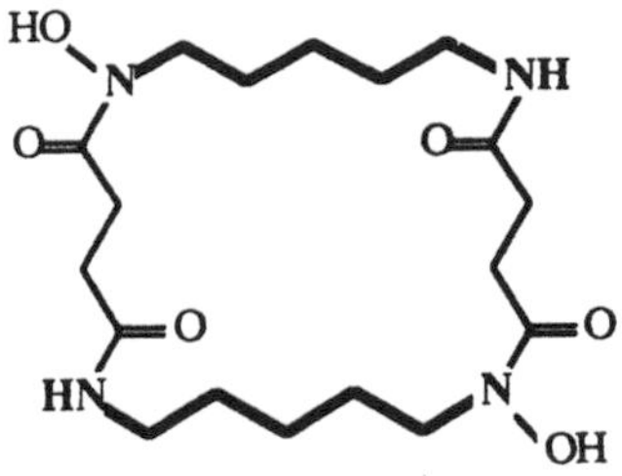

Bisucaberin

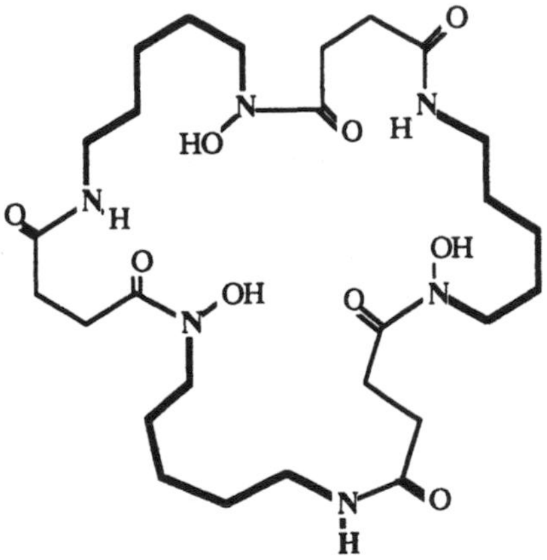

Desferrioxamine E, Nocardamine,

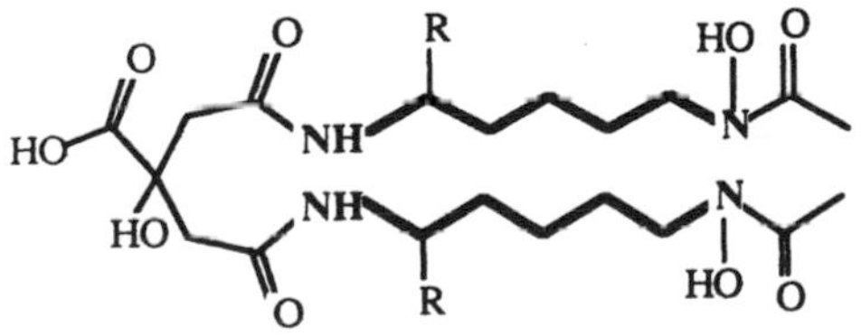

R=CO₂H   Aerobactin

R=H   Arthrobactin

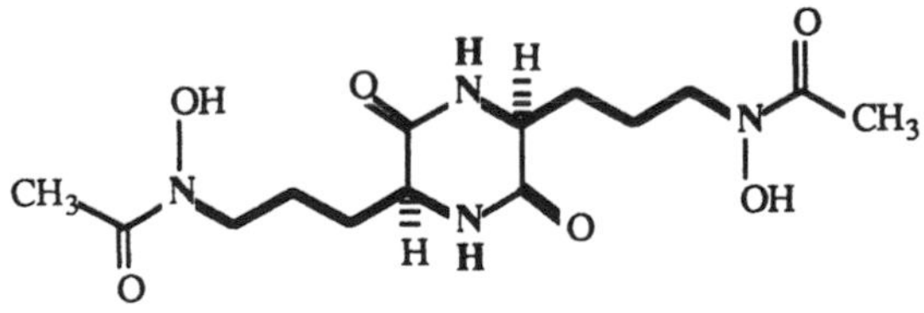

Rhodotorulic acid

A

FIGURE 2.   (A and B) Hydroxamate siderophores. Polyamine backbones are highlighted by darkened bonds.

ferrioxamine G, bisucaberin, nocardamine (desferrioxamine E), anthrobactin, rhodotorulic acid, desferri-ferrichrome, aerobactin, schizokinen, and mycobactin S2. In an attempt to demonstrate the general utility of the synthetic methods employed, we will note how the methods have been applied to the production of various homologues and analogues.

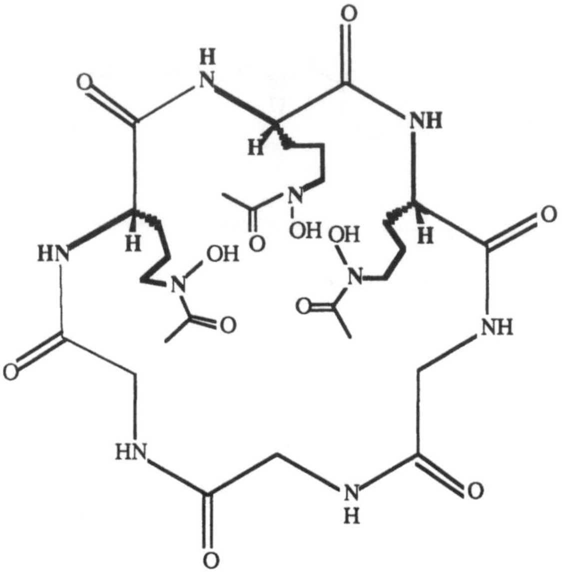

Desferri-ferrichrome

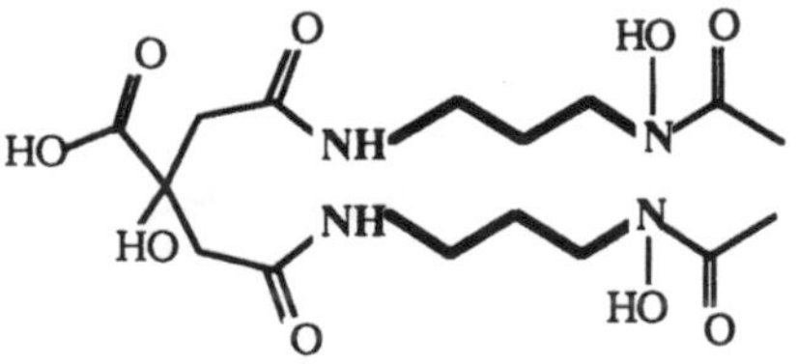

Schizokinen

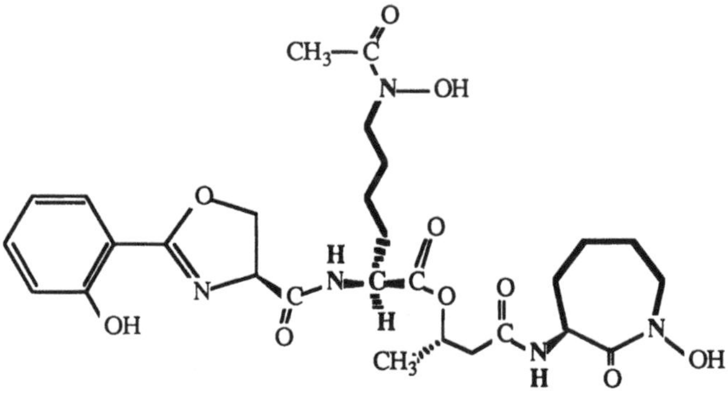

Mycobactin S2

FIGURE 2B

COMMONALITY IN BIOSYNTHETIC ORIGINS

Most of the hydroxamate and catecholamide chelators to be described in this review have common structural denominators. They can be regarded as either predicated directly on the polyamine backbones, putrescine, cadaverine, norspermidine, or spermidine or on their biochemical precursors ornithine or lysine. The catecholamide chelators, bis-N¹N⁸-(2,3-dihydroxybenzoyl)spermidine, parabactin,[26] agrobactin,[27] and vibriobactin[28] (Figure 1), all have 2,3-dihydroxybenzoyl groups fixed directly to either a spermidine or norspermidine backbone. Clearly enterobactin, a siderophore based on a macrocyclic serine backbone, is an exception to this observation. The cadaverine units of the hydroxamates (Figure 2), desferrioxamine B,[29] desferrioxamine G,[30] nocardamine,[29] bisucaberin,[31] and arthrobactin,[32] are obvious, while in mycobactin S2 [33] and aerobactin,[34] the cadaverine biochemical precursor, lysine, can be seen. Of course in the case of desferri-ferrichrome[35] and rhodotorulic acid,[36] both of which contain putrescine components, one immediately sees ornithine, the biochemical precursor to putrescine, as the actual synthon. Finally, the diaminopropane segment of schizokinen[37] is equally apparent.

We will begin our discussion of the total synthesis of siderophores with the polyamine catecholamides followed by a description of methods developed for the synthesis of enterobactin and several analogues. Next the total synthesis of a number of hydroxamate chelators will be described. The hydroxamate section will begin with the cadaverine containing siderophores, desferrioxamine B, desferrioxamine G, desferrioxamine E, bisucaberin, and arthrobactin, followed by the ornithine-derived ligands rhodotorulic acid and desferriferrichrome. The synthesis of the 1,3-diaminopropane-based chelator schizokinen will be described, and this section will also include the syntheses of aerobactin and mycobactin S2, which are predicated on lysine. Where appropriate, we will indicate how the methods developed for total synthesis of the natural product can be utilized for the generation of various analogues and homologues.

CATECHOLAMIDE CHELATORS

The mechanism by which bacteria process catecholamide siderophore iron complexes, recent discoveries regarding parabactin, agrobactin, and vibriobactin inhibition of tumor cell growth, coupled with the ligand's potential in the treatment of iron overload, have compelled investigators to synthesize additional catecholamides. An understanding of the mechanism of iron removal from the catecholamide iron complex will be facilitated through access to various analogues and homologues of the siderophores. Although initial experiments in animals treated with iron dextran and infected with L1210 cells implied that "excess iron" was serving as a micronutrient and was responsible for increased cell growth and early animal death,[38] more recent studies suggest that other factors contributed at least in part to the early deaths. The precise role of excess iron in tumor growth remains unclear.[39] However, it is certainly true that "iron deprivation" does inhibit tumor growth.[4,19-21,40,41] In fact, the catecholamide chelators have been shown to be extremely potent cell synchronization agents, holding the cells at the G1/S border. Compound II, parabactin, and vibriobactin have all been shown to strongly inhibit the growth L1210 cells, while both Compound II and parabactin inhibit HSV-1 viruses. Evident is consistent with the idea that the ligands operate at the ribonucleotide reductase level by removing iron from the enzyme. These observations have encouraged workers to synthesize additional chelators for study.

In each of the catecholamide chelators to be described (Figure 1)—bis-N^1,N^8-(2,3-dihydroxybenzoyl)spermidine, parabactin, agrobactin, vibriobactin, enterobactin, and chrysobactin—the catechol function is in the form of a 2,3-dihydroxybenzoyl group or its oxidative equivalent. The 2,3-dihydroxybenzoyl unit is an excellent bidentate ligand for Fe(III). In fact, 2,3-dihydroxybenzoic acid itself forms very tight, three to one, high spin, hexacoordinate, octahedral complexes[17] with Fe(III), $K_f = 10^{38}\ M^{-1}$. When the 2,3-dihydroxybenzoyl group is fixed to polyamine backbones as in the cases of parabactin, agrobactin, or vibriobactin, or to a triserine macrocycle as in enterobactin, the formation constants of the Fe(III) complexes become even more impressive, as entropy plays its hand. The most outstanding example can be seen with the enterobactin-Fe(III) complex,[17] which has a formation constant of $10^{52}\ M^{-1}$. In this instance the 2,3-dihydroxybenzoyl groups are coupled to a macrocyclic serine backbone. Synthetically, then, with each siderophore the directive becomes: identify methods for fixing the 2,3-dihydroxybenzoyl functionality to the appropriate anchor, e.g., polyamine or serine macrocycle.

The three catecholamide siderophores predicated on the spermidine backbone, bis-N^1,N^8-

(2,3-dihydroxybenzoyl)spermidine (Compound II), parabactin, and agrobactin (Figure 1), all present the same synthetic obstacle, that is, selective functionalization of a triamine. Compound II involves the fixing of a 2,3-dihydroxybenzoyl group to each of the primary nitrogens of spermidine.[42-44] The selectivity problem is obviously associated with avoiding functionalization of the secondary nitrogen. Syntheses of parabactin[45] and agrobactin[46] present the same kind of selectivity complication, acylation of the primary vs. secondary nitrogens of spermidine. In the case of parabactin there are a number of alternatives, e.g., the 2-(2-hydroxyphenyl)-4-carboxyl-5-methyl-2-oxazoline group could first be secured to the secondary nitrogen (Scheme 1) of spermidine followed by addition of 2,3-dihydroxybenzoyl groups. Alternatively, the terminal primary nitrogens could first be acylated with the 2,3-dihydroxybenzoyl group to give its biosynthetic precursor, Compound II, followed by introduction of the 2-(2-hydroxyphenyl)-4-carboxyl-5-methyl-2-oxazoline functionality at the secondary nitrogen. The synthetic design of agrobactin involves similar considerations, although in this case the 2-(2,3-dihydroxyphenyl)-4-carboxyl-5-methyl-2-oxazoline group is involved (Scheme 1).

Vibriobactin, which is predicated on norspermidine, requires an alternative synthetic design.[47] The polyamine backbone of vibriobactin, unlike the unsymmetrical spermidine backbone of parabactin and agrobactin, consists of a symmetrical norspermidine unit. However, vibriobactin has two different groups fixed to the terminal primary amino nitrogens of the symmetrical amine: a 2,3-dihydroxybenzoyl function on one end, and a 2-(2,3-dihydroxyphenyl)-4-carboxyl-5-methyl-2-oxazoline group on the other end, as well as a second 2-(2,3-dihydroxyphenyl)-4-carboxyl-5-methyl-2-oxazoline fixed to the central secondary nitrogen. Synthetically the problem is again reduced to selective acylation of polyamine nitrogens. The alternatives include bis-acylation of norspermidine with the two oxazoline segments followed by introduction of the 2,3-dihydroxybenzoyl group, or the inverse. The common denominator in the synthesis of all of the polyamine catecholamide siderophores is the availability of appropriately protected triamines.

POLYAMINE REAGENTS

The synthesis of each of the polyamine catecholamide chelators to be described below weighs heavily on the availability of the appropriately protected polyamines norspermidine, spermidine, and homospermidine. Each of these linear triamines has two primary terminal amines as well as a secondary central nitrogen. Although it is possible to realize some selectivity with regards to primary vs. secondary nitrogen acylation utilizing *N*-hydroxysuccinimide or imidazole activated acids,[48,49] selectively protected polyamines are of far greater utility. With, for example, the secondary nitrogens of the triamines appropriately protected, any acylating agent can be utilized for primary nitrogen acylation. Furthermore, in the case of vibriobactin it is clear that selective acylation of one primary and one secondary nitrogen of norspermidine is an unreasonable expectation in the absence of protecting groups. With these considerations in mind, a brief discussion of polyamine reagents is in order.[50-55]

Although methods are available for accessing the secondary *N*-protected triamines, N^4-benzylnorspermidine, N^4-benzylspermidine, and N^5-benzylhomospermidine for parabactin, agrobactin, and the corresponding homologue and analogue syntheses, we will describe a group of more versatile reagents which can also be used to generate vibriobactin and its homologues (Scheme 2). These protected triamines have three different protecting groups, trifluoroacetyl, benzyl, and *tert*-butoxycarbonyl, each of which is removed under different conditions.[53] Selective annealing of the orthogonal protecting groups onto the parent triamines is unrealistic. Although it is possible to selectively acylate the terminal nitrogens of the polyamines if one carefully chooses the acylating agent, the polyamine reagents to be described allow for the use of virtually any acylating agent. Most importantly, one can easily fix three different groups to the polyamine backbone in any order by utilizing these reagents.

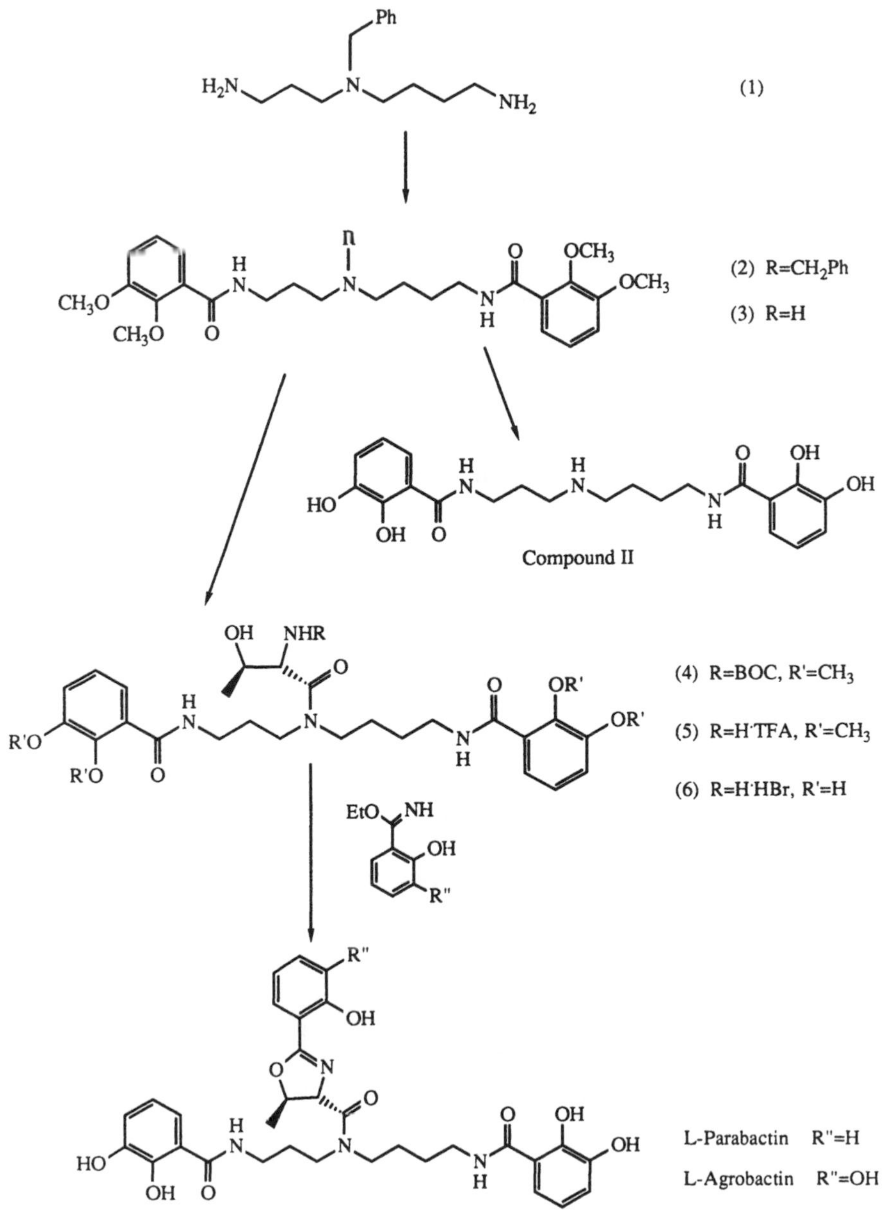

SCHEME 1. Total synthesis of Compound (II), L-parabactin and L-agrobactin.

The synthesis of the triprotected polyamine reagents begins with a suitable *N*-benzyl diamine (Scheme 2). Nitrile (1), obtained by cyanoethylation of benzylamine, is reduced to diamine (2a, n = 3) utilizing a Raney nickel catalyst.[51,53] However, the analogous diamine (2b, n = 4) is not as easily accessible. Attempts to monoalkylate benzylamine with 4-chlorobutyronitrile met with only limited success. Alternatively, monobenzoylation of 1,4-diaminobutane (putrescine) followed by reduction was considered. Unfortunately, both amino groups of 1,4-diaminobutane react with acylating agents even when the diamine is present in large excess. For example, acylation of 1,4-diaminobutane (10 mmol) with benzoyl chloride (2 mmol) gave a 95% yield of the bisacylated product. Furthermore, reaction of putrescine (90.7 mmol) with benzaldehyde (22.7 mmol) generated the corresponding bis(imine) in a 67% yield. However, putrescine can be monobenzylated with benzaldehyde under

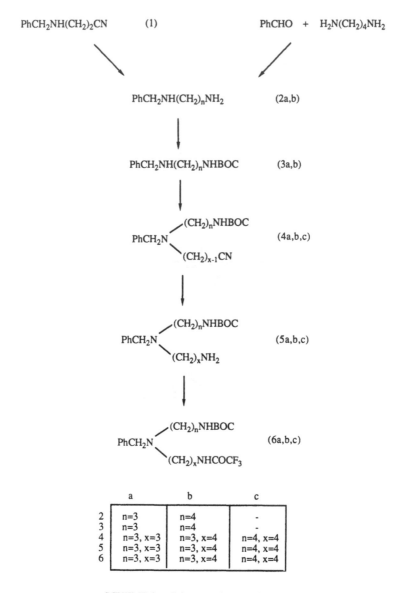

SCHEME 2.   Triprotected polyamine reagents.

reductive amination conditions in formic acid to produce (2b) in high yields (81%). The excess 1,4-diaminobutane required for monofunctionalization in this step is easily recovered (70% after distillation).

The diamines (2) can be regioselectively mono-*t*-butoxycarbonylated with one equivalent of 2-(*tert*-butoxycarbonyloxyimino)-2-phenylacetonitrile (BOC-ON). Both of the amino groups can be protected by using this reagent; however, with one equivalent of BOC-ON at 0°C, monoacylation occurs quantitatively at the primary amine of (2). The monoamides (3) from this reaction can be purified by vacuum distillation at less than 0.5 mmHg. At higher pressures, the high temperature required promotes thermal decomposition of the BOC protecting group.

The amide (3a) is alkylated with acrylonitrile resulting in (4a) in quantitative yield. The homologous nitriles (4b) and (4c) are obtained in 95% yield by alkylation of (3a) and (3b), respectively, with 4-chlorobutyronitrile. These compounds do not require further purification and are taken directly to the corresponding triamines (5a to c) by Raney nickel reduction.

The amines (5a), (5b), or (5c) can also be easily converted to the corresponding secondary *N*-benzylated triamines by brief exposure to trifluoroacetic acid. This has the advantage of overcoming the problems associated with removing small amounts of contaminating N^5-benzylhomospermidine found in our alternative N^4-benzylspermidine synthesis.[50] Alternatively, (5a), (5b), or (5c) can be reacted with BOC-ON followed by hydrogenolysis in methanolic HCl leading to the previously described terminal bis(BOC) reagents.[52] One note of interest is the inertness of the carbamate and the N-benzyl moieties to the nitrile reduction conditions employed. Finally, acylation of the free amine in (5a to c) with trifluoroacetic anhydride gives the versatile triprotected spermidine reagent (6b) and its symmetrical analogues (6a and c) in 91 to 96% yield. These reagents make it possible to access the catecholamide chelators of interest and were utilized in the synthesis of the siderophores, N^1,N^8-bis(2,3-dihydroxybenzoyl)spermidine (Cpd II), parabactin, agrobactin (Scheme 1), and vibriobactin (Scheme 3), as well as a number of homologues and analogues.

N^1,N^8-BIS(2,3-DIHYDROXYBENZOYL)SPERMIDINE (CPD II), PARABACTIN, AND AGROBACTIN

The syntheses of Cpd II,[42-44] parabactin,[45] and agrobactin[46] are initiated with the polyamine reagent N^4-benzylspermidine (1) (Scheme 1). The benzylated spermidine is first terminally bis-*N*-acylated with 2,3-dimethoxybenzoyl chloride generating diamide (2) in 87% yield. Next the N^4-benzyl group is removed by hydrogenolysis over palladium at atmospheric pressure to afford (3) in 90% yield. This intermediate can be taken to parabactin's biological precursor, Cpd II, by exposure to boron tribromide in methylene chloride, thus removing the methyl protecting groups. Alternatively, (3) can be acylated at the free secondary nitrogen with L-*N*-(*tert*-butoxycarbonyl)threonine, activated as the *N*-hydroxysuccinimide ester, producing (4) in 84% yield.

On brief exposure of (4) to trifluoroacetic acid, the *tert*-butoxycarbonyl group collapses to carbon dioxide, isobutylene, and the corresponding amine salt (5) in 95% yield. Again the methoxy methyl groups are removed by treatment of (5) with boron tribromide in methylene chloride providing (6) in 90% yield. Either parabactin or agrobactin can now be accessed depending on which ethyl imidate is condensed with the threonyl segment of the molecule. Stereospecific formation of the acid-sensitive trans oxazoline ring of parabactin was achieved in high yield by heating amino alcohol (6) with ethyl 2-hydroxybenzimidate. The imidate was derived from 2-cyanophenol and ethanolic hydrogen chloride.[45] Alternatively, the condensation of (6) with ethyl 2,3-dihydroxybenzimidate results in agrobactin. Unfortunately, the benzimidate required for agrobactin is not accessible from the reaction of 2,3-dihydroxybenzonitrile with ethanolic hydrogen chloride.

A synthesis of the imidate ester is based upon the *O*-alkylation of amides.[56] The catechol hydroxyls of 2,3-dihydroxybenzaldehyde were first benzylated to prevent unwanted hydroxyl alkylation. Aminolysis of 2,3-bis(benzyloxy)benzoyl chloride, available by a known route from the aldehyde,[57] furnished the required amide in 92% yield. Finally, selective amide *O*-alkylation with triethyloxonium hexafluorophosphate in CH_2Cl_2, followed by basification, provided the dibenzylated imidate ester in 81% yield.

The imidate ester is debenzylated to the less hindred catechol imidate and is condensed with (6), providing agrobactin in 61% yield. It is interesting to note that hydrogenolysis (10% Pd-C, atmospheric pressure) of the dibenzyl imidate resulted in a 73% yield of the benzyl-cleaved product with little or no imidate reduction. Utilizing the protected polyamines, a large number of parabactin and agrobactin analogues were also prepared in high yield, in addition to a variety of structurally related octacoordinate catecholamide actinide-binding ligands (Figure 3).

An alternative approach to the above parabactin synthesis (Scheme 4) involves coupling of the central 2-(2-hydroxyphenyl)-4-carboxyl-5-methyl-2-oxazoline group to the secondary

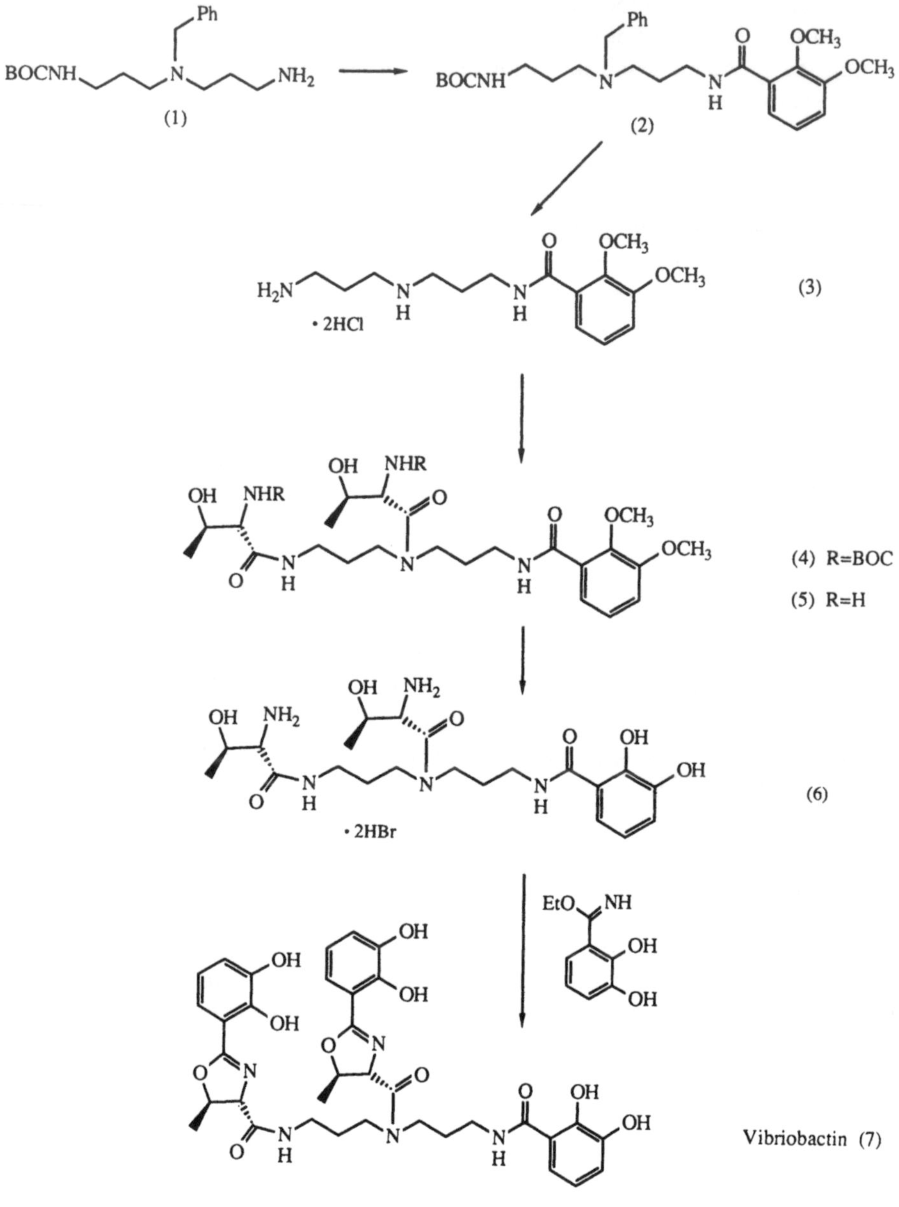

SCHEME 3.   Total synthesis of vibriobactin.

nitrogen of bis(N^1,N^8-carbobenzoxy)spermidine.[49] The scheme begins with the benzylation of methyl 2-hydroxybenzoate (1) followed by hydrolysis of the methyl ester (2) to produce the acid (3). This compound is coupled with 1,3-thiazolidine-2-thione, utilizing dicyclohexyl-carbodiimide in the presence of 4-(dimethylamino)pyridine to afford 3-[2-(benzyloxy) benzoyl]-1,3-thiazolidine-2-thione (4). Threonine is *N*-acylated with the thione (4) to generate (5). Esterification of (5), debenzylation of intermediate (6) to the phenol-ester (7), and cyclization with SOCl$_2$ gives oxazoline (8). The epimerizable hydrogen in (8) is isomerized to provide the trans oxazoline (9), and this intermediate is condensed with N^1,N^8-bis(benzyloxycarbonyl)spermidine (10) in the presence of a phosphonamide coupling agent and *N*,*N*-diisopropylethylamine to produce (11). The N^1,N^8-protecting groups are removed under acidic conditions to yield diamine (12), and the free terminal amino groups reacylated with 2,3-diacetoxybenzoyl chloride to give (13). This intermediate is finally deacylated, providing "crude parabactin".

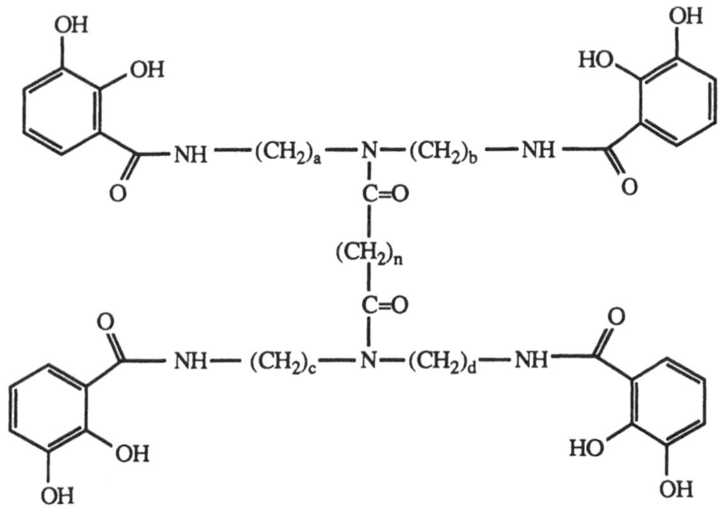

Cpd #	n	a	b	c	d
1	2	3	3	3	3
2	2	4	4	4	4
3	2	3	4	3	4
4	3	3	3	3	3
5	3	4	4	4	4
6	3	3	4	3	4
7	2	3	3	3	4
8	2	3	3	4	4
9	2	3	4	4	4
10	3	3	3	3	4
11	3	3	3	4	4
12	3	3	4	4	4

FIGURE 3. Octacoordinate catecholamide ligands.

VIBRIOBACTIN

This siderophore, N-[3-(2,3-dihydroxybenzamido)propyl]-1,3-bis[2-(2,3-dihydroxy-phenyl)-*trans*-5-methyl-2-oxazoline-4-carboxamido]propane,[28] has no symmetry with respect to the terminal acyl groups and thus presents a new synthetic challenge. The norspermidine backbone has a 2-(2,3-dihydroxyphenyl)-*trans*-5-methyl-2-oxazoline-4-carboxamido group fixed to both a primary and a secondary amine nitrogen. The unsymmetrical structure mandates beginning with a mono primary amino-protected norspermidine or a primary, secondary amino-diprotected norspermidine. The monoprotected triamine would allow for initial fixing of the threonyl groups, while the diprotected triamine would allow for initial attachment of the 2,3-dihydroxybenzoyl group. The polyamine reagents described allow for either approach. The latter route was adopted in order to avoid the problems associated with acylation of the intermediate bis(threonyl) or bis(oxazoline) compounds.

The synthesis of vibriobactin (Scheme 3) begins with N^4-benzyl-N^1-(*tert*-butoxycarbonyl)norspermidine (1). The free primary amine is acylated with 2,3-dimethoxybenzoyl chloride in the presence of triethylamine providing the trisubstituted norspermidine (2) in quantitative yield.[47] Both of the amine protecting groups are next removed, producing monoacylated norspermidine (3). The order of deprotection is unimportant: the *tert*-butoxycarbonyl group is removed by brief exposure to trifluoroacetic acid and the N^4-benzyl group by hydrogenolysis over palladium chloride in methanolic HCl, furnishing the dihydrochloride

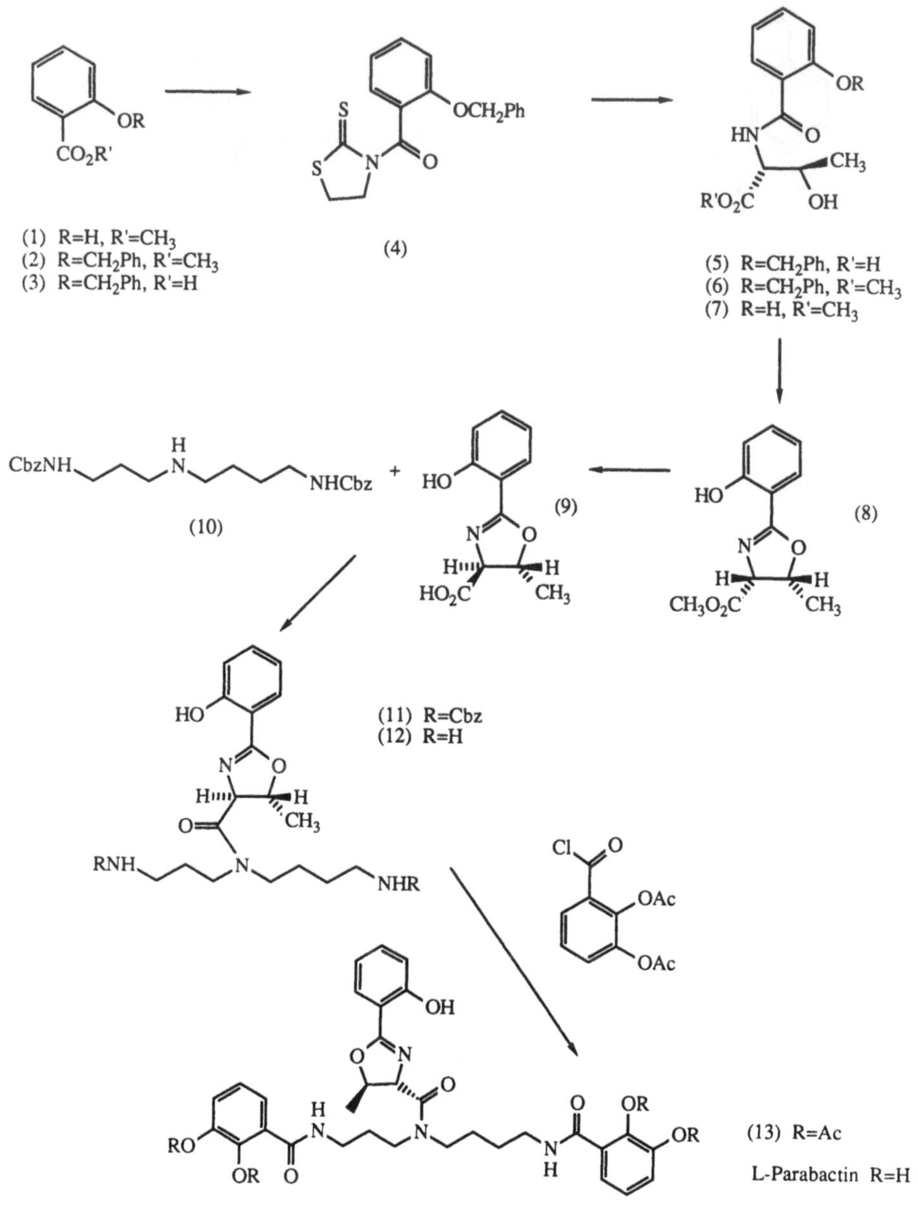

(1) R=H, R'=CH$_3$
(2) R=CH$_2$Ph, R'=CH$_3$
(3) R=CH$_2$Ph, R'=H

(4)

(5) R=CH$_2$Ph, R'=H
(6) R=CH$_2$Ph, R'=CH$_3$
(7) R=H, R'=CH$_3$

(10)

(9)

(8)

(11) R=Cbz
(12) R=H

(13) R=Ac

L-Parabactin R=H

SCHEME 4.   Total synthesis of parabactin.

salt of the monoacylated norspermidine (3) in 93% overall yield. Exposure of (3) to aqueous base produces the free diamine, which is bis-acylated in DMF with the activated ester of L-*N*-(*tert*-butoxycarbonyl)threonine. The unsymmetrically substituted norspermidine (4) is obtained in quantitative yield. The BOC protecting groups of (4) are easily removed by brief exposure to trifluoroacetic acid and N^1,N^4-bis(L-threonyl)-N^7-(2,3-dimethoxy benzoyl)norspermidine (5) is obtained in 93% yield. The catechol methyl protecting groups are removed with BBr$_3$ in methylene chloride, affording the dihydrobromide salt of the completely deprotected triamide (6) in 63% yield. The final step, condensation of the threonyl groups with an imidate, is accomplished by reaction of (6) with excess of ethyl 2,3-dihydroxybenzimidate in refluxing methanol, producing vibriobactin (7) in 58% yield.

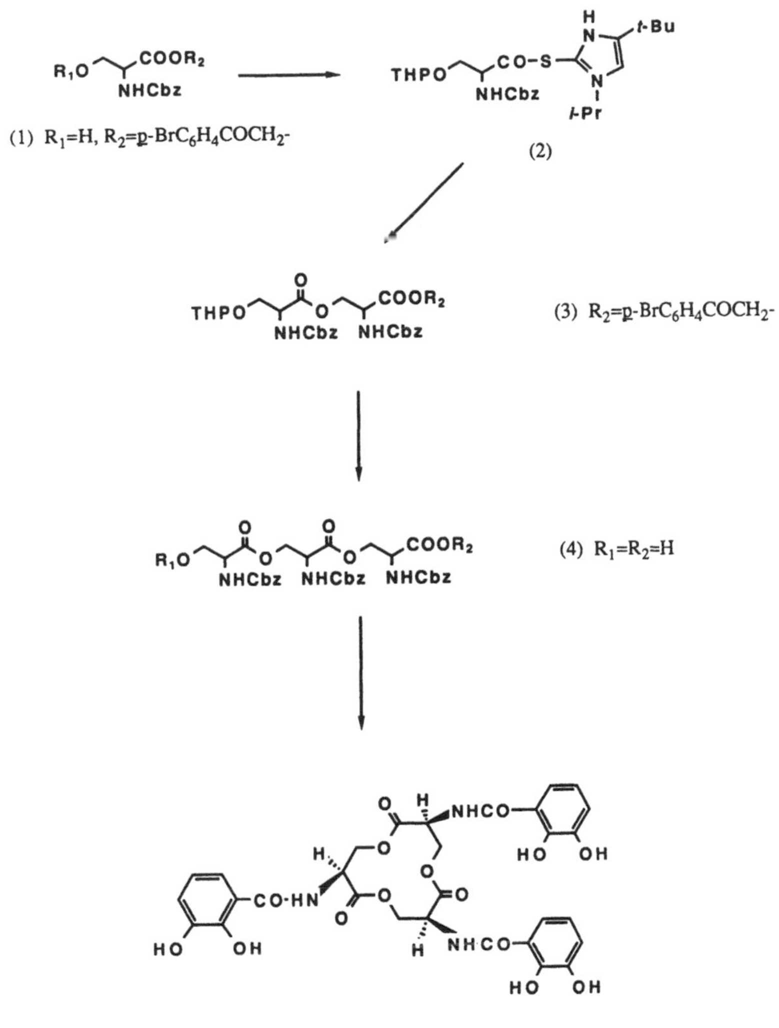

SCHEME 5. Total synthesis of enterobactin (1).

ENTEROBACTIN

This siderophore was first synthesized[58] employing double activation for closure to the macrocyclic lactone (Scheme 5). The key intermediates are the *p*-bromophenacyl ester of L-*N*-(benzyloxycarbonyl)serine (1) and the *O*-(tetrahydropyranyl)-L-*N*-(benzyloxycarbonyl)serine thioester (2). The thioester is condensed with protected serine (1) producing the ester (3) in excellent yield. The phenacyl protecting group is next removed with zinc and acetic acid and the free carboxyl group converted to the thioester by reaction with bis(4-*tert*-butyl-1-isopropylimidazol-2-yl)disulfide and triphenylphosphine in benzene. This thioester is coupled with (1) and the tetrahydropyranyl group of the triserine product removed by exposure to acetic acid and methanol in THF. Next the *p*-bromophenacyl group is removed by treatment with zinc and acetic acid and the intermediate (4) cyclized by carboxyl activation again with bis(4-*tert*-butyl-1-isopropylimidazol-2-yl)disulfide and triphenylphosphine in benzene. The benzyloxycarbonyl protecting groups are cleaved by hydrogenolysis and the triamine trihydrochloride reacted with 2,3-dihydroxybenzoyl chloride in the presence of triethylamine in tetrahydrofuran to generate enterobactin. One could utilize this method to produce small quantities of enantioenterobactin simply by replacement of the L-serine with

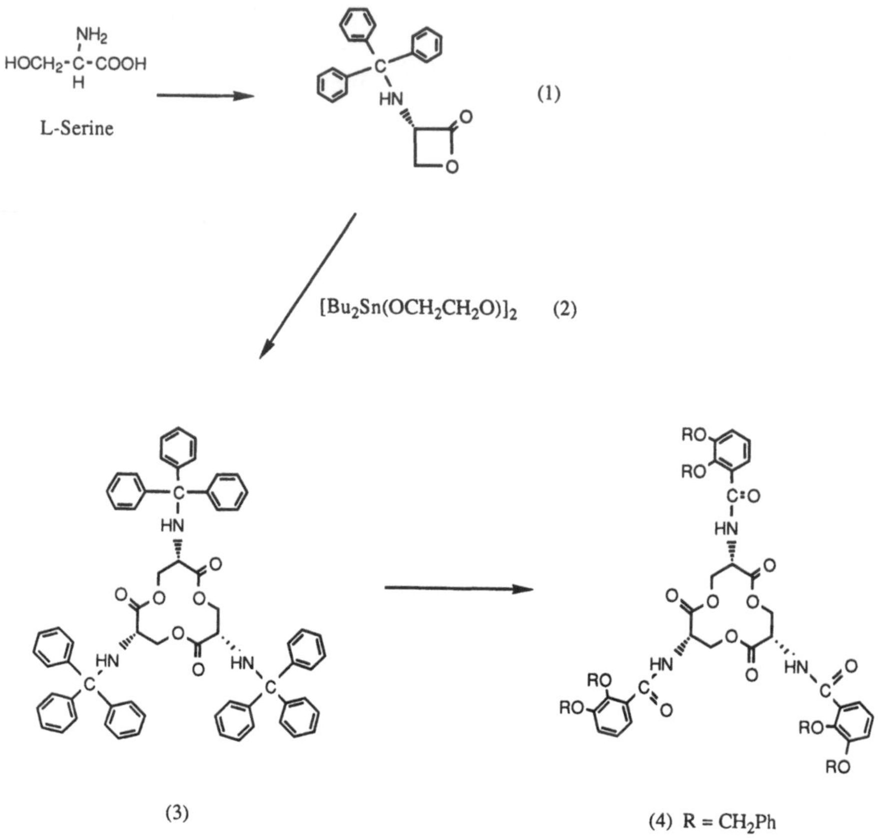

SCHEME 6.   Total synthesis of enterobactin (2).

D-serine, and, in fact, a similar synthesis of enantioenterobactin has been described.[57] The difference in the two approaches is associated with the serine carboxyl and the catechol hydroxyl protecting groups. In the alternate scheme, the carboxyl is protected with the photochemically labile 2-methylanthraquinone, and the 2,3-dihydroxybenoyl groups are introduced by coupling the amines with 2,3-bis(benzyloxy)benzoyl chloride. The benzyl groups are removed under hydrogen olysis conditions. Enantioenterobactin, the target system, was accessed in this 12-step synthesis in approximately a 20% yield.

A somewhat less tedious method of accessing enterobactin[59] employs tin as a metal template (Scheme 6). *N*-tritylated-L-serine is cyclized to its β-lactone (1) utilizing 4-(dimethylamino)pyridine and diisopropylcarbodiimide. The stannoxane (2) is employed to cyclize (1) to the enterobactin skeleton (3) and the trityl protecting groups are removed with HC1 in ethanol. The product is acylated with the *p*-nitrophenyl ester of 2,3 bis(benzyloxy)-benzoic acid to produce (4) and the *O*-benzyl protecting groups removed by hydrogenolysis generating enterobactin.

ENTEROBACTIN ANALOGUES

In an attempt to further understand how microorganisms process the enterobactin iron complex and to prevent the kind of aqueous or nonspecific serum esterase hydrolysis which enterobactin is subject to, several analogues were synthesized (Figure 4). The first of these was a carbocyclic system[60] predicated on all *cis*-cyclododecane-1,5,9-triol. The triol is first tosylated and converted to the corresponding tris(azide) by reacting the tritosylate with sodium azide in dimethylformamide. The tris(azide) is catalytically reduced to the corresponding

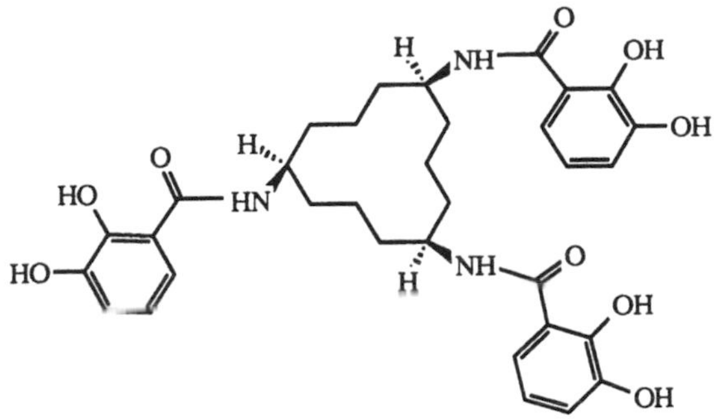

FIGURE 4. Enterobactin analogues.

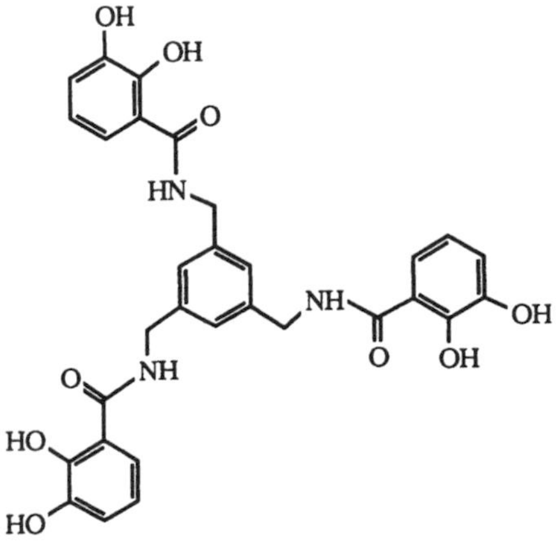

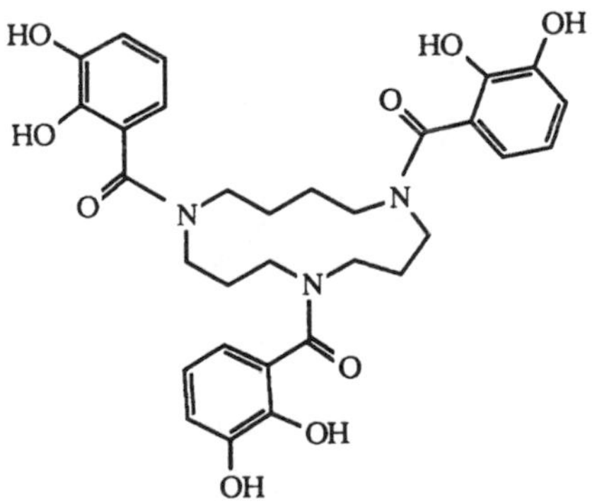

triamine and acylated with the acetonide of 2,3-dihydroxybenzoyl chloride. In the last step, the isopropylidene protecting groups are removed by heating the tris(amide) in aqueous acetic acid. The product hexacoordinate ligand (1) was shown to form very stable complexes with iron(III) and also to promote iron transport in microbial systems.

The next isosteric equivalent of enterobactin synthesized, 1,3,5,tris[[(2,3-dihydroxy-benzoyl)amino]methyl]benzene (2),[61,62] is predicated on a rigid 1,3,5-trisubstituted benzene platform. The synthesis of this enterobactin analogue is accomplished very effectively by coupling 1,3,5-tris(aminomethyl)benzene trihydrochloride, which is obtained by catalytic reduction of 1,3,5-benzenetrialdoxime, with 2,3-bis(benzyloxy)benzoyl chloride, affording the triacylated amine in 81% yield. The benzyl protecting groups are easily removed by hydrogenolysis over palladium in ethanolic acetic acid to provide (2) in 90% yield. This analogue is also very effective at delivering iron to a number of *Escherichia coli* mutants.

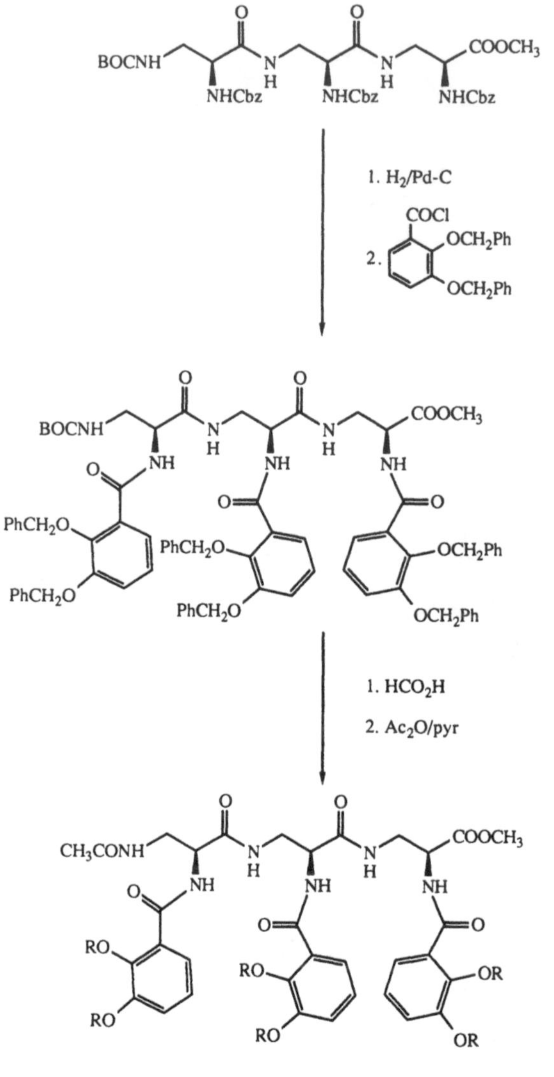

SCHEME 7.   Total synthesis of enterobactin analogue.

An alternative scheme for the synthesis of this system is initiated by the reaction of "trimzoyl" chloride with concentrated ammonium hydroxide to produce the corresponding amide. The amide is reduced by reduction with diborane in THF to generate the triamine, which is acylated with 2,3-dimethoxybenzoyl chloride. The methyl protecting groups are easily removed with boron tribromide in methylene chloride to again give hexacoordinate (2).

In further exploring the role of the platform in biological activity, another cyclic enterobactin analogue was synthesized, an N,N',N''-tris(2,3-dihydroxybenzoyl) derivative of 1,5,9-triazacyclotridecane (3)[62] (Figure 4). The parent macrocycle can be generated by the reaction of N,N',N''-tris(p-toluenesulfonyl)-1,5,10-triazadecane dianion with 1,3-ditosyloxypropane. The p-toluenesulfonyl protecting groups are removed by treatment with concentrated sulfuric acid. The resulting cyclic triamine is condensed with 2,3-dimethoxybenzoyl chloride, followed by removal of the methyl protecting groups with BBr$_3$ and methylene chloride to yield analogue (3).

Finally, an acyclic enterobactin model system was also synthesized (Scheme 7). The synthesis of this compound requires some 22 steps.[57] The key intermediated (1), a diamido ester with three internal carbobenzyloxy-protected nitrogens and a terminal *tert*-butoxycarbonyl-protected nitrogen, is hydrogenolyzed removing the carbobenzyloxy groups, followed by acylation of the free nitrogens with 2,3-bis(benzyloxy)benzoyl chloride to produce the polyamide (2). The *tert*-butoxycarbonyl protecting group of (2) is next removed with formic acid followed by acetylation of the free nitrogen generating (3), and finally the catechol benzyl protecting groups were removed under hydrogenolysis conditions resulting in the linear tris(catechol)enterobactin analogue (4).

It is interesting that the iron(III) acyclic amido analogue (4) and the enterobactin Fe(III) complex CD curves are essentially identical and that the stability constant for the linear iron(III) complex, 10^{46} M^{-1}, is also very high.

CHRYSOBACTIN

This siderophore, isolated from the phytopathogenic bacterium *Erwinia chrysanthemi*, is structurally one of the simplest catecholamide chelators. The synthesis of this siderophore, N-[N^2-(2,3-dihydroxybenzoyl)-D-lysyl]-L-serine (6),[63] begins with the condensation of N^6-Cbz-D-lysine (1) with the *p*-nitrophenyl ester of 2,3-dibenzyloxybenzoic acid (2) (Scheme 8). The resulting N^2-(2,3-dibenzyloxybenzoyl)-N^6-Cbz-D-lysine (3) is next coupled with L-serine benzyl ester (4) in the presence of EEDQ and triethylamine. The benzyl protecting groups of the condensation product (5) are removed by hydrogenation over palladium to make chrysobactin (6). Similar peptide methods will be nicely applicable to related siderophores, e.g., amonobactin.

HYDROXAMATE CHELATORS

The major functional difference between hydroxamate siderophores and the catecholamides is related to environmental iron concentration.[64] The hydroxamates are generated by the microorganism in a high iron environment, while the catecholamide "backup" system is activated when iron concentrations are low. It follows that the catecholamide chelators typically bind iron far more tightly than hydroxamates.

In this section the total synthesis of a number of hydroxamate chelators will be described. This will begin with the cadaverine containing siderophores, desferrioxamine B, desferrioxamine G, desferrioxamine E, bisucaberin, and arthrobactin, followed by the ornithine-derived ligands, rhodotorulic acid, and desferri-ferrichrome. The synthesis of the 1,3-diaminopropane-based chelator schizokinen will be described, and the section will also include the syntheses of aerobactin and mycobactin S2, which are predicated on lysine.

The key to the synthesis of the hydroxamate siderophores is the availability of the appropriately protected N-hydroxydiamines and amino acids. Where appropriate, we will indicate how such reagents developed for the total synthesis of the natural products can be utilized for the generation of various analogues and homologues.

DESFERRIOXAMINE B — NITRO SYNTHESIS

Desferrioxamine B is a linear trihydroxamate ligand isolated from *Streptomyces pilosus* (Figure 2), which forms a very stable hexacoordinate, octahedral[65] complex with iron(III), $K_f = 1 \times 10^{30}$ M^{-1}. Interestingly, a series of iron(III) chelators, desferrioxamines A-1,[66] have been isolated from the same soil bacterium. Although desferrioxamine B will bind a number of different +3 cations, e.g., Al(III), Ga(III), and Cr(III) (as do a number of the catecholamide siderophores), it exhibits a high specificity for iron(III). Desferrioxamine B, as its mesylate salt Desferal, has been employed in the treatment of several iron overload diseases, e.g., thalassemia;[67] however, its short half-life in the body requires that patients be maintained on constant infusion therapy. This failing has compelled investigators to continue the search for better therapeutic iron chelators. However, approaches have not

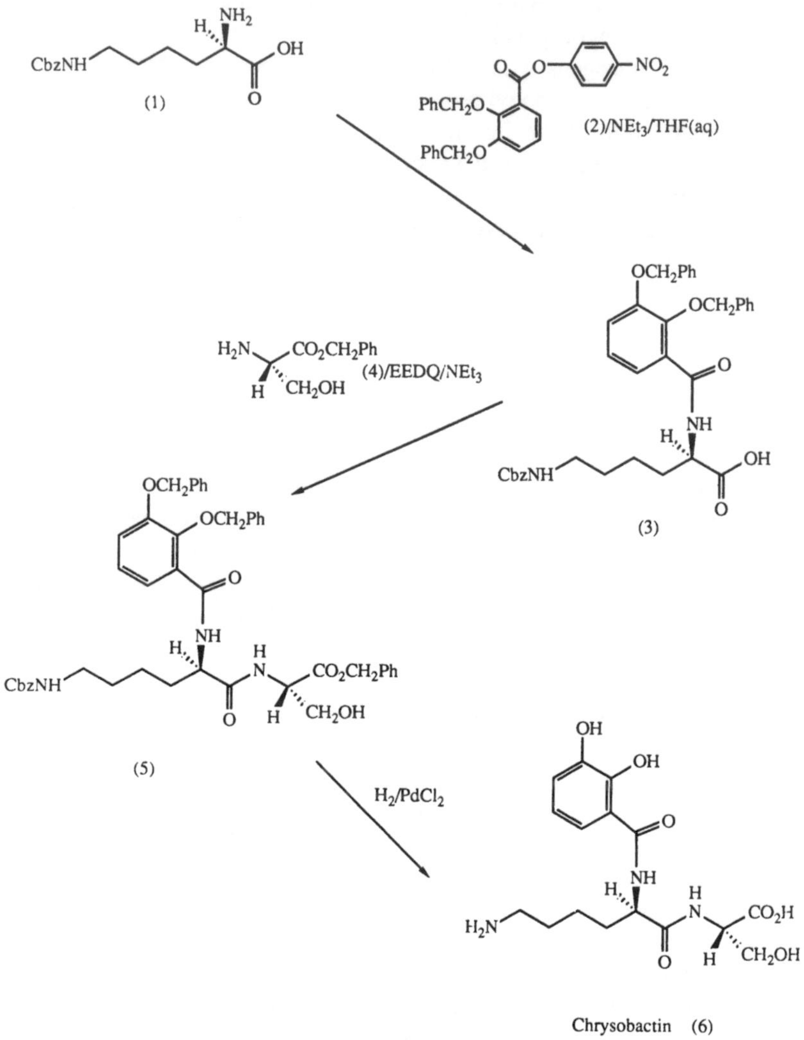

SCHEME 8. Total synthesis of chrysobactin.

included modification of the desferrioxamine B molecule itself, probably because of the lack of a high yield, facile entry into the desferrioxamine B skeleton.

Desferrioxamine B is predicated on two fundamental synthons, 1-amino-5-(N-hydroxy-amino)pentane and succinic acid, and the key to its synthesis is the production and regiospecific condensation of these units. The initial synthesis of this hydroxamate (Scheme 9) begins with 1-amino-5-nitropentane (1), a starting material which is accessible in only 46% yield.[68] This amine is protected with a carbobenzyloxy group, generating (2). Reduction of the terminal nitro group in (2) produces the hydroxyamino compound (3), which is condensed with either succinic anhydride providing (4) or with acetic anhydride giving (5). The half-acid amide (4) is cyclized to (6), while the carbobenzyloxy protecting group of (5) is removed by hydrogenolysis. The amine (7) is condensed with (6) and the resulting amide (8) deprotected and coupled with a second equivalent of (6) to produce (10). Catalytic reduction of (10) results in a 6% overall yield of desferrioxamine B (11).

N-HYDROXYDIAMINE REAGENT

Since an *N*-hydroxydiamine containing three to five carbons is the basic unit of many

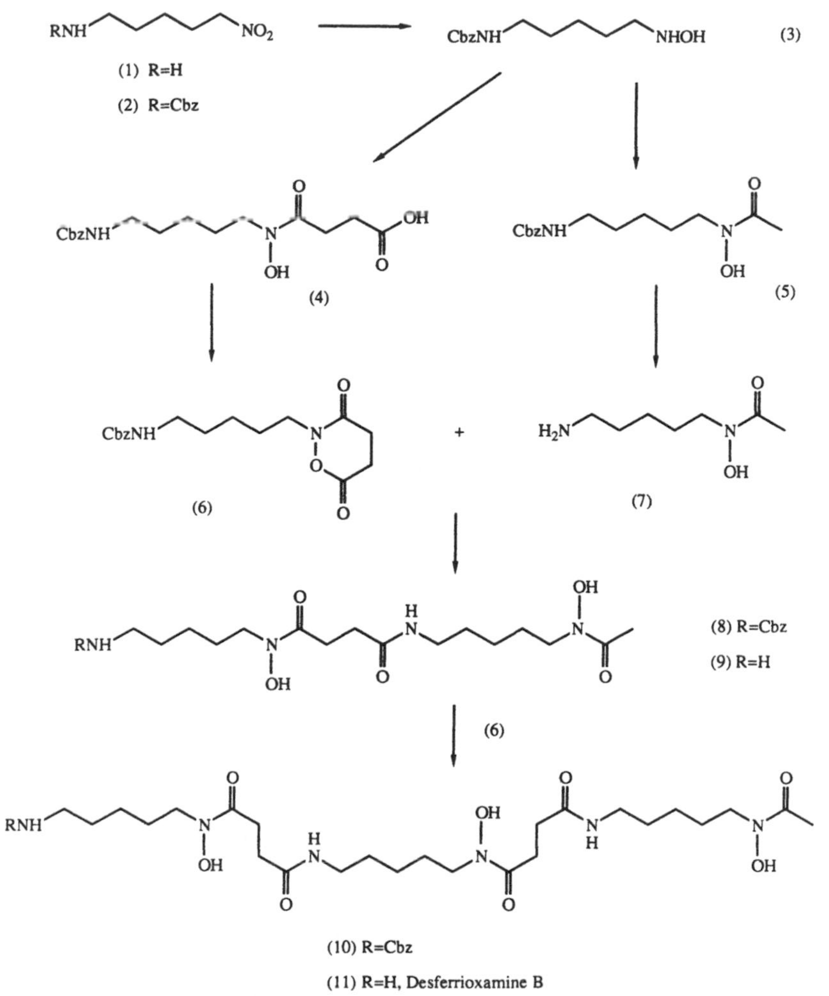

SCHEME 9. Total synthesis of desferrioxamine B (1).

hydroxamate chelators, a method of differentiating the two nitrogens and protecting the oxygen is desirable. The generation and selective deprotection of triprotected N-hydroxy-cadaverine (3) is presented in Scheme 10. In this versatile reagent the primary amine is masked as a nitrile, while the hydroxylamine is N-tert-butoxycarbonyl, O-benzyl diprotected. The synthesis of reagent (3)[69] begins with the conversion of O-benzylhydroxylamine hydrochloride (1) to its crystalline N-(tert-butoxycarbonyl) derivative (2). This is accomplished by reaction of (1) with di-tert-butyl dicarbonate (NEt$_3$, aqueous THF). The carbamate (2), obtained in 97% yield, is next N-alkylated with 5-chlorovaleronitrile (NaH, DMF, NaI) to give O-benzyl-N-(4-cyanobutyl)hydroxylamine (3) in 87% yield. The production of reagent (3) is not only efficient but is also flexible; N-(tert-butoxycarbonyl)amine (2) could also be monoalkylated with commercially available Cl(CH$_2$)$_n$CN, n = 1 to 3, to shorten the N-hydroxydiamine chain or with 7-bromoheptanenitrile to lengthen it. Thus, a large number of natural hydroxamate chelators and their homologues are accessible. Brief exposure of N-(tert-butoxycarbonyl) nitrile (3) to trifluoroacetic acid (TFA) results in collapse to carbon dioxide, isobutylene, and O-benzyl-N-I(4-cyanobutyl)hydroxylamine (4) in 75% yield, thus freeing up the hydroxylamine nitrogen. Alternatively, nitrile (3) is selectively hydrogenated in the presence of the benzyl group to generate primary amine (5) in 83% yield with W-2

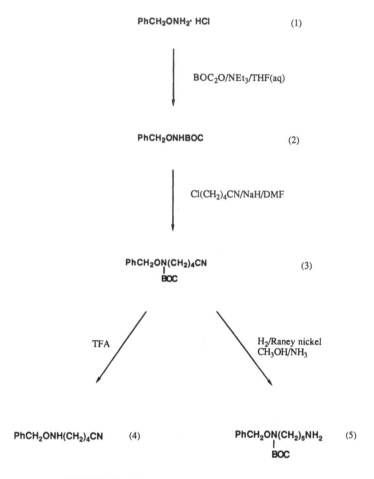

PhCH₂ONH₂· HCl   (1)

BOC₂O/NEt₃/THF(aq)

PhCH₂ONHBOC   (2)

Cl(CH₂)₄CN/NaH/DMF

PhCH₂ON(CH₂)₄CN   (3)
|
BOC

TFA   H₂/Raney nickel
                       CH₃OH/NH₃

PhCH₂ONH(CH₂)₄CN   (4)   PhCH₂ON(CH₂)₅NH₂   (5)
                                |
                                BOC

SCHEME 10.   Triprotected N-hydroxydiamine reagent.

grade Raney nickel in methanolic ammonia. The hydroxylamine oxygen in (3) remains protected by the benzyl moiety until catalytic (Pd-C) unmasking of the hydroxamic acid chelator in the last step of the total synthesis.

DESFERRIOXAMINE B

An alternative and more efficient synthesis of desferrioxamine B (Scheme 11) is predicated on the convenient production of *O*-benzyl-*N*-(4-cyanobutyl)hydroxylamine (2) from the *N*-hydroxycadaverine reagent (1). The nitrile (2) is condensed with either succinic anhydride to produce the half-acid amide (3) in 88% yield or with acetic anhydride, resulting in a quantitative yield of cyano acetyl amide (4). The amide (4) is next hydrogenated to primary amine (5) in 82% yield in methanolic ammonia utilizing prewashed nickel catalyst. The amine (5) is condensed with the half-acid amide (3) in an 88% yield utilizing dicyclohexylcarbodiimide. The resulting nitrile (6) is then reduced with Raney nickel to the corresponding amine (7) in 82% yield.[70] The amine (7) is coupled with the half-acid amide (3), again with dicyclohexylcarbodiimide, and the resulting nitrile (8) isolated in 88% yield. Finally the nitrile (8) is reduced to desferrioxamine B (9) in 84% yield utilizing 10% Pd/C in 0.1 *N* HCl and methanol.

This scheme, because of its flexibility, lends itself nicely to the preparation of various analogues and homologues. For example, the synthesis can be terminated at the nitrile (6) and this reduced to the tetracoordinate ligand simply by hydrogenation over palladium. Alternatively, the octacoordinate ligand can be synthesized by reduction of the nitrile group

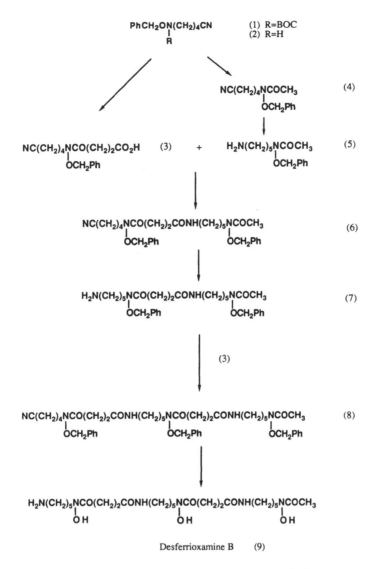

PhCH$_2$ON(CH$_2$)$_4$CN   (1) R=BOC
  | (2) R=H
  R

NC(CH$_2$)$_4$NCOCH$_3$ (4)
         |
         OCH$_2$Ph
         |

NC(CH$_2$)$_4$NCO(CH$_2$)$_2$CO$_2$H   (3)   +   H$_2$N(CH$_2$)$_5$NCOCH$_3$ (5)
         | |
         OCH$_2$Ph   OCH$_2$Ph

NC(CH$_2$)$_4$NCO(CH$_2$)$_2$CONH(CH$_2$)$_5$NCOCH$_3$ (6)
         | |
         OCH$_2$Ph   OCH$_2$Ph

H$_2$N(CH$_2$)$_5$NCO(CH$_2$)$_2$CONH(CH$_2$)$_5$NCOCH$_3$ (7)
          | |
          OCH$_2$Ph   OCH$_2$Ph

(3)

NC(CH$_2$)$_4$NCO(CH$_2$)$_2$CONH(CH$_2$)$_5$NCO(CH$_2$)$_2$CONH(CH$_2$)$_5$NCOCH$_3$ (8)
         | | |
         OCH$_2$Ph   OCH$_2$Ph   OCH$_2$Ph

H$_2$N(CH$_2$)$_5$NCO(CH$_2$)$_2$CONH(CH$_2$)$_5$NCO(CH$_2$)$_2$CONH(CH$_2$)$_5$NCOCH$_3$
          | | |
          O H   O H   O H

Desferrioxamine B   (9)

SCHEME 11.   Total synthesis of desferrioxamine B (2).

in (8), followed by condensation of the amine with another equivalent of (3) and reduction of the condensate with hydrogen over palladium. Utilizing this method, the desaminodesferrioxamine (6) (Scheme 12) can be prepared.

The desamino analogue synthesis begins with conversion of heptanal (1) to an oxime by condensation with the hydrochloride salt of O-benzylhydroxylamine. The resulting oxime is reduced with sodium cyanoborohydride to the corresponding O-benzylhydroxylamine (2) in 22% yield. The hydroxylamine (2) is condensed with succinic anhydride to generate the half-acid amide (3) in 85% yield, and this amide is condensed with amine (7) (synthesized in Scheme 11). The condensation product (5) is finally reduced with hydrogen over palladium, resulting in the desamino analogue (6) in 86% yield.

DESFERRIOXAMINE G

The synthetic methodology based upon the hydroxydiamine reagent also offers entry into other acyclic as well as cyclic hydroxamates, including the linear hexacoordinate ω-amino acid desferrioxamine G, also isolated from *Streptomyces pilosus* (Figure 2). Easy

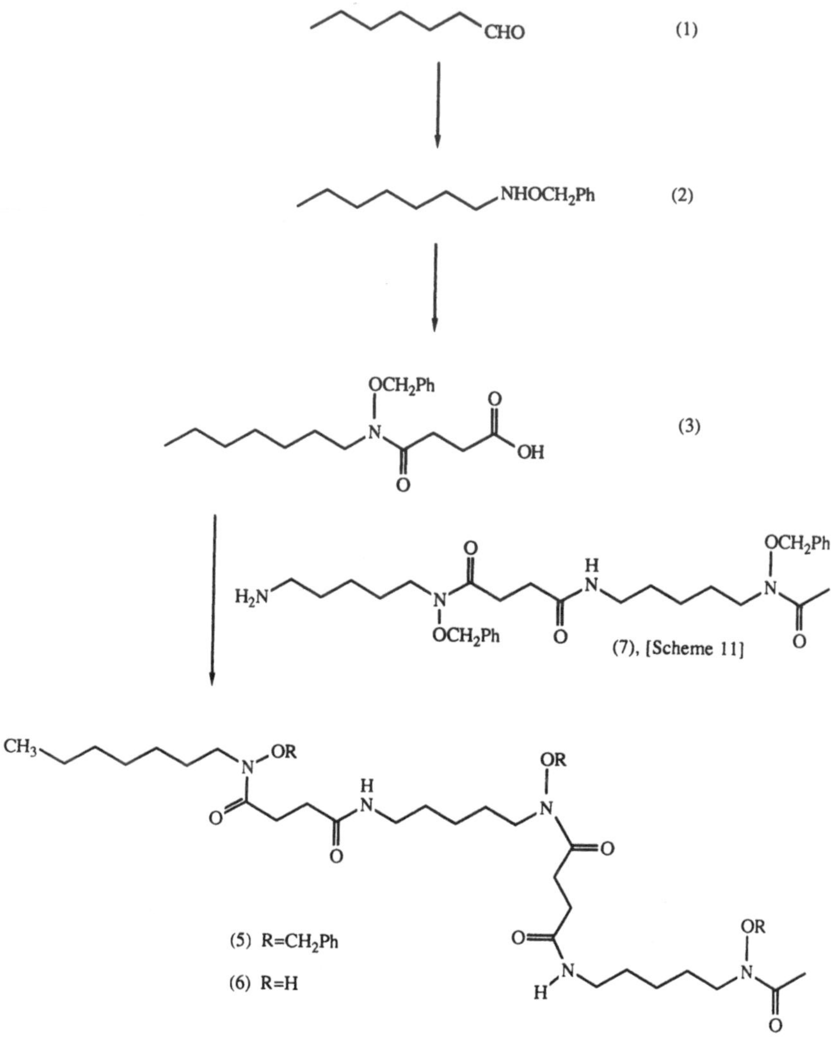

SCHEME 12.   Synthesis of desferrioxamine B analogue.

access to desferrioxamine G, which contains alternating *N*-hydroxycadaverine and succinic acid segments, would provide a versatile molecule which could be funtionalized at either the amine or carboxylic acid terminus to generate a wide variety of prodrugs and analogues for biological evaluation.[71]

Primary amine (1), which is derived from the *N*-hydroxycadaverine reagent (Scheme 10), is coupled with nitrile-acid (2), a desferrioxamine B segment (Scheme 11), in the presence of dicyclohexylcarbodiimide and catalytic DMAP to afford nitrile (3) in 65% yield (Scheme 13). The *tert*-butoxycarbonyl group of (3) is cleaved by brief exposure of the compound to trifluoroacetic acid, providing the secondary amine (4) in 83% yield. This amine is next acylated with succinic anhydride in pyridine affording nitrile acid (5) in 96% yield, which is a masked tetracoordinate amino acid.

Nitrile acid (5) and *N*-(5-aminopentyl)hydroxylamine (1) are coupled again using dicyclohexylcarbodiimide to form the cyano compound (6) in 44% yield. Addition of *tert*-(butoxycarbonyl)nitrile (6) in methylene chloride to trifluoroacetic acid at 0°C gave *N*-benzyloxyamine (7) in quantitative yield along with carbon dioxide and isobutylene. Condensation of the third succinate unit was accomplished with succinic anhydride in pyridine,

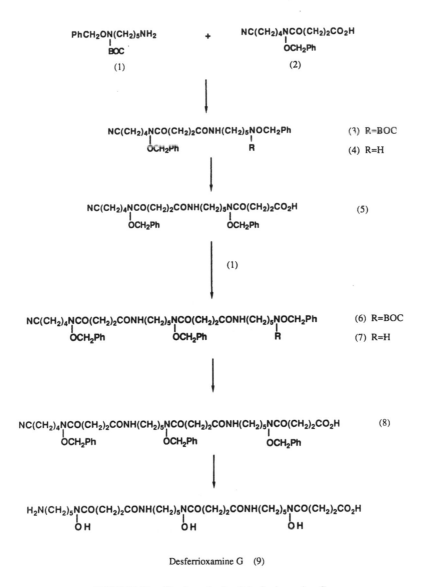

Desferrioxamine G   (9)

SCHEME 13.   Total synthesis of desferrioxamine G.

as before, to produce nitrile acid (8) in 78% yield. Hydrogenation of the trihydroxamate ester (8) under the mildly acidic conditions which were used to generate desferrioxamine B (Pd catalyst at atmospheric pressure) removed the three benzyl groups and saturated the nitrile to produce desferrioxamine G as its hydrochloride salt (92%).

BISUCABERIN

The above methodology also provides for entry into cyclic hydroxamate analogues including bisucaberin, a 22-membered ring dihydroxamate ligand (Figure 2).[69] Retrosynthetic analysis of this cyclic tetracoordinate chelator reveals that the molecule can also be segmented into the two repeating units, succinic acid and 1-amino-5-(hydroxyamino)pentane, (N-hydroxycadaverine). The interest in this molecule, isolated from the marine bacterium *Alteromonas haloplanktis*, is related to the fact that it was found to slow the growth of both L-1210 and IMC carcinoma cells and to sensitize tumor cells to macrophage-promoted cytolysis. Interestingly, the biological activity of bisucaberin is absent in nocardamine or desferrioxamine E, the analogous 33-membered trihydroxamate cyclic siderophore (Figure 2).[71] Both

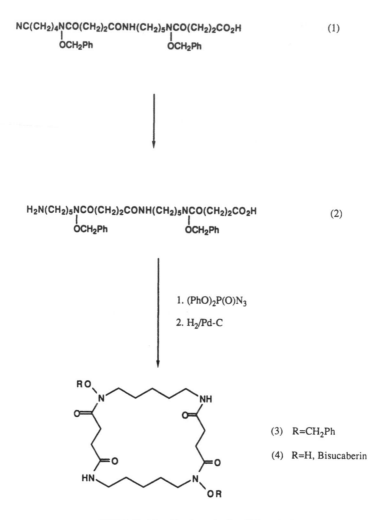

NC(CH₂)₄NCO(CH₂)₂CONH(CH₂)₅NCO(CH₂)₂CO₂H   (1)

SCHEME 14.   Total synthesis of bisucaberin.

bisucaberin and nocardamine bind with ferric ion to form a red, water-soluble complex. *N*-(*t*-butoxycarbonyl)nitrile ([3], Scheme 10) is efficiently converted to nitrile (1), which is a desferrioxamine G intermediate (Scheme 14). Selective hydrogenation of the nitrile acid (1), which contains the basic molecular framework of bisucaberin (4), with Raney nickel in methanolic ammonia furnishes the ω-amino acid (2) in 65% yield. At this point all that remains is the cyclization of (2) and the unmasking of the hydroxamates. Diphenylphosphoryl azide, the Yamada reagent,[72,73] which has been utilized for the formation of large lactams, is employed in this system with success. Specifically, the azide reagent (1.2 equivalent) is added to a solution of amino acid (2) in DMF and stirred for 3 d at 0°C. The hydroxamate-protected macrocycle, 6,17-dibenzylbisucaberin (3), is obtained in 43% yield. Finally, the benzyl groups of (3) are removed under a hydrogen atmosphere (10% Pd-C, CH₃OH, 1 atm) to give bisucaberin (4) in 89% yield.

This synthetic route permits the efficient production of homologues of bisucaberin (4). The size of the *N*-hydroxydiamine chain can be altered, depending on the halonitrile used to prepare the key reagent (3) (Scheme 10). Moreover, replacement of succinic anhydride with glutaric anhydride in one or both of the benzyloxyamine acylations will give homologues of nitrile acid (1) (Scheme 14). Thus, the length of each methylene chain in bisucaberin (4) can be adjusted to provide a variety of both symmetrical and unsymmetrical bisucaberin homologues in order to tailor the chelator's size to a given metal and to determine structure-activity relationships.

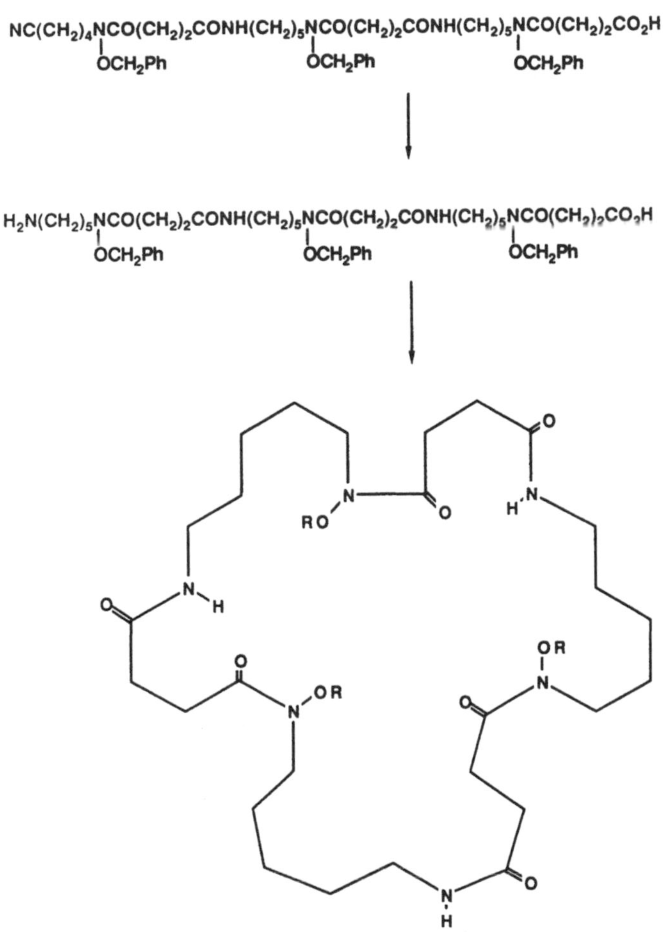

$$NC(CH_2)_4NCO(CH_2)_2CONH(CH_2)_5NCO(CH_2)_2CONH(CH_2)_5NCO(CH_2)_2CO_2H \qquad (1)$$
$$\underset{OCH_2Ph}{|} \qquad\qquad \underset{OCH_2Ph}{|} \qquad\qquad \underset{OCH_2Ph}{|}$$

$$H_2N(CH_2)_5NCO(CH_2)_2CONH(CH_2)_5NCO(CH_2)_2CONH(CH_2)_5NCO(CH_2)_2CO_2H \qquad (2)$$
$$\underset{OCH_2Ph}{|} \qquad\qquad \underset{OCH_2Ph}{|} \qquad\qquad \underset{OCH_2Ph}{|}$$

(3) R=CH₂Ph

(4) R=H DFO E, Nocardamine

SCHEME 15.   Total synthesis of desferrioxamine E.

DESFERRIOXAMINE E (NOCARDAMINE)

Nocardamine (Figure 2) was determined to be a 33-membered macrocycle, which contains three hydroxamate coordination sites evenly spaced around the ring. This large ring iron chelator, one of the series of siderophores isolated from *S. pilosus*, is the hexacoordinate homologue of bisucaberin (Figure 2). The total synthesis of nocardamine (4) (Scheme 15) further exemplifies the generality of the methods utilized to prepare linear desferrioxamines B and G and macrocyclic bisucaberin. In fact, cyclization of the amino acid desferrioxamine G would give desferrioxamine E directly.

In Scheme 15 the hydroxamic acids remain protected as their benzyl esters until the last step of the synthesis. The key nitrile acid (1), the penultimate desferrioxamine G intermediate, is subjected to mild catalytic conditions (Raney nickel, methanolic ammonia) to give tribenzyl amino acid (2) in 21% yield. Acyclic precursor (2) in DMF (2.6 mM) is cooled to 0°C, and diphenylphosphoryl azide, the Yamada reagent, is added and stirred for 4 d at 0°C to product O,O',O''-tribenzylnocardamine (3) in 54% percent yield. The generation of such a large ring, 33 membered, by routine reaction conditions further illustrates the usefulness of this lactamization method. Finally, catalytic removal of the O-benzyl protecting groups of (3) under mild conditions gives nocardamine (4) in quantitive yield.[71]

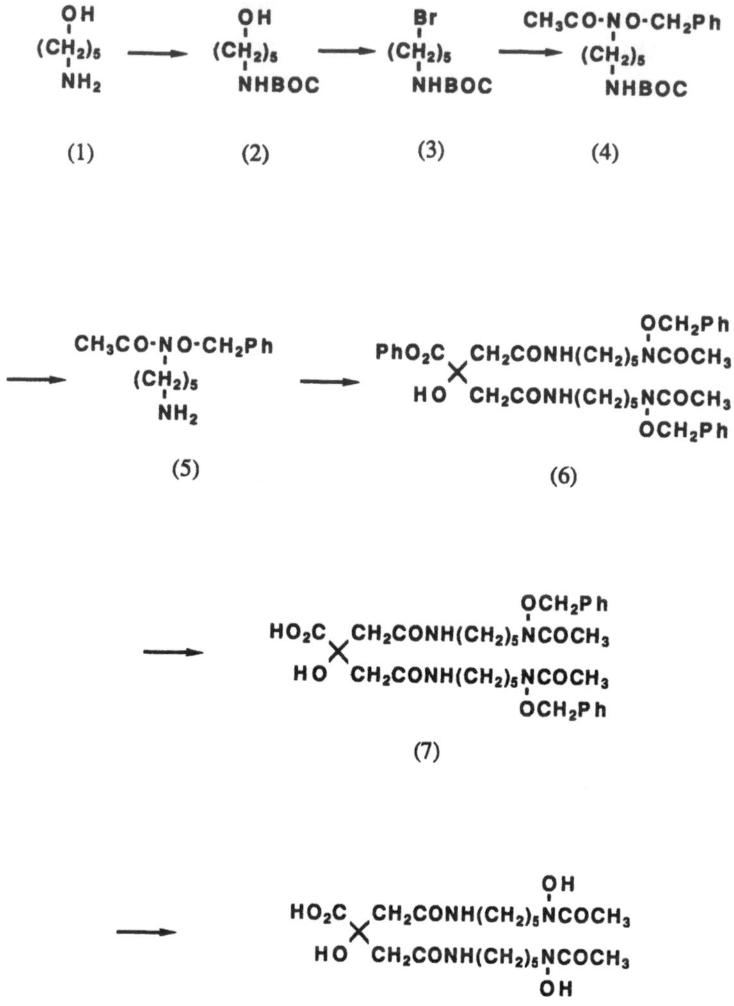

$$\underset{\text{NH}_2}{\overset{\text{OH}}{\underset{|}{(\text{CH}_2)_5}}} \longrightarrow \underset{\text{NHBOC}}{\overset{\text{OH}}{\underset{|}{(\text{CH}_2)_5}}} \longrightarrow \underset{\text{NHBOC}}{\overset{\text{Br}}{\underset{|}{(\text{CH}_2)_5}}} \longrightarrow \underset{\text{NHBOC}}{\overset{\text{CH}_3\text{CO-N O-CH}_2\text{Ph}}{\underset{|}{(\text{CH}_2)_5}}}$$

(1)   (2)   (3)   (4)

$$\longrightarrow \underset{\text{NH}_2}{\overset{\text{CH}_3\text{CO-N O-CH}_2\text{Ph}}{\underset{|}{(\text{CH}_2)_5}}} \longrightarrow \underset{\underset{\text{OCH}_2\text{Ph}}{\text{HO} \quad \text{CH}_2\text{CONH(CH}_2)_5\text{NCOCH}_3}}{\overset{\text{OCH}_2\text{Ph}}{\text{PhO}_2\text{C} \quad \text{CH}_2\text{CONH(CH}_2)_5\text{NCOCH}_3}}$$

(5)   (6)

$$\longrightarrow \underset{\underset{\text{OCH}_2\text{Ph}}{\text{HO} \quad \text{CH}_2\text{CONH(CH}_2)_5\text{NCOCH}_3}}{\overset{\text{OCH}_2\text{Ph}}{\text{HO}_2\text{C} \quad \text{CH}_2\text{CONH(CH}_2)_5\text{NCOCH}_3}}$$

(7)

$$\longrightarrow \underset{\underset{\text{OH}}{\text{HO} \quad \text{CH}_2\text{CONH(CH}_2)_5\text{NCOCH}_3}}{\overset{\text{OH}}{\text{HO}_2\text{C} \quad \text{CH}_2\text{CONH(CH}_2)_5\text{NCOCH}_3}}$$

Arthrobactin   (8)

SCHEME 16.   Total synthesis of arthrobactin.

ARTHROBACTIN

This siderophore is predicated on a 1,5-diaminopentane, or cadaverine, fragment (Scheme 16), and its synthesis employs methods very similar to those found in the aerobactin synthesis.[74] The amine in 1-amino-5-hydroxypentane (1) is *tert*-butoxycarbonylated, and the hydroxyl of (2) converted to the halide (3) by reaction with triphenylphosphine and carbon tetrabromide. The resulting bromide is utilized to alkylate the anion of *O*-benzyl acetohydroxamate. This alkylation product (4) is de-*t*-butoxycarbonylated by brief exposure to trifluoroacetic acid, and the resulting amine salt (5) is then condensed with the terminal bis(*p*-nitrophenyl) ester-activated, internal benzyl ester-protected citric acid to produce the corresponding diamide (6). The benzyl ester (6) is cleaved by reaction with sodium hydroxide in THF, generating (7) and the benzyl protecting groups cleaved by hydrogenation over palladium to generate arthrobactin (8). The basic cleavage of the benzyl ester is necessary because attempted removal of all three benzyl protecting groups by hydrogenolysis of (6) in methanol failed. The corresponding methyl ester resulted and attempts to convert it to the final product in alkaline THF/water resulted in cyclized analogues, e.g., arthrobactin A or simply mixtures of undesired material.

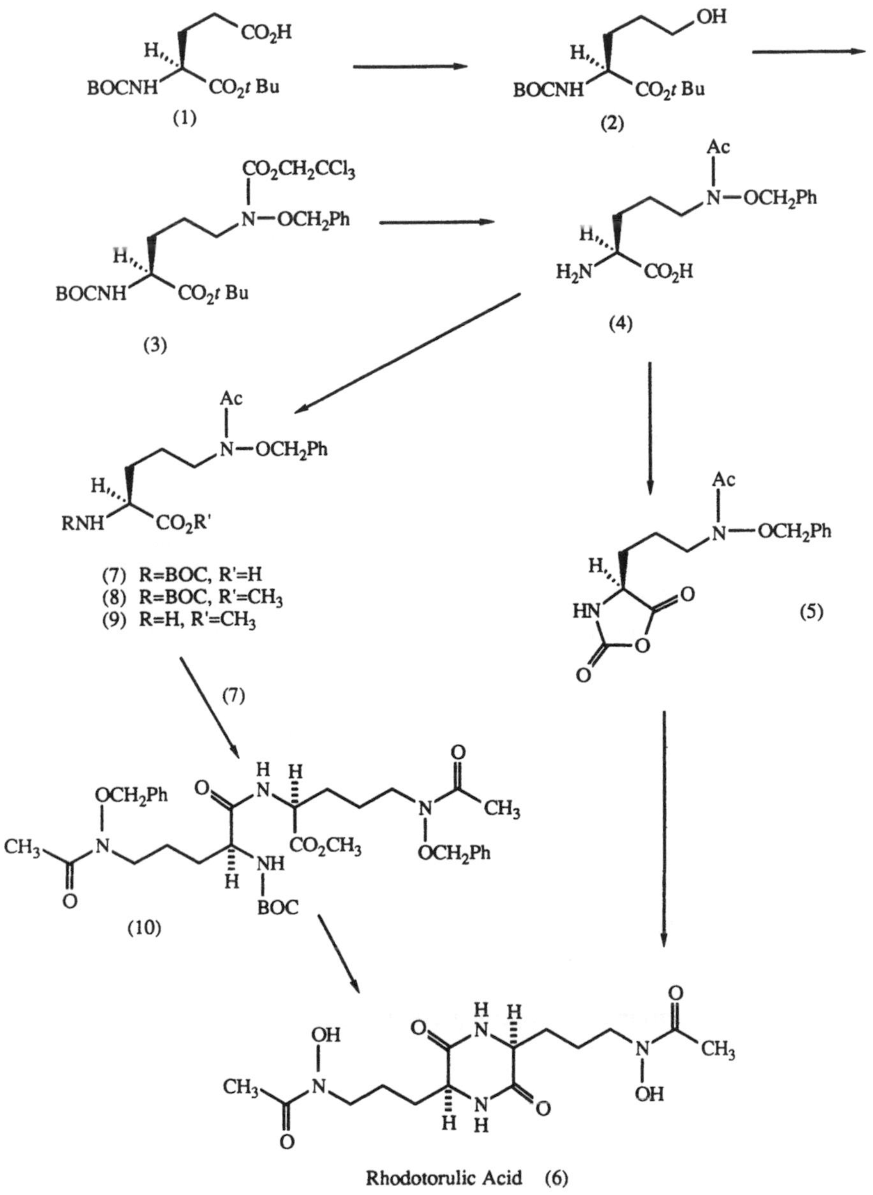

SCHEME 17.   Total synthesis of rhodotorulic acid.

RHODOTORULIC ACID

Unlike desferrioxamine, rhodotorulic acid (Figure 2) is not predicated on cadaverine but on the putrescine precursor ornithine. Most of the synthetic approaches to this molecule focus on altering the ornithine backbone prior to dimerization to the diketopiperazine, rhodotorulic acid. One of the more recent and interesting approaches[75] begins with a readily available protected glutamic acid derivative (1) (Scheme 17). The free acid is converted to the corresponding alcohol (2) by first reacting it with ethyl chloroformate and reducing the mixed anhydride with sodium borohydride. The resulting alcohol (2) is then coupled to N-[(2,2,2-trichloroethoxy)carbonyl]-O-benzylhydroxylamine with DEAD/triphenylphosphine to produce fully protected δ-N-hydroxyl-L-ornithine (3). The (2,2,2-trichloroethoxy)carbonyl group of (3) is removed and the resulting free nitrogen acetylated with zinc and acetic anhydride in acetic acid. Next the *N-t*-butoxycarbonyl and the carboxyl *t*-butyl protecting groups are cleaved by exposure to trifluoroacetic acid producing amino acid (4).

Two different routes to rhodotorulic acid (6) were taken, the more direct of which involved conversion of (4) with phosgene to Leuch's anhydride (5), followed by dimerization and protective group removal. In an alternative route the free amine (4) is again *N-t*-butoxycarbonylated and the free carboxyl of the product (7) is esterified. The resulting methyl ester (8) is again de-*t*-butoxycarbonylated with trifluoroacetic acid to (9), which is coupled with (7) utilizing EEDQ. The condensation product (10) is de-*t*-butoxycarbonylated, the resulting amino ester is cyclized to the diketopiperazine with sodium carbonate, and the benzyl groups are removed under hydrogenolysis conditions resulting in rhodotorulic acid (6). The intermediates (7) and (9) also allow for a rather nice entry into the ferrichrome system utilizing standard peptide methods.

DESFERRI-FERRICHROME

Two fundamental approaches to the synthesis of desferri-ferrichrome and its analogues have been utilized. Both approaches have generally involved introduction of the acetyl group of the hydroxamate in the final steps of the procedure. One method employs the generation of acyclic, linear peptides with the *N*-hydroxy ornithine fragment implicit in the backbone.[76,77] A total synthesis of ferriochrome by this route is shown in Scheme 18. Standard peptide synthetic techniques are employed to join N^δ-benzyloxy-N^δ-tosyl-L-ornithine and glycine units to form linear hexapeptide (1). The key step of the chelator synthesis involves cyclization of peptide (1) to the macrolide (2), in which each hydroxylamine is *N,O*-diprotected. Detosylation under acidic conditions, *N*-acetylation, and catalytic debenzylation afford desferri-ferrichrome (3).

The second entry into this system involves the production of the appropriate macrocycle in which the δ nitrogens of the ornithines are free amines. The crucial steps in the second method focus largely on the conversion of the δ primary amine in each ornithine of the macrocycle to its acetohydroxamate, as in the total synthesis of enantioferrichrome (9) (Scheme 19). Specifically, oxidation of the tris(γ-aminopropyl) macrocycle (3) involves the rather clever formation of a tri-oxaziridine ring-bearing macrocycle (5) and its cleavage to furnish hydroxylamine (6); however, there has been some suggestion in the literature that the yields of the tri-oxaziridine ring opening are not reproducible.[78]

AEROBACTIN

The key intermediate in this synthesis, N^6-acetyl-N^6-(benzyloxy)-L-lysine methyl ester (6) (Scheme 20), is accessed from ε-hydroxynorleucine (1), a previously prepared amino acid.[79] The amino acid is first *N-t*-butoxycarbonylated by reaction with di-*tert*-butyl dicarbonate in THF/water and the resulting carbamate (2) converted to the methyl ester (3). The hydroxyl group of this ester is next brominated with triphenylphosphine and carbon tetrabromide, and the resulting bromide (4) utilized to alkylate *O*-benzyl acetohydroxamate in anhydrous acetone in the presence of potassium carbonate and potassium iodide to produce (5). Brief exposure of the alkylated product to trifluoroacetic acid collapsed the *tert*-butoxycarbonyl protecting group to carbon dioxide, isobutylene, and the amine salt. The free amine (6), which is obtained by treatment of the salt with sodium carbonate, is coupled with anhydromethylenecitryl chloride, providing the diamide (7) in excellent yield. The anhydromethylene citric acid derivative (7) is exposed briefly to sodium hydroxide in water/THF followed by an acid workup, resulting in carboxylic acid (8). Finally, aerobactin (9) is generated by reductive removal of the benzyl protecting groups with an overall yield for the nine steps of about 35%.

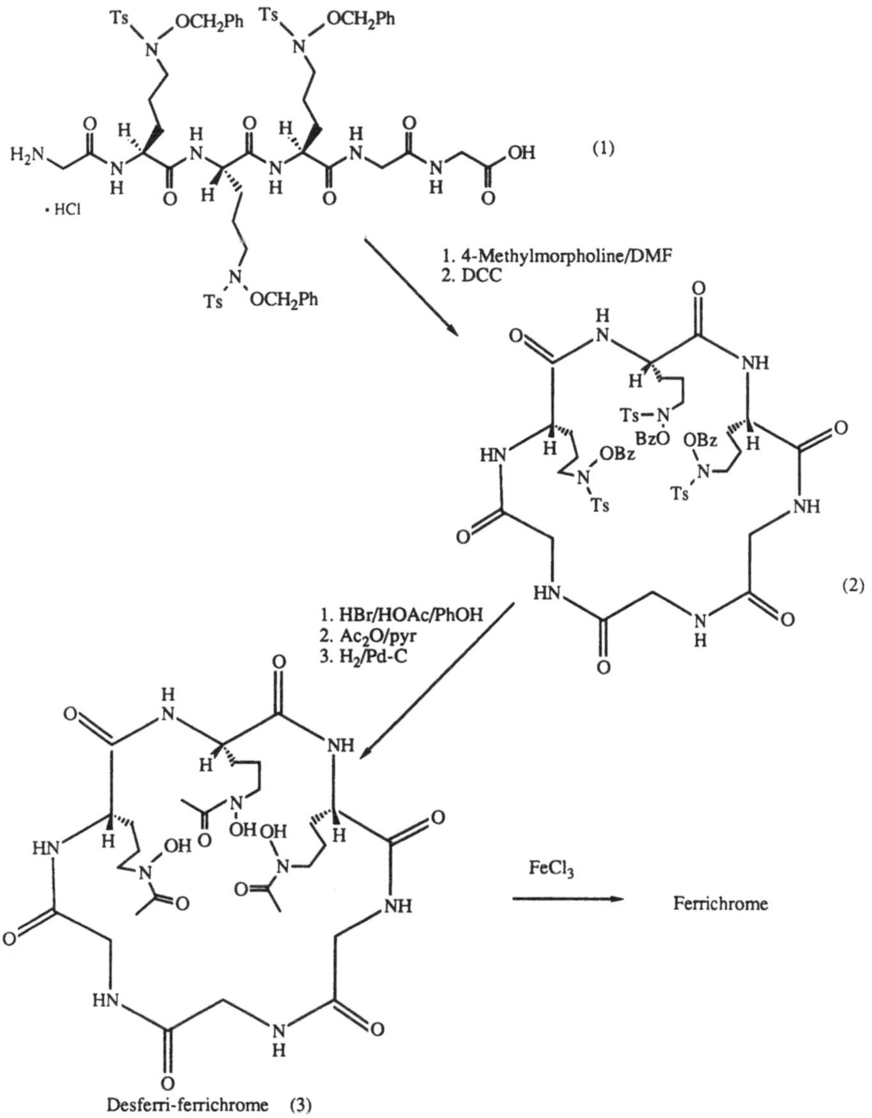

SCHEME 18. Total synthesis of ferrichrome.

SCHIZOKINEN

This ligand is predicated on the shorter polyamine 1,3-diaminopropane, and its synthesis begins with the masked diamine, 1-amino-3-[(benzyloxy)amino]propane (1).[74] The benzyloxy compound is coupled with the terminal bis(p-nitrophenyl), internal isopropyl ester of citric acid to give (2) (Scheme 21). After coupling, acetic anhydride is added prior to workup to acetylate the hydroxylamine nitrogen. As with arthrobactin, final protective group removal is accomplished by treatment of the ester (2) with base followed by hydrogenolysis, providing schizokinen (3).

MYCOBACTIN S2

The mycobactins incorporate both the hydroxamate funtionality of ferrichrome, ferrioxamine, etc. as well as the oxazoline of parabactin. The synthesis is a well-designed convergent one.[80] In fact, many of the synthetic methods described for synthesis of both the hydroxamates and the catecholamides are utilized in accessing S2. The synthesis of this mycobactin serves

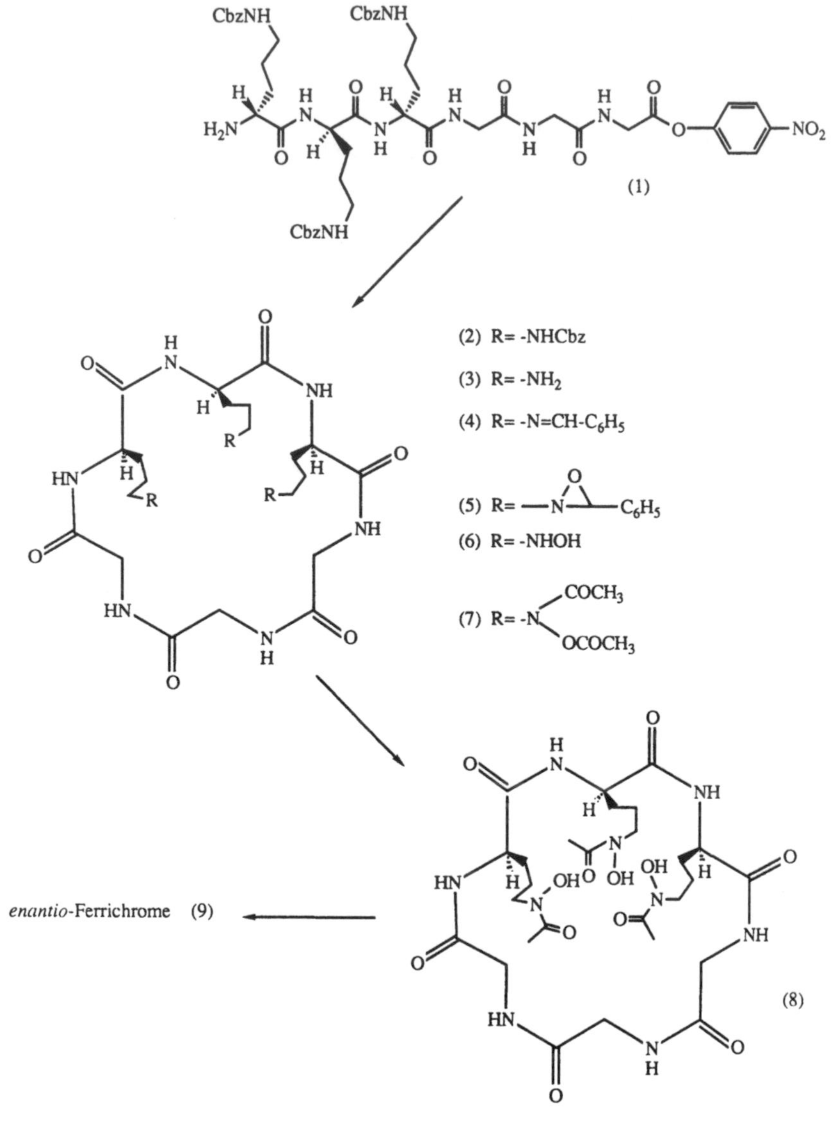

(2) R= -NHCbz

(3) R= -NH$_2$

(4) R= -N=CH-C$_6$H$_5$

(5) R= —N$\overset{O}{\diagdown}$—C$_6$H$_5$

(6) R= -NHOH

(7) R= -N$\diagdown\overset{COCH_3}{\diagdown OCOCH_3}$

enantio-Ferrichrome   (9)

SCHEME 19.   Total synthesis of enantio-ferrichrome.

as an excellent example of the general utility of the methods described. The strategy (Scheme 22) begins with the *N-t*-butoxycarbonylation of ε-hydroxynorleucine (1), utilizing di-*tert*-butyl dicarbonate, followed by condensation of the carboxyl group in (2) with *O*-benzyl-hydroxylamine to produce the corresponding benzyloxyamide (3). This amide is cyclized with triphenylphosphine and diethyl azodicarboxylate (DEAD) to lactam (4) in 43% yield. The *tert*-butoxycarbonyl protecting group of the lactam is next removed by exposure to trifluoroacetic acid followed by a basic workup to generate the amine (5). The amino nitrogen is coupled with D-β-hydroxybutyric acid utilizing EEDQ to produce the key intermediate, benzyl-protected cobactin (6).

The oxazoline portion of the mycobactin molecule must now be constructed. L-Serine is first coupled with the *p*-nitrophenyl activated ester of *O*-benzyl salicylic acid providing acylated serine derivative (7), which is coupled with N^6-acetyl-N^6-(benzyloxy)-L-lysine methyl ester ([6], Scheme 20) using EEDQ to afford dipeptide (8). The salicylamide (8) is cyclized to the oxazoline utilizing thionyl chloride followed by treatment with aqueous base to give

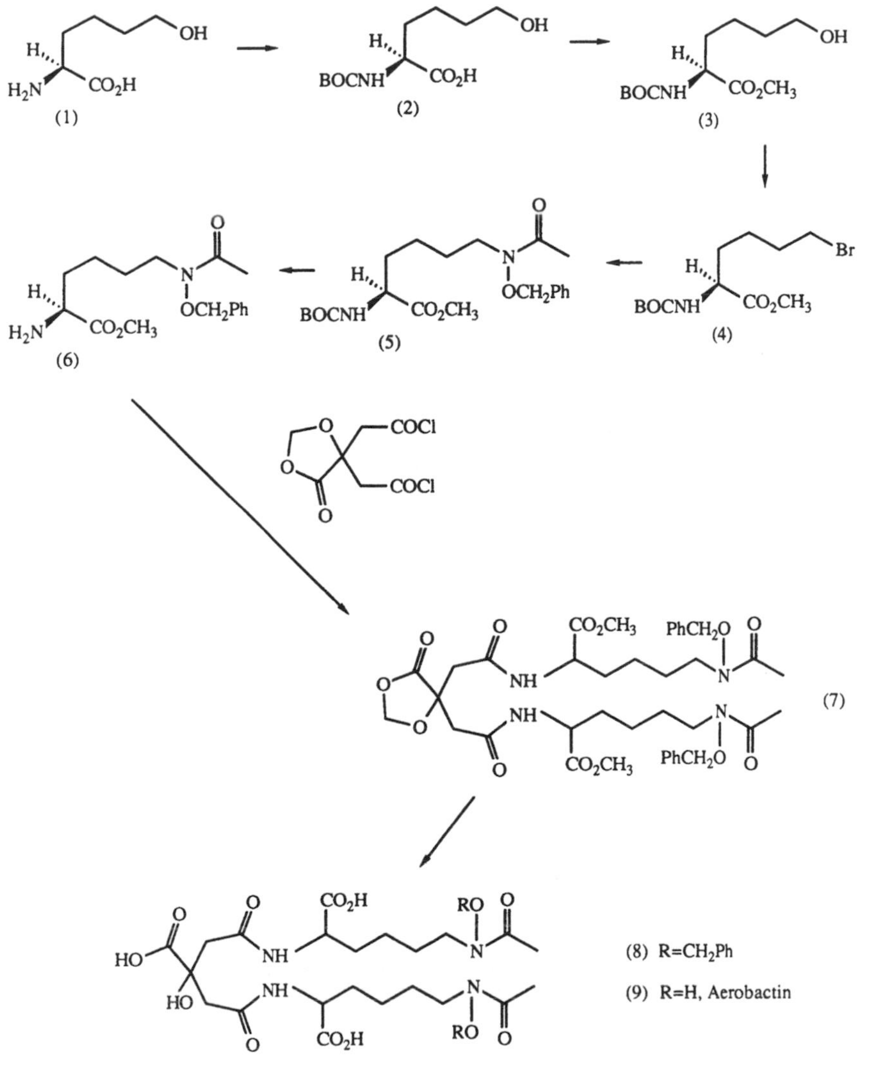

SCHEME 20. Total synthesis of aerobactin.

methyl ester (9), which is hydrolyzed to the corresponding carboxylic acid (10) under mild alkaline conditions. Oxazoline acid (10) is coupled with the benzyl-protected cobactin (6) employing triphenylphosphine and DEAD to generate tribenzylmycobactin (11). Finally the adduct (11) is catalytically debenzylated to produce mycobactin S2 (12).

CONCLUDING REMARKS

Because of their interesting biological properties, it is clear that investigators will continue to develop additional methods for the synthesis of new siderophores and siderophore analogues. Such synthetic methods will provide researchers with the opportunity to establish valuable structure-activity relationships and to identify ligands of therapeutic value. The synthons described in this review for the production of both hydroxamate and catecholamide siderophores should provide workers with many of the building blocks necessary to fabricate even more complicated molecules.

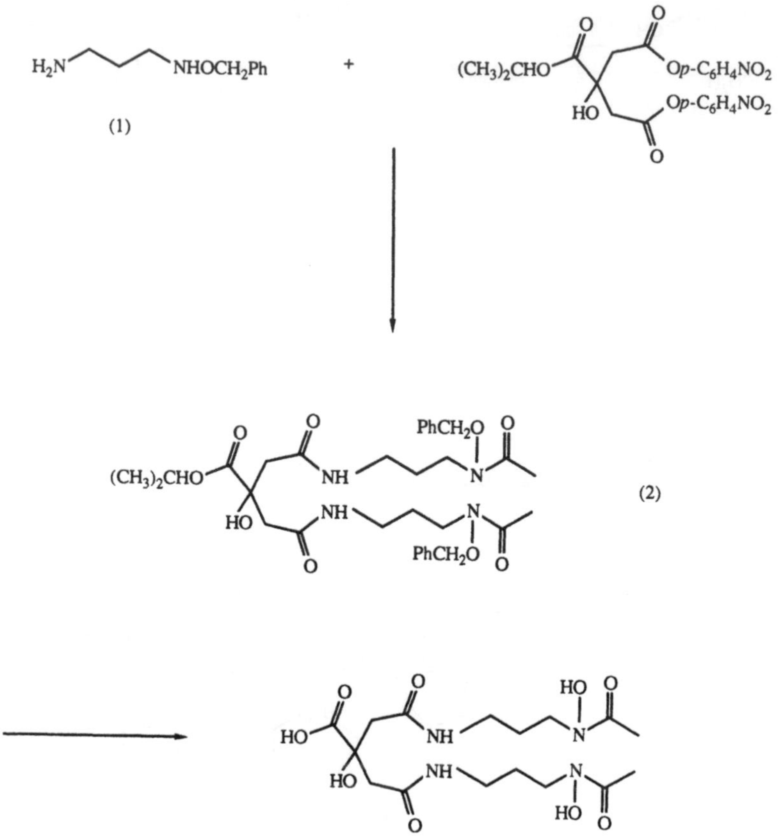

SCHEME 21.   Total synthesis of schizokinen.

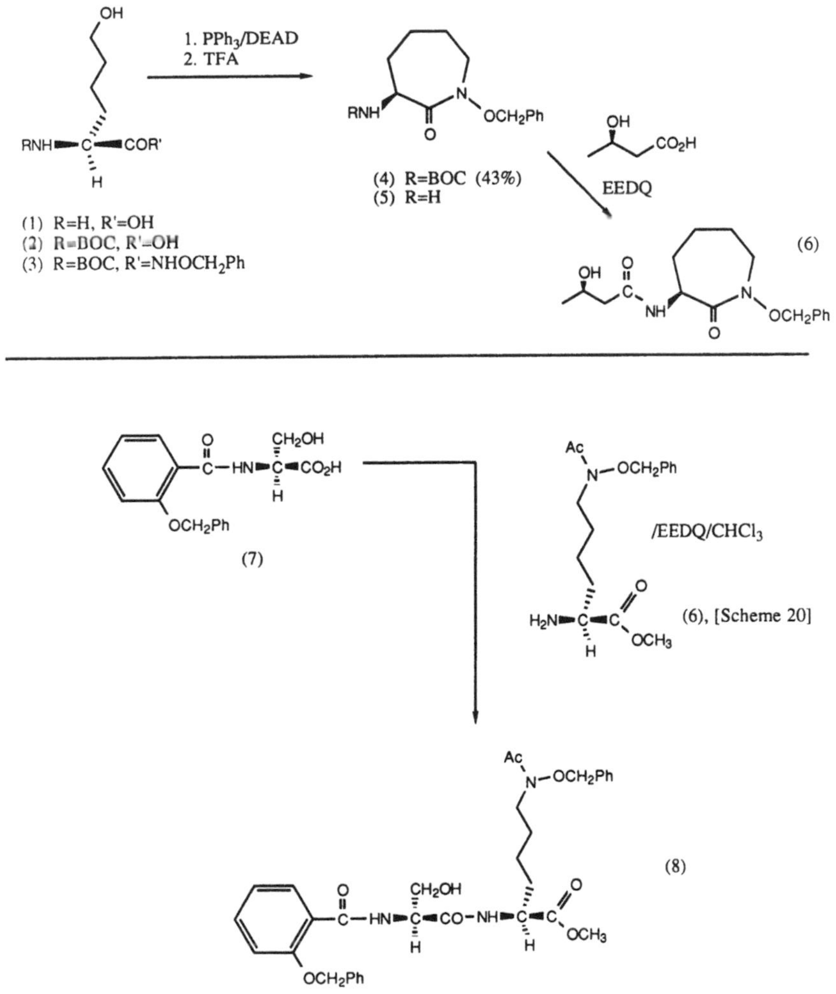

SCHEME 22. (A and B) Total synthesis of mycobactin S2.

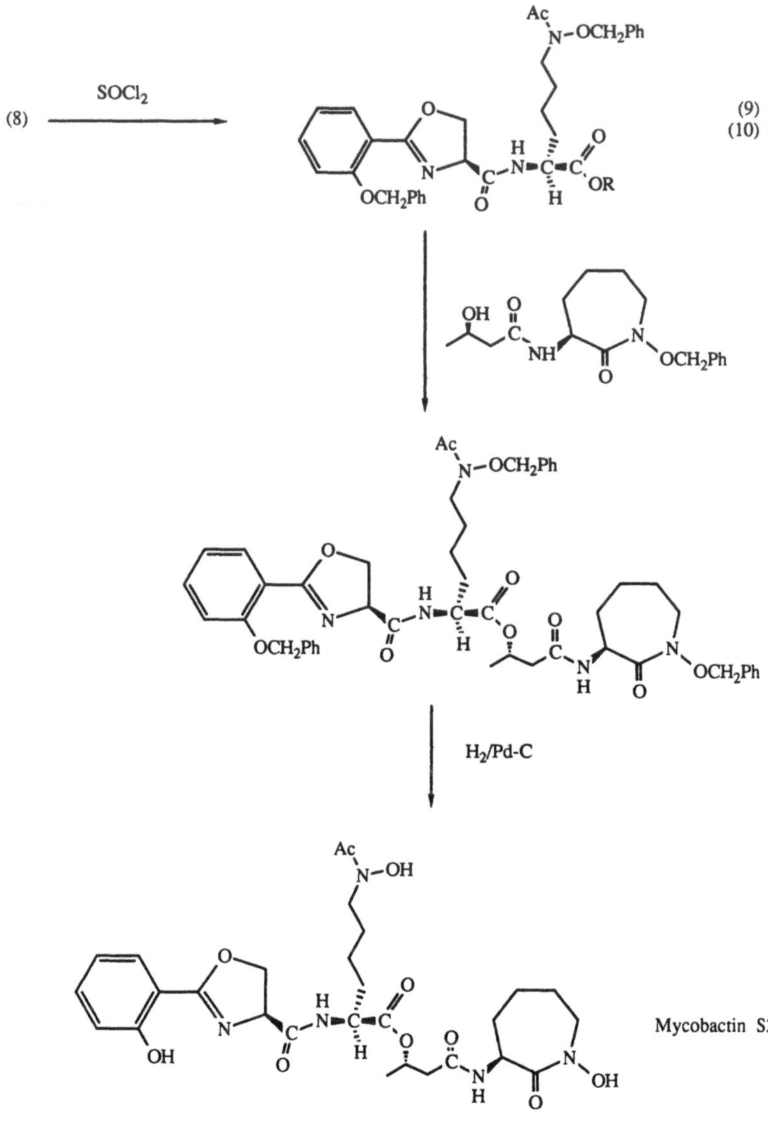

(9) $R=CH_3$
(10) R=H

(6)

(11)

Mycobactin S2 (12)

SCHEME 22B

REFERENCES

1. **Cooper, S. R., McArdle, J. V., and Raymond, K. N.,** Siderophore electrochemistry: relation to intracellular iron release mechanism, *Proc. Natl. Acad. Sci. U.S.A.*, 75, 3551, 1978.
2. **Ortiz de Montellano, P. R.,** *Cytochrome P450-Structure, Mechanism, and Biochemistry,* Ortiz de Montellano, P. R., Ed., Plenum Press, New York, 1986.
3. **Sahlin, M., Petersson, L., Graslund, A., Ehrenberg, A., Sjoberg, B.-M., and Thelander, L.,** Magnetic interaction between the tyrosyl free radical and the antiferromagnetically coupled iron center in ribonucleotide reductase, *Biochemistry,* 26, 5541, 1987.
4. **Bergeron, R. J.,** Iron: a controlling nutrient in proliferative processes, *Trends Biochem. Sci.*, 11, 133, 1986.
5. **Bergeron, R. J., Weimar, W. R., and Dionis, J. B.,** Demonstration of ferric L-parabactin-binding activity in the outer membrane of *Paracoccus denitrificans, J. Bacteriol.*, 170, 3711, 1988.

6. **Bergeron, R. J., Dionis, J. B., Elliott, G. T., and Kline, S. J.,** Mechanism and stereospecificity of the parabactin-mediated iron transport system in *Paracoccus denitrificans, J. Biol. Chem.,* 260, 7936, 1985.

7. **Neilands, J. B.,** Microbial iron compounds, *Annu. Rev. Biochem.,* 50, 715, 1981.

8. **Payne, S. M.,** Iron and virulence in the family enterobacteriaceae, *CRC Crit. Rev. Microbiol.,* 16, 81, 1988.

9. **Winkelmann, G., van der Helm, D., and Neilands, J. B., Eds.,** *Iron Transport in Microbes, Plants and Animals,* VCH Publishers, Weinheim, West Germany, 1987.

10. **Hider, R. C.,** Siderophore-mediated absorption of iron, *Struct. Bonding (Berlin),* 58, 25, 1984.

11. **Neilands, J. B., Konopka, K., Schwyn, B., Coy, M., Francis, R. T., Paw, B. H., and Bagg, A.,** Comparative biochemistry of microbial iron assimilation, *Iron Transport in Microbes, Plants and Animals,* Winkelmann, G., van der Helm, D., and Neilands, J. B., Eds., VCH Publishers, Weinheim, West Germany, 1987, chap. 1.

12. **Raymond, K. N. and Carrano, C. J.,** Coordination chemistry and microbial iron transport, *Acc. Chem. Res.,* 12, 183, 1979.

13. **Swinburne, T. R.,** Iron siderophore and plant diseases, Nato ASI series, *Series A: Life Sciences 117,* Plenum Press, New York, 1986.

14. **Demange, P., Wendenbaum, S., Boteman, A., Dell, A., and Abdollah, M. A.,** *Iron Transport in Microbes, Plants and Animals,* Winkelmann, G., van der Helm, D., and Neilands, J. B., Eds., VCH Publishers, Weinheim, West Germany, 1987, chap. 10.

15. **Meyer, J.-M., Halle, F., Hohnadel, D., Lemanceu, P., and Ratefiarivelo, H.,** *Iron Transport in Microbes, Plants and Animals,* Winkelmann, G., van der Helm, D., and Neilands, J. B., Eds., VCH Publishers, Weinheim, West Germany, 1987, chap. 11.

16. **Neilands, J. B.,** Some aspects of microbial iron metabolism, *Bacteriol. Rev.,* 21, 101, 1957.

17. **Avdeef, A., Sofen, S. R., Bregante, T. L., and Raymond, K. N.,** Coordination chemistry of microbial iron transport compounds. IX. Stability constants for catechol models of enterobactin, *J. Am. Chem. Soc.,* 100, 5362, 1978.

18. **Weatherall, D. J.,** *Development of Iron Chelators for Clinical Use,* Martell, A. E., Anderson, W. F., and Badman, D. G., Eds., Elsevier/North-Holland, New York, 1981, 3.

19. **Bergeron, R. J., Braylan, R., Goldey, S., and Ingeno, M.,** Effects of the *Vibrio cholerae* siderophore vibriobactin on the growth characteristics of L1210 cells, *Biochem. Biophys. Res. Commun.,* 136, 273, 1986.

20. **Cavanaugh, P. F., Jr. Porter, C. W., Tukalo, D., Frankfurt, O. S., Pavelic, Z. P., and Bergeron, R. J.,** Characterization of L1210 cell growth inhibition by the bacterial iron chelators parabactin and Compound II, *Cancer Res.,* 45, 4754, 1985.

21. **Bergeron, R. J. and Ingeno, M. J.,** Microbial iron chelator-induced cell cycle synchronization in L1210 cells: potential in combination chemotherapy, *Cancer Res.,* 47, 6010, 1987.

22a. **Neilands, J. B. and Valenta, J. R.,** *Metal Ions in Biological Systems,* Sigel, H., Ed., Marcel Dekker, New York, 19, 1985, chap. 11.

22b. **Rogers, H. J.,** *Iron Transport in Microbes, Plants and Animals,* Winkelmann, G., van der Helm, D., and Neilands, J. B., Eds., VCH Publishers Weinheim, West Germany, 1987, chap. 13.

23. **Braun, V., Gunther, K., Hantke, K., and Zimmerman, L.,** Intracellular activation of albomycin in *Escherichia coli* and *Salmonella typhimurium, J. Bacteriol.,* 156, 308, 1983.

24. **Zahner, H., Diddens, H., Keller-Schierlein, W., and Nageli, H.-U.,** *Jpn. J. Antibiot.,* 30, S-201, 1977.

25. **Ohi, N., Aoki, B., Shinozaki, T., Moro, K., Kourki, T., Noto, T., Nehashi, T., Matsumoto, M., Okazaki, H., and Matsunaga, I.,** Semisynthetic β-lactam antibiotics. IV. Synthesis and antibacterial activity of new ureidocephalosporin and ureidocephamycin derivatives containing a catechol moiety or its acetate, *Chem. Pharm. Bull.,* 35, 1903, 1987.

26. **Tait, G. H.,** The identification and biosynthesis of siderochromes formed by *Micrococcus denitrificans, Biochem. J.,* 146, 191, 1975.

27. **Peterson, T., Falk, K.-E., Leong, S. A., Klein, M. P., and Neilands, J. B.,** Structure and behavior of spermidine siderophores, *J. Am. Chem. Soc.,* 102, 7715, 1980.

28. **Griffiths, G. L., Sigel, S. P., Payne, S. M., and Neilands, J. B.,** Vibriobactin, a siderophore from *Vibrio cholerae, J. Biol. Chem.,* 259, 383, 1984.

29. **Bickel, H., Hall, G. E., Keller-Schierlein, W., Prelog, V., Vischer, E., and Wettstein, A.,** Stoffwechselprodukte von actinomyceten uber die konstitution von ferrioxamin B, *Helv. Chim. Acta,* 43, 2129, 1960.

30. **Keller-Schierlein, W. and Prelog, V.,** Stoffwechselprodukte von actinomyceten ferrioxamin G, *Helv. Chim. Acta,* 45, 590, 1962.

31. **Takahashi, A., Nakamura, H., Kameyama, T., Kurasawa, S., Naganawa, H., Okami, Y., Takeuchi, T., and Umezawa, H.,** Bisucaberin, a new siderophore, sensitizing tumor cells to macrophage-mediated cytolysis, *J. Antibiot.,* 40, 1671, 1987.

32. **Linke, W. D., Crueger, A., and Diekmann, H.,** Metabolic products of microorganisms. CVI. Structure of terregens-factor, *Arch. Mikrobiol.*, 85, 44, 1972.

33. **Snow, G. A.,** Mycobactins: iron-chelating growth factors from mycobacteria, *Bacteriol. Rev.*, 34, 99, 1970.

34a. **Harris, W. R., Carrano, C. J., and Raymond, K. N.,** Coordination chemistry of microbial iron transport compounds. XVI. Isolation, characterization, and formation constants of ferric aerobactin, *J. Am. Chem. Soc.*, 101, 2722, 1979.

34b. **Gibson, F. and Magrath, D. I.,** The isolation and characterization of a hydroxamic acid (aerobactin) formed by *Aerobacter aerogenes* 62-I, *Biochim. Biophys. Acta*, 192, 175, 1966.

35. **Keller-Schierlein, W., Prelog, V., and Zahner, H.,** Siderochromes. (Natural Fe(III)-trihydroxamate complexes), *Fortscher. Chem. Org. Naturst.*, 22, 279, 1964.

36. **Atkin, C. L. and Neilands, J. B.,** Rhodotorulic acid, a diketopiperazine dihydroxamic acid with growth-factor activity. I. Isolation and characterization, *Biochemistry*, 7, 3734, 1968.

37. **Mullis, K. B., Pollack, J. R., and Neilands, J. B.,** Structure of schizokinen, an iron-transport compound from *Bacillus megaterium*, *Biochemistry*, 10, 4894, 1971.

38. **Bergeron, R. J., Streiff, R. R., and Elliott, G. T.,** Influence of iron on *in vivo* proliferation and lethality of L1210 cells, *J. Nutr.*, 115, 369, 1985.

39. **Weinberg, E. D.,** Iron in neoplastic disease, *Nutr. Cancer*, 4, 223, 1983.

40. **Hann, H.-W. L., Stahlhut, M. W., and Blumberg, B. S.,** Iron nutrition and tumor growth: decreased tumor growth in iron-deficient mice, *Cancer Res.*, 48, 4168, 1988.

41. **Blatt, J. and Stiteley, S.,** Antineuroblastoma activity of desferrioxamine in human cell lines, *Cancer Res.*, 47, 1749, 1987.

42. **Bergeron, R. J., McGovern, K. A., Channing, M. A., and Burton, P. S.,** Synthesis of N^4-acylated N^1,N^8-bis(acyl)spermidines: an approach to the synthesis of siderophores, *J. Org. Chem.*, 45, 1589, 1980.

43. **Bergeron, R. J., Burton, P. S., Kline, S. J., and McGovern, K. A.,** Biomimetic synthesis of a *Paracoccus denitrificans* siderophore analogue, *J. Org. Chem.*, 46, 3712, 1981.

44. **Bergeron, R. J., Kline, S. J., Stolowich, N. J., McGovern, K. A., and Burton, P. S.,** Flexible synthesis of polyamine catecholamides, *J. Org. Chem.*, 46, 4524, 1981.

45. **Bergeron, R. J. and Kline, S. J.,** Short synthesis of parabactin, *J. Am. Chem. Soc.*, 104, 4489, 1982.

46. **Bergeron, R. J., McManis, J. S., Dionis, J. B., and Garlich, J. R.,** An efficient total synthesis of agrobactin and its gallium (III) chelate, *J. Org. Chem.*, 50, 2780, 1985.

47. **Bergeron, R. J., Garlich, J. R., and McManis, J. S.,** Total synthesis of vibriobactin, *Tetrahedron*, 41, 507, 1985.

48. **Joshua, A. V. and Scott, J. R.,** A simple method for the direct bis-acylation of the primary amino groups in spermidine and other linear triamines, *Tetrahedron Lett.*, 25, 5725, 1984.

49. **Nagao, Y., Miyasaka, T., Hagiwara, Y., and Fujita, E.,** Total synthesis of parabactin, a spermidine siderophore, *J. Chem. Soc. Perkin Trans. 1*, p. 183, 1984.

50. **Bergeron, R. J.,** Methods for the selective modification of spermidine and its homologues, *Acc. Chem. Res.*, 19, 105, 1986.

51. **Bergeron, R. J. and Garlich, J. R.,** Amines and polyamines from nitriles, *Synthesis*, p. 782, 1984.

52. **Bergeron, R. J., Stolowich, N. J., and Porter, C. W.,** Reagents for the selective secondary N-acylation of linear triamines, *Synthesis*, p. 689, 1982.

53. **Bergeron, R. J., Garlich, J. R., and Stolowich, N. J.,** Reagents for the stepwise functionalization of spermidine, homospermidine and bis(3-aminopropyl)amine, *J. Org. Chem.*, 49, 2997, 1984.

54. **Bergeron, R. J., Burton, P. S., McGovern, K. A., and Kline, S. J.,** Reagents for the selective acylation of spermidine, homospermidine, and bis(3-aminopropyl)amine, *Synthesis*, p. 732, 1981.

55. **Bergeron, R. J. and McManis, J. S.,** Reagents for the stepwise functionalization of spermine, *J. Org. Chem.*, 53, 3108, 1988.

56. **Borch, R. J.,** A new method for the reduction of secondary and tertiary amides, *Tetrahedron Lett.*, p. 61, 1968.

57. **Rastetter, W. H., Erickson, T. J., and Venuti, M. C.,** Synthesis of iron chelators. Enterobactin, enantioenterobactin, and a chiral analogue, *J. Org. Chem.*, 46, 3579, 1981.

58. **Corey, E. J. and Bhattacharyya, S.,** Total synthesis of enterobactin, a macrocyclic iron transporting agent of bacteria, *Tetrahedron Lett.*, p. 3919, 1977.

59. **Shanzer, A. and Libman, J.,** Total synthesis of enterobactin via an organotin template, *J. Chem. Soc. Chem. Commun.*, p. 846, 1983.

60. **Corey, E. J. and Hurt, S. D.,** Synthesis of the carbocyclic analog of enterobactin, *Tetrahedron Lett.*, p. 3923, 1977.

61. **Venuti, M. C., Rastetter, W. H., and Neilands, J. B.,** 1,3,5-Tris(N,N′,N″-2,3-dihydroxyben-zoyl)aminomethylbenzene, a synthetic iron chelator related to enterobactin, *J. Med. Chem.*, 22, 123, 1979.

62. **Weitl, F. L. and Raymond, K. N.**, Ferric ion sequestering agents. I. Hexadentate O-bonding N,N′,N″-tris(2,3-dihydroxybenzoyl) derivatives of 1,5,9-triazacyclotridecane and 1,3,5-triaminomethylbenzene, *J. Am. Chem. Soc.*, 101, 2728, 1979.

63. **Persmark, M., Expert, D., and Neilands, J. B.**, Isolation, characterization, and synthesis of chrysobactin, a compound with siderophore activity from *Erwinia chrysanthemi, J. Biol. Chem.*, 264, 3187, 1989.

64. **Martell, A. E., Anderson, W. F., and Badman, D. G.**, *Development of Iron Chelators for Clinical Use*, Martell, A. E., Anderson, W. F., and Badman, D. G., Eds., Elsevier/North-Holland, New York, 1980.

65. **Modell, B. and Berdoukas, V.**, *The Clinical Approach to Thalassaemia*, Grune & Stratton London, 1984, 217.

66. **Aksoy, M. and Birdwood, G. F. B., Eds.**, *Hypertransfusion and Iron Chelation in Thalassaemia*, Hans Huber Publishers, Berne, 1985, 80.

67. **Anderson, W. F.**, *Inorganic Chemistry in Biology and Medicine*, Martell, A. E., Ed., American Chemical Society, Washington, D.C., 1973, chap. 15.

68. **Prelog, V. and Walser, A.**, Stoffwechselprodukte von actinomyceten uber die synthese der ferrioxamine B and D_1, *Helv. Chim. Acta*, 45, 631, 1962.

69. **Bergeron, R. J. and McManis, J. S.**, The total synthesis of bisucaberin, *Tetrahedron*, 45, 4939, 1989.

70. **Bergeron, R. J. and Pegram, J. J.**, An efficient total synthesis of desferrioxamine B, *J. Org. Chem.*, 53, 3131, 1988.

71. **Bergeron, R. J. and McManis, J. S.**, The total synthesis of desferrioxamines E and G, *Tetrahedron*, 46, 5881, 1990.

72. **Shioiri, T., Ninomiya, K., and Yamada, S.**, Diphenylphosphoryl azide. A new convenient reagent for a modified Curtius reaction and for the peptide synthesis, *J. Am. Chem. Soc.*, 94, 6203, 1972.

73. **Boger, D. L. and Yohannes, D.**, Studies on the total synthesis of bouvardin and deoxybouvardin: cyclic hexapeptide cyclization studies and preparation of key partial structures, *J. Org. Chem.*, 53, 487, 1988.

74. **Lee, B. H. and Miller, M. J.**, Natural ferric ionophores: total synthesis of schizokinen, schizokinen A, and arthrobactin, *J. Org. Chem.*, 48, 24, 1983.

75. **Lee, B. H., Gerfen, G. J., and Miller, M. J.**, Constituents of microbial iron chelators. Alternate syntheses of δ-N-hydroxy-L-ornithine derivatives and applications to the synthesis of rhodotorulic acid, *J. Org. Chem.*, 49, 2418, 1984.

76. **Keller-Schierlein, W. and Maurer, B.**, Stoffwechselprodukte von mikroorganismen synthese des ferrichroms. II. Teil, *Helv. Chim. Acta*, 52, 603, 1969.

77. **Isowa, Y., Ohmori, M., and Kurita, H.**, Total synthesis of ferrichrome, *Bull. Chem. Soc. Jpn.*, 47, 215, 1974.

78. **Naegeli, H.-U. and Keller-Schierlein, W.**, Stoffwechselprodukte von mikroorganismen eine neue synthese des ferrichroms, enantio-ferrichrom, *Helv. Chim. Acta,.* 61, 2088, 1978.

79. **Maurer, P. J. and Miller, M. J.**, Microbial iron chelators: total synthesis of aerobactin and its constituent amino acid, N^6-acetyl-N^6-hydroxylysine, *J. Am. Chem. Soc.*, 104, 3096, 1982.

80. **Maurer, P. J. and Miller, M. J.**, Total synthesis of a mycobactin: mycobactin S2, *J. Am. Chem. Soc.*, 105, 240, 1983.

BIOMIMETIC SIDEROPHORES

Abraham Shanzer and Jacqueline Libman

SUMMARY

Biomimetic chemistry is the art of chemical modeling and aims at simulating with the simplest possible synthetic molecules some properties of highly complex, natural compounds. The success of biomimetics in the study of molecular recognition phenomena justifies its application to microbial iron uptake, and particularly siderophore-mediated iron uptake, which is in essence a problem of multiple molecular recognition.

In this article we describe the principles of the biomimetic design and its implementation for the synthesis of three families of siderophore analogues: enterobactin, ferrichrome, and the ferrioxamines. We will then elaborate on the utilization of these analogues (1) in establishing the structural requirements for biological activity and (2) in providing biological probes to trace iron uptake routes. Finally, we will show how biomimetic siderophores may be applied to achieve two inherently opposite effects; growth promotion of desired cultures by enhancing iron uptake, or growth inhibition of pathogenic parasites by causing iron deprivation.

INTRODUCTION

Biomimetic chemistry[1] is essentially the art of chemical modeling, and ultimately aims at mimicking the properties of complex biological systems with all-synthetic molecules. Biomimetics is guided by the rational, that there is more than a single chemical solution to exert a given biological function, and that it should therefore be possible to mimic the performance of biological compounds by reproducing their essential structural features with synthetic molecules.

The area of biomimetic chemistry emerged in the late 1960s as an outgrowth of several scientific developments. These developments were initiated by the discovery of Pedersen[2] that macrocyclic polyethers, the so-called "crown ethers", selectively bind a specific alkali or alkaline earth metal ion. These compounds proved to also transport the bound metal ions across biological membranes, and to thereby mimic the behavior of naturally occurring ion carriers, or ionophores.[3,4] This discovery drew attention to one of the most fundamental phenomena, the phenomenon of *molecular recognition,* by providing synthetic host molecules (or receptors) that are able to recognize a specific metal ion from a mixture of many. It further confirmed the assumption that synthetic compounds may simulate the performance of natural molecules without copying their structures.

As an outgrowth of these developments the problem of molecular recognition became one of the most prevalent topics in the area of biomimetics. These research efforts were further nourished by the fact that molecular recognition phenomena are at the heart of the most vital biological processes, as they govern specific molecular interactions such as enzyme-substrate, hormone-receptor, and antigen-antibody binding. One of the most relevant conclusions of these interdisciplinary endeavors was the realization that molecular recognition is controlled by the principles of complementarity and necessitates the synchronized action of a multitude of weak forces. This insight led to the successful preparation of synthetic compounds that are not only capable of discriminating between metal ions of different size,[2,5,6] but also of distinguishing between molecular ions of enantiomeric relationships[6-8] and of differentiating between closely related noncharged molecules.[7-9]

It should be emphasized that the biomimetic approach is inherently different from the classical trial-and-error approach in the search for a good match between "substrate" and "receptor". The biomimetic approach relies on identifying the principle structural features of a family of bioactive molecule, on reproducing these very features with the simplest, possible synthetic compounds, and on thereby providing synthetic analogues of a whole class of bioactive compounds. Biomimetics may thereby be regarded as a generally applicable system approach, which promises entrance to a multitude of solutions, rather than to a single solution for a specific case.

Biomimetic chemistry has thus become one of the prevalent areas at the interface between chemistry and biology. The interest in this area was further stimulated by the fact that the biomimetic approach is not limited to the problem of recognition, but may readily be applied to other biological processes such as information transfer,[5] enzymatic catalysis,[7,10-13] or energy conversion,[14] although the latter processes are composites of a sequence of steps. Information transfer, for example, involves recognition and signaling, enzymatic catalysis, necessitates recognition, stabilization of the transition state, and fast release of the product. In these cases the methodology of biomimetics involves first reproduction of each of the steps and finally integration of the individual solutions to a single synthetic hybrid molecule.

Recent progress in the studies of biological iron(III) uptake phenomena[15] and particularly microbial iron(III) uptake[16-20] provided insight into these processes to such a degree that their exploration by the tools of biomimetics promises to become a rewarding undertaking. This article is not intended to provide a comprehensive account of current knowledge; for this purpose the reader is referred to a recent book[15] and several excellent reviews by Neilands,[16] Emery,[19] Raymond et al.,[17] Hider,[18] Miller,[21] and others[22] that cover various aspects of siderophore-mediated iron uptake and iron metabolism. The following account deliberately highlights the biomimetic approach in the study of iron uptake phenomena, while concentrating on enterobactin, ferrichrome, and the ferrioxamine siderophores as representative examples.

GENERAL ASPECTS OF MICROBIAL IRON(III) UPTAKE

Iron is essential for the well-being and growth of every living system, as it is involved in several fundamental enzymatic reactions such as oxygen metabolism, electron transfer processes, and DNA and RNA synthesis. The dependence of these metalloenzymes on iron is so pronounced that no viable substitute for this metal has been found. Even when soluble iron became scarce on earth due to the transition from a reducing to an oxidating atmosphere, and the simultaneous conversion of soluble ferrous salts to sparingly soluble ferric salts, microorganisms did not replace iron by any other metal ion. Instead, they adapted to the changing environment by developing highly sophisticated means to ensure adequate iron supply. One of these means is based on the use of molecular iron carriers, termed siderophores.[16-19]

Siderophores are low molecular weight molecules that are excreted by microorganisms when grown in iron-deficient media. Released to the environment, these compounds effectively bind and solubilize iron, and serve as molecular vehicles by transporting the bound metal ion across the bilayer membrane into the cell.

The study of microbial iron(III) uptake proceeded initially along two almost independent lines. The first aimed primarily at the isolation and characterization of microbial siderophores,[16,23,24] the second at elucidating the underlying biological machinery.[16,19,20]

The first line led to the identification of more than 200 siderophores that fall into two major classes: those that are based on catecholates as ion binding sites and those that make use of hydroxamates. Although encompassing a large range of structural varieties, these siderophores were found to have several features in common: they all scavenge effectively and selectively iron(III) by embedding it into an octahedral ion binding cavity, and are all hydrophilic, rather than hydrophobic. The latter observation implied that siderophores cannot affect ion transport via a diffusion-controlled process like ionophores,[3,4] but have to rely on a different mechanism. Further support for this possibility was obtained when only the natural isomers, but not their enantiomers, proved to be biologically active.[25-27] It thus appeared that siderophore-mediated iron transport necessitates specific interactions with membranal components.

Progress in genetics helped to unravel the details of this iron uptake machinery by locating the chromosomal genes that govern siderophore-mediated iron uptake.[16,20] These studies resulted in the identification of at least four proteins that include membranal receptors and transport systems, induce iron release, and act as repressors which regulate the whole iron uptake process. At last many of the puzzles encountered when studying the chemistry of siderophores could be put at rest. Receptor-driven iron uptake could account for the hydrophilic, rather than hydrophobic, character of siderophores, and for the dependence of their activity on their absolute configuration. In addition, this mechanism provides means for iron accumulation against unfavorable concentration gradients and for the control of intracellular iron levels within a narrow window.

Applying two different methodologies, those of chemistry and those of biology, the conclusions derived from each converged to provide a coherent picture of the intricate features of microbial iron uptake. Siderophore-mediated iron(III) uptake thus emerged to be in essence a problem of *molecular recognition*, involving recognition of ferric ions by the siderophore and recognition of the siderophore-ferric complex by the receptor. Therefore, siderophore-mediated iron uptake is eminently suited to be studied by the tools of biomimetics. When applied *in vivo* and in conjunction with the products provided by genetic engineering, biomimetic siderophores are anticipated to help sorting out the structural parameters that govern each of the iron uptake steps, and to shed light on molecular aspects of iron acquisition that so far escaped examination by pure chemical or biological methods.

Although siderophore-mediated iron uptake is unique to microorganisms, it has direct bearing on the well-being of higher plants, which may profit from microbial metabolites products, and on the well-being of mammalians, who may fight microbial invasions by reducing available iron levels.[22] Some of these aspects will become apparent, as we will screen possible applications of biomimetic siderophores that have been modeled after microbial iron(III) carriers.

IRON BINDING PROPERTIES OF SIDEROPHORIC COMPOUNDS

The biomimetic approach toward synthetic, bioactive compounds relies on identifying the essential structural features of the natural molecules and subsequently reproducing these features with the simplest possible synthetic compounds.

Our following representation will accordingly proceed along these lines. We will first describe the roads that have been taken to identify the essential structural characteristics of the natural siderophores, and the conclusions that could be drawn as to the origin of the natural compounds' specific iron binding properties. Subsequently, we will turn to the design and synthesis of several biomimetic analogues that satisfy the first of the requirements: specific recognition of ferric ions. As a first test case we will examine enterobactin, the most powerful iron binder known.

ENTEROBACTIN AND SYNTHETIC ANALOGUES

The Structures of Enterobactin and of Its Ferric-Complex

Enterobactin is a catecholate siderophore which is produced by enteric bacteria when grown in Fe^{3+}-deficient media.[28] Structurally, enterobactin is a tripode-like molecule derived from three L-seryl residues. The seryl residues are condensed to a 12-membered trilactone ring which serves as an anchor, while their symmetrically projecting amino groups are extended by catecholates. When binding iron(III), the ion binding catecholates generate an octahedral cavity of Δ-cis configuration.[29]

It has been found that for the case of enterobactin as well as for most naturally occurring siderophores, only the natural isomers are biologically active, while their enantiomers fail to facilitate cellular iron(III) uptake and growth. Enterobactin,[28] for example, is only bio-logically active as its right-handed, Δ-cis ferric complex,[25,29] while ferrichrome,[30,31] a rep-resentative of the hydroxamate family of siderophores, is active as its left-handed, Λ-cis complex.[32,33]

The high chiral discrimination between enantiomeric ferric-siderophore complexes can only fully be appreciated by considering their detailed configuration. When binding iron(III) in an octahedral geometry, *a priori*, a right-handed, Δ-cis or left-handed Λ-cis helicity may be adopted. In the absence of asymmetric centers in the ligands the two structures are enantiomeric, energetically equivalent, and thereby equally populated. However, in the presence of chiral centers, as occurring in most siderophores, the two alternative structures become diastereomeric, energetically nonequivalent, and therefore either of them predom-inantly or even exclusively formed (Figure 1).

The coordinative purity of enterobactin and its exceptionally high complexation constant (log K = +52)[34] attracted extensive attention and stimulated extensive research efforts aimed at establishing the origin of enterobactins' unique iron(III) binding properties[35] and at synthesizing artificial analogues that would stimulate enterobactins' binding properties[36-38] (Table 1).

Enterobactin Analogues with Catecholate Binding Sites

The first systematic work in the search for synthetic enterobactin analogues was carried out by Raymond et al.,[36,37] who introduced the tripodal triscatecholate MECAM. This molecule exceeded the ion binding efficiency of any triscatecholate previously prepared with a remarkable ion binding constant of log K = +46. Yet, this iron(III) binder still lagged behind the natural siderophore by six orders of magnitude (Figure 2).

The structurally related Me₃MECAM and TRIMCAM were found to be significantly less efficient.[37] Comparison between the latter three binders, MECAM, Me₃MECAM, and TRIMCAM, that all make use of the same mesitylene base as anchor, but differ in the nearest environment around the catecholate binding site, indicates the essentiality of the 2,3-dihydroxybenzoyl-amido group.

Replacement of the –NH–C=O– group by –NMe–C=O–, as occurring in Me₃MECAM, or dislocation of the –NH–C=O– group by one methylene from the ion binding catecholate, as occurring in TRIMCAM, proved to be detrimental to ion binding. The spacial proximity of the amide NH group to the catecholate group thus appears to be essential for efficient binding.

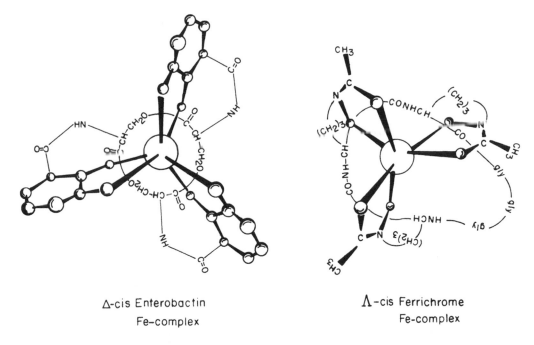

Δ-cis Enterobactin
Fe–complex

Λ-cis Ferrichrome
Fe-complex

FIGURE 1.   Enterobactin (left and ferrichrome (right).

TABLE 1
Ion Binding Properties of Catecholates[a]

	pM (Fe^{3+})	pM (Ga^{3+})	pM (In^{3+})	log β_{110} (Fe^{3+})
Enterobactin	35.5			52.0
Tiron	19.9	19.4		
DMB	15.0			43.9
DMBS	19.2	16.1	15.1	
MECAM	29.1			46.0
MECAMS	29.4	26.3	27.4	41.0
ME₃MECAMS	26.9			40.6
TRIMCAMS	25.1			
3,4-LICAMS	28.5	26.0	26.5	41.0

[a]   From References 17 and 37.

The possibility that the proximal NH groups contribute to the stability of the ferric-enterobactin complex by forming H-bonds to the ferric-catecholate oxygens was first suggested by Neilands.[39] This view has been confirmed in more general terms by Force Field Calculations which demonstrated the presence of such H-bonds in the lowest energy conformations of the ferric-enterobactin complex as well as in other ferric-catecholates possessing suitably positioned amide protons.[35] Direct evidence for H-bonded ferric-catecholates was most recently obtained by X-ray diffraction analysis of bicapped complexes. These data showed distances of 1.89 Å, in good agreement with the presence of strong H-bonds H··O–Fe.[40]

What is the Origin of Enterobactins' High Binding Efficiency?

The MECAM model thus relies on the optimal ion binding site, 2,3-dihydroxy benzamide, which is identical with that present in enterobactin and allows stabilization of the iron(III) complex by H-bonding. The inferiority of MECAM relative to enterobactin must thus derive from stereochemical factors: enhanced strain of the metal complex, more entropy

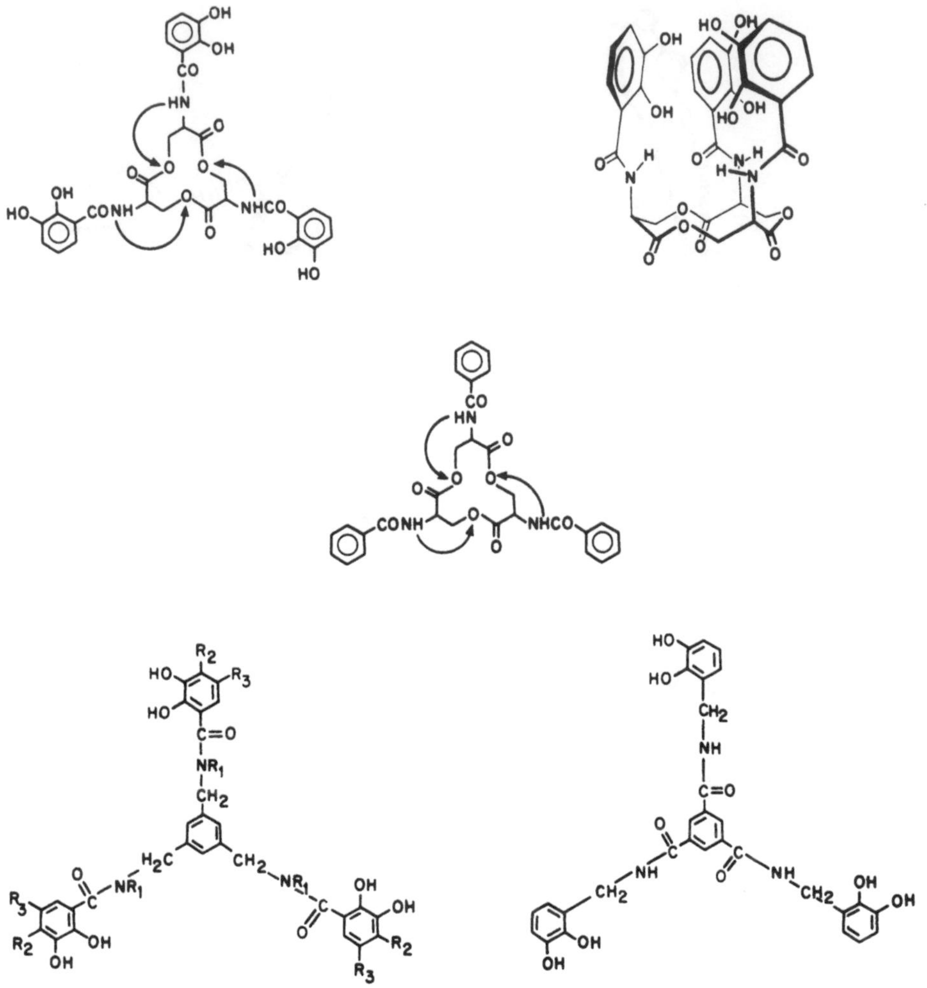

MECAM ($R_1 = R_2 = R_3 = H$)
MECAM-Me ($R_1 = R_3 = H$, $R_2 = CH_3$)
MECAMS ($R_1 = R_2 = H$, $R_3 = SO_3^-$)
Me_3MECAM ($R_2 = R_3 = H$, $R_1 = CH_3$)

TRIMCAM

FIGURE 2. Enterobactin (top), structural enterobactin analogue (center), and ligating enterobactin analogues (bottom).

loss upon ion binding, or both. In an attempt to identify these stereochemical factors, we undertook a thorough study which relied on (1) comparing the structures of ferric-enterobactin with that of ferric-MECAM, and on (2) identifying the essential conformational features of metal-free enterobactin.[35] Since both ferric-enterobactin and ferric-MECAM, as well as free enterobactin, had consistently failed to provide crystalline samples for X-ray diffraction analysis, we resorted to a combined theoretical-experimental study. In this study the structures of ferric-enterobactin and of ferric-MECAM were determined by EFF (empirical force field calculations), which are highly reliable when applied to conformationally restricted molecules or metal complexes. The conformation of free enterobactin, on the other hand, was scrutinized by a combination of theoretical and experimental tools, using the structural analogue to enterobactin, the trisbenzamide TBA, as link between theory and experiment. Trisbenzamide TBA is an ideal structural analogue, as it is identical to enterobactin in its molecular backbone, but has the catecholate groups replaced by phenyl groups. This replacement enables reliable experimental examination because of both significantly enhanced solubility in polar as well as apolar solvents, and lack of hydroxyl groups that hamper interpretation of spectroscopic data.

EFF calculations on ferric-MECAM in comparison with ferric-enterobactin revealed that the synthetic metal-ion complex is more strained than the enterobactin-ion complex and that most of the strain relies in the mesitylene anchor and its connections to the side chains. However, the estimated strain energy of ferric-MECAM was found to be only 2.0 kcal/mol higher than that of ferric-enterobactin. The measured binding constant of MECAM is, however, six orders of magnitude lower, equivalent to about 8 kcal/mol difference in binding free energy. This gap could *a priori* be closed by considering the differences in the conformational entropy of the two molecules, enterobactin vs. MECAM.

In order to evaluate the possible role of conformational entropy, EFF calculations were performed on free enterobactin and on its structural analogue, trisbenzamide TBA, in combination with experimental analysis of TBA. Spectroscopic examination demonstrated that TBA adopts a well-defined C_3-symmetric propeller-type conformation with its side chains axially positioned and its amide-NH groups H-bonded to the lactone oxygen of the ring. *A priori*, these H-bonds may involve the oxygen of the same seryl unit or of its adjacent one. CD analysis allowed to differentiate between these two possibilities by demonstrating predominance of the right-handed orientation of the benzoate groups implying H-bonds to the same seryl unit.

EFF calculations supported the findings derived from experiments on TBA and demonstrated that free enterobactin adopts essentially the same conformation as the tribenzamide model, TBA. Enterobactin becomes thus ideally predisposed for ion binding by adopting a propeller-like conformation in the free state which is close to that of its iron(III) complex. Upon ion binding merely a rearrangement of the H-bonding network occurs. The H-bonds between the amide-NHs and the ring oxygens become weakened and replaced by H-bonds between the amide-NHs and the catecholate anions. Enterobactins' H-bonding network thus shapes the molecule in the free state to adopt a conformation prone to ion binding and rearranges to stabilize the complex once formed. This H-bonding network is believed to contribute significantly to both the high ion binding efficiency of enterobactin and the high optical purity of its ferric complex. We suggest to coin the term "wiper shift" for this type of H-bond (Figure 3).

It may be argued that in an aqueous environment where enterobactin exerts its high ion binding efficiency, the H-bonds of the free ligand might be weakened or even altogether broken. This argument could seriously challenge the validity of the suggested model where the conformation of enterobactin is predisposed for ion binding by adopting a propeller-like structure. Yet, experimental data on both the tribenzamide model and enterobactin are in agreement with the presence of conformations where the side chains adopt an axial orientation similar to the one in its ferric-complex, even in polar solvents such as DMSO (NMR data of both enterobactin[39] and TBA[35]) or ethanol (CD data of TBA[35]).

Chiral Enterobactin Analogues with Restricted Conformational Freedom?

Aiming at artificial ion binders with superior ion binding properties, we therefore searched for molecules that (1) were strain-free when binding ferric ions, and that (2) simulated enterobactin by generating noncovalently defined ion binding cavities already prior to binding and by undergoing the wiper shift upon binding.

The first step in this undertaking aimed at synthesizing tripode-like molecules that adopt propeller-like conformations by virtue of a network of interstrand H-bonds.[41,42] We selected for the synthesis of such molecules C_3-symmetric residues as "anchors" and extended these anchors symmetrically by amino acid residues. Amide linkages were selected as H-bond formers because of their well-established tendency to form strong H-bonds and because of

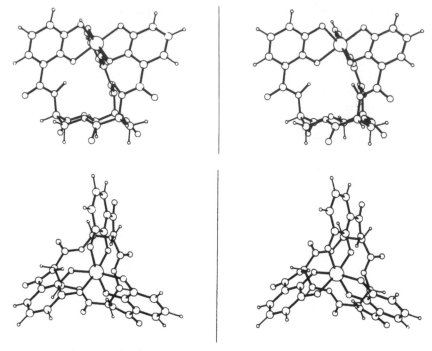

FIGURE 3. Computed conformation of ferric enterobactin (top, side view; bottom, top view).

the ready availability of amino acid residues in optically pure form (Figure 4). This approach proved successful with the synthesis of a large variety of C_3-symmetric trispeptides that showed interstrand H-bonds between the amide linkages of adjacent chains. Such H-bonds cause a tilt of the chains to generate propeller-like arrangements. In the absence of chiral centers, the right- and left-handed arrangements are enantiomeric, energetically identical, and equally populated. However, in the presence of chiral centers the right- and left-handed arrangements become diastereomeric, energetically nonidentical, and therefore either of the arrangements preferentially populated to form propellers of defined helical sense. The strength of the H-bond pattern proved to thereby depend on the nature of the anchor used. While the trispeptides based on mesitylene as anchor lost their propeller-like arrangements in protic media, those based on TREN (tris[2-aminoethyl]amine) as anchor retained some of their propeller-like arrangements even in protic media.[42]

Extension of such tripodal structures with catecholate ion binding sites was expected to provide iron(III) binders that adopt propeller-like conformations of defined chiral sense in the free state and form iron(III) complexes of the same absolute configuration as enterobactin. In addition, these binders were anticipated to exceed the ion binding efficiencies and selectivities of the parent ion binders, MECAM and TRENCAM, that lack the intermittent amino acid residues because of two factors: (1) first, these structural modifications were anticipated to relax the iron(III) complexes by adding two additional degrees of freedom to each ion binding chain; and lower their strain energies (2) second, the presence of intermittent amino acid residues was hoped to shape the arrangement of the free molecules toward conformations that are prone to ion binding by virtue of a belt of interstrand H-bonds.

All these expectations were borne out by experiment.[38,42] The protected chiral triscatecholates, similarly to the nonfunctionalized trispeptides, were all found to adopt propeller-like arrangements by virtue of interstrand H-bonds in apolar media. Depending on the anchor employed, these arrangements were partially retained in polar solvents such as methanol (in the case of TREN-based ligands), or broken (in the case of mesitylene-based ligands). When

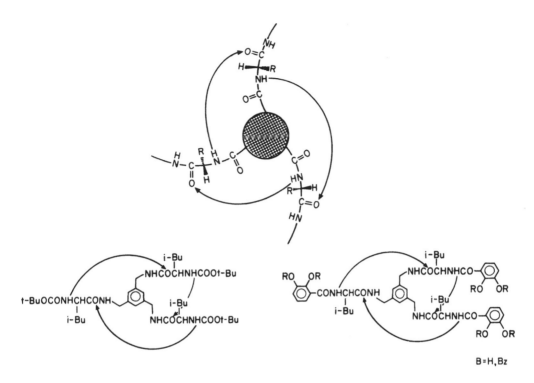

FIGURE 4. Principle design of preorganized chiral, tripodal enterobactin analogues with interstrand H-bonds ("circular H-bonds").

titrated with ferric ions, these ligands provided complexes of 1:1 iron/ligand stoichiometry. These complexes adopted preferentially the Δ-cis configuration, as does ferric-enterobactin, when using L-amino acids as bridges (Figure 5). However, when the NH groups adjacent to the catecholate were replaced by NMe groups, the preferred configuration of the ferric complex was inverted toward the left-handed one (Table 2).

The effect of these structural features on binding was examined by comparing the iron binding efficiencies of the chiral triscatecholates with those of the parent, achiral trisca-techolates MECAM and TRENCAM. Both families of binders, the chiral derivatives of MECAM as well as those of TRENCAM, proved to be more efficient and more selective for ferric ions than the corresponding achiral parent molecules. At this point it is, however, difficult to estimate the contribution of each effect, reduced strain of the ferric-complex, or reduced entropy loss upon binding. The significantly larger gain when introducing amino acid residues to the TRENCAM-type ligand than to the MECAM-type ligand is in line with the stronger H-bond network in the former which provides structures that are conformationally predisposed for ion binding.

The origin of the detrimental effect when replacing amide-NH with amide NMe is similarly difficult to deduce, as the amide-NH may exert a double function: (1) creation of interstrand H-bonds before binding and (2) stabilization of the ferric-complex once formed by H-bonding to the catecholate oxygen. Separation of these effects is not a trivial task. One possibility would be replacement of the internal amide (close to the anchor) by an ester linkage. This replacement would substantially reduce interstrand H-bonds because of the lower charge density of ester carbonyls relative to amide carbonyls, while H-bond stabili-zation of the ferric complex would not be curtailed by these changes.

The Remaining Puzzle of Enterobactin

Although much has been learned about the intricate features of enterobactin, its structural

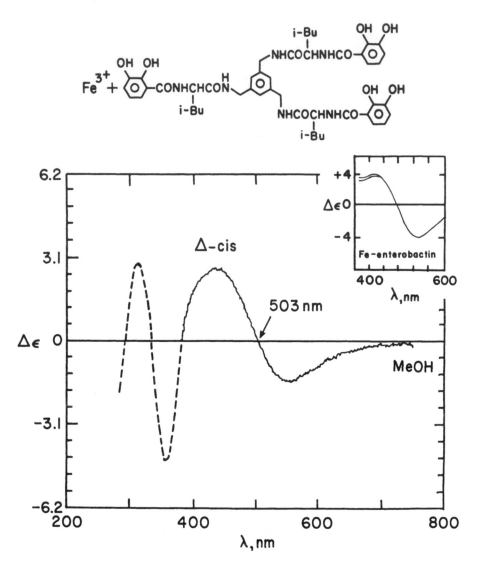

FIGURE 5.   CD-spectrum of chiral ferric-enterobactin analogue and of ferric enterobactin (insert).

TABLE 2
Iron Binding Properties of Chiral Catecholates

Ligand	R	R′	UV $\lambda_{max}(\epsilon)$	CD $\lambda_{max}(\Delta\epsilon)$	Preferred absolute configuration	Relative binding efficiency
1	H	L-iBu	502 (4650)	550 (−2.3), 432 (+4.2)	Δ-cis	7.3
1	H	L-Me	502 (4370)	560 (−1.0), 435 (+3.3)	Δ-cis	—
1	H	D-Me	502 (4460)	560 (+1.0), 435 (−3.3)	Λ-cis	6.7
1	H	L-Bz	502 (5040)	540 (−2.4), 435 (+2.8)	Δ-cis	2.0
1	Me	L-iBu	496 (4790)	540 (+2.1), 430 (−3.0)	Λ-cis	0.3
2	—	L-iBu	500 (5030)	540 (−2.9), 428 (+3.8)	Δ-cis	>50
2	—	L-Me	496 (5040)	548 (−1.6), 430 (+3.5)	Δ-cis	18
MECAM			496 (4700)	—	—	1.0

TABLE 3
Iron Binding Properties of Hydroxamates

	pM (Fe^{3+})	log β_{110}	Ref.
Ferrichrome	25.2	29.1	17
Ferrichrome A		32.0	17
Ferrioxamine E	27.7	32.5	17
Ferrioxamine B	26.6	30.5	17
Coprogen	27.5	30.2	17
Retroferrichrome		>29.1	45
C_3-sym. ferrichrome analogue		31.8	46
Acetyl hydroxamate		28.33	44
TRENDROX			
N-$(CH_2CH_2NHCOCH_2CH_2CONHOH)_3$	27.8	32.9	48
$1,3,5$-$Ar[CH_2NHCO(CH_2)_2NOHCOPh]_3$		28.0	47

simplicity, and outstanding iron(III) binding properties, it is still difficult to fully account for its remarkable properties. Most striking is the fact that catecholate iron(III) binders span a large range of iron(III) binding efficiencies even when making use of one of the best catecholate binding sites, the 2,3-dihydroxy-benzamide moiety. It is tempting to speculate that this phenomenon might derive from the highly rigid geometrical requirements of the ferric-dihydroxybenzamide chromophore, which should ideally adopt coplanarity in order to gain maximal resonance energy. This rigid geometric requirement is difficult to satisfy, and compromise solutions have to be adopted necessitating substantial deviations from coplanarity, deviations from octahedral iron(III) binding geometry, or strain in the ligand's backbone. This possibility is strongly supported by a recent X-ray diffraction study of a bicapped ferric-triscatecholate where the iron(III) binding geometry is prismatic, rather than octahedral, in order to retain coplanarity of the catecholate binding sites.[40]

HYDROXAMATE SIDEROPHORES AND SYNTHETIC ANALOGUES

Hydroxamate-based siderophores span a large family of compounds that differ in their symmetry and topology.[16-19,43] The hexadentate hydroxamate-based siderophores include linear structures like ferrioxamine B, cyclic structures like ferrioxamine E, and tripodal ones like the ferrichromes. The ferrichromes, although being tripodal-like enterobactin, differ from enterobactin by lacking its perfect C_3-symmetry. Moreover, while enterobactin forms complexes of predominant Δ-cis configuration,[29] in the ferrichromes the Λ-cis configuration prevails[32] (Table 3 and Figure 6).

In contrast to the catecholates, the span of the ferric ion binding efficiencies within the hydroxamate binders is much smaller and lies within less than five orders of magnitude.[17,44-48] Moreover, the natural hydroxamate siderophores are not significantly more efficient than the simplest bidentate hydroxamate, and several synthetic trishydroxamates have already been prepared by that approach,[21,45-47] and even surpass the efficiency of the natural compounds.[48]

Synthetic Ferrichrome Analogues of C_3-Symmetry

The search for synthetic hydroxamate analogues developed along two almost independent lines. One line concentrated on the synthesis of achiral C_3-symmetric trishydroxamates[47,49] quite similar in structure to the C_3-symmetric catecholates developed by Raymond. This line provided derivatives that surpassed ferrichrome in its ion binding efficiency. The other line was drawn from the structure of the natural ferrichrome and aimed at its modifications. The first modifications involved inversion of the hydroxamate directionality to provide retro-hydroxamate and its de-methylated derivative.[45] The binding efficiencies of both of these compounds proved inferior to that of natural ferrichrome, although the retro-ferrichrome

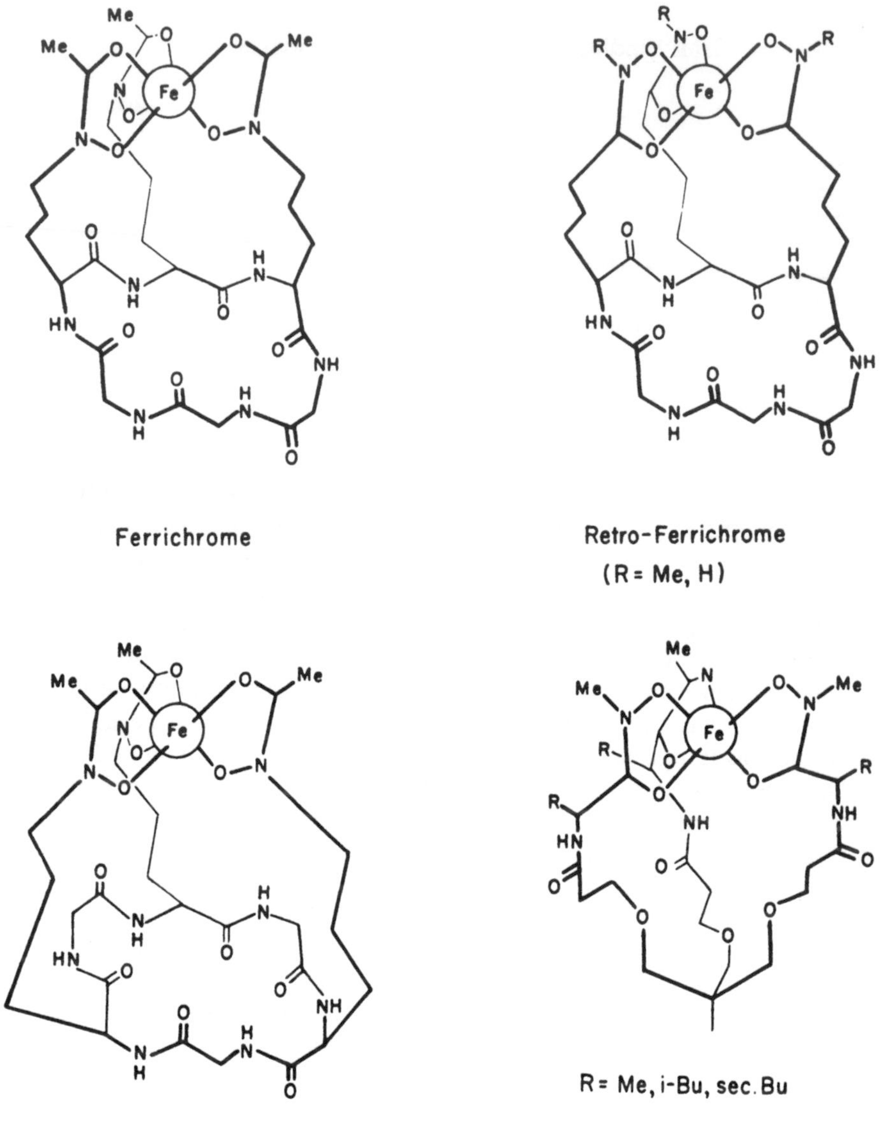

FIGURE 6.   Ferrichrome (top left) retro-ferrichrome (top right), and C$_3$-symmetric ferrichrome analogues (bottom).

analogue equaled ferrichrome as microbial growth promoter. The second modification involved the macrocyclic C$_3$-symmetric peptide analogue of ferrichrome which was estimated to be two orders of magnitude more efficient than the natural siderophore as iron(III) binder.[46]

Our approach toward biomimetic ferrichrome analogues relied on replacing the nonsymmetric structure of the natural siderophore by a C$_3$-symmetric one while making use of the modular design previously applied for the synthesis of biomimetic enterobactin analogues.[35,50] It was based on the use of tripodal molecules that generate propeller-type conformations of defined chiral sense by virtue of a network of interstrand H-bonds.[41] In these structures the chiral information contents are located in the amino acid bridges, instead of the macrocyclic peptide ring used in ferrichrome. It should be emphasized that the use of amino acids in these structures not only induces helicity of preferred chiral sense, but also allows systematic modifications of the molecule's cavity size and of its shape by incorporating different amino acid residues. In order to obtain ligands that form complexes of a predominant Λ-cis configuration like in ferrichrome, rather than a Δ- configuration, we selected triscar-

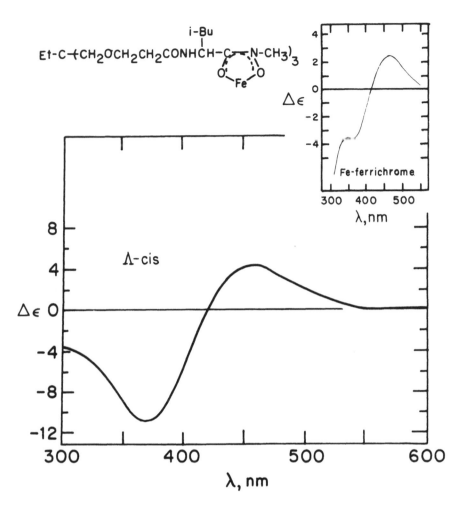

FIGURE 7. CD-spectrum of biomimetic ferrichrome and of ferrichrome (insert).

boxylates as anchors and extended them by attachment of amino acids via their N terminal, rather than via their C terminal. Inverting the directionality of the terminal hydroxamate groups relative to that of the natural ferrichrome was sought to have no significant effect on the *in vivo* properties, since ferrichrome and its retro-isomer have been shown to have comparable biological activity, but to greatly facilitate synthesis.

Following this principle design several chiral trishydroxamates were synthesized and examined. The leucyl-, alanyl-, and ileucyl-derivatives proved to adopt in the free state propeller-like conformations that are stabilized by interchain H-bonds. When binding Fe^{3+}, the Fe^{3+} complexes adopted Λ-cis configuration according to their CD spectra. The remarkably high intensities of the Cotton effects for this complex (which surpassed any values so far published for chiral Fe^{3+} hydroxamates) suggested high configurational purity[50] (Figure 7).

The high configurational purity of the chiral tris-hydroxamates was confirmed by the NMR spectra of their Ga^{3+} complexes. Ga^{3+} has been shown to form complexes that are very close to those of Fe^{3+}, but to be amenable to NMR analysis.[51] The NMR pattern of the Ga^{3+} complexes of the chiral derivatives showed a single set of signals confirming the presence of a single configurational isomer.

Comparison of the free hydroxamate ligands and their complexes by a combination of IR and NMR spectroscopy confirmed the occurrence of an H-bond shift upon binding to stabilize the complexes formed by H-bonds from the amide-NH to the ferric-hydroxamate Fe-O (Figures 8 and 9).

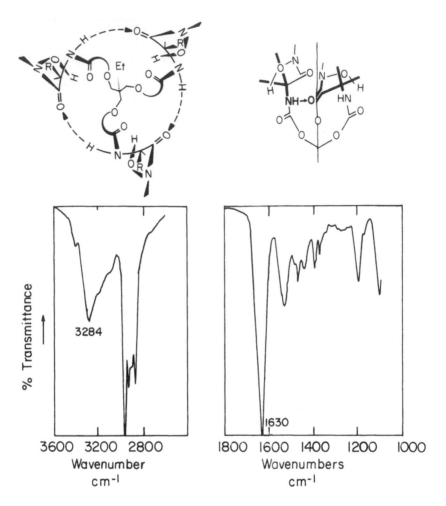

FIGURE 8.   Schematic representation of chiral ferrichrome analogue (top) and its IR-spectrum (bottom).

These biomimetic hydroxamates thereby reproduce the "wiper mode" of ion binding first suggested to occur in natural enterobactin. The chiral hydroxamate which makes use of proline, instead of primary amino acids, and which therefore cannot form interchain H-bonds, provides isomeric mixtures of Ga^{3+} complexes and forms Fe^{3+} complexes that give rise to Cotton effects of significantly lower intensities. Interchain H-bonds thus appear inducive to the formation of configurationally pure metal complexes (Table 4).

In order to provide an estimate of the effect of H-bonding on complex stability, the relative stabilities of the Fe^{3+} complexes of the leu derivative and of the pro derivative were determined in comparison with the parent hydroxamate by competition experiments in apolar, aprotic media. These measurements indicated threefold higher binding efficiency for the leucyl derivative relative to the prolyl derivative. In protic media, on the other hand, the prolyl derivative was comparable with the leucyl derivative. Independent evidence that the Fe^{3+} ion is more strongly bound to the leu derivative than to the pro derivative was obtained by examining the temperature effect on their Moessbauer spectra.[52] The temperature effect was thus smaller for the ferric complex of the leucyl derivative than that of the prolyl derivative, indicating higher force constants for the Fe-O bond and thereby stronger binding to the former.

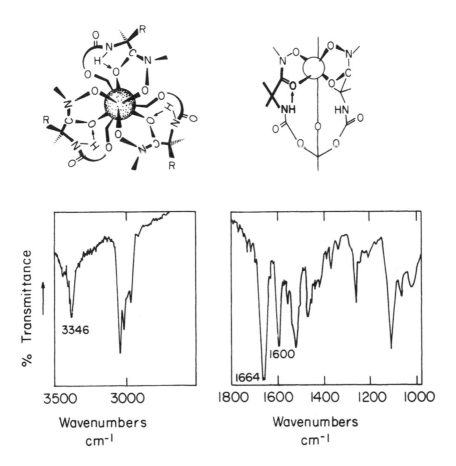

FIGURE 9. Schematic representation of chiral ferric-ferrichrome analogue (top) and of its IR-spectrum (bottom).

TABLE 4
Iron Binding Properties of Chiral Hydroxamates

$$Et-C-[-CH_2O(CH_2)_nCON-CH-C-N-CH_3]_3$$

(with R R' above CH and below: O OH)

	Compound		CD spectra of ferric hydroxamates		Preferred absolute
n	R	R'	λ_e	$\Delta\epsilon$	configuration
1	H	L-iBu	368, 410, 450	−7.1, 0.0, +4.04	Λ
2	H	L-Me	365, 420, 455	−7.5, 0.0, +2.5	Λ
2	H	D-Me	365, 420, 455	+7.5, 0.0, −2.5	Λ
2	H	L-iBu	365, 415, 450	−6.8, 0.0, +3.43	Λ
2	H	L-sec.Bu	368, 408, 444	−7.4, 0.0, +4.4	Λ
2	L-(CH₂CH₂CH₂)		378, 430, 465	−4.3, 0.0, +1.27	Λ

Synthetic Ferrioxamine Analogues

The ferrioxamines,[16-18,23,24] similar to the ferrichromes described above, make use of hydroxamate groups as ion binding sites. Yet, at variance with the ferrichromes, the ferrioxamines are not tripode-like molecules, but macrocyclic and linear structures that lack chiral centers. They all possess three hydroxamate groups on a string which are bridged by

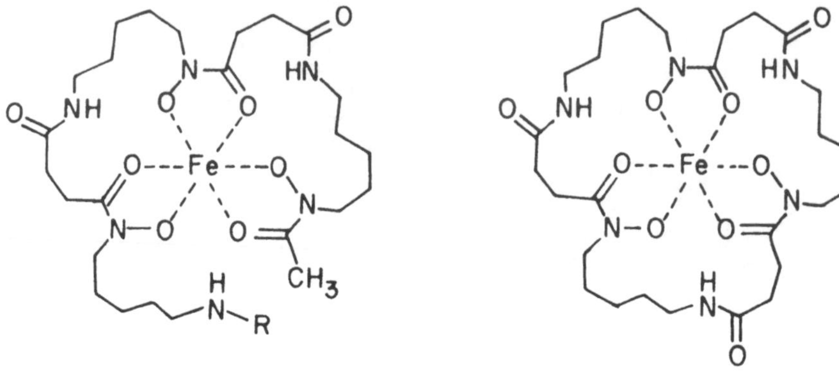

FIGURE 10.   Representatives of the ferrioxamine family of siderophores.

amide-containing methylene chains. This motive is created by a linear assembly of repeating alpha-amino-omega-hydroxyamino alkanes and succinic or acetic acid residues. The most studied representatives are the macrocyclic ferrioxamine E and linear ferrioxamine B. The latter siderophore represent an extreme case of configurational variability, in that it can form a total of five isomers when binding three valent metal ions, each as a racemic mixture[53] (Figure 10).

Replacing kinetically labile ferric ions with rather inert ions of chromium, Raymond et al. demonstrated that ferrioxamine B complexes consist of a mixture of 65% cis and 35% trans coordination isomers. In an attempt to establish whether the ferrioxamine receptor is capable of geometric discrimination, the isomeric chromium complexes were separated and their efficiency in inhibiting ferrioxamine-mediated iron uptake into *streptomyces piloses* examined.[54] Both the trans and cis isomers were found to have comparable activity, indicating little geometric differentiation by the ferrioxamine receptor (Figure 11).

Inspecting the structures of the ferrioxamine siderophores, it occurred to us that they might originally have been built from lysine residues, but lost their substituents and thereby their asymmetry during the evolutionary process.[61] Should this hypothesis be correct, the "archaic" ferrioxamines would have differed from their "modern" forms by containing carboxy side chains adjacent to the hydroxamate nitrogen or the amide nitrogen. As a consequence, archaic ferrioxamines would have preferentially formed either right- or left-handed ferric complexes. Considering the high chiral discrimination exhibited by the receptors of enterobactin,[25] ferrichrome,[26] and rhodotorulic acid,[48] it appeared plausible that the ferrioxamine receptor might also exhibit high chiral discrimination, and geometric discrimination, when appropriately substituted.

In an attempt to examine the possibility of geometric and chiral discrimination by the ferrioxamine receptor, and possibly also identify the domains of recognition, we aimed at designing biomimetic ferrioxamine B analogues that (1) form exclusively cis complexes, (2) adopt a minimal number of conformational isomers when binding ferric ions, and (3) exhibit significant chiral preference. Toward this target we reduced the pentamethylene chain of Ferrioxamine B by a shorter methylene bridge, and introduced asymmetric centers. These modifications were envisioned to not only induce chiral preference of either the right- or left- handed ferric complex, but also to have the additional advantage of highlighting the differences between the two faces of the ferrioxamine complexes (Figure 12).

The detailed design of these molecules relied on linear tris-hydroxamate derivatives assembled from natural amino acids (either L or D) and 3-hydroxy amino acids. The use of

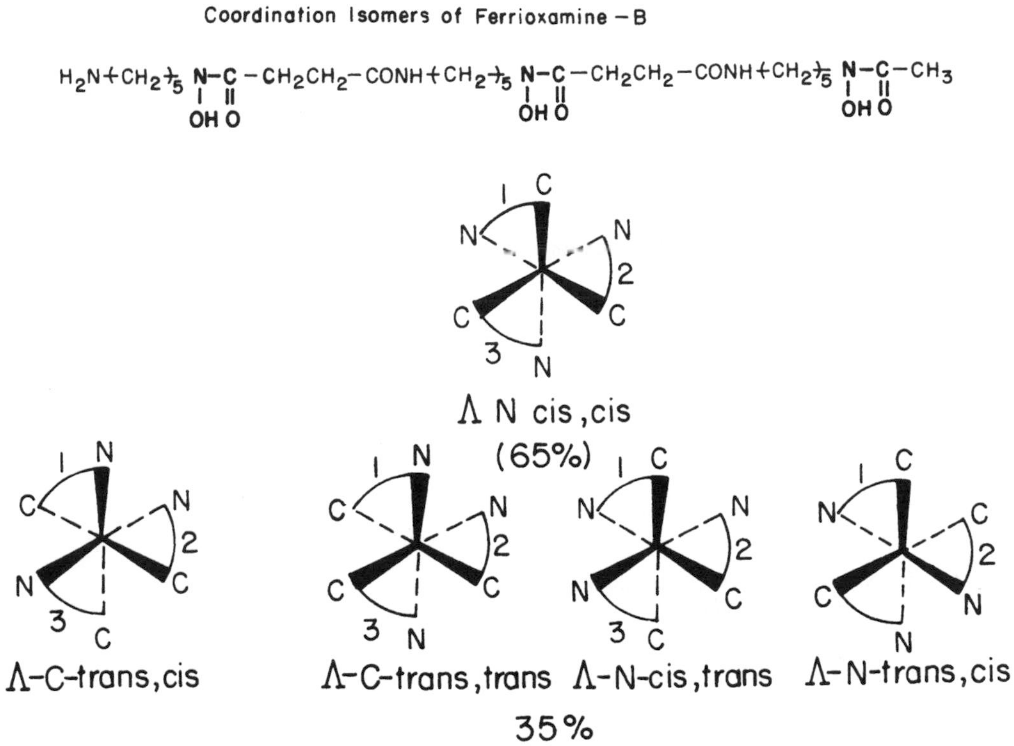

Coordination Isomers of Ferrioxamine — B

$$H_2N+CH_2\overset{}{\rightarrow_5} \overset{N-C}{\underset{OH\ O}{|\ ||}} - CH_2CH_2 - CONH + CH_2\overset{}{\rightarrow_5} \overset{N-C}{\underset{OH\ O}{|\ ||}} - CH_2CH_2 - CONH + CH_2\overset{}{\rightarrow_5} \overset{N-C}{\underset{OH\ O}{|\ ||}} - CH_3$$

Λ N cis,cis
(65%)

Λ-C-trans,cis Λ-C-trans,trans Λ-N-cis,trans Λ-N-trans,cis

35%

FIGURE 11. Geometric isomers of ferrioxamine B.

$$R'(NH(CH_2)_5NOHCOCH_2CH_2CO)_2NH(CH_2)_5NOHCOCH_3$$

Ferrioxamine B (R' = H)

$$R'(NH-CHR(CH_2)_n-CONOH-CH_2CH_2-CO)_3OMe$$

Biomimetic Ferrioxamine B

n = 0, 1, 2

R = H, CH$_3$; CH$_2$CH(CH$_3$)$_2$; CONR$_2$

R' = H, Me$_3$C–OCO–, Ar

FIGURE 12. Ferrioxamine B (top) and biomimetic analogues (bottom).

amino acids as building blocks was thought to facilitate the preparation of such binders by relying on the large arsenal of available natural amino acids as well as their enantiomers. These binders represent "retro" analogues of ferrioxamine B in regard to the relative directionality of the amide and hydroxamate groups. Although this inversion may reduce binding efficiency, it was anticipated to be of little biological relevance. The recently prepared

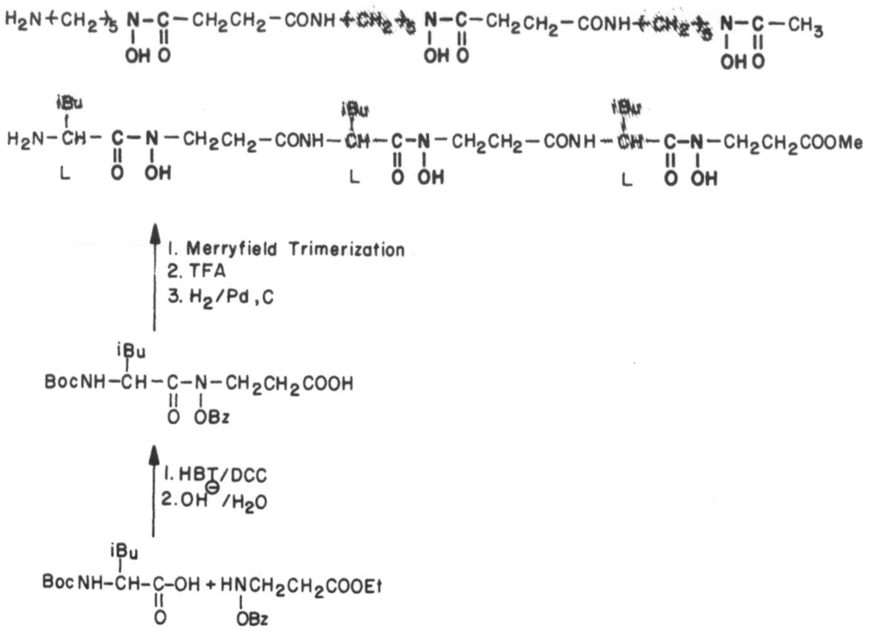

$$H_2N \{CH_2\}_5 \underset{OH}{N}-\underset{O}{\overset{\parallel}{C}}-CH_2CH_2-CONH \{CH_2\}_5 \underset{OH}{N}-\underset{O}{\overset{\parallel}{C}}-CH_2CH_2-CONH \{CH_2\}_5 \underset{OH}{N}-\underset{O}{\overset{\parallel}{C}}-CH_3$$

$$H_2N-\underset{L}{\overset{iBu}{\overset{\mid}{C}H}}-\underset{O}{\overset{\parallel}{C}}-\underset{OH}{N}-CH_2CH_2-CONH-\underset{L}{\overset{iBu}{\overset{\mid}{C}H}}-\underset{O}{\overset{\parallel}{C}}-\underset{OH}{N}-CH_2CH_2-CONH-\underset{L}{\overset{iBu}{\overset{\mid}{C}H}}-\underset{O}{\overset{\parallel}{C}}-\underset{OH}{N}-CH_2CH_2COOMe$$

↑ 1. Merryfield Trimerization
2. TFA
3. H$_2$/Pd ,C

$$BocNH-\overset{iBu}{\overset{\mid}{C}H}-\underset{O}{\overset{\parallel}{C}}-\underset{OBz}{N}-CH_2CH_2COOH$$

↑ 1. HBT/DCC
2. OH$^{\ominus}$/H$_2$O

$$BocNH-\overset{iBu}{\overset{\mid}{C}H}-\underset{O}{\overset{\parallel}{C}}-OH + H\underset{OBz}{N}CH_2CH_2COOEt$$

FIGURE 13.   Synthesis of chiral, biomimetic ferrioxamine B analogs from L-leucine and 3-hydroxyamino propionic acid.

retroferrioxamine B showed ferric ion binding properties similar to those of ferrioxamine B, although some differences in its redox potential.[55] Retroferrichrome, for example, has shown to equal the biological activity of ferrichrome, although being a weaker binder.[45]

In order to select for synthesis and testing the most promising structures that adopt preferentially the cis configuration in a minimal number of conformations, we joined theoretical with experimental tools. EFF calculations on three families of biomimetic ferrioxamine analogues where R = H and n = 0, 1, or 2 demonstrated that binders with n = 0 or n = 2 from a small number of closely related low energy conformations, while compounds with n = 1 yield a large number of substantially different conformations. Accordingly, the latter series of compounds was dropped for further consideration.

The first compounds synthesized were the alanyl, leucyl, and glutamyl derivatives, where n = O, R = CH$_3$ or CH$_2$CH(CH$_3$)$_2$, and n = 2, R = CON(CH$_2$CH$_3$)$_2$. The synthesis relied on the Merryfield method as illustrated below for the leucyl derivative (Figure 13). All these ligands were found to bind ferric ions in a 1:1 stoichiometry, and to exhibit some preference for the Δ configuration when binding ferric ions. The two isomeric configurations of the α-amino acid derivatives are schematically illustrated in Figure 14.

Inspection of the above structures reveals that the presence of side chains renders the two faces of the complex significantly different in their stereochemical requirements. Moreover, the orientations of the side chains relative to the carbonyl amides are opposite in the diastereomeric complexes. While in the Λ configuration the side chains and carbonyl amides point to opposite faces of the chelate rings; in the Δ configuration they point to the same face. To which extent these ferrioxamine analogues may replace ferrioxamine in receptor-driven ferric ion uptake, and to which extent the biological receptors differentiate between the two faces or even between the two configurations, is under current investigation.

COMPARISON OF CATECHOLATE AND HYDROXAMATE-BASED SIDEROPHORES

Comparing the hydroxamate siderophores with the catecholate siderophores two very basic differences become apparent. First, hydroxamate siderophores form electrically neutral

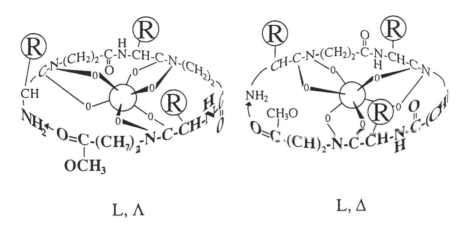

L, Λ   **L, Δ**

FIGURE 14.   Diastereomeric ferric complexes of biomimetic ferrioxamine B analogues derived from α amino acids (R = CH$_3$, R = CH$_2$CH[CH$_3$]$_2$).

complexes when binding iron(III) and related trivalent metal ions, while catecholates form electrically charged complexes. Second, the ion binding efficiencies of hydroxamates fall within a rather narrow range, while those of the catecholates cover a wide span. The latter differences are likely to derive from the more stringent geometrical requirements of the catecholate binding site.

In spite of these differences, the two types of binding sites share common features. These include the dependence of their ion binding properties on proximal groups. The stability and optical purity of both metal-trishydroxamates and metal-triscatecholates are enhanced by the presence of adjacent amide-NH groups that can form H-bonds to the metal complex (although the effects on the hydroxamates are less pronounced than on the catecholates that bear a higher charge density), Even more so, the very principles of preorganization and the "wiper shift" that have first been postulated to occur in the catecholate siderophore enterobactin could successfully be reproduced in synthetic siderophores, and in both biomimetic catecholate and hydroxamate siderophores. It could also be demonstrated that biomimetic hydroxamate siderophores designed in accordance with these principles show enhanced iron(III) binding efficiency, iron (III) binding selectivity, and optical purity of the metal complexes. This successful translation of a principle from one type of iron(III) binder to a different type of iron(III) binder illustrates the power of the biomimetic approach. Identifying the functional principles of a given natural compound, i.e., catecholate siderophore, not only allows the synthesis of biomimetic analogues of this very compound or family of compounds, but also of related ones, i.e., hydroxamate siderophores.

IN VIVO PROPERTIES OF BIOMIMETIC SIDEROPHORES

GENERAL CONSIDERATIONS

Siderophore-mediated iron(III) uptake is in essence governed by chemical recognition. For an iron(III) carrier to be biologically active it has to meet two requirements: (1) to specifically bind iron(III), and (2) to favorably interact with membranal proteins. This implies the presence of a double recognition phenomenon: iron binding and receptor matching. The challenge of the latter task is further highlighted by the lack of structural information on the recognition sites of the membranal proteins, and implies the search for a key to a largely unknown lock (Figure 15).

In vivo tests of biomimetic siderophores should not only provide information on their overall growth promotion activity, but also shed light on the occurrence of specific ferric-

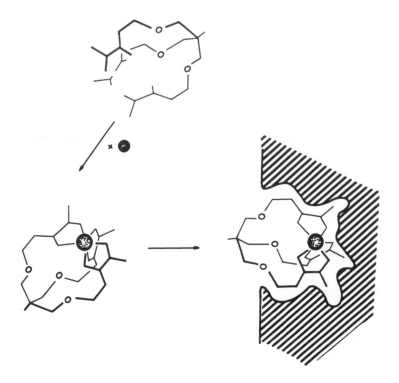

FIGURE 15.   Schematic representation of receptor-driven siderophore-mediated
iron uptake.

siderophore-protein interactions and the extent of these interactions in comparison with those
of the natural counterparts. The selection of the biological model and of the type of activity
to be examined is crucial in this undertaking. *E. coli* and its mutants have long been the
systems of choice[20] because of the large array of mutants available that differ in the intactness
of their iron(III) uptake systems. However, with the pronounced progress made in genetic
engineering, other microbial models and several of their mutants have become available.
Comparing *in vivo* properties of ferric ion carriers on mutants deficient in a very specific
gene product, i.e., the outer membrane receptor, the siderophore translocation protein or a
degradation enzyme allows to deduce the extent of fit of the biomimetic analogues to either
of the transport proteins. In addition, two alternative *in vivo* activities may be screened:
growth promotion or iron(III) uptake. These activities are not necessarily proportional. While
measurements of growth promotion activities are generally more sensitive, iron(III) uptake
experiments are more easily quantitized and prone to competition experiments with the
natural siderophores. Such competition experiments enable to establish to which degree the
biomimetic analogues are viable surrogates of the natural systems when interacting with
membranal proteins and enzymes.

The pioneering efforts in this field concentrated on the study of enterobactin, the most
powerful natural iron(III) binder known.[34,56]

BIOMIMETIC ENTEROBACTIN ANALOGUES FOR FACILITATED IRON UPTAKE AND GROWTH IN MICROORGANISMS

Major contributions in both the synthesis of biomimetic enterobactin analogues and the
analysis of their *in vivo* performance came from Raymond's laboratory.[34,56,57] The biomimetic
enterobactin analogues were screened on *E. coli* strains that differ in the intactness of their
enterobactin uptake systems, and both growth promotion and iron(III) uptake were tested.

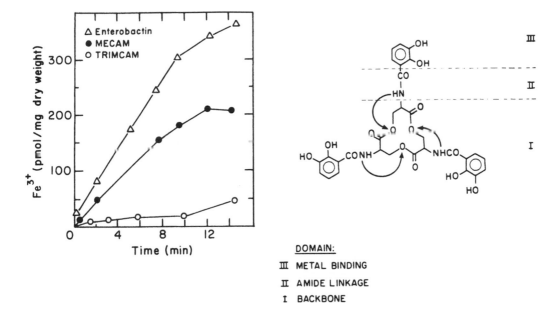

FIGURE 16. Promotion of iron uptake into *E. coli* by enterobactin analogues (left), and domains of receptor recognition of enterobactin (right).

Competition experiments between the biomimetic iron(III) carriers and enterobactin on *E. coli* allowed to establish the extent of fit of the synthetic molecules to the enterobactin receptor, the FepA protein. These experiments identified MECAM as the only triscatecholate of the series that binds to the receptor; the methylated analogue, Me_3CAM, as well as the positional analogue, TRIMCAM, failed to be recognized by the microbial receptor, as did MECAM derivatives containing substituents on their aromatic nuclei (Figure 16).

These results led to the conclusion that the exposed part of ferric-catecholate is the fragment relevant for recognition, while the trilactone of the natural enterobactin does not play a major role in this step.[56] Combing the tools of biomimetics with those of genetic engineering, it was also possible to trace the structural requirements for recognition by other membrane components such as the fepB protein, which is responsible for translocation. All four enterobactin analogues, MECAM, Me_3MECAM, TRIMCAM, and LCAMS, were found to act as growth promoters in *E. coli* gene types with an intact transport system. Yet, there were drastic differences in the efficiency of these compounds to transport iron(III) into the cell. These observations appear to suggest that growth promotion is still occurring with ferric(III) ion carriers of very low iron(III) transport activity. The discrepancy between growth promotion and iron(III) uptake efficiency seems not to be an isolated curiosity, but will come back again in other microbial systems in relation with ferrichrome.

BIOMIMETIC FERRICHROME ANALOGUES FOR FACILITATED IRON UPTAKE AND GROWTH IN MICROORGANISMS AND PLANTS

The first synthetic analogues of ferrichrome, and structurally the closest ones, have been prepared and examined by Emery et al.[45] These analogues, termed retrohydroxamate ferrichromes, differ from the natural compound by the directionality of the hydroxamate groups. The binding efficiencies of the retro-isomer where the terminal acetyl group had been replaced by N-methyl-hydroxyl amine was lower than that of the natural counterpart. Yet its biological activity as growth promoter of *Arthrobacter flavescens*,[58] and its potency in antagonizing albomycin was indistinguishable from ferrichrome. Also, the activity of this isomer in promoting iron(III) uptake in *Ustilago sphaerogena* was practically identical with that of ferrichrome. On the other hand, desmethylretrohydroxamate ferrichrome, where the terminal acetyl group had been replaced by hydroxyl amine, showed no significant growth promotion activity toward *A. flavescens,* and only one third of the iron(III) transport efficiency toward *U. sphaerogena.*

These observations demonstrate that the directionality of the hydroxamate groups does not play a major role in receptor recognition, while replacement of the terminal methyl groups by hydrogen may drastically reduce recognition.

More remote from the structure of natural ferrichrome, both C_3-symmetric and nonsymmetric derivatives were prepared and examined *in vivo* by the groups of Miller et al.[49] Several of these synthetic iron(III) carriers were found to act as growth promoters of *E. coli* strains having intact transport systems for hydroxamate-type siderophores, but were inactive in those mutants that lack the outer membrane ferrichrome receptor or gene fhuB products. These observations suggest that the activity of these carriers simulates that of ferrichrome qualitatively (although not quantitatively), by making use of the ferrichrome iron uptake system. Structurally, the active compounds included C_3-symmetric and nonsymmetric derivatives, derivatives based on tetrahedral carbon, trigonal nitrogen, or isocyanuric acid as anchors, and derivatives containing the terminal hydroxamate group in either of the two directionalities. On the other hand, two very similar compounds proved either not active at all or active in all mutants. Both compounds were based on cyanuric acid as stem, but differed in the directionality of the terminal hydroxamate groups and the length of the methylene bridges that connect the binding sites to the anchor. This extreme sensitivity of the biological system to minute structural changes seems rather surprising. Further studies, including iron(III) transport experiments, will have to be performed to trace the origin of this phenomenon.

Our own approach to biomimetic ferrichrome analogues drew from both the study of Emery's retrohydroxamate ferrichromes[45] and rather simple, C_3-symmetric trishydroxamates.[50] It, however, differed from the former in replacing the hexapeptide ring by a tetrahedral carbon as anchor, and from the latter by introducing chiral amino acids as bridges. Chiral amino acids were used (1) to enable systematic modifications of the molecules' envelope and (2) to impart chiral preference to the iron(III) carriers' ferric complexes. Systematic modification of the molecules' envelope was conceived to be of prime importance as it allows to trace the receptor surface. Chiral preference, and the possibility to prepare either of the two optical isomers, were anticipated to provide a powerful probe to distinguish between specific and nonspecific interactions with the microbial systems.

In order to test the biological performance of these compounds, *A. flavescens* was chosen as the model.[50,58] This bacterium possesses ferrichrome receptors, but does not produce ferrichrome itself. It therefore is completely dependent on externally added ferrichrome, or ferrichrome substitutes, for growth. Addition of the synthetic ferrichrome analogues to the culture medium of this bacterium and measurement of its resulting growth provide a sensitive indicator for the biological effectiveness of these compounds. The tricarboxylate-based L-leu derivative (Figure 17) showed 1% of the growth promotion activity of the natural ferrichrome. Should the observed activity be of any significance, it should be possible to optimize these structures by further modifications. Considering the fact that the ion binding chains in ferrichrome do not possess substituents, we proceeded to reduce the bulkiness of the projecting amino acid residues and started by replacing L-leu by either L-pro or sar (Figure 17). Indeed, the activity increased from 1 to 80 and 85%, respectively. In order to establish whether the observed effect derived from the "shrinking" of the amino acid, or from a conformational change caused by replacing secondary amide bonds by tertiary amide bonds, the L-pro and sar were substituted by L-ala. The L-ala derivative was found to fully equal ferrichrome in its growth promotion activity toward *A. flavescens.* Although the details of the observed growth promotion activity are still under investigation,[60] the considerations given below strongly support the involvement of the outer membrane receptor.

In-vitro and in-vivo properties of
Biomimetic Ferrichrome Analogs

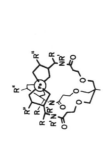

	L-ileu,Λ	L-ileu,Λ	L-pro,Λ	Sar	L-ala,Λ	D-ala,Δ	L-ala,Λ
Rel. binding efficiencies (Desferal=100)	25	10	14		110		
Fe^{3+} - extraction efficiencies	52	35	38		88		
Rel. growth promotion in Arthrobacter Flavescens (Ferrichrome =100)		1	80	85	100	<1	<1
Rel. Fe^{3+} uptake in Corn Roods (citric acid =1)	1.5	1.3	1.2		1.5	1.5	
Rel. Ga^{3+} uptake in Corn Roods (citric acid =1)	0.92				1.05		

FIGURE 17.   Biological activity of ferrichrome analogues as growth promoters of *Arthrobacter flavescens* and as iron carriers into corn roots.

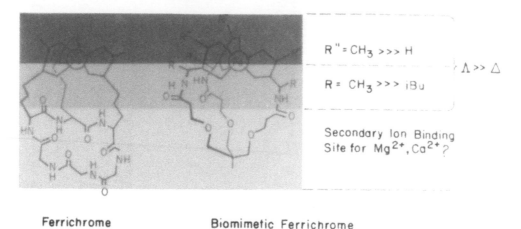

R" = CH$_3$ >>> H

Λ >> Δ

R = CH$_3$ >>> iBu

Secondary Ion Binding
Site for Mg^{2+}, Ca^{2+}?

Ferrichrome   Biomimetic Ferrichrome

FIGURE 18.   Domains of receptor recognition of ferrichrome.

The higher activity of the L--ala derivative relative to the more lipophilic L-leu derivative excludes the possibility of mere diffusion. Moreover, the similar biological activity of the L-pro and L-ala derivative is in line with the observation that L-ala has most frequently been found to replace L-pro during the evolutionary process. Further support for the occurrence of a receptor-driven process was obtained when the D-ala derivative, which forms iron(III) complexes of Δ-cis configurations instead of Λ-cis configuration, was found to lack any activity. This pronounced chiral discrimination is analogous to that observed with ferrichrome in fungi, where the antipode of the natural isomer, enantio-ferrichrome, lacked any activity.[26] Moreover, replacement of the terminal methyl group in the synthetic L-ala derivative by hydrogen decreased the molecule's effectiveness, as did analogous replacements of the terminal methyl group by hydrogen in retro-ferrichrome[45] (Figure 18).

It is appealing to use the observed scale of activity for defining the domains relevant for recognition. Of major importance is the stereochemistry around the metal center, where Λ configuration is a prerequisite for biological activity and where a terminal methyl group is significantly superior to a terminal hydrogen. The lateral portion of the molecule allows fine tuning to optimally match the biological receptor. The anchor, on the other hand, seems to allow significant modifications. These findings are in line with those reported for enterobactin,[56,62] where the exposed part of the iron(III) binding region was suggested of relevance for biological activity.

Further insight into the mode of action of these ferric ion carriers was obtained by examining their capability to enhance iron(III) uptake.[60] The hydrophilic L-alanyl derivative, which is an effective growth promoter, was found to also enhance iron(III) uptake into *Arthrobacter*, although three times less efficiently than ferrichrome. On the other hand, the enantiomeric D-alanyl derivative that forms ferric complexes of the unfavorable Δ-*cis* configuration and failed to stimulate growth, also failed to enhance iron(III) uptake. However, growth promotion and facilitated iron(III) uptake do not seem linearly related. Thus the L-pro derivative which is almost as good a growth promoter as the L-alanyl derivative did cause only minute intracellular iron(III) uptake. It might accordingly be speculated that growth promotion in *Arthrobacter* might not necessitate cytoplasmic iron uptake, but merely recognition of the ferric siderophore by the receptor and possibly iron(III) release on the outside of the cytoplasmic membrane. Although many alternative models might be put forward, we are confident that the use of biomimetic siderophores in combination with genetically modified mutants will enable to unravel the intricate facets of this problem.

With a systematic series of biomimetic ferrichrome analogues at hand, we had produced a ''kit'' that offered itself as probe to examine iron uptake mechanisms in other biological systems. Not only microorganisms, but plants also need iron. We therefore chose to examine this possibility on *Zea-Mays* (corn roots), which has been shown to make effective use of ferrichrome as an iron(III) carrier in a highly specific iron(III)-uptake process.[63]

Iron uptake experiments (using radioactive $^{59}Fe^{3+}$) established facilitated iron(III) uptake with the L-ileu and L-ala derivatives, and to a smaller extent with the corresponding L-leu and L-pro derivatives.[64]

In order to establish whether iron(III) uptake by the synthetic compounds is specific and involves the same pathway as that promoted by ferrichrome, we tested its sensitivity to stereochemical effects. Most indicatively, the D-ala derivative proved practically inactive. This pronounced chiral discrimination in favor of the left-handed isomer supports the involvement of a highly specific, chiral recognition site.

A priori, the observed high chiral discrimination may imply either the involvement of (1) a chiral receptor, (2) a highly stereoselective enzyme, or (3) a chiral receptor prior to a nonselective enzyme. In order to examine the possible involvement of a reductive, enzymatic process, $^{59}Fe^{3+}$ was replaced by $^{69}Ga^{3+}$ and its uptake monitored. Gallium(III) simulates iron(III) in its coordination to hydroxamate binders and forms hydroxamate complexes of the same overall shape.[51] However, gallium(III) hydroxamates cannot undergo reduction and therefore only be taken up via nonreductive processes. None of the most active synthetic iron(III) carriers facilitated gallium(III) uptake. This result suggests that the synthetic ferrichrome-mediated iron(III) uptake processes involve either pathway (2) or (3).

BIOMIMETIC FERRICHROME ANALOGUES AS GROWTH INHIBITORS OF PLASMODIUM FALCIPARUM

Siderophores and their analogues enhance cellular iron(III) uptake and growth through irreversible, transmembranal iron(III) transport. It was therefore anticipated that biomimetic iron(III) carriers designed to scavenge cytosolic iron(III) and deprive the cell of iron(III) could function as growth inhibitors and thereby provide powerful tools against pathogenic microorganisms and parasites.

For iron(III) carriers to exert these functions, their actions need to be opposite, or reversed, to those of the natural siderophores. "Reversed siderophores" have to (1) penetrate cellular membranes in the uncomplexed state by diffusion, (2) scavenge intracellular iron(III), and finally (3) evade the cell as iron(III) complexes. We suggest to coin for these compounds the term "reversed siderophores".

For assessing our siderophoric hypothesis we have chosen to apply the agents to *P. falciparum* as potential antimalarial drugs. These malaria parasites complete their erythrocyte cycle of growth in 48 h and are highly dependent on an immediate and substantial supply of iron. Therefore, scavenging iron promises to be an effective tool for arresting parasite growth and opens new avenues for chemotherapy. This is of particular importance, since currently used drugs are steadily losing their efficacy because of increased drug resistance displayed by parasites.[65] It is most likely, and preliminary evidence confirms it, that such resistance is not developed against reversed siderophores. Moreover, the ostensible difference in the permeability properties of the host cell membrane in infected cells as compared to noninfected cells opens the possibility of the design of drugs with selective access to infected cells.[66,67]

Indeed, both natural and synthetic iron binders have been found to function as growth inhibitors of intraerythrocytic *P. falciparum*.[68-73] Yet, each of the compounds examined so far suffered from one or another limitation. While the hydroxamate-based natural siderophores were found to act by iron(III) deprivation, their efficiency was curtailed by their poor penetration across erythrocytic membranes.[68-70,72] Synthetic hydroxyquinolines and related chelators, on the other hand, proved satisfactory in respect to their transmembranal penetration characteristics, but acted by the formation of toxic metal complexes, rather than by iron deprivation.[73] In contrast synthetic catecholates did act as iron scavengers and showed satisfactory transmembranal penetration characteristics, but simultaneously removed iron from serum.[17,71]

Reversed hydroxamate siderophores that simulate the iron(III) binding properties of the natural siderophores, but are of hydrophobic, rather than hydrophilic, character, were anticipated to provide iron(III) scavengers that would overcome the limitations of both the natural and synthetic iron(III) chelators examined so far. Being hexacoordinating ferric ion binders, the reversed siderophores were expected not to form complexes with metal ions other than iron(III). In addition, being based on the hydroxamate group as an ion binding site, these compounds were anticipated not to remove iron from transferrin at any appreciable rate.[37] Moreover, due to their hydrophobicity, such iron(III) carriers were anticipated to readily transverse erythrocyte membranes by diffusion and thereby overcome the major limitation of the natural siderophores as intracellular iron(III) scavengers.

These expectations were fully realized[74] when the hydrophobic ferrichrome analogs, reversed siderophores, were found to inhibit growth of intraerythrocytic *P. falciparum*. The extent of growth inhibition was found to depend on the relative iron(III) binding or iron(III) extraction efficiency of the siderophores, but mainly on their hydrophobicity, as determined by their partition coefficients between saline and octanol (Table 5). This conclusion was supported by the fact that in open systems where the erythrocytic membrane does not represent a hydrophobic barrier, the less hydrophobic chelators proved as effective as the hydrophobic ones.

The observed growth inhibition was demonstrated to derive from iron(III) deprivation. Application of the reversed siderophores as their iron(III) complexes instead of as their free ligands completely reversed the inhibitory effect on parasite growth. Since the activity of these chelators derive from iron(III) deprivation, they were expected to have similar activity on both quinine-sensitive as well as quinine-resistant strains of *P. falciparum*. And indeed, the most resistant as well as the most sensitive strains proved to be equally effected by the reversed siderophores. Although these hydroxamate binders were found to exhibit their antiplasmodial action by iron(III) deprivation, they did not deplete serum from iron(III). Preincubation of whole serum with the most effective of the reversed siderophores did not lower its iron(III) contents.

Although these results appear promising in relation to possible iron scavengers as antimalarials, several very basic questions are still unanswered. These include the source of intraerythrocytic iron(III) which is scavenged by the ''reversed siderophores'' and the nature of the metabolic process which is curtailed by iron(III) deprivation. Elucidation of these problems is hoped to shed light on the iron(III) metabolism of the malaria-causing parasite and allow the development of ever superior chelators of potential antimalarial activity.

CONCLUSIONS AND OUTLOOK

Concentrating on three siderophores, enterobactin, ferrichrome, and the ferrioxamines, we have outlined the biomimetic approach toward the preparation of synthetic analogues that simulate the performance of the natural compounds *in vitro* and *in vivo*. We thereby followed the biomimetic methodology that relies on identifying the essential features of the natural compounds and on reproducing these features with the simplest possible synthetic structures. In the course of this undertaking the origin of enterobactins' outstanding iron binding selectivity and efficiency was demonstrated to derive from the nature of the dihydroxybenzamido binding site and the presence of H-bonding networks. The latter shape the molecule to adopt an ion binding configuration prior to binding and to stabilize the complex once formed. The principle of preorganization by the use of H-bonding networks was successfully reproduced in chiral enterobactin analogues to provide binders that exceed the

TABLE 5
Properties of Reversed Siderophores

$$Et-C-(CH_2-O-(CH_2)_n-CO-(NRCHR'CO)_m-NOHCH_3)_3$$

n	m	R	R'	Relative Binding[a]	Extraction eff.[b]	Partition Coefficients[c]	Hydrophobicity[d]	IC-50[e]
2	0	—	—	47	83	1.7		4
2	1	H	L-Me	110	88	0.53	1.0	40
2	1	H	D-Me					45
2	1	L	(CH₂)₃-	14	38	0.65	1.5	>50
2	1	H	L-iBu	10	35	12.5	3.5	17
2	1	H	L-sc.iBu	25	52	14.0	5.0	<2
1	1	H	L-iBu	20	62	28.6	3.5	4
Desferrioxamine B				100				43

[a] Relative binding efficiencies were determined by competition with EDTA in aq MeOH (MeOH—0.1 N NaOAc) using UV as an analytical tool.

[b] Extractions were performed by equilibrating solutions of ligand in $CHCl_3$ with aq solutions of $FeCl_3$, citric acid, 40 mM Tris at pH 6.9. Percent are given relative to total ligand present.

[c] Partition experiments were performed by equilibrating the free ligands between equal volumes of n-octanol and saline.

[d] The hydrophobicity scale given is that of the amino acid residues according to Tanford/Segret.

[e] Ligand concentration at which the growth of *Plasmodium falciparum* is reduced to 50%.

iron(III) binding efficiency of synthetic tris-catecolate analogues that lack this feature. The generation of H-bonding networks proved to also favorably effect iron(III) binding in ferrichrome analogues. These latter points illustrate the difference and the advantage of biomimetics relative to classical trial-and-error approaches in the design of biologically active molecules. Following the rational of biomimetics, identifying the functional principles of one system allows to successfully mimic another, related system.

When applied to biological systems, biomimetic iron(III) carriers serve as probes to trace the structural requirements for receptor binding, and also binding possibly to other membrane components that play a role in cellular iron uptake or growth. Biomimetic iron(III) carriers, which may systematically be varied, thereby provide "biological kits" to not only probe recognition to different membrane components of a single organism, but also to trace the differences between different organisms. Due to their modular design, they also allow the attachment of labels, possibly fluorescent or radiative labels, to further facilitate mapping of cellular iron(III) uptake processes.

In addition to the usefulness of biomimetic iron(III) carriers as mechanistic probes, such carriers promise application as growth promoters of agronomically important cultures. Even more so, biomimetic siderophores may be designed to exert a role opposite to, or reversed to that of the natural counterparts. While the natural molecules are to promote cellular iron(III) uptake and thereby growth, the synthetic analogues may be designed to scavenge intracellular iron(III) and thereby inhibit growth of unwanted parasites. The spectrum of possibilities appears unlimited, even with the few examples discussed above, and much more is envisioned to follow as the intricate facets of iron(III) metabolism will be unraveled.

ACKNOWLEDGMENT

The authors thank Drs. B. Bar-Josef, Z. I. Cabantchick, T. Emery, C. E. Felder, R. Ganmore, S. Lifson, Y. Tor, and G. Winkelmann for their enthusiastic collaboration in different aspects of this work, Mrs. R. Lazar for her outstanding technical assistance, and the MINERVA foundation and US-ISRAEL Binational Science Foundation for financial support.

REFERENCES

1. **Breslow, R.,** Biomimetic chemistry, *Chem. Soc. Rev.,* 1, 553, 1972. **Breslow, R.,** Biomimetic control of chemical selectivity, *Acc. Chem. Res.,* 13, 170, 1980.
2. **Pedersen, C. J.,** *Synthetic Multidentate Macrocyclic Compounds,* Vol. 1, Izatt, R. M. and Christensen, J. J., Eds., Wiley ‒‒‒‒‒‒‒c 1978, 1.
3. **More, C. and Pressman, B. C.,** Mechanism of action of valinomycin on mitochondria, *Biochem. Biophys. Res. Commun.,* 15, 562, 1964; **Pressman, B. C.,** Properties of ionophores with broad range cation selectivity, *Fed. Proc.,* 32, 1698, 1973, **Pressman, B. C.,** Antibiotic-mediated transport of alkali ions across lipid barriers, *Proc. Natl. Acad. Sci. U.S.A.* 58, 1949, 1967.
4. **Ovchinninkov, Yu.A. and Ivanov, V. T.,** The cyclic peptides, structure, conformation and function, in *The Proteins,* Vol. 5, 3rd ed., Neurath, H. and Hill, R. L., Ed., Academic Press, New York, 1982, 310.
5. **Lehn, J. M.,** Supramolecular chemistry — scope and perspectives, *Angew. Chem.,* 100, 91, 1988, **Lehn, J. M.,** Supramolecular chemistry: receptors, catalysts and carriers *Science,* 227, 849, 1985.
6. **Cram, D. J.,** The design of molecular hosts, guests and their complexes, *Science,* 240, 760, 1988; **Cram, D. J.,** Cavitands: organic hosts with enforced cavities, *Science,* 219, 1177, 1983.
7. **Kellogg, R. M.,** Bioorganic modelling — stereoselective reactions with chiral neutral ligand complexes as model systems for enzyme catalysis, *Top. Curr. Chem.,* 101, 111, 1982; **Kellogg, R. M.,** Chiral macrocycles as reagents and catalysts, *Angew. Chem. Int. Ed. Eng.,* 23, 782, 1984.
8. **Rebek, J.,** Model studies in molecular recognition, *Science,* 235, 1478, 1987; **Rebek, J.,** Recent progress in molecular recognition, *Top. Curr. Chem.,* 149, 189, 1988.
9. **Voegtle, F., Sieger, H., and Mueller, W. M.,** Complexation of uncharged molecules and anions by crown-type host molecules, *Top. Curr. Chem.,* 98, 107, 1982.
10. **Lehn, J. M. and Sirlin, C.,** Molecular catalysis, *J. Chem. Soc. Chem. Commun.,* p. 949, 1978.
11. **Breslow, R.,** Artificial enzymes: studies relevant to reaction mechanisms of enzymes and enzyme models, *Proc. Robert A. Welch Found. Conf. Chem. Res.,* 31, 72, 1987, **Breslow, R.,** Approaches to artificial enzymes, *Ann. N.Y. Acad. Sci.,* 471 (Int. Symp. Bioorg. Chem., 1985), 60, 1986.
12. **Tabushi, I.,** Design and synthesis of artificial enzymes, *Tetrahedron,* 40, 269, 1984.
13. **Brown, R. S., Huguet, J., and Curtis, N. J.,** Models for Zn(II) binding sites in enzymes, *Met. Ions Biol.,* 15, 55, 1984.
14. **Balzani, V., Ed.,** *Supramolecular Photochemistry,* Reidel, Dordrecht, 1987.
15. **Winkelmann, G., van der Helm, D., and Neilands, J. B., Eds.,** *Iron Transport in Microbes, Plants and Animals,* VCH Verlagsgesellschaft mbH, D-6940 Weinheim, West Germany, 1987.
16. **Neilands, J. B.,** Methodology of siderophores, *Struct. Bonding,* 58, 1, 1984; **Bagg, A. and Neilands, J. B.,** Molecular mechanism of regulation of siderophore-mediated iron assimilation, *Microbiol. Rev.,* 51, 509, 1987.
17. **Raymond, K. N., Mueller, G., and Matzanke, B. F.,** Complexation of iron by siderophores. A review of their solution and structural chemistry and biological function, *Top. Curr. Chem.,* 123, 49, 1984.
18. **Hider, R. C.,** Siderophore mediated absorption of iron, *Struct. Bonding,* 58, 25, 1984.
19. **Emery, T.,** The storage and transport of iron, *Met. Ions Biol. Syst.,* 7, 77, 1978.
20. **Braun, V.,** *Enzymes Biol. Membranes,* 3, 617, 1985.
21. **Miller, J. M.,** Synthesis and therapeutic potential of hydroxamic acid based siderophores and analogs, *Chem. Rev.,* 89, 1563, 1989.
22. **Weinberg, E. D.,** Cellular regulation of iron assimilation, *Q. Rev. Biol.,* 64, 61, 1989.
23. **Maehr, H.,** Antibiotics and other naturally occurring hydroxamic acids and hydroxamates, *Pure Appl. Chem.,* 28, 603, 1971.
24. **Prelog, V.,** The role of certain microbial metabolites as specific complexing agents, *Pure Appl. Chem.,* 6, 327, 1963.
25. **Neilands, J. B., Erickson, T. J., and Rastetter, W. H.,** Stereospecificity of the ferric enterobactin receptor of *E. coli, J. Biol. Chem.,* 256, 3831, 1981.
26. **Winkelmann, G.,** Evidence for stereospecific uptake of iron chelates in fungi, *FEBS Lett.,* 97, 43, 1979, **Winkelmann, G. and Braun, V.,** Stereoselective recognition of ferrichrome by fungi and bacteria, *FEMS Microbiol. Lett.,* 11, 237, 1981.
27. **Mutzanke, B. F., Muller, G., and Raymond, K. N.,** Hydroxamate siderophore mediated iron uptake in *E. coli;* stereospecific recognition of ferric rhodotorulic acid, *Biochem. Biophys. Res. Commun.,* 121, 922, 1984.
28. **O'Brian, I. G. and Gibson, F.,** The structure of enterochelin and related 2,3-dihydroxybenzoylserine conjugates from *E. coli, Biochim. Biophys. Acta,* 215, 393, 1970.
29. **Isied, S. S., Kuo, G., and Raymond, K. N.,** Coordination isomers of biological iron transport compounds, *J. Am. Chem. Soc.,* 98, 1763, 1976; **McArdle, J. V., Sofen, S. R., Cooper, R. S., and Raymond, K. N.,** Coordination chemistry of microbial iron transport compounds. III, *Inorg. Chem.,* 17, 3075, 1978.

30. **Emery, T.**, Role of ferrichrome as a ferric ionophore in *Ustilago spherogena, Biochemistry,* 10, 1483, 1971.
31. **Leong, J. and Neilands, J. B.**, Mechanisms of siderophore iron transport in enteric bacteria, *J. Bacteriol.,* 126, 823, 1976.
32. **Emery, T.**, Initial steps in the biosynthesis of ferrichrome. Incorporation of d N-hydroxyornithine and d N-acetyld N-hydroxyornithine, *Biochemistry,* 5, 3694, 1966, **Van der Helm, D., Baker, J. R., Eng-Wilmot, D. L., Hossain, M. B., and Loghry, R. A.**, Crystal structure of ferrichrome and a comparison with the structure of ferrichrome A, *J. Am. Chem. Soc.,* 102, 4224, 1980.
33. **Leong, J., Neilands, J. B., and Raymond, K. N.**, Coordination isomers of biological iron transport compounds. III, *Biochem. Biophys. Res. Commun.,* 60, 1066, 1974.
34. **Raymond, K. N. and Carrano, C. J.**, Coordination chemistry and microbial iron transport, *Acc. Chem. Res.,* 12, 183, 1979.
35. **Shanzer, A., Libman, J., Lifson, S., and Felder, C. E.**, Origin of the Fe^{3+}-binding and conformational properties of enterobactin, *J. Am. Chem. Soc.,* 108, 7609, 1986.
36. **Harris, W. R. and Raymond, K. N.**, Ferric ion sequestering agents. III, *J. Am. Chem. Soc.,* 101, 6534, 1979.
37. **Pecoraro, V. L., Weitl, F. L., and Raymond, K. N.**, Ferric ion specific sequestering agents. VII, *J. Am. Chem. Soc.,* 103, 5133, 1981; **Harris, W. R., Raymond, K. N., and Weitl, F. L.**, Ferric ion specific sequestering agents. VI, *J. Am. Chem. Soc.,* 103, 2667, 1981.
38. **Tor, Y., Libman, J., Shanzer, A., and Lifson, S.**, Biomimetic ferric ion carriers. A chiral analog of enterobactin, *J. Am. Chem. Soc.,* 109, 6517, 1987.
39. **Llinas, M., Wilson, D. M., and Neilands, J. B.**, Effect of metal binding on the conformation of enterobactin, *Biochemistry,* 12, 3836, 1972.
40. **McMurry, T. J., Hosseini, M. W., Garrett, T. M., Hahn, F. E., Reyes, Z. E., and Raymond, K. N.**, Macrobicyclic iron(III) sequestering agents, *J. Am. Chem. Soc.,* 109, 7196, 1987.
41. **Tor, Y., Libman, J., Shanzer, A., Felder, C. E., and Lifson, S.**, A trispeptide circularly organized through inter-chain hydrogen bonds, *J. Chem. Soc. Chem. Commun.,* p. 749, 1987.
42. **Tor, Y., Ph.D.**, thesis, Weizmann Institute of Science, Israel, 1990.
43. **Crumbliss, A. L., Brink, C. P., and Fish, L.**, Mechanistic studies of synthetic and naturally occurring hydroxamic acid complexes of iron(III) relating to biological transport, *Inorg. Chim. Acta Bioinorg. Chem.,* 79, 218, 1983.
44. **Anderegg, G., L'Eplattenier, F., and Schwarzenbach, G.**, *Helv. Chim. Acta,* 46, 1409, 1963.
45. **Emery, T., Emery, L., and Olsen, R. K.**, Retrohydroxamate ferrichrome; a biomimetic analogue of ferrichrome, *Biochem. Biophys. Res. Commun.,* 119, 1191, 1984.
46. **Akiyama, M., Katoh, A., and Muto, T.**, N-hydroxy amides. VII. Synthesis and properties of linear and cyclic hexapeptides as models for ferrichrome, *J. Org. Chem.,* 53, 6089, 1988.
47. **Mitchell, M. S., Walker, D. L., Whelan, D. L., and Bosnich, B.**, Biological analogs of synthetic, iron(III) specific chelators based on the natural siderophores, *Inorg. Chem.,* 26, 396, 1987.
48. **Ng, Y. C., Rodgers, S. J., and Raymond, K. N.**, Ferric ion sequestering agents, *Inorg. Chem.,* 28, 2062, 1989.
49. **Lee, B. H., Miller, M. J., Prody, C. A., and Neilands, J. B.**, Artificial siderophores. I. Synthesis and microbial iron transport properties, *J. Med. Chem.,* 28, 317, 1985.
50. **Tor, Y., Libman, J., and Shanzer, A.**, Biomimetic ferric ion carriers. A chiral analog of ferrichrome, *J. Am. Chem. Soc.,* 109, 6518, 1987.
51. **Borgias, B. A., Barclay, S. J., and Raymond, K. N.**, Structural chemistry of gallium(III), *J. Coord. Chem.,* 15, 109, 1986.
52. **Bauminger, R.**, personal communications.
53. **Leong, J. and Raymond, K. N.**, Coordination isomers of biological iron transport compounds. IV, *J. Am. Chem. Soc.,* 97, 293, 1975.
54. **Mueller, G. and Raymond, K. N.**, Specificity and mechanism of ferrioxamine-mediated iron transport in *Streptomyces pilosus, J. Bacteriol.,* 160, 304, 1984.
55. **Shimzu, K., Nakayama, K., and Akiyama, M.**, N-hydroxy amides. IV. Synthesis and properties of a trihydroxamic acid anilide as a model for ferrioxamines, *Bull. Chem. Soc. Jpn.,* 59, 2421, 1986; 1975.
56. **Ecker, D. J., Matzanke, B. F., and Raymond, K. N.**, Recognition and transport of ferric enterobactin in *Escherichia coli, J. Bacteriol.,* 167, 666, 1986.
57. **Heidinger, S., Braun, V., Pecoraro, V. L., and Raymond, K. N.**, Iron supply to *E. coli* by synthetic analogs of enterochelin, *J. Bacteriol.,* 153, 109, 1983.
58. **Burnham, B. F. and Neilands, J. B.**, Studies on the metabolic function of the ferrichrome compounds, *J. Biol. Chem.,* 236, 554, 1960.
59. **Shanzer, A., Libman, J., Lazar, R., Tor, Y., and Emery, T.**, Synthetic ferrichrome analogues with growth promotion activity for *Arthrobacter Flavescens, Biochem. Biophys. Res. Commun.,* 157, 389, 1988.
60. **Winkelmann, G.** personal communication.

61. **Dayhoff, M. O., Schwartz, R. M., and Orcutt, B. C.,** *Atlas of Protein Sequence and Structure,* Vol. 5 (Suppl. 3), Dayhoff, M. O., Ed., National Biomedical Research Foundation, 1978, 345.

62. **Ecker, D. J., Loomis, L. D., Cass, M. E., and Raymond, K. N.,** Substituted complexes of enterobactin and synthetic analogs as probes of the ferric-enterobactin receptor in *E. coli, J. Am. Chem. Soc.,* 110, 2457, 1988.

63. **Crowley, D. E., Reid, C. P. P., and Szaniszlo, P. J.,** Microbial siderophores as iron sources for plants, in *Iron Transport in Microbes, Plants and Animals,* Winkelmann, G., van der Helm, D., and Neilands, J. B., Eds., VCH Publishers, West Germany, 1987, 365.

64. **Ganemore-Neumann, R., Bar-Josef, B., Shanzer, A., and Libman, J.,** submitted for publication.

65. **Clyde, D. F.,** *Epidemiol., Rev.,* 9, 250, 1986.

66. **Cabantchik, Z. I.,** Editorial review, *Blood, 1989.* **Cabantchik, Z. I., Silfen, J., and Glickstein, H.,** *J. Subc. Biochem.* in press.

67. **Rodriguez, H. M. and Jungery, M.,** A protein on *Plasmodium falciparum* infected erythrocytes functions as a transferrin receptor, *Nature,* 324, 388, 1986.

68. **Fritsch, G., Sawatzki, G., Treumer, J., Jung, A., and Spira, D. T.,** *Plasmodium falciparum:* inhibition in vitro with lactoferrin, desferriferrithiocin, and desferricrocin, *Exp. Parasitol.,* 1, 63, 1987; **Fritsch, G. and Jung, A.,** ^{14}C-desferrioxamine B: uptake into erythrocytes infected with *Plasmodium falciparum, Z. Parasitenkd.,* 72, 709, 1986. **Fritsch, G., Treumer, J., Spira, D. T., and Jung, A.,** *Plasmodium vinckei:* suppression of mouse infections with desferrioxamine B, *Exp. Parasitol.,* 60, 171, 1985.

69. **Pollack, S., Rassan, R. N., Davidson, D. E., and Escajadillo, A.,** Desferrioxamine suppresses *Plasmodium falciparum* in Aotus monkeys, *Proc. Soc. Exp. Biol. Med.,* 184, 162, 1987, **Pollack, S.,** Effects of iron and desferrioxamine on the growth of *Plasmodium falciparum* in vitro (letter), *Br. J. Haematol.,* 65, 256, 1987; **Pollack, S. and Schnelle, V.,,** Br. J. Haematol., 68, 125, 1988.

70. **Raventos-Suarez, C., Pollack, S., and Nagel, R. L.,** *Plasmodium falciparum:* inhibition of in vitro growth by desferrioxamine, *Am. J. Trop. Med. Hyg.,* 31, 919, 1982.

71. **Heppner, D. G., Hallaway, P. E., Kontoghiorghes, G. J., and Eaton, J. W.,** Antimalarial properties of orally active iron chelators, *Blood,* 72, 358, 1988.

72. **Hershko, C. and Peto, T. E.,** Desferrioxamine inhibition of malaria is independent of host iron status, *J. Exp. Med.,* 168, 375, 1988.

73. **Scheibel, L. W. and Stanton, G. G.,** Antimalarial activity of selected aromatic chelators, IV. Cation uptake by *Plasmodium falciparum* in the presence of oxines and siderochromes, *Mol. Pharmacol.,* 30, 364, 1986.

74. **Cabantchick, Z. I., Shanzer, A., Lytton, S., and Libman, J.,** in preparation.

THERAPEUTICALLY USEFUL IRON CHELATORS

John B. Dionis, Hans-Beat Jenny, and Heinrich H. Peter

INTRODUCTION

Desferrioxamine B (DFO, Desferal®) is a microbial siderophore containing three hydroxamate groups as iron binding sites. It is being produced industrially by large-scale fermentation of *Streptomyces pilosus* strain A 21748 and isolated in crystalline form as the methanesulfonate salt.

Desferal® is still the only approved drug for treatment of patients suffering from acute iron intoxication or from chronic iron overload as a result of recurrent blood transfusions.[1,2] In conjunction with regular blood transfusions chelation therapy with Desferal® has proved to be a life-saving treatment for patients affected with thalassemia (or Cooley's anemia), a formerly lethal inheritable disease. The long-term treatment with Desferal® is generally well tolerated but it has two major drawbacks: the hydrophilic compound is not absorbed when given orally and its serum half-life is quite short as a result of rapid clearance via the kidneys. In order to achieve sufficiently high serum levels over a long period of time the chelator is usually administered by slow subcutaneous infusion using a portable pump. This tedious mode of administration which has to be performed almost daily leads to serious compliance problems. Therefore there is an urgent need for an orally active chelator for the life-long treatment of thalassemia patients. The availability of an oral chelator could also lead to a reduction of the high costs of the current treatment.[3]

More recently Desferal® has also found therapeutic application for various pathological conditions due to aluminum overload. Accumulation of this toxic metal is frequently observed in chronically dialyzed patients who have lost the ability to clear it via renal excretion.[4,5] Systemically administered Desferal® is able to mobilize pathogenic aluminum from bone and nervous tissue deposits. The aluminum complex is subsequently cleared from the body by the dialysis treatment. Desferal® is not only being used as a therapeutic agent, but has also been recommended for the diagnosis of such an overloaded state. The increase of serum aluminum levels after a single dose of the chelator provides semiquantitative information on the degree of the overload in a patient.[6]

In the following paragraphs we will first give an overview of some physicochemical aspects of iron chelators and the established analytical methods used for the determination of both the free ligand as well as its complex with iron or aluminum. A second paragraph will deal with the problems associated with the development of new drugs for chelation therapy, mainly orally effective ones. The last paragraph will provide a short outlook on new applications of chelating drugs for use in different therapeutic and diagnostic areas.

CHEMICAL AND ANALYTICAL METHODS FOR THE DETERMINATION OF DFO AND ITS COMPLEXES

Different aspects of the physicochemical behavior of DFO and its iron complex have been investigated with great effort over the past few years. Since DFO is a potent ligand for trivalent transition metal ions the determination of complex formation constants is of considerable interest. In the area of clinical chemistry and biomedical research the main task was to develop reliable analytical methods for the quantitative determination of DFO and its complexes, mainly with iron and aluminum present in different kinds of biological samples. The following paragraph will give an overview of the different techniques used today.

PHYSICOCHEMICAL PROPERTIES

DFO as the free ligand and its complexes formed with either Fe(III) or Al(III) exhibit various characteristic properties based on which analytical methods have been worked out.

Complex Formation

DFO, like other trihydroxamic acids, shows a very high affinity for ferric ions. According to different authors, a complex forming constant with Fe(III) in the range of 10^{30} to 10^{32} [7,8] seems to be well established. The complex-forming constant with aluminum ion is reported to be in the range of 10^{23} to 10^{25}.[9] The affinity for divalent ions, including the biologically relevant ions Fe(II) (K = 10^{10}), Cu(II) (K = 10^{14}), Zn(II) (K = 10^{11}), and Mn(II), is significantly lower.[8]

By reduction of Fe(III) to Fe(II) the complex can therefore be easily dissociated. As a reducing agent ascorbic acid or thioglycolic acid is often used in analytical methods.

Electronic Absorption Spectra

The photoelectronic spectrum of the iron complex ferrioxamine (FO) exhibits a maximum at a wavelength of 430 nm, with a molar extinction coefficient of about 2700.[10,11] Other authors reported 440 nm for the maximum and an extinction coefficient of 2640.[12] The free ligand DFO, on the other hand, exhibits useful photometric absorption only at wavelengths below 240 nm.

Electrochemical Properties

The iron complex FO exhibits electrochemical activity caused by the redox-pair Fe(III)/ Fe(II). This property has been exploited by using electrochemical techniques like polarography for the determination of the concentration of the complex in a given sample.[5]

Other Spectrophotometric Methods

The ions Fe(III) and Al(III) can easily be detected directly by different atomic spectroscopy methods like AAS[13] or ICP.[14,15]

Fe(II), on the other hand, can only be detected after reduction from Fe(III) using specific methods, usually based on photometric detection of a specific complex formed.[16-19]

Solubility

In spite of its solubility in water FO can be extracted from complex aqueous sample matrices with benzyl alcohol. In order to achieve a quantitative extraction a high salt concentration in the aqueous phase must be established. Usually this is accomplished by saturating the sample with sodium chloride prior to extraction.[15,20-22] Some other authors used saturated ammonium nitrate solution buffered to pH 8 instead.[13]

Based on these different physicochemical properties a number of analytical methods for specific applications have been developed and will be described in the following paragraph.

ANALYTICAL METHODS FOR BIOLOGICAL SAMPLES

The choice of the analytical method depends very much on the type of sample in question. Three of them are mainly of clinical interest: serum, urine, and dialysis fluids. Biological and chemical research samples like fermentation broths or reaction mixtures are in most cases easier to deal with and can usually be analyzed using, in principle, the same methods as for the clinical samples.

Serum analysis is mostly done for research purposes, where information on the pharmacokinetic profile has to be gained, e.g., on the excretion rate of ligand and complex, organ specificity, ratio of free ligand and its complexes with Fe(III) and Al(III), or studies on the mechanism of iron mobilization from different biological iron pools.[13,20-26]

Urine or dialysis fluids are usually analyzed in order to determine the therapeutic efficacy of DFO treatment applied to patients with iron and/or aluminum overload. The method of administering single standard doses of free ligand to a patient and monitoring excretion of iron and, more recently, aluminum is also used as a diagnostic tool to detect such an overloaded status.[5,6,13,14,16,18,19,27,28]

Direct Photometric Determination of the Complex

This group of test methods takes advantage of the inexpensive photometric determination combined with a usually rather simple sample pretreatment. The first example was published by Meyer-Brunot and Keberle.[20] The authors extracted FO from serum and determined its concentration directly at a wavelength of 430 nm. The amount of free ligand was determined by adding an excess of Fe(III) ions to the sample before extraction. The disadvantage of this method is its relatively high detection limit of about 5 mg/l.

Indirect Determination via Ferrous Iron

Several other methods are based on the determination of iron in its reduced form and therefore give direct access to the total amount of iron present in the sample. Since in most biological samples iron is present in ferric form, it has to be reduced to ferrous iron before the selective complex can be formed. To perform this reduction from Fe(III) to Fe(II) in a quick and quantitative manner, either thioglycolic acid (= mercaptoacetic acid)[19,29] or ascorbic acid[17] is used. Different bidentate ligands can then be used to form colored complexes with ferrous iron. Three of them mentioned below are most often used in clinical chemistry. Total ligand concentration can again be measured by addition of excess ferric iron to the sample before extraction and reduction of the iron.

BPS = Bathophenantroline disulfonic acid [4,7-diphenyl-1,10-phenanthroline-disulfonic acid disodium salt]; the ferrous complex with this ligand is also called Ferroin, absorbance wavelength: 527 to 535 nm[18,27]

PPST = Ferrozine [3-(2-pyridyl)-5,6-bis(4-phenyl-sulfonic acid)-1,2,4-triazine disodium salt]; absorbance wavelength: 562 to 570 nm[16,17]

TPTZ = Tripyridyltriazine [2,4,6-tri(2'-pyridyl)-1,3,5-triazine]; absorbance wavelength: 600 nm[26]

These methods have been optimized for serum and urine samples. If TPTZ is used serum samples have to be deproteinated[17] or dialyzed[26] before measurement. With BPS[18,27] or PPST[16,17] urine and serum samples are usually simply mixed with the reagents and incubated before photometric determination. DuPont's *aca* autoanalyzer method has also been modified to give good results even in the presence of DFO and FO.[29] In certain cases it can be necessary to wet ash the samples before reduction and determination.[19]

Spectroscopic Methods for Iron and Aluminum Determination

The direct determination of ferric iron by atomic emission spectroscopy using the inductively coupled plasma technique (ICP) can be applied to urine samples after dilution[14] or serum samples after extraction with benzyl alcohol.[15] The latter method furthermore allows the determination of free ligand by adding an excess of ferric iron to the sample prior to extraction.

A similar approach is used for determinations by atomic absorption spectrometry (AAS). After an extraction step both iron and aluminum can be measured directly in plasma or urine samples.[13]

Electrochemical Methods

The concentration of complex and free ligand in dialysis fluids can be determined by the pulsed polarographic technique. The complex can be directly determined, while the free ligand is titrated by adding ferric ion in a stepwise manner until the readout of the signal reaches a maximum value.[5,28]

Chromatographic Separation

These most powerful methods have only recently been developed. Several examples are based on normal phase chromatography. Serum samples can be directly injected for the determination of the concentration of FO.[25] If an extraction step with benzyl alcohol is included, the free ligand and the aluminum complex can be determined as well.[22] The extraction step can also be performed using liquid-solid extraction, for example, SepPak C_{18}.[21] Ion-exchange chromatography can be used to separate ferrous from ferric iron. Postcolumn treatment with ascorbic acid and BPS allows detection of the separated peaks at 530 nm for both components.

Reversed-phase chromatography allows the direct separation of different ligands and their ferric complexes as well as their aluminum complexes.[24,30,31] Precolumn and column switching techniques give access to even more sophisticated separation schemes.[23,32] The development and availability of nonsilica-based reversed-phase materials will further enhance the performance of the technique in the near future.[30,33]

Other Methods

The presence of siderophores in screening samples can easily be detected by observing the disappearance of the blue color of the Fe(III)-Chrome-azurol S complex added previously to the culture medium.[34]

Mono-, di-, and trihydroxamates can be distinguished by observing the change of the color of the iron complexes in the presence of CN^- anions.[35]

DISCUSSION

The determination of DFO, FO, or iron can be achieved by using a number of different methods based on either photometric or atomic spectroscopic principles as described above. The techniques allowing direct or indirect determinations seem to be well established. For routine monitoring in the field of clinical chemistry selectivity as well as sensitivity are in most cases satisfactory.

The ongoing rapid development of the more elaborate chromatographic methods, however, will provide a very powerful tool to solve more complex problems. Especially the development of new stationary phases for reversed chromatography will contribute to further improvement of the reliability of chromatographic separations of this class of compounds. In contrast to the colorimetric methods, chromatography is able to distinguish easily between different types of trihydroxamates. This is of great importance in production monitoring and quality control of the ligand and is already widely used. More recently pronounced interest in determining the metabolic degradation products of the ligand itself in biological fluids has increased. Such investigations will provide a more profound insight into the mechanism of the interaction of chelators with the iron metabolism.

DEVELOPMENT OF NEW DRUGS FOR THE TREATMENT OF IRON OVERLOAD

GENERAL REQUIREMENTS

It is now widely acknowledged that a decisive breakthrough in chelation therapy can only be achieved by the discovery and development of a truly orally effective iron chelating has proven extremely difficult. This is due in part to the stringent requirements which must be fulfilled by an iron chelating agent intended for therapeutic use.[36] Such a compound must have a high affinity for ferric ions and a high degree of selectivity relative to other trace metals. Additionally, the ability to penetrate membranes and to compete with other natural iron chelating systems is of considerable importance. As a matter of course the drug must be devoid of acute toxic effects and have good long-term tolerability. Most importantly, treatment with the drug must maintain a negative iron balance in overloaded patients and prevent iron accumulation in patients at risk.

The requirements are even more difficult to meet if an oral mode of application is desired. First the compound must be resistant to hydrolysis at the low pH of the stomach. In addition, drug solubility and dissolution in this acidic environment may also be a limiting factor in the drug's potential efficacy. Another point which should be emphasized is the potential for increased iron absorption following the oral administration of an iron chelator, especially at low doses.

Weakly acidic drugs which are predominantly in the undissociated form are more readily absorbed from the stomach. Conversely, weakly basic drugs will be poorly absorbed from this compartment due to their increased charge and lipid insolubility. For many compounds absorption from the gastrointestinal epithelium is a passive diffusion process with a rate proportional to the lipophilicity of the drug.[37] The physicochemical characteristics of the absorbed drug will also influence heavily its distribution into various compartments. The toxicity associated with iron overload is related not only to the absolute quantities of body iron, but also in its distribution and severity among the vital organs.[38] Therefore, in assessing the therapeutic potential of new iron chelators information regarding the site and mechanism of action will be required in addition to demonstrating a reduction in total body iron.

It has been repeatedly stated that the ideal chelator should also be inexpensive. The high cost of a life-long treatment with DFO is a factor which precludes its use in many less-developed countries of the world. Unfortunately the iron-loading anemias, in particular thalassemia, are prevalent in many such areas throughout the Mediterranean region, the Middle East, India, and Southeast Asia. The drugs evaluated to date encompass a wide range of chemical structures including the naturally occurring microbial siderophores. The common feature to many of these ligands is their ability to bind ferric ions in a hexacoordinate, octahedral complex with formation constants in the range of 10^{30} to 10^{50}.[39]

The performance of any new iron chelator intended for clinical use will have to compare favorably with subcutaneously infused DFO, the current treatment of choice for transfusional iron overload. The enormous amount of information which has been collected since the introduction of the drug in the early 1960s documents its beneficial effects. Long-term drug treatment in patients has been shown to maintain a negative iron balance, resulting in diminished hepatic and cardiac toxicity with a concomitant increase in survival rate.[1] The drug is well tolerated with minimal toxicity in iron-loaded patients when administered in the short or long term in doses up to 25 mg/kg·d.[2] In fact, DFO fulfills virtually all of the essential requirements mandated for an iron chelator intended for therapeutic use. Therefore, the selection of an appropriate orally effective agent as a replacement for DFO has turned out to be a difficult and time-consuming task. Extensive documentation will be required of a potential candidate demonstrating an efficacy and long-term safety at least equal to DFO.

TEST SYSTEMS FOR NEW CHELATING AGENTS

The test systems employed in selecting and evaluating chelators as potentially useful therapeutic agents have to take into account possible interactions with iron transport and storage proteins, the effects on cellular iron uptake, incorporation and mobilization of iron in various cultured cell lines, and, above all, the ability to enhance iron excretion in different animal models. The reader is referred to a recent overview for a more thorough discussion.[40]

There are several animal test systems in current use for comparative evaluation of the relative efficiencies of iron chelating drugs. Almost all of them are based on the use of normal or iron-loaded rodents such as rats, mice, or rabbits. However, since there exist fundamental differences between the iron metabolism in rodents and in man the results obtained in these species could not always be reproduced in patients with transfusional iron overload.

Preliminary information is now available on a most promising new development which utilizes a primate model of iron overload to measure iron excretion in response to orally administered iron chelators.[41] Owing to the difficulties associated with the handling of the animals and the expenses involved this screen will be limited to selected compounds which have shown promise in previous biological screens. This emphasizes the importance of the continued use of *in vitro* as well as rodent test models in the primary assessment of iron chelators. The primate model will serve as a secondary screen for promising candidates and thus complement existing methods. The combined use of these models is expected to enable an accelerated development of a new, orally effective iron chelator for therapeutic use.

It is appropriate to discuss the merits of the various primary screens used in the evaluation of biological activity. The early animal models were developed to simulate the condition of iron overload in man. Iron overload in a rat and mouse model has been accomplished using a hypertransfusion regimen with heat-denatured red blood cells.[42-44] Evaluation of the test compounds involved metabolic balance studies, combined in the mouse model with hepatic and splenic iron determination. Since these early studies, numerous modifications and refinements of the test methods have been described. These include the use of selective radioiron probes which are used to label the parenchymal and reticuloendothelial iron stores as well as bile duct cannulation in the rat. Unfortunately, it is not possible to make a direct comparison of a chelator's efficacy from one study to the next, mainly because of the many variations in the experimental protocols. This includes factors such as iron loading procedures and equilibration time after administration of radioactive isotope and/or senescent red blood cells. In addition, the mode of iron excretion and the metabolism of drugs between species can vary considerably, thus complicating the situation even further.

EVALUATION OF POTENTIAL IRON CHELATING AGENTS

Using a rat model Grady et al. investigated a series of benzoic acid derivatives, hydroxamic acids, and other naturally occurring agents with iron binding capacity.[43,44] Drugs were administered i.p. or p.o. at a dose of 100 mg/kg·d for 5 d to rats which had been iron loaded via a hypertransfusion regimen which was continued throughout the experiment. Fecal and urinary iron excretion were determined every 24 h and the relative effectiveness of the various drugs reported in μg/kg·d. In terms of total amount of iron excreted in both urine and stool, *rhodotorulic acid* (Figure 1, 1-1), a tetradentate hydroxamic acid, was found to be the most efficacious chelating agent, promoting nearly twice the iron excretion as that induced by DFO. However, like DFO, this microbial siderophore was not effective via the oral route. Several compounds also demonstrated activity after oral administration in this model, including two bidentate ligands, *cholylhydroxamic acid* (Figure 1, 1-2) and *2,3-dihydroxybenzoic acid* (Figure 1, 1-3). The former compound was designed with the intent of mimicking the bile acids which are readily reabsorbed and transported to the liver by the enterohepatic circulation. After additional positive results in animal and *in vitro* studies these three promising candidates were subsequently tested in clinical trials with varying degrees of success. While an initial trial with 2,3-dihydroxybenzoic acid showed promising results, clinical efficacy could not be demonstrated after a 1-year evaluation as evidenced from liver biopsies.[45] The study did demonstrate that the drug was very well tolerated at a dose of 25 mg/kg four times a day. An initial clinical study with cholylhydroxamic acid revealed that net iron excretion in response to the drug approached the desired levels. However, further studies confirmed that therapeutically useful doses were not well tolerated, causing diarrhea

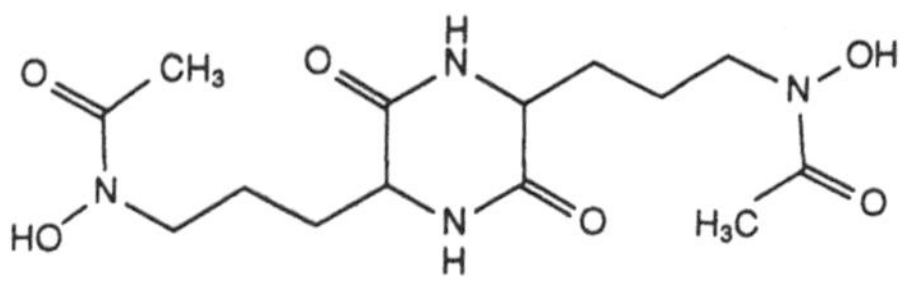

1-1 Rhodotorulic acid

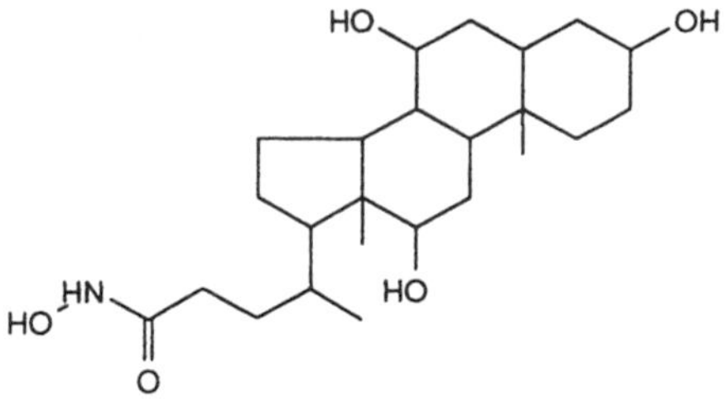

1-2 Cholylhydroxamic acid

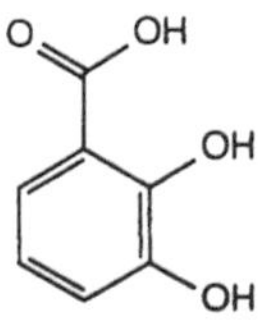

1-3 2,3-dihydroxy benzoic acid (DBH)

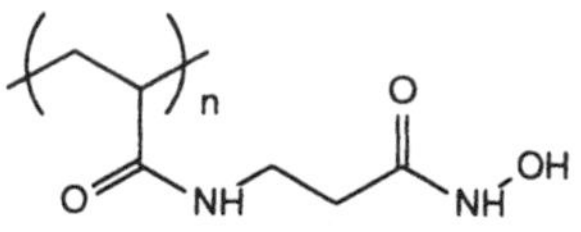

1-4 Poly(N-methacryl-β-alanine
hydroxamic acid)

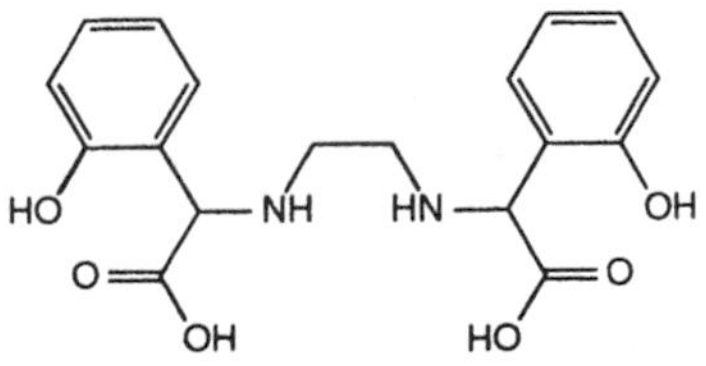

1-5 Ethylenediamine-N,N'-bis-o-
hydroxyphenylglycine (EHPG)

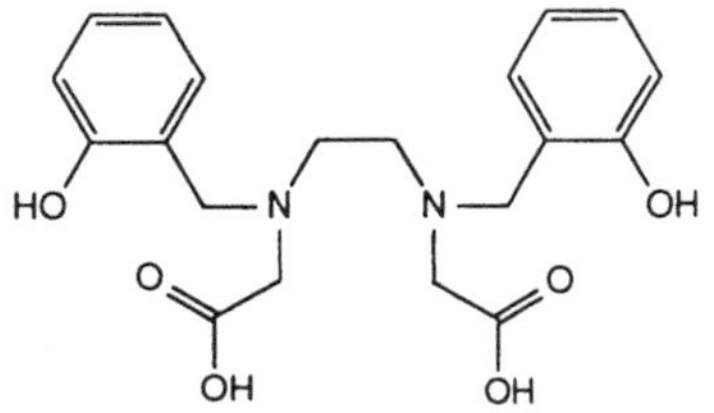

1-6 (N,N'-bis(2-hydroxybenzyl)ethylene-
diamine-N,N'-diacetic acid (HBED)

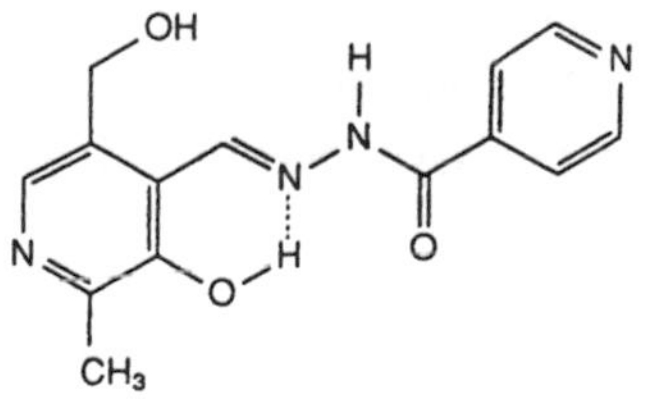

1-7 Pyridoxylidene isonicotinoyl
hydrazone (PIH)

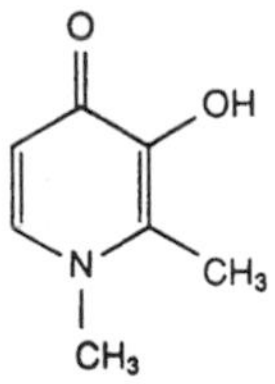

1-8 1,2-dimethyl-3-hydroxy-
pyrid-4-on (L₁)

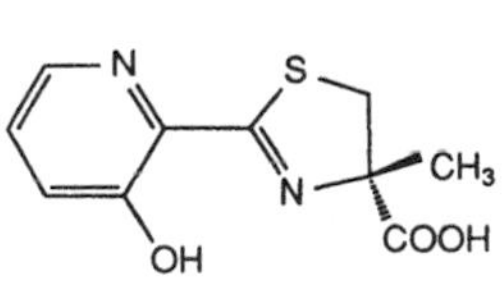

1-9 Desferrithiocin

FIGURE 1. (1-1) Rhodotorulic acid; (1-2) cholylhydroxamic acid; (1-3) 2,3-dihydroxy benzoic acid (DBH); (1-4) poly(*N*-methacryl-β-alanine hydroxamic acid); (1-5) ethylenediamine-*N,N'*-bis-o-hydroxyphenylgly-cine (EHPG); (1-6) (*N,N'*-bis[2-hydroxybenzyl]ethylenediamine-*N,N'*-diacetic acid (HBED); (1-7) pyri-doxylidene isonicotinoyl hydrazone (PIH); (1-8) 1,2-dimethyl-3-hydroxypyrid-4-on (L₁); (1-9) desferrithiocin.

as the most common side effect.[46] It was hoped that rhodotorulic acid would offer advantages relative to DFO in that the drug could be administered as a depot preparation due to its very low water solubility. In addition, the compound was at that time thought to have the added advantage of a much lower production cost, as fermentation titers were in the range of 1 g/

1. However, when suspensions of the drug were injected via the i.m. or the s.c. route to patients, pain and inflammation were observed at the site of injection which persisted for more than 5 d, a symptom not seen in dogs receiving the drug for up to 17 weeks.[47] This unacceptable side effect offset the significant increase in iron excretion observed and prevented further clinical evaluation of rhodotorulic acid.

While these results were disappointing they demonstrated quite clearly that extrapolation of data from an animal test system to humans with their more highly conservative iron metabolism is quite problematic. They also showed that the search for a suitable replacement of DFO was going to be an expensive and time-consuming process.

A widely used test system for the screening of iron chelators is the hypertransfused mouse model originally developed by Gralla.[48] Pitt et al. evaluated over 70 chelating agents including natural and synthetic hydroxamic acids, phenols, catechols, and tropolones utilizing this test system.[42] In this screening model, the drug was administered for 7 d as a single daily i.p. injection given 2 d after the last transfusion. The measurement of urinary iron excretion in combination with hepatic and splenic iron depletion was used to assess the efficacy of the various test compounds. Those chelators which were inefficient in decreasing hepatic iron levels were considered less likely to be of clinical value. Splenic iron measurements were shown to be a useful parameter in demonstrating iron redistribution among other tissues subsequent to hepatic iron depletion. Of the microbial siderophores investigated only triacetylfusarinine·C, a cyclic hexadentate trihydroxamic acid, exhibited activity almost comparable to DFO, as demonstrated by an increase in urinary iron excretion and significant depletion of hepatic iron stores. Rhodotorulic acid was much less effective in this animal model enhancing only urinary iron excretion. Numerous synthetic bidentate hydroxamic acids were shown to be capable of mobilizing iron internally; however, subsequent metabolic degradation of the iron complexes occurred as evidenced by increased splenic and/or hepatic iron levels with little or no increase in urinary iron excretion.

A fully synthetic polydentate polymer, *poly(N-methacryl-β-alanine hydroxamic acid)* (Figure 1, 1-4) has been prepared with the objective to prolong the retention time in the plasma, thus leading to an increased efficacy in comparison to DFO which is rapidly excreted via the kidneys.[49] Based on a thorough evaluation of a series of polymeric hydroxamic acid derivatives using the hypertransfused mouse model, it was shown that systemic administration of the above-mentioned optimized polymer caused a comparable or even higher iron excretion than equal doses of DFO. This favorable result could not be reproduced when polymeric N-methacryl-β-alanine hydroxamic acid was tested in our own primary test system, i.e., in normal rats with a cannulated bile duct. This compound produced no change in iron excretion levels when given p.o. More surprisingly, only a very small and delayed increase in biliary and urinary output could be observed after s.c. administration, even when the relatively high dose of 300 mg/kg was employed.

The same appealing concept of increasing residence time of chelators by linking suitable ligands to biologically well-tolerated polymer backbones has also been used in the recent work of Mahoney et al.[50]

Several other chelating agents which had been shown effective in the rat model were also investigated in this model. The most striking disparity between the animal models was the lack of activity and toxicity associated with 2,3-dihydroxybenzoic acid when administerd to mice i.p. at 100 mg/kg. This and other discrepancies may be related to the concurrent drug/transfusion therapy in the rat screen. As the time interval between transfusion and drug administration is increased the amount of iron accessible to chelation is expected to become progressively lower.[51] While this may in part explain the more discriminative nature of the mouse screen, it is obvious that inherent metabolic differences of the mouse and rat can also play an important role.

One of the most effective chelators evaluated in the hypertransfused mouse model and subsequently confirmed in the rat model was *ethylenediamine-N,N'-bis(2-hydroxyphenyla-cetic acid) (EDHPA),* now referred to as *EHPG (ethylenediamine-N,N'-bis-o-hydroxyphen-ylglycine)* (Figure 1, 1-5). This commercially available synthetic phenolic hexadentate chelator and several of its derivatives were shown to be more efficacious than DFO when administered i.p. and also retained activity after oral administration, corresponding to 33% of the i.p. dose.[46] A detailed evaluation of EHPG in rats was also carried out by Hershko and co-workers.[38] Using the hypertransfused rat model in conjunction with radioiron probes these researchers demonstrated that EHPG-induced iron excretion was enhanced eightfold relative to identical doses of DFO injected i.m. Moreover, the study revealed that EHPG was able to interact with both parenchymal and reticuloendothelial iron stores and that the iron excretion was limited mainly to the gut. DFO, in comparison, is also interacting with these iron pools; however, the mode of iron excretion is somewhat different and the magnitude of iron excretion from parenchymal stores is significantly lower. Another promising finding was that hepatic ferritin iron and nonhaem iron stores were considerably reduced in animals treated with i.m. EHPG. The potential of this drug for removing excess iron in man was documented already in 1963.[52] In three patients administration of the drug (3 g) i.v. resulted in the net urinary iron excretion of 18 to 25 mg/d. More recently EHPG was tested clinically in four patients, the drug being administered orally at a daily dose of 25 mg/kg for 1 week.[46] Only a slight increase in urinary iron excretion was observed, with no change in fecal iron levels. This apparent lack of oral activity in man prompted investigations into the preparation of various, more lipophilic esters. While several analogues were shown to be more efficacious in the rat and mouse screen after oral administration, concerns over long-term toxicity studies in rats of the parent molecule resulted in a loss of interest in EHPG and its derivatives.[53]

The structurally related hexadentate chelator *HBED (N,N'-bis(2-hydroxyben-zyl)ethylenediamine-N,N'-diacetic acid)* (Figure 1, 1-6) has been investigated even more thoroughly. Pitt et al. showed that the drug was comparable to DFO when administered i.p. in hypertransfused mice.[42] Further studies reported by Grady and Jacobs in the hypertrans-fused rat model revealed that the drug was as effective orally as EHPG and that net iron excretion was nearly three times that of DFO when given i.p.[46] A further twofold increase in iron excretion was effected by conversion of the carboxylic acid groups into methyl esters. This and several other esters were shown to have greatly improved intestinal absorption properties as evidenced by their substantial activity after oral administration corresponding to 80% of the i.p. dose. In fact, the dimethyl ester of HBED was the most active chelating agent identified in animal test systems when given orally or by injection. Pitt et al.[54] reported an investigation of several phenolic amino carboxylic acid esters and lactones as prodrugs for iron chelation therapy utilizing their mouse screen. This study confirmed the results previously found using the rat screen in that HBED and several ester derivatives were very effective after oral administration. Although these chelators did not display signs of toxicity in the rats, significant CNS activity was observed in the hypertransfused mouse screen. Oral administration of HBED resulted in increased toxicity in which severe CNS inhibition and anorexia were observed. A full toxicological evaluation of HBED and its derivatives has been completed recently by Rosenkrantz and Metterville.[53]

In order to resolve conflicting claims concerning the oral effectiveness of EHPG as compared to HBED, a detailed study was carried out by Hershko and co-workers utilizing the hypertransfused rat model in conjunction with selective radioiron probes.[55] They demonstrated that at a standard dose of 40 mg/kg i.m. EHPG, HBED, and their respective methyl esters were superior to DFO in their ability to promote radioiron excretion, regardless of the source of the radioiron given. Moreover, these compounds more effectively removed iron from parenchymal iron stores with the chelated iron being predominantly excreted in the feces. In a study of dose-response relationships these workers concluded that EHPG, HBED, and dimethyl HBED were 9, 12, and 15 times more active than DFO, respectively, at a dose of 5 mg per animal. As expected none of these phenolic ethylenediamine derivatives was able to increase urinary iron excretion after oral administration. In addition EHPG and HBED effected only negligible excretion of radiolabel in the feces. In contrast, the dimethyl esters were capable of enhancing fecal iron excretion after oral administration. The dimethyl ester of HBED was found to be the most potent compound, retaining two thirds of the activity of an equal dose given i.m.

The use of diethylenetriamine pentaacetic acid (DTPA) to reduce iron overload in thalassemic patients has been under investigation for several years.[56-59] The main problem associated with this chelator relates to its limited selectivity, in particular its ability to form stable complexes with zinc *in vivo*. In order to counteract serious side effects due to zinc depletion, chronic administration of this drug must be combined with zinc supplementation. Moreover, it is also not active when given orally. Therefore, its use can only be considered in a few specific cases, e.g., in patients developing symptoms which have been attributed to high doses of DFO. It has been reported that such symptoms could be reversed when DTPA was used for chelation therapy of these patients.[60]

During the past decade several interesting new types of compounds have been investigated including pyridoxylidene isonicotinoyl hydrazone (PIH) (Figure 1, 1-7) , the keto-hydroxy pyridine (Figure 1, 1-8), as well as its derivatives and desferrithiocin (Figure 1, 1-9). While structurally very dissimilar they all have the common feature of increased lipid solubility and improved intestinal absorption. Hoy and colleagues first demonstrated that PIH was able to enhance iron excretion in the rat following oral administration.[61] Hershko et al. also investigated the mechanism of iron chelation by PIH in normal and hypertransfused rats using selective radioiron probes.[62] It was found that fecal excretion of radiolabeled iron was nearly doubled in animals which had been hypertransfused. In a confirmation of the earlier studies radiolabeled iron excretion in response to orally administered PIH was predominantly confined to the gut and compared favorably with parenterally administered DFO. It was also shown that PIH is capable of interacting with both parenchymal and reticuloendothelial iron stores from which the chelated iron is effectively transported via the biliary route. Subsequent studies with PIH analogues in rats following i.v. administration have revealed that the efficacy of the drug can be enhanced by structural modification.[63] Reports on the results of long-term oral treatment with PIH in rats have been inconsistent. Williams et al.[64] reported that after 10 weeks of treatment no apparent reduction of iron overload occurred. However, Kim and co-workers[65] demonstrated a one third reduction in hepatic and splenic iron stores and a greater than twofold increase in fecal and urinary iron excretion in animals treated for 4 weeks. While several other PIH analogues have also shown interesting activity in animal screens the abundance of *in vitro* and *in vivo* studies on PIH itself and the encouraging toxicity data have recently resulted in limited clinical studies of the drug. While these studies are still in course initial reports tend to indicate that iron excretion in response to the drug is apparently not high enough to induce negative iron balance.

The keto-hydroxy pyridones are another class of compounds which have been widely investigated. Most of the studies were carried out with the *1,2-dimethyl-3-hydroxy pyrid-4-one* which has been designated L_1 by some authors (Figure 1, 1-8). This bidentate ligand has been shown to be effective orally and parenterally in various rodents including mice, rats, and rabbits. In rats having received ^{59}Fe-labeled ferritin subsequent to loading with iron dextran, administration of an oral or parenteral dose (40 mg) of L_1 induced a mobilization of iron from hepatocellular stores comparable to parenteral DFO.[66] A more extensive survey of L_1 and its analogues has been carried out in iron-loaded mice. The loading procedure involved weekly i.p. injections of iron dextran followed by labeling of iron stores with ^{59}Fe lactoferrin.

Kontoghiorghes and others have studied several *N*-substituted 3-hydroxy pyridones in this model following the i.p. and i.g. administration at a dose of 200 mg/kg.[67] By far the most effective compounds tested were the *N*-alkyl-substituted 3-hydroxy pyrid-4-ones containing the methyl (L_1), ethyl, and propyl side chains. These compounds which were nearly as active or even more active i.g. were as effective as an identical dose of DFO given i.p. in increasing excretion of the radiolabel. The *N*-substituted 3-hydroxypyrid-2-ones were generally less effective.[68]

In a more systematic study Gyparaki et al. have evaluated several N-alkyl-substituted 3-hydroxypyrid-4-one chelators in a mouse model and correlated their potencies with the respective partition coefficients for the free ligand and the iron complex.[69] Employing dose-response curves these researchers demonstrated marked differences in activity when comparing the methyl, ethyl, propyl, isopropyl, butyl, and hexyl side chains. The ethyl, propyl, and butyl derivatives were shown to be significantly more effective orally than the methyl and isopropyl compounds in total ^{59}Fe excretion. It was concluded that the more hydrophilic compound L_1 is less able to penetrate membranes, in particular those of the hepatocyte in comparison to the more lipophilic chelators. As predicted the most lipophilic chelators tested were indeed the most active, but they also showed progressive signs of toxicity. A possible explanation for the more discriminative nature of this study may relate to the dosages used. As the permeation of most drugs through membranes is a passive diffusion process, it is conceivable that at higher doses a saturation phenomenon occurs which limits the amount of hepatic ferritin iron which can be mobilized by the chelator.

In order to probe the mechanism of action of these N-alkyl hydroxypyridones an elegant study with hepatocytes in culture was initiated.[70] Hepatocellular iron mobilization was determined by measuring release of ^{59}Fe after incubation with the test compounds at a concentration of 100 μM each. As already seen in the mouse study, L_1 was the least active chelator evaluated and in addition it was considerably less effective than DFO. Iron release from the hepatocyte monolayers was highest in the presence of the N-propyl derivative which may relate to the favorable partition coefficient in its free and iron-bound forms. Even more lipophilic compounds were less effective, presumably because they partitioned preferentially into the lipid phase in both the free and complexed form.[1]

Although several reports suggest that L_1 may not be the ideal candidate from this group of chelators in terms of potency, its decreased lipophilicity may be advantageous with regard to drug tolerability. While long-term tolerability studies in animals have not yet been reported, the available data from short-term toxicity studies look promising. Due to its ease of preparation, effectiveness in various *in vitro* and *in vivo* test systems, and low acute toxicity, L_1 has received widespread attention. Initial reports on a limited clinical evaluation have appeared and demonstrated its lack of short-term toxicity at relatively high doses which significantly enhance urinary iron excretion.[71,72] At this time data on fecal excretion are still not available. Therefore it remains to be demonstrated whether net negative iron balance can be achieved using dose levels which are well tolerated in the long term. There is also considerable concern among clinical experts that no significant reduction of serum ferritin levels has been observed even after many months of treatment. Moreover the tolerability of the compound in human patients is still in debate.[73-76]

Desferrithiocin (Figure 1, 1-9) is a novel tridentate siderophore isolated from the culture broth of *Streptomyces antibioticus* DSM 1865.[77] This ligand has also proven orally effective in the mobilization of iron in various animal models. Longueville and Crichton investigated the ability of desferrithiocin to mobilize hepatic ferritin iron in an animal model of iron overload in which rats had received 3,5,5-trimethylhexanoyl ferrocene (HOE 117) in their diet.[78] Oral administration of this ferrocene derivative was shown to substantially increase both hepatic ferritin levels and ferritin iron content in the treated animals. Chelators were administered either by i.p. injections or by gavage every 2 d over a period of 14 d followed by analysis of liver ferritin and hepatic ferritin iron content. At higher doses (100 mg/kg) desferrithiocin was shown to be more effective in mobilizing hepatic ferritin iron than an equivalent dose of DFO given i.p. Its efficacy in chelating hepatocellular iron was shown to increase exponentially as a function of decreasing cumulative dose. The efficacy approached 29% following i.p. injection of desferrithiocin at a dose of 10 mg/kg. Unfortunately desferrithiocin was not without side effects. Only 7 animals out of 12 survived the treatment with repeated doses of 100 mg/kg. More extensive toxicological investigations have subsequently been carried out in several animal species with unsatisfactory results. Therefore, further development of this drug for the treatment of iron overload has been stopped.

Bergeron et al. have also examined desferrithiocin and several of its analogues and derivatives for their ability to promote iron clearance in a noniron overloaded bile duct-cannulated rat model.[79] Biliary iron excretion was monitored in intervals of 3 h over a 24-h period, thus allowing the determination of the complete time course of the chelator-induced iron clearance. In addition, total iron output in both the bile and urine in response to the chelators was measured. The main route of iron excretion observed was biliary regardless of the chelators used. Desferrithiocin administered orally at a dose of 150 μmol per animal was found to be 1.2 times as effective in promoting total iron excretion as an identical dose of DFO given s.c. Further, the rates of iron excretion differed considerably for the two compounds. Whereas biliary iron excretion in response to DFO was rapid and essentially back to baseline after 10 h, desferrithiocin-promoted iron excretion was more protracted, reaching baseline values only after 24 h. The longer residence time of desferrithiocin may in part account for the increased toxicity observed at higher drug concentrations. None of the synthetic analogues or derivatives tested were shown to be as effective as desferrithiocin following oral administration.

Up to now *in vivo* tests have been performed almost exclusively in rodent models involving normal or iron-loaded mice or rats. Several candidates which showed potential in these investigations were eventually tested in human trials. Unfortunately, the favorable results obtained in these models could not be reproduced in humans with their specific and highly conservative iron metabolism. Some of the compounds proved to be ineffective in promoting iron excretion in man, while others produced inacceptable side effects. Therefore, increased emphasis has been placed on the development of more predictive test models. A particularly interesting new animal model based on iron-loaded Cebus monkeys has been proposed recently.[41] This primate model relates much more closely to the situation in iron-overloaded humans. However, it also requires special facilities and know-how, especially with respect to animal handling and care, and to trace metal analysis. It is extremely critical to use suitable methods for measuring the iron content in samples of urine and feces, and to accurately determine trace levels of iron in the components of the diet. Moreover, it is important to avoid inadvertent exposure of the food and the excreta of the animals during the experiments.[80] It is expected that the fully optimized version of this model, which has been worked out in collaboration with a research group at the University of Florida in Gainesville, will greatly contribute to a rapid and predictive evaluation of promising new chelating drugs and help to reduce the risks involved in premature clinical trials in humans.

NEW USES OF CHELATORS AND CHELATES

In view of the many crucial biochemical functions of iron in cells and organisms it is not surprising that various other therapeutic uses for iron chelators have been proposed. These topics have been reviewed earlier and will therefore not be presented here extensively.[81]

DIALYSIS AND ALUMINUM OVERLOAD

In spite of its much lower affinity for aluminum ions (see Complex Formation.) DFO has also been found to be of therapeutic use in various pathological conditions caused by aluminum overload. Ackrill et al.[4] were the first investigators reporting a curative effect of long-term DFO treatment in patients disabled by aluminum-induced dialysis encephalopathy or osteomalacia. Due to the efforts of numerous other investigators chelation therapy with DFO is now a well-accepted treatment for all manifestations of aluminum accumulation in chronically dialyzed patients.[82,83] These patients are unable to clear this toxic metal via renal excretion. Moreover, they are exposed to additional sources, such as aluminum contamination of the dialysis water and aluminum hydroxide-based phosphate binding agents.

DFO is also used as a diagnostic tool for the detection of systemic aluminum overload in patients. A single dose is administered and the subsequent increase of aluminum levels in the serum or in the dialysis fluid is determined. The chelator is causing a ligand exchange of the aluminum ions from serum-proteins, thereby producing a hydrophilic low molecular weight complex which can easily cross the dialysis membrane.[84] The repeated treatment eventually leads to a depletion of toxic deposits in bone and nervous tissue.

PARASITES

Since iron is a vital element for the growth of almost all living organisms the effect of iron on infective microorganisms has been extensively investigated (for a recent review see Reference 85). Iron withholding is indeed a general host-defense mechanism against infections by rapidly proliferating microorganisms and protozoa. The inhibitory effect of iron chelators on the growth of different types of parasites in cell cultures and animal models has been reported in a number of recent publications. The pathogenicity of infections by *Trypanosoma cruzi* in mice can be reduced by depleting host iron stores with DFO and an iron-deficient diet.[86] Different species of *Plasmodium* are sensitive to low iron levels in the host organism. The inhibitory effect of DFO on *P. falciparum in vitro* was first reported by Raventos-Suarez et al.[87] The detailed mechanism of action, however, is not yet known,[88] but it has been shown that the iron complex FO is not active in this test model. A suppression of the infection *in vivo* could be induced by administration of DFO to mice infected with *P. vinckei*[89] and to Aotus monkeys infected with *P. falciparum*.[90] Other ligands have also been investigated for possible use as antimalaria drugs.[91] The positive results of short-term chelation treatment of malaria infections with DFO in various animal test models have been considered to be sufficiently promising to warrant initiation of limited clinical trials in patients. If these trials confirm the therapeutic usefulness of iron chelation against infections by malaria parasites, more widespread application of this therapeutic concept would have to await the availability of an approved oral iron chelating drug.

INFLAMMATION

Inflammatory disease could be another potential indication for iron chelation therapy. It has been shown by various authors that free ferric ions are involved in the development of inflammatory processes by catalyzing the formation of free radicals. These radicals are promoting lipid peroxidation which results in membrane breakdown and subsequent inflammatory tissue damage. Complexation of the free ferric ions should therefore have a beneficial effect on the progress of inflammatory diseases.[92-96] The complex interactions of iron and free radicals in generating oxidative tissue damage and the effects of iron chelators on these mechanisms have already been reviewed in detail.[81] Some recent findings reported by Fridovich and co-workers[97,98] documented an interesting biological activity of the manganese(IV) complex of DFO which apparently mimics the protective effect of the enzyme superoxide-dismutase (SOD).

ANTIPROLIFERATIVE EFFECTS

DFO has also been shown to be a potent inhibitor of DNA synthesis by human B and T lymphocytes *in vitro*.[99] The inhibition of ribonucleotide-reductase activity mediated by the binding of iron by DFO prevents cells from completing the S phase of the cell-proliferation cycle.[100,101] The effect of DFO on DNA synthesis, DNA repair, cell proliferation, and differentiation of HL-60 cells has been described in detail by a group of Canadian researchers.[102] The same authors have also reported a synergistic antiproliferative effect of DFO when cultures of the same cell line were treated with various antileukemic agents with a cytotoxic mechanism of action.[103] These initial findings have generated a lot of transient enthusiasm. Preliminary treatment of a single patient suffering from neonatal acute leukemia confirmed the antiproliferative properties of DFO *in vivo*. However, the fatal outcome could at best be slowed down but not prevented.[104] Much more preclinical and clinical work will be needed in order to find out whether DFO or other iron chelators will eventually find a place in the therapy of proliferative diseases.

MRI WITH COMPLEXES

Recent developments in clinical diagnostics open up a new area for the use of biologically well-tolerated transition metal complexes. Magnetic resonance imaging (MRI) provides a powerful tool for noninvasive *in vivo* diagnostics. For improved contrast enhancing different paramagnetic ions such as Mn^{2+}, Fe^{3+}, and Gd^{3+} have been used. However, salts of these ions have shown undesirable toxicity when used in *in vivo* experiments. The toxic effects can be reduced by using chemically stable complexes of the ions. Complexes of gadolinium are now used clinically, while FO and some of its derivatives are still in the preclinical evaluation stage.[105-109]

REFERENCES

1. **Modell, B., Letsky, E. A., Flynn, D. M., Peto, R., and Weatherall, D. J.,** Survival and desferrioxamine in thalassemia major, *Br. Med. J.*, 284, 1081, 1982.
2. **Modell, B. and Berdoukas, V.,** The clinical approach to thalassaemia, *Br. Med. J. Clin. Res.*, 289, 996, 1984.
3. **Hoffbrand, A. V. and Wonke, B.,** Results of long-term subcutaneous desferrioxamine therapy, *Baillieres Clin. Haematol.*, 2, 345, 1989.
4. **Ackrill, P., Ralston, A. J., Day, J. P., and Hodge, K. C.,** Successful removal of aluminium from patient with dialysis encephalopathy, *Lancet*, 2, 692, 1980.
5. **Romero, R. A. and Day, J. P.,** Polarographic determination of desferrioxamine B in dialysis samples, *Trace Elem. Med.*, 2, 1, 1985.
6. **De Broe, M. E.,** private communication, University of Antwerp, Belgium, 1990.
7. **Anderegg, G. and Schwarzenbach, G.,** Hydroxamatkomplexe I. Die Stabilitaet der Eisen(III)-Komplexe einfacher Hydroxamsaeuren und des Ferrioxamins B, *Helv. Chim. Acta*, 96, 1390, 1963.
8. **Keberle, H.,** The biochemistry of desferrioxamine and its relation to iron metabolism, *Ann. N.Y. Acad. Sci.*, 119, 758, 1964.
9. **Evers, A., Hancock, R. D., Martell, E. A., and Motekaitis, R. J.,** Metal ion recognition in ligands with negatively charged oxygen donor groups. Complexation of Fe(III), Ga(III), Al(III), and other highly charged metal ions, *Inorg. Chem.*, 28, 2189, 1989.
10. **Barry, M. and Cartei, G. C.,** Estimation of ferrioxamine in jaundiced urine, *J. Clin. Pathol.*, 21, 169, 1968.
11. **Fielding, J. and Brunström, G. M.,** Estimation of ferrioxamine and desferrioxamine in urine, *J. Clin. Pathol.*, 17, 395, 1964.
12. **Anderegg, G., L'Eplattenier, F., and Schwarzenbach, G.,** Hydroxamatkomplexe. III. Eisen(III)-Austausch zwischen Sideraminen und Komplexonen. Diskussion der Bindungkonstanten der Hydroxamatkomplexe, *Helv. Chim. Acta*, 96, 1409, 1963.
13. **Allain, P., Mauras, Y., Beaudeau, G., and Hingouet, P.,** Indirect micro-scale method for the determination of desferrioxamine and its aluminium and iron chelated forms in biological samples by AAS with electrothermal atomisation, *Analyst*, 111, 531, 1986.
14. **Leflon, P. and Plaquet, R.,** Rapid determination of iron in urine, in the presence of deferoxamine, by inductively coupled plasma emission spectrometry, *Clin. Chem.*, 32, 521, 1986.

15. **Bourdon, S., Houze, P., and Bourdon, R.,** Quantification of desferrioxamine in blood plasma by inductively coupled plasma atomic emission spectrometry, *Clin. Chem.,* 33, 132, 1987.

16. **Grisler, R., Ferrari, M., Rajnoldi, A. C., and Ravazzani, V.,** Determinazione colorimetrica del ferro urinario, chelato con deferossmina B, mediante impiego di un monoreattivo, *Quad. Sclavo Diagn.,* 18, 383, 1982.

17. **Van Stekelnburg, G. J., Valk, C., and Deboer, G. J.,** Determination of deferoxamine chelated iron, *Clin. Chem.,* 28, 2328, 1982.

18. **Arts, J. W. M. M. and Hafkenscheid, J. C. M.,** Determination of ferrioxamine bound iron in urine, *Clin. Chem.,* 30, 155, 1984.

19. **Barry, M.,** Determination of chelated iron in the urine, *J. Clin. Pathol.,* 21, 166, 1968.

20. **Meyer-Brunot, H. G. and Keberle, H.,** The metabolism of desferrioxamine B and ferrioxamine B, *Biochem. Pharmacol.,* 16, 527, 1967.

21. **Kruck, T. P. A., Teichert-Kuliszewska, K., Fisher, E., Kalow, W., and McLachlan, D. R.,** High performance liquid chromatography analysis of desferrioxamine. Pharmacokinetic and metabolic studies, *J. Chromatogr.,* 433, 207, 1988.

22. **Kruck, T. P. A. and Kalow, W.,** Determination of desferrioxamine and a major metabolite by high performance liquid chromatography. Application to the treatment of aluminium-related disorders, *J. Chromatogr.,* 341, 123, 1985.

23. **Hughes, H., Hagen, L. E., Cameron, E. C., and Sutton, R. A. L.,** Estimation of aluminoxamine and ferrioxamine in plasma by high performance liquid chromatography, *Clin. Chim. Acta,* 157, 115, 1986.

24. **Mattiello, G., Rizzo, F., Zoccolan, R., and Andriani, M.,** High performance liquid chromatography determination of low levels of free desferrioxamine and its iron and aluminium chelates, in *Aluminium and Other Trace Metals in Renal Disease,* Tylor, A., Ed., Balliere Tindall, London, 1986, 241.

25. **Catalan, R. B., Leung, F. Y., Hodsman, A. B., and Henderson, A. R.,** Determination of desferrioxamine in serum by reversed-phase-high performance liquid chromatogrpahy, in *Aluminium and Other Trace Metals in Renal Disease,* Taylor, A., Ed., Balliere Tindall, London, 1986, 235.

26. **Evans, J. R. and Shepherd, A. M. M.,** The effect of desferrioxamine on the colorimetric determination of iron in human serum and plasma, *Clin. Chim. Acta,* 60, 401, 1975.

27. **Pippard, M. J. and Strat, S.,** Simple assay for urinary iron after desferrioxamine therapy, *Am. J. Clin. Pathol.,* 77, 324, 1982.

28. **Romero, R. A. and Day, J. P.,** Polarographic determination of desferrioxamine B in dialysis samples, *Trace Elem. Med.,* 2, 1, 1985,.

29. **Steinmetz, W. L., Glick, M. R., and Oei, T. O.,** Modified aca method for determination of iron chelated by deferoxamine and other chelators, *Clin. Chem.,* 26, 1593, 1980.

30. **Jenny, H. B. and Peter, H. H.,** Determination of desferrioxamine B, its complexes with iron and aluminium and selected derivatives by high performance liquid chromatography using a polymer column, *J. Chromatogr.,* 438, 433, 1988.

31. **Cramer, S. M., Nathanael, B., and Horvath, C.,** High performance liquid chromatography of deferoxamine and ferrioxamine: interference by iron present in the chromatographic system, *J. Chromatogr.,* 295, 405, 1984.

32. **Van Der Horst, A., De Goede, P. N. F. C., Willems, H. J. J., and Van Loenen, A. C.,** Determination of desferrioxamine and ferrioxamine by high performance liquid chromatography with direct serum injection and pre-column enrichment, *J. Chromatogr.,* 381, 185, 1986.

33. **Palmieri, M. D. and Fritz, J. S.,** Determination of metal ions by high performance liquid chromatography separation of their hydroxamic acid chelates, *Anal. Chem.,* 59, 2226, 1987.

34. **Schwyn, B. and Neilands, J. B.,** Universal chemical assay for the detection and determination of siderophores, *Anal. Biochem.,* 160, 47, 1987.

35. **Emery, T.,** Reaction of cyanide with hydroxamic acid iron complexes to distinguish trihydroxamates from simple monohydroxamates, *Anal. Biochem.,* 139, 301, 1984.

36. **Peter, H. H.,** Industrial aspects of iron chelators: pharmaceutical applications, in *Proteins of Iron Storage and Transport,* Spik, G., Montreuil, J., Crichton, R. R., and Mazurier, J., Eds., Elsevier, Amsterdam, 1985, 293.

37. **Fingl, E. and Woodburg, D. M.,** General principles, in *The Pharmacological Basis of Therapeutics,* 5th ed., Goodman, L. S. and Gilman, A., Eds., Macmillan, New York, 1975, 1.

38. **Hershko, C., Grady, R. W., and Link, G.,** Evaluation of iron-chelating agents in an *in vivo* system: potential usefulness of EHPG, a powerful iron-chelating drug, *Br. J. Haematol.,* 51, 251, 1982.

39. **Hider, R. C.,** Siderophore mediated absorption of iron, in *Structure and Bonding,* Vol. 85, *Siderophores from Microorganisms and Plants,* Springer-Verlag, Berlin, 1984, 25.

40. **Baker, E.,** Biologic screens for iron chelators, *Birth Defects Orig. Artic. Ser.,* 23, 49, 1988.

41. **Wolfe, L. C., Nicolosi, R. J., Renaud, M. M., Finger, J., Hegsted, M., Peter, H., and Nathan, D. G.,** A non-human primate model for the study of oral iron chelators, *Br. J. Haematol.,* 72, 456, 1989.

42. **Pitt, C. G., Gupta, G., Estes, W. E., Rosenkrantz, H., Metterville, J. J., Crumbliss, A. L., Palmer, R. A., Nordquest, K. W., Sprinkle Hardy, K. A., Whitcomb, D. R., Byers, B. R., Arceneaux, J. E. L., Gaines, C. G., and Sciortino, C. V.,** The selection and evaluation of new chelating agents for the treatment of iron overload, *J. Pharmacol. Exp. Ther.,* 208, 12, 1979.

43. **Grady, R. W., Graziano, J. H., Akers, H. A., and Cerami, A.,** The development of new iron-chelating drugs, *J. Pharmacol. Exp. Ther.,* 196, 478, 1976.

44. **Grady, R. W., Graziano, J. H., White, G. P., Jacobs, A., and Cerami, A.,** The development of new iron-chelating drugs. II, *J. Pharmacol. Exp. Ther.,* 205, 757, 1978.

45. **Peterson, C. M., Graziano, J. H., Grady, R. W., Jones, R. L., Markenson, A., Lavi, U., Canale, V., Gray, G. F., Cerami, A., and Miller, D. R.,** Chelation therapy in β-thalassemia major: a one-year double blind study of 2,3-dihydroxybenzoic acid, *Exp. Hematol.,* 7, 74, 1979.

46. **Grady, R. W. and Jacobs, A.,** The screening of potential iron chelating drugs, in *Development of Iron Chelators for Clinical Use,* Martell, A. E., Anderson, W.-F., Badman, D. G., Eds., Elsevier/North-Holland, Amsterdam, 1981, 133.

47. **Grady, R. W., Peterson, C. M., Jones, R. L., Graziano, J. H., Bhargava, K. K., Berdoukas, V. A., Kokkini, G., Loukopoulos, D., and Cerami, A.,** Rhodotorulic acid — investigation of its potential as an iron-chelating drug, *J. Pharmacol. Exp. Ther.,* 209, 342, 1979.

48. **Gralla, E. J.,** Efforts to develop a bioassay system for detecting compounds which actively deplete tissue iron stores, in Proc. Symp. on the Development of Iron Chelators for Clinical Use, Anderson, W. F. and Hiller, M. C., Eds., National Institute of Health, Bethesda, 1977, 229.

49. **Winston, A., Varaprasad, D. V. P. R., Metterville, J. J., and Rosenkrantz, H.,** Evaluation of polymeric hydroxamic acid iron chelators for treatment of iron overload, *J. Pharmacol. Exp. Ther.,* 232, 644, 1985.

50. **Mahoney, J. R., Hallaway, P. E., Hedlund, B. E., and Eaton, J. W.,** Acute iron poisoning, *J. Clin. Invest.,* 84, 1362, 1989.

51. **Pitt, C. G. and Martell, A. E.,** The design of chelating agents for the treatment of iron overload, in *Inorganic Chemistry in Biology and Medicine,* Vol. 140, Martell, A. E., Ed., American Chemical Society, Washington, D. C., 1980, 279.

52. **Cleton, F., Turnbull, A., Finch, C. A., Thompson, L., and Martin, J.,** Synthetic chelating agents in iron metabolism, *J. Clin. Invest.,* 42, 327, 1963.

53. **Rosenkrantz, H. and Metterville, J. J.,** Preliminary toxicity data on new ethylenediamine derivatives designed for iron chelation, *J. Am. Coll. Toxicol.,* 7, 617, 1988.

54. **Pitt, C. G., Bao, Y., Thompson, J., Wani, M. C., Rosenkrantz, H., and Metterville, J.,** Esters and lactones of phenolic amino carboxylic acids: prodrugs for iron chelation, *J. Med. Chem.,* 29, 1231, 1986.

55. **Hershko, C., Grady, R. W., and Link, G.,** Phenolic ethylenediamine derivatives: a study of orally effective iron chelators, *J. Lab. Clin. Med.,* 103, 337, 1984.

56. **Sephton-Smith, R.,** Iron excretion in thalassaemia major after administration of chelating agents, *Br. Med. J.,* 2, 1577, 1962.

57. **Bannerman, R. M., Callender, S. T., and Williams, D. L.,** Effect of desferrioxamine and DTPA in iron overload, *Br. Med. J.,* 2, 1573, 1962.

58. **Fahey, J. L., Roth, C. E., Princiotto, J. V., Brick, I. B., and Rubin, M.,** Evaluation of trisodium calcium diethylenetriamine pentaacetate in iron storage disease, *J. Lab. Clin. Med.,* 57, 436, 1961.

59. **Pippard, M. J., Jackson, M. J., and Modell, C. B.,** Subcutaneous diethylene triamine pentaacetic acid: comparison with desferrioxamine in thalassaemic patients with iron overload, *Birth Defects,* 23, 105, 1988.

60. **Wonke, B., Hoffbrand, A. V., Aldouri, M., Wickens, D., Flynn, D., Stearns, M., and Warner, P.,** Reversal of desferrioxamine induced auditory neurotoxicity during treatment with Ca-DTPA, *Arch. Dis. Child.,* 64, 77, 1989.

61. **Hoy, T., Humphrys, J., Jacobs, A., Williams, A., and Ponka, P.,** Effective iron chelation following oral administration of an isoniazid-pyridoxal hydrazone, *Br. J. Haematol.,* 43, 443, 1979.

62. **Hershko, C., Avramovici-Grisaru, S., Link, G., Gelfand, L., and Sarel, S.,** Mechanism of in vivo iron chelation by pyridoxal isonicotinoyl hydrazone and other imino derivatives of pyridoxal, *J. Lab. Clin. Med.,* 98, 99, 1981.

63. **Johnson, D. K., Pippard, M. J., Murphy, T. B., and Rose, N. J.,** An *in vivo* evaluation of iron-chelating drugs derived from pyridoxal and its analogs, *J. Pharmacol. Exp. Ther.,* 221, 399, 1982.

64. **Williams, A., Hoy, T., Pugh, A., and Jacobs, A.,** Pyridoxal complexes as potential chelating agents for oral therapy in transfusional iron overload, *J. Pharm. Pharmacol.,* 34, 730, 1982.

65. **Kim, B.-K., Huebers, H. A., and Finch, C. A.,** Effectiveness of oral iron chelators assayed in the rat, *Am. J. Hematol.,* 24, 277, 1987.

66. **Kontoghiorghes, G. J.,** Orally active α-ketohydroxypyridine iron chelators: studies in mice, *Mol. Pharmacol.,* 30, 670, 1986.

67. **Kontoghiorghes, G. J., Sheppard, L., Hoffbrand, A. V., Charalambous, J., Tikerpae, J., and Pippard, M. J.,** Iron chelation studies using desferrioxamine and the potential oral chelator, 1,2-dimethyl-3-hydroxypyrid-4-one, in normal and iron loaded rats, *J. Clin. Pathol.,* 40, 404, 1987.

68. **Kontoghiorghes, G. J. and Hoffbrand, A. V.**, Orally active α-ketohydroxypyridine iron chelators intended for clinical used: in-vivo studies in rabbits, *Br. J. Haematol.*, 62, 607, 1986.
69. **Gyparaki, M., Porter, J. B., Hirani, S., Streater, M., Hider, R. C., and Huehns, E. R.**, In vivo evaluation of hydroxypyridone iron chelators in a mouse model, *Acta Haematol.*, 78, 217, 1987.
70. **Porter, J. B., Gyparaki, M., Huehns, E. R., and Hider, R. C.**, The relationship between lipophilicity of hydroxypyrid-4-one iron chelators and cellular iron mobilization, using an hepatocyte culture model, *Biochem. Soc. Trans.*, 14, 1180, 1986.
71. **Kontoghiorhes, G. J., Aldouri, M. A., Sheppard, L., and Hoffbrand, A. V.**, 1,2-Dimethyl-3-hydroxypyrid-4-one, an orally active chelator for treatment of iron overload, *Lancet*, ii, 1294, 1987.
72. **Kontoghiorghes, G. J., Aldouri, M. A., Hoffbrand, A. V., Barr, J., Wonke, B., Kourouclaris, T., and Sheppard, L.**, Effective chelation of iron in β thalassaemia with the oral chelator 1,2-dimethyl-3-hydroxypyrid-4-one, *Br. Med. J.*, 295, 1509, 1987.
73. **Hoffbrand, A. V., Bartlett, A. N., Veys, P. A., O'Connor, N. T. J., and Kontoghiorghes, G. J.**, Agranulocytosis and thrombocytopenia in patient with blackfan-diamond anaemia during oral chelator trial, *Lancet*, ii, 457, 1989.
74. **Kontoghiorghes, G. J., Nasseri-Sina, P., Goddard, J. G., Barr, J. M., Nortey, P., and Sheppard, L. N.**, Safety of oral iron chelator L1, *Lancet*, ii, 457, 1989.
75. **Editor**, Oral iron chelators, *Lancet*, ii, 1016, 1989.
76. **Kontoghiorghes, G. J. and Hoffbrand, A. V.**, Clinical trials with oral iron chelator L1, *Lancet*, ii, 1516, 1989.
77. **Naegeli, H. U. and Zaehner, H.**, Metabolites of microorganisms. CXCIII. Ferrithiocin, *Helv. Chim. Acta*, 63, 1400, 1980.
78. **Longueville, A. and Crichton, R. R.**, An animal model of iron overload and its application to study hepatic ferritin iron mobilization by chelators, *Biochem. Pharmacol.*, 35, 3669, 1986.
79. **Bergeron, R. J., Crist, C. M., Prudencio, S., LaGraves, E. S., Peter, H. H., and Dionis, J. B.**, Evaluation of desferrithiocin and its synthetic derivatives as orally effective iron chelators, *J. Med. Chem.*, in press.
80. **Bergeron, R. J., Streiff, R. R., and collaborators**, private communication University of Florida, Gainesville, 1989.
81. **Hershko, C. and Weatherall, D. J.**, Iron-chelating therapy, *CRC Clin. Lab. Sci.*, 26, 303, 1988.
82. **Ackrill, P.**, Aluminium removal by desferrioxamine: clinical practice, in *Aluminium and Other Trace Elements in Renal Disease*, Taylor, A., Ed., Balliere Tindall, London, 1986, 193.
83. **Nebeker, H. G. and Coburn, J. W.**, Aluminium and renal osteodystrophy, *Annu. Rev. Med.*, 37, 79, 1986.
84. **Day, J. P.**, Chemical aspects of aluminium chelation by desferrioxamine, in *Aluminium and Other Trace Elements in Renal Disease*, Taylor, A., Ed., Balliere Tindall, London, 1986, 184.
85. **Weinberg, E. D.**, Iron, infection, and neoplasia, *Clin. Physiol. Biochem.*, 4, 50, 1986.
86. **Lalonde, R. G. and Holbein, B. E.**, Role of iron in *Trypanosoma cruzi* infection of mice, *J. Clin. Invest.*, 73, 470, 1984.
87. **Raventos-Suarez, C., Pollack, S., and Nagel, R. L.**, *Plasmodium falciparum*: inhibition of in vitro growth by desferrioxamine, *Am. J. Trop. Med. Hyg.*, 31, 919, 1982.
88. **Peto, T. E. A. and Thompson, J. L.**, A reappraisal of the effects of iron and desferrioxamine on the growth of *Plasmodium falciparum* in vitro: the unimportance of serum iron, *Br. J. Haematol.*, 63, 273, 1986.
89. **Fritsch, G., Treumer, J., Spira, D. T., and Jung, A.**, *Plasmodium vinckei*: suppression of mouse infections with desferrioxamine B, *Exp. Pathol.*, 60, 171, 1985.
90. **Pollack, S., Rossan, R. N., Davidson, D. E., and Escajadillo, A.**, Desferrioxamine suppresses *Plasmodium falciparum* in aotus monkeys, *Proc. Soc. Exp. Biol. Med.*, 184, 162, 1987.
91. **Yinnon, A. M., Theanacho, E. N., Grady, R. W., Spira, D. T., and Hershko, C.**, Antimalarial effect of HBED and other phenolic and catechic iron chelators, *Blood*, 74, 2166, 1989.
92. **Blake, D. R., Hall, N. D., Bacon, P. A., Dieppe, P. A., Halliwell, B., and Gutteridge, J. M.**, The importance of iron in rheumatoid disease, *Lancet*, ii, 1142, 1981.
93. **Blake, D. R. and Lunec, J.**, Copper, iron, free radicals and arthritis, *Br. J. Rheumatol.*, 24, 123, 1985.
94. **Sedgwick, A. D., Blake, D. R., Winwood, P., Moore, A. R., Al Duaij, A., and Willoughby, D. A.**, Studies into the effects of the iron chelator desferrioxamine on the inflammatory process, *Eur. J. Rheumatol. Inflamm.*, 7, 87, 1984.
95. **Andrews, F. J., Morris, C. J., Kondratowicz, G., and Blake, D. R.**, Effect of iron chelation on inflammatory joint disease, *Ann. Rheum. Dis.*, 46, 327, 1987.
96. **Hewitt, S. D., Hider, R. C., Sarpong, P., Morris, C. J., and Blake, D. R.**, Investigation of the anti-inflammatory properties of hydroxypyridinones, *Ann. Rheum. Dis.*, 48, 382, 1989.
97. **Beyer, W. F. and Fridovich, I.**, Characterization of a superoxide dismutase mimic prepared from desferrioxamine and MnO_2, *Arch. Biochem. Biophys.*, 271, 149, 1989.

98. **Darr, D., Zarilla, K. A., and Fridovich, I.,** A mimic of superoxide dismutase activity based upon desferrioxamine B and manganese(IV), *Arch. Biochem. Biophys.,* 258, 351, 1987.
99. **Lederman, H. M., Cohen, A., Lee, J. W. W., Freedman, M. H., and Gelfand, E. W.,** Deferoxamine: a reversible S-phase inhibitor of human lymphocyte proliferation, *Blood,* 64, 748, 1984.
100. **Hoffbrand, A. V., Ganeshaguru, K., Hooton, J. W. L., and Tattersall, M. H. N.,** Effect of iron deficiency and desferrioxamine on DNA synthesis in human cells, *Br. J. Haematol.,* 33, 517, 1976.
101. **Bomford, A., Isaac, J., Roberts, S., Edwards, A., Young, S., and Williams, R.,** The effect of desferrioxamine on transferrin receptors, the cell cycle and growth rates of human leukaemic cells, *Biochem. J.,* 236, 243, 1986.
102. **Kaplinsky, C., Estrov, Z., Freedman, M. H., Gelfand, E. W., and Cohen, A.,** Effect of deferoxamine on DNA synthesis, DNA repair, cell proliferation, and differentiation of HL-60 cells, *Leukemia,* 1, 437, 1987.
103. **Estrov, Z., Cohen, A., Gelfand, E. W., and Freedman, M. H.,** Synergistic antiproliferative effects on HL-60 cells: deferoxamine enhances cytosine arabinoside, methotrexate, and daunorubicin cytotoxicity, *Am. J. Ped. Hematol. Oncol.,* 10, 288, 1988.
104. **Estrov, Z., Tawa, A., Wang, X.-H., Dube, I. D., Sulh, H., Cohen, A., Gelfand, E. W., and Freedman, M. H.,** In vitro and in vivo effects of deferoxamine in neonatal acute leukemia, *Blood,* 69, 757, 1987.
105. **Tzika, A. A., Thurnher, S., Hricak, H., Price, D. C., Arrive, L., Aboseif, S., Engelstad, B. L., and Rector, F. C.,** Rapid, contrast-enhanced, diuretic magnetic resonance imaging of unilateral partial ureteral obstruction. An experimental study in micropigs, *Invest. Radiol.,* 24, 37, 1989.
106. **Worah, D., Berger, A. E., Burnett, K. R., Cockrill, H. H., Kanal, E., Kendall, C., Leese, P. T., Lyons, K. P., Ross, E., Wolf, G. L., et al.,** Ferrioxamine as a magnetic resonance contrast agent. Preclinical studies and phase I and II human clinical trials, *Invest. Radiol.,* 23, S281, 1988.
107. **Niedrach, W. L., Tonetti, F. W., Katzberg, R. W., Morris, T. W., Ventura, J. A., Totterman, S., and Cos, L. R.,** Effects of the magnetic resonance contrast medium ferrioxamine methanesulfonate on systemic and renal hemodynamics in the anesthetized dog, *Invest. Radiol.,* 23, 687, 1988.
108. **Von Schulthess, G. K., Duewell, S., Jenny, H.-B., Wüthrich, R., and Peter, H. H.,** Polyethylene-Glycol-Ferrioxamine: A new magnetic resonance contrast agent, *Invest. Radiol.,* 25, 548, 1990.
109. **Carr, D. H., Brown, J., Leung, A. W. L., and Pennock, J. M.,** Iron and gadolinium chelates as contrast agents in NMR imaging: preliminary studies, *J. Comput. Assist. Tomogr.,* 8, 385, 1984.

INDEX

siderophore uptake in, 28
transport of substrates into
across sytoplasmic membrane, 109—110
across outer membrane, 108—109
Exochelins, 70
exb mutations, 122

F

Fast Atom Bombardment Mass Spectrometry (FAB-
MS), 139, 147
of pyoverdins, 143
pyoverdin sequence determined by, 149
fec IR genes, 115
Fenton reagents, 36
Ferric chloride reaction, 2
Ferric complex, of enterobactin, 311—319
Ferrichrome A, 185
ascorbate reduction of, 228
citrate-promoted reductive removal of iron from,
214
and competing legands, 216
electron transfer reactivities of, 227
isolation of microbial iron chelates with, 7—8
in siderophore ligand exchange reaction, 217
siderophore redox potential for, 223
stability constants for, 189
Ferrichrome analogues, 197
of C_3-symmetry, 319—322
for facilitated iron uptake, 329—333
as growth inhibitors of *P. falciparum,* 333—334
as growth promoters of *Arthrobacter flavescens,*
330, 331
IR-spectrum of, 322
spectrum of, 323
structure of, 320
Ferrichromes, 185
CD spectral parameters of, 25
CD-spectrum of, 321
characteristics of, 235, 236
citrate-promoted reductive removal of iron, from,
214
deferri forms of, 263—264
electron transfer reactivities of, 227
and FhuBCD proteins, 124
growth factor activity in, 66
identification of, 1
modification of, 112
in *N. crassa,* 87
NMR spectroscopy of, 258—262
protonation constant for, 182
pseudo first-order rate constants for, 218
receptor recognition of, 332
retrohydroxamate, 330
in siderophore ligand exchange raction, 217
stability constants for, 189
structure of, 8, 83
structural variations of, 18
transport systems for, 111
Ferrichrysin
CD spectral parameters of, 25

FAB mass spectra of, 248
stability constants for, 189
Ferricrocin, 185
in *Aspergillus nodulans,* 89
and competing ligands, 216
in *Conidia,* 86
FAB mass spectra of, 248
metal transfer from coprogens to, 37
in *N. crassa,* 37
in siderophore ligand exchange reaction, 217
siderophore redox potential for, 223
stability constants for, 189
Ferrimycin A, structures of, 69
Ferimycins, 67, 69
agar-diffusion bioassays for, 67
potential application of, 130
Ferrioxamine analogues, synthetic, 323—326
Ferrioxamine B, 183—184, 187, 188
binding site for, 113
and biomimetric analogues, 325
CD spectral parameters of, 25
citrate-promoted reductive removal of iron from,
214
compared with model systems, 209—213
dechelation of, 208, 210—211
electron transfer reactivities of, 227
exchange with ferrichrome A, 215
and FhuBCD proteins, 124
geometric isomers of, 325
protonation constants for, 182
siderophore redox potential for, 223
stability constant for, 189
Ferrioxamine B analogues
biomimetric, 324
distereometric ferric complexes of, 327
Ferrioxamine D, stability constant for, 189
Ferrioxamine E
preorganization of, 188
stability constant for, 189
Ferrioxamine E uptake, Mossbauer spectra of, 41
Ferrioxamines, 1, 16—17
"archaic vs. modern," 324
family of, 324
iron transport mediated by, 68
structures of, 67, 68
Ferrirubin, FAB mass spectra of, 248, 249
Ferrithiocin, 20
Ferropyrimine, 21
Ferroverdin, 21
Ferrozine, 72
fes gene, 117
FE^{3+} uptake, via specific receptors, 166—169
FE uptake, pseudobactin 358-mediated, 171
FhuA gene, 120
FhuA mutants, 119
Fhu genes, 112
Formation constants
calculations of, 178
for monocatechols ligands, 201
use of, 177
Fungal siderophores

Milton Keynes UK
Ingram Content Group UK Ltd.
UKHW050259161024
449569UK00043B/1444